DATE			

Soybeans:

Chemistry and Technology

VOLUME 1

Proteins

other AVI books

SOYBEANS:
CHEMISTRY AND TECHNOLOGY

Volume 1
Proteins

Edited by **ALLAN K. SMITH, Ph.D.**
  Oilseeds Protein Consultant,
  New Orleans, La.

and **SIDNEY J. CIRCLE, Ph.D.**
  Director, Protein Research,
  W. L. Clayton Research Center,
  Anderson Clayton Foods,
  Richardson, Texas

WESTPORT, CONNECTICUT

THE AVI PUBLISHING COMPANY, INC.

1972

Contributors

BILLY E. CALDWELL, Ph.D., Leader, Soybean Investigations, USDA Plant Science Research Division, Beltsville, Maryland

CLYDE M. CHRISTENSEN, Ph.D., Professor, Department of Plant Pathology, University of Minnesota, St. Paul, Minnesota

SIDNEY J. CIRCLE, Ph.D., Director, Protein Research, W. L. Clayton Research Center, Anderson Clayton Foods, Richardson, Texas

ARTHUR C. ELDRIDGE, Chemist, Meal Products Investigations, Oilseed Crops Laboratory, USDA Northern Regional Research Laboratory, Peoria, Illinois

CLIFFORD W. HESSELTINE, Ph.D., Chief, Fermentation Laboratory, USDA Northern Regional Research Laboratory, Peoria, Illinois

DANIEL E. HOOTON, Assistant to the Vice President, Agricultural & Industrial Operations, Anderson, Clayton & Co., Houston, Texas

ROBERT W. HOWELL, Ph.D., Chief, Oilseed and Industrial Crops Research Branch, USDA Plant Science Research Division, Beltsville, Maryland

DALE W. JOHNSON, Ph.D., Executive Vice President, Crest Products, Inc., Park Ridge, Illinois

HENRY H. KAUFMANN, Director, Grain Research Laboratory, Cargill, Inc., Minneapolis, Minnesota

IRVIN E. LIENER, Ph.D., Professor, Department of Biochemistry, University of Minnesota, St. Paul, Minnesota

JOSEPH J. RACKIS, Ph.D., Principal Chemist, Oilseed Crops Laboratory, USDA Northern Regional Research Laboratory, Peoria, Illinois

ALLAN K. SMITH, Ph.D., Oilseeds Protein Consultant, New Orleans, Lousiana

HWA L. WANG, Ph.D., Research Chemist, Fermentation Laboratory, USDA Northern Regional Research Laboratory, Peoria, Illinois

WALTER J. WOLF, Ph.D., Head, Meal Products Investigations, Oilseed Crops Laboratory, USDA Northern Regional Research Laboratory, Peoria, Illinois

Foreword

The appearance of this timely volume dealing with underlying problems and the basic science and technology of soybean utilization coincides with growing concern for the seriousness of the world protein problem. At this time also, after years of speculation and some effective research on perfecting for food use the so-called unconventional sources of protein, such as fish protein concentrates, single cell (microbiological) protein and cottonseed protein, a more realistic appraisal is now possible of the time scale and the human and material investments required to achieve their effective use. The harvest of these recent years of effort has been, in general, a disheartening array of new or unanticipated problems which have yet to be solved before these novel resources can begin to make any important impact.

The outstanding exception to this rather somber assessment is the remarkable progress which has been achieved in the industrialization of the soybean, which now provides on a large and increasingly effective scale a host of highly acceptable food and industrial products whose impact on consumer needs is already great. By means of what the late Dr. M. L. Anson called the "new" soybean technologies, this ancient and traditional cornerstone of nutrition in Southeast Asia has re-emerged in the form of an impressive new array of products whose effect on world protein needs is growing rapidly.

It must be emphasized that the other novel sources of protein can expect to achieve similar success only after investment of comparable resources for research and development. Soybean utilization is now enjoying, as it were, a 20-yr head start.

A greater commitment of research to other proteins seems urgently needed, as appears obvious from the rapidly growing impact of the Green Revolution, which, while having produced a dramatic increase in cereal supplies in formerly food-deficient countries such as India, Pakistan, and the Philippines, is not solving the increasing protein deficit quickly enough. This highly important emphasis on the production of wheat and rice has unfortunately led to neglect of important protein crops; legumes, the primary supplementary protein food of

India, have dropped in per capita availability to a significant degree. Obviously, more attention must be given to legume production but an additional challenge exists in applying more effectively the "new" technologies to production of human food not only from locally-produced soybeans (now increasing in India) but also from large resources of oilseeds such as peanut and cottonseed.

This observer has seen in some countries which are seeking to increase soybean production and utilization a virtually uncritical euphoria toward the apparently limitless beneficial potentials of this crop. This lack of realism was frequently unaccompanied by any adequate understanding of the basic and difficult technological problems which had to be overcome. The profound lesson of this book is that success in soybean utilization has been achieved only because of the research accomplishments of dedicated human talent and years of steadfast financial and institutional support for its efforts. A particularly fascinating aspect of this volume is the fact that the editors and a number of the chapter authors are themselves pioneers in much of the research which has brought soybean technology to its present sophisticated and effective level.

MAX MILNER
Executive Secretary
Protein Advisory Group
    of the United Nations System
United Nations, New York

Preface

Soybeans are the most important cash crop in the United States. They have found an extremely wide area of utilization in animal feeds, human foods, and industrial applications. The U.S. soybean crop contributes more protein and fat to our food economy than any other single source. For example, milk protein contributes 4.7 billion pounds or about 1/3 of the total food protein consumed in the United States, whereas the soybean crop provides approximately 27 billion pounds of protein, which is nearly twice the amount consumed by our population in the form of ingested food. It is unrealistic to presume that we can continue to feed our expanding population with the food industry as presently constituted; in the future we will have to depend more on bypassing the animal and producing acceptable foods directly from vegetable crops. Thus, the recent surge of interest in developing protein foods from soybeans and other oil seeds for the United States, as well as developing countries, is one practical approach to the solution of our food problems.

The present volume reviews the research and development on food uses of soybean protein products in the past 20 yr, primarily in occidental type foods, and also the fundamental chemistry and associated research important in improving these products for utilization by the food industry. Occidental type uses of soybeans differ markedly from traditional oriental methods. Although in recent years the Japanese have initiated an extensive research program to improve their diet, yet there is much that Western scientists can learn from the ancient oriental methods of food preparation and current research in that area; thus a brief review of oriental foods is included. In turn, it should be noted that the Japanese are much interested in increasing their use of Western type foods. There is a substantial use of soy protein in foods in Britain, Germany, and some of the other European countries. There is also a growing interest in India, Brazil, Mexico, and other countries of South America.

The chemistry of the proteins, because of their high molecular weight and large number of chemically reactive groups, is very difficult to master; and despite extensive and intensive research in recent years, there is still much to learn. Only since about 1950 have the automatized tools desirable for the purification and characterization of the proteins become available, yet accrual of knowledge in this field is still time-consuming. However, very definite progress has been made in understanding the proteins with respect to their chemistry, nutrition, and biologically reactive components as well as their functional uses in foods. Recent research leading to a better understanding of the chemistry of the soybean flavor problem suggests that its solution is very near. This will open the gates for a major expansion of the food uses of soybean protein.

The editors wish to express their thanks to the many people in government and industry who have generously supplied information, photographs, and drawings for our use. The W. L. Clayton Research Center of Anderson Clayton Foods, and Anderson, Clayton & Co. have been especially generous in their support and services during preparation of this manuscript for publication. The editors wish also to express their gratitude to the chapter authors who have expended so much of their time and energy outside their regular responsibilities in the preparation of their respective chapters. Their contributions denote a sincere dedication to their chosen profession as well as to the advancement of the soybean industry. The editors also appreciate the assistance of Richard R. Fergle and Donald L. Cook of the W. L. Clayton Research Center in preparation of some of the Figures.

ALLAN K. SMITH
SIDNEY J. CIRCLE

May, 1971

Contents

CHAPTER 1

A. K. Smith
S. J. Circle

Historical Background

INTRODUCTION

Although soybeans in the Orient go back to ancient times, their history in the United States covers a very recent period. Soybeans in the United States, including the whole vine, were used as hay and silage for a number of years before the value of their oil and protein for food and feed was clearly recognized; commercial processing of the seed was not established until 1922. From a farm production in 1922 of less than 4 million bushels they have increased, as shown in Fig. 1.1, to more than 1.17 billion bushels in 1971, which is an average yearly increase of more than 25 million bushels. It is not likely that such a tremendous increase in production will ever again be duplicated by soybeans or any other crop.

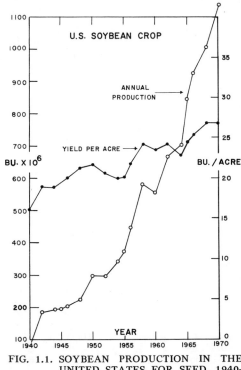

FIG. 1.1. SOYBEAN PRODUCTION IN THE UNITED STATES FOR SEED, 1940-1970

According to the National Soybean Processors Association, soybeans are now the leading U.S. cash crop, for which farmers received 2.6 billion dollars in 1970. Also, they are the number one agricultural product in our export market. Government acreage restrictions on other plantings have been a stimulant to the increasing acreage of soybeans. Because of their phenomenal success soybeans have been called the "miracle crop." C. V. Piper (Dies 1942) has referred to soybeans as "gold from the soil." After only 50 yr of commercial operations, the United States is now producing approximately 75% of the world supply. The world production of soybeans and other important commercial edible oilseeds is reported in Table 1.1.

The low cost of producing high quality oil and protein is largely responsible for the soybean success story. The total U.S. crop presently yields about 13 million tons of protein and 13 billion pounds of oil. Domestic soybean production is approximately 5.4 bu per capita of which 3.6 bu are processed into oil and meal and consumed in the United States (with part exported); the total export of meal, oil, and beans is equivalent to almost ½ of the crop. In comparison, the consumption of soybeans in Japan, which are mostly imported, are about 0.75 bu per capita and in China a rough approximation is 0.33 bu. The high consumption in the United States is the result of our extensive use of the meal in animal feeds.

In 1968 soybean utilization in the United States started to level off which led some authorities to believe that our soybean production in the United States had reached a climax. The yearly soybean carry-over had increased from a previous normal level of about 30 million bushels in 1965 to more than 230 million bushels for 1969–1970, but dropped to 99 million bushels in September, 1971.

TABLE 1.1

U.S. AND WORLD PRODUCTION OF IMPORTANT OILSEEDS
(METRIC TONS × 1000)

Year	Soybeans U.S.	Soybeans World	Cottonseed U.S.	Cottonseed World	Peanuts U.S.	Peanuts World	Sunflower World	Rape World	Sesame World
1971	32727	43636							
1970	30397	40562	3877	21764	1192	16720	9071	5425	1717
1969	30023	39752	4209	22325	1153	14880	9218	5161	1612
1968	26564	36531	2912	20319	1122	17236	9374	5319	1587
1967	25269	34785	3593	20291	1093	16234	8583	5019	1653
1966	23014	32246	5522	22405	1081	15670	7533	4344	1520
1965	19076	28074	5649	21935	952	16174	7806	4792	1544
1964	19028	28290	5617	21818	881	15256	6123	3576	1665
1963	18212	28233	5569	20829	780	14711	6804	3431	
1962	18046	28438	5423	19654	752	14124	6393	3745	
1961	15107	25060	5340	20298	779	13747	5953	3780	
1960	14503	25865	5434	20362	691	12945	4692	3609	

Source: USDA Foreign Agr. Serv.

T. H. Hieronymus (1969) stated that although soybeans "have been the Cinderella crop and the brightest part of U.S. agriculture for 30 yr" they had reached a critical stage in U.S. agriculture and because of increasing competition with other oilseeds, such as cottonseed, peanuts, sunflower, safflower, and others they faced an uncertain future.

More recent developments seem to indicate that the soybean crop will be able to sustain further expansion successfully since our excess carry-over has diminished and the 1970–1971 prices for oil and meal are at their highest levels in several years, partly because of the shortage of fish meal, sunflower meal, and other oil seeds. Still more important for the future of soybeans are the recent reports which indicate an increasing demand for soybeans in our European and Asian export markets. Perhaps of equal importance for the future of the soybean crop is its potential for increasing yields, indicated by Cooper (1970), through the efforts of our plant breeding program. It is expected that within the next few years soybean yields will increase substantially over the present average of about 26 bu per acre. Many test plots have produced in the range of 60–100 bu per acre and new research suggests that average yields approaching 40 bu per acre may be possible. If these increased yields materialize it will give soybeans an advantage over competitive oilseeds in the food and feed markets for years to come. It should be noted, also, that the food and feed technology developed in the United States in recent years can be used to good advantage in increasing our exports to developing countries.

U.S. HISTORY

Introduction of Soybeans

The United States is fortunate in having large areas of land in the Mississippi valley and in the eastern and southeastern parts of the country which are well adapted by climate and soil for soybean production.

Soybeans were introduced into the United States at a time when horse power was being replaced by motor power and U.S. agriculture was reaching a high state of mechanization. The development of mechanical equipment for harvesting soybeans was an essential factor in reducing labor costs and U.S. soybeans soon became competitive with beans grown in the Orient. By the 1950's, it was possible for the farmer to produce a bushel of beans with less than 10 min of labor, and recently, soybeans have become the leading source of oil and high quality protein for our feed and food industries.

The first mention of soybeans in American literature was in 1804 by James Mease in Willics Domestic Encyclopedia, first American Edition, in which he suggested their production in Pennsylvania. The Perry Expedition in 1854 brought back two varieties of soybeans which were distributed to interested people. Thereafter a number of literature references on the introduction of soybeans into the United States were reported by Piper and Morse (1943), but there was no interest indicated in the production of soybeans for commercial purposes until after the turn of the century.

Some European countries, especially England, started importing soybeans from Manchuria in 1908 to supplement short supplies of cottonseed and flax-seed. The beans were processed into oil and meal; the oil was used mostly for the manufacture of soap and the cake or meal for feeding dairy cattle. The success in the utilization of soybean cake and oil in Europe was an inspiration for similar experimentation in the United States on oil and cake imported from Manchuria.

Soon interest began to develop in the United States for domestic processing of soybeans, and in 1911, Herman Meyer operated a small crushing plant in Seattle, Washington on beans imported from Manchuria (Dies 1942). The short-age of industrial and food grade oils and of protein for animal feeds, partly aggravated by World War I, further encouraged the processing of domestic soybeans and in 1915 the Elizabeth City Oil and Fertilizer Company of Eliza-beth, N. Carolina was the first to crush domestic soybeans grown in that area. However, because of the difficulty in obtaining a suitable supply of soybeans and lack of experience in processing, these early ventures were unsuccessful.

At about that same period the growing of soybeans spread from the Carolinas to Illinois and surrounding states, and A. E. Staley, a corn processor located at Decatur, Ill., visualized the advantages to the midwestern states of a domestic supply of oil and meal and became a missionary among Illinois farmers on growing more soybeans. In 1921, he announced that his company would open a soybean processing plant by harvest time in 1922 (Anon. 1959).

Following the lead of the A. E. Staley Co., other oil mill processing plants were built. They were Funk Brothers, 1924; William Goodrich, 1926, acquired by Archer Daniels Midland in 1928; American Milling Co., 1927; Shellabarger Grain Products, 1929; Central Soya, 1934; and Spencer Kellogg Co., 1935. This rapid increase in soybean processing plants indicated the accelerating interest in soybeans by the farmers of the midwest.

As the number of processing plants increased it became difficult for them to secure adequate quantities of soybeans for their operations. To overcome this problem the so-called "Peoria plan" was developed whereby the farmer was guaranteed a price of $1.35 per bu for beans delivered at the plant. Also at this time, the farmers and processors were facing competition from imported soy-beans and soybean products from Manchuria. With the assistance of the Amer-ican Soybean Association, the Smoot-Hawley Tariff regulation was passed which stipulated a tariff of 2c per lb on soybeans, 3½¢ on soybean oil and $6.00 per ton on soybean meal. With the help of these tariff regulations the production and processing of soybeans increased very rapidly. Illinois became the leading soybean producing state, and soybean processing expanded in Decatur until it became known as the soybean processing capital of the world.

Processing for Oil

The early oil mill processing of soybeans was on a small scale using hydraulic and screw presses and was often combined as a part time operation with the processing of flaxseed or cottonseed (Goss 1944). Gradually the screw press replaced the less efficient hydraulic press and in 1934 the first countercurrent solvent extraction process was introduced. Some of the equipment for the early solvent plants was imported from Germany. Since the soybean has a relatively high ratio of protein to oil, it is easily formed into thin but strong flakes which are excellent for solvent extraction; thus soybeans set the pattern for countercurrent solvent extraction of oil seeds in the United States.

Nakamura and Hieronymus (1965) reported that by 1954 the total number of plants processing soybeans in the United States had increased to 261, of which 18 were hydraulic plants, 158 screw press plants, and 85 solvent extraction plants. The total number of plants processing soybeans exclusively was 88, which was a sizable decline from the 128 plants in 1951–1952 season. Of the 85 solvent plants, 57 were located in the 4 major soybean producing states of Illinois, Iowa, Indiana, and Ohio.

The early solvent plants ranged in size from 50 to 100 tons daily capacity but as soybean production increased and mechanization of the plants improved, starting about 1960, the size of the plants that processed soybeans exclusively increased in size to a range of 1000 to over 2000 tons daily capacity, for a single extraction unit.

Because of the surplus of cotton in the United States and the resulting government controls, the acreage planted to cotton has steadily decreased. Much of the vacated acreage has been replaced with soybeans, and in many of the southern states soybean acreage now exceeds that of cotton. In keeping with this movement, many of the cottonseed processing plants now process soybeans part of the year. According to the Soybean Blue Book (Anon. 1971) there were 117 mills processing soybeans in 1970; more than 50 of these plants are in the cotton belt and process both cottonseed and soybeans. Kromer (1970) stated the annual soybean crush was approaching approximately 900 million bushels.

Soybean Oil

The growth of the soybean industry in the United States was influenced more by the shortage of oil and its relatively high price than the need for protein. During the soybean introduction period the value of protein for poultry, swine, and other animal feeds and the world shortage of food protein had not been recognized. The price of oil was as high as 20¢ per lb or higher whereas the 44% protein meal or cake was as low as $20.00 per ton, or lower. The oil mill processor, by solvent extraction, would obtain approximately 10.5 lb of oil and 47.8 lb of 44% protein meal from 1 bu of soybeans, thus the oil was the more valuable product from soybeans and the meal was referred to as a by-product. The soybean processors urged the plant breeders to develop new varieties of soybeans with high oil content.

Soybean oil is highly unsaturated and is classified as a semidrying oil. In the 1920's, it was used mostly in soaps, paints, and varnishes; and considerable research was carried out to increase the industrial uses of the oil. The use of soybean oil in foods was restricted for a considerable period because of its unsaturation and high content of linolenic acid which created a flavor stability problem. This problem had to be solved before the oil became acceptable to the food industry.

By the 1930's, the industrial uses of soybean oil, because of competition with synthetic resins and detergents, started to decline. At the same time, there was an increasing demand for edible fats and oil which encouraged research on soybean oil for food uses. The research by government and industrial laboratories on oil refining, on the problem of oil flavor reversion, and on hydrogenation solved the flavor stability problem and the oil became acceptable for use in shortenings, margarines, salad dressings, and in cooking oils. Because of the abundance of soybean oil and its low cost, it has become the most important domestic vegetable food oil. Nonfood uses have declined to less than 10% of the food uses.

The plant breeding program, among other objectives, placed strong emphasis on developing new varieties of soybeans with high oil content. This program has been so successful that U.S. soybeans have a higher oil content than beans from other countries and they often receive a premium in foreign markets.

SOYBEAN MEAL AND PROTEIN

Animal Feed Industry

During the early years of soybean production, the meal or cake, as it was called, was generally regarded as a by-product and had little value; it was used as cattle feed or occasionally as a fertilizer. The use of the protein for poultry, swine, and other animal feeds was not developed until the late 1930's; and the world shortage of food protein was not generally recognized until after World War II.

The importance of the protein in animal feeds stems from preliminary observations by J. W. Hayward in feeding tests on poultry at Purdue University. He observed erratic nutritional results while feeding soybean meal, which he attributed to the variations in the methods of processing soybeans for oil; the principal variables were the time, temperature and moisture content of the beans or bean meal during processing. These observations encouraged further nutritional research by Hayward *et al.* (1936A) at the University of Wisconsin concerning the effects of heat and moisture on the nutritive value of the protein during processing. They also investigated the effects of cystine and casein supplements (Hayward *et al.* 1936B) upon the nutritive value of the heat-treated soybean meal. The results ultimately demonstrated that moist heat treatment is necessary for developing the optimum nutritional value of the protein and that the limiting amino acid in soybean protein is methionine.

Osborne and Mendel (1917) had previously reported a low nutritional value for raw soybeans when fed to rats and a good nutritional value when the beans were cooked. While this was valuable information it was not sufficient for defining the heat and moisture requirements in processing to produce optimum nutritional value in the meal for use in animal feeds.

Ham and Sandstedt (1944) discovered the presence of proteolysis inhibitors (trypsin inhibitors) in soybeans and Liener (1969) reported on the presence of a toxic hemagglutinating compound. However it was soon found that the growth inhibitors in raw soybeans were easily inactivated by moist heat treatment and thus did not affect the nutritional value of properly processed soybean meal.

Nutritional investigations with poultry demonstrated that the most efficient growth rate was obtained for broilers by feeding a ration containing 20% protein and for turkeys, 27% protein. The importance of a high protein diet for swine was demonstrated also. Thus soybean meal was established as an important ingredient for mixed feeds for the meat and poultry industries.

Poultry Industry

The technology and growth of the poultry industry has been thoroughly reviewed recently by Schaible (1970). Poultry nutrition, rations, breeding, housing, anatomy, physiology, metabolism, disease control, and economics have all played a part in the development of this tremendous industry. Schaible has summarized all of the factors responsible for increasing the feed efficiency in poultry production which is based on a high protein diet containing the dehulled and defatted soybean meal. The efficiency in broiler production increased from a gain of about 4 lb of feed per pound of bird in 1948 to the present rate of 2.2 lb of feed per pound of bird. Turkey hens and toms now can be produced with an experimental feed conversion of 1.85 and 2.25 lb of feed per pound of turkey, respectively.

Poultry and egg production, formerly the part-time occupation of the farm wife, has been relocated in the more efficient large scale poultry houses which depend mostly on the mixed feed industry for a scientifically formulated feed supply. The U.S. mixed feed industry produces annually more than 28 million metric tons of poultry feeds which is about 53% of the world production and is generally reported to consume more than 65% of the soybean meal processed in the United States.

Schaible (1970) reports that in a recent 10-yr period, broiler production increased from 1.7 to 3.0 billion birds and the consumption of chicken and turkeys increased from 15.7 lb per person in 1909 to 46 lb in 1967. Poultry production in the United States is about 1/3 of the world production.

Although poultry production in the United States has now reached a much slower growth rate, which will reduce the rate of increase in the domestic use of soybean meal, there is still a great potential for the meal in the export markets of the world. Since the cost of production of poultry is substantially below that of other sources of meat, the use of poultry is now expanding rapidly in Europe, Japan, and other countries and this industry should make a great contribution to the world protein food problem. These foreign markets will be an important factor in preventing a burdensome surplus of soybeans in the United States.

Industrial Uses

Soybeans gained favor as an agricultural crop in the United States during a period when other crops, such as corn, wheat, cotton, tobacco, were being produced in surplus quantities and soybeans took over much of the acreage vacated by these crops. At that early period it was the hope of many leaders of agriculture, government, and industry that much of the oil and protein of the soybean could be diverted from the food and feed industries into industrial products such as paints, varnishes, soap stock, plastics, adhesives, plywood glue, paper coating and lamination, paper sizing, textile fibers, and other uses (Brother et al. 1940; Smith and Max 1942). In 1936, the US organized the Regional Soybean Industrial Products Laboratory for this purpose. These new industrial uses were expected to help relieve the problem of farm surpluses. The market potential of oilseed proteins for industrial uses has been reviewed by USDA (1951). In 1935, the Glidden Company built the first plant for the isolation of industrial grade soybean protein (transferred to Central Soya in 1958). The largest use of industrial grade protein is in the paper-making industry, for coating and sizing of paper and paper board.

After World War I, soybean meal, because of its low cost, replaced casein as an adhesive for Douglas fir plywood glue, where it still retains a substantial part of the market for the interior grade product. Good glues using a combination of phenolic resin with protein concentrates were developed by McKinney et al. (1943), but they were somewhat less stable for outdoor use than the glues made with synthetic resins and failed to gain a profitable market. It now appears that use of protein glues will be limited to interior grade products and there is no reason to anticipate any substantial expansion in the use of protein in this market. The soybean glues will retain a share of this market only as long as their price remains attractive.

Protein-resin plastics were developed by Babcock and Smith (1947) and manufactured on a small scale for a few years, but were unable to compete with synthetic resins. The water absorption of protein plastics cannot be reduced to the very low level obtained with synthetic resin products, and their use in plastics, originally deemed promising, was soon eliminated from further consideration.

Commercial textile fibers which had wool-like properties were produced for several years from casein, peanut, and corn products and Boyer et al. (1945) developed fibers from soy protein but these did not reach commercial production. In retrospect, the failure of the protein fibers is attributed to insufficient knowledge of the basic chemistry of proteins and structure of protein molecules which was necessary for developing fiber wet strength. The protein fibers which had wool-like properties could have supplemented the short supply of U.S. domestic wool.

The rapid development of synthetic textile fibers combined with shifts in clothing habits have changed textile use requirements so that it does not appear possible for vegetable proteins to regain any share of this large market.

While the soybean proteins have several important industrial applications, especially in the paper industry for coating and sizing paper (Bain *et al.* 1961), which are expected to continue for years to come, the original dream of an ever-expanding industrial market has faded. In the polymer market it appears that for most applications the proteins cannot be made competitive with the increasing number of low cost, high quality synthetic resins. In any search for new industrial markets for proteins it will be necessary to find applications in which a small degree of water absorption is a desirable characteristic. It is generally recognized that the increasing demands for proteins for feed and food will greatly surpass the anticipated industrial uses.

SOYBEAN PRODUCTION

Production of soybeans in the United States has not been subjected to the government restraints that have been imposed on many other crops such as corn, wheat, cotton, and tobacco. To a large extent soybeans have taken over the acreage vacated by the surplus crops and have supplied the increasing demand for domestic oil and protein in the food and feed industries. Soybeans have become such a popular crop that in recent years their production has greatly exceeded domestic requirements and nearly 50% of the crop in the form of oil, meal, and whole beans must be exported to prevent the accumulation of a serious surplus.

With the steady increase in world population the demand for both oil and protein is still increasing but the demand for protein is increasing more rapidly than that for oil. The imbalance arises because most of the meal is consumed in animal feeds whereas most of the oil goes into human food. This imbalance has had an unfavorable influence on the domestic price of oil.

The agronomic program on soybeans, carried out by USDA and many state experiment stations has played a very important role in the soybean develop-ment program. Since a domestic shortage of oil was the most important factor stimulating the introduction of soybeans, the plant breeders have included in their program the selection and development of soybeans with a high oil content. However, they found that when the percentage of oil is increased there is usually a decrease in the protein content; 1% increase in oil may result in a 2% loss in protein. Fortunately, they found certain varieties for which this limita-tion does not hold true. Thus it now appears possible to develop soybean varieties in which the percentage of protein varies widely; the practical limits have not been defined, but it appears that eventually a level of 46% may be attained without serious loss of oil or of yield.

Of the thousands of varieties and strains which have been investigated for farm production in the United States (Cartter and Hopper 1942) only about 40 are officially recommended to the farmer for commercial production. This list is constantly changing because of the development of new and improved varieties to replace the less desirable ones; the currently recommended varieties and their area of adaptation are reported in the Soybean Blue Book, published each year by the Soybean Digest, Hudson, Iowa. The composition of presently grown varieties, on a moisture free basis, averages about 20% oil and 40% protein. The high protein soybeans will have a special advantage in processing protein isolates, concentrates, and in the mixed feed industry; and also they will have an advantage in the Orient for processing products which yield best from high protein soybeans such as tofu, miso, and shoyu.

ORIENTAL HISTORY

Ancient History

Soybeans are reported to have originated in Eastern Asia and were used as food long before the existence of written records. The wild soybean (Fig. 1.2), a vine-like plant that grows prone on the ground, is regarded as the ancestor of present-day soybean varieties.

Courtesy of USDA Regional Soybean Laboratory

FIG. 1.2. WILD SOYBEANS — PRONE ON THE GROUND
COMPARED TO MODERN VARIETIES

Legend tells us of the first use of soybeans for food when a caravan laden with gold and gems, returning to an eastern Chinese city was attacked by bandits. The caravan took refuge in a cave and was nearing starvation when a servant, upon eating beans from a vine-like plant, recovered his vigor and induced others also to eat the bean, which sufficed as food until the bandits were driven away. This bean, supposedly the wild soybean, was thus established as a food in China.

The first Chinese records which mention soybeans date back to about the time of the building of the Egyptian Pyramids. In 2838 B.C., Emperor Shang-Nung published the books of Pen Ts'ao Kong Mu, which described the plants of China including a description of the soybean, (Shih 1918). Soybeans were mentioned frequently in later Chinese records and have been considered China's most important legume. Soybeans were reported to be one of the "Mu Ku" or sacred grains of China, which included also rice, wheat, barley, and millet and were considered essential for the existence of the Chinese race.

The great interest in growing soybeans as well as the advanced state of agriculture in China at that early period are indicated by writings published as early as 2207 B.C. in which the agricultural experts gave advice on raising soybeans. The experts advised on the selection of proper soil types, the best varieties, proper time of planting, method and rate of planting, time of harvest, method of storage, and the utilization of the many varieties for different purposes. It appears that even in ancient times the soybean occurred in various sizes and in several colors including black, brown, yellow, and mottled.

The soybean has many different names depending on the country where it is grown and used. It is generally reported that its name was derived from the Chinese "chiang-yiu" which means soy sauce; in Japanese it would be pronounced "sho-yu." Rather recent names include soya bean, soja bean, soy, so-ya, Chinese pea, Manchurian bean, and soia. Piper and Morse (1943) have recorded more than 50 names which occur throughout the Orient.

A materia medica published in China in about A.D. 450 recommended the use of soybeans as a drug. They were regarded as a specific remedy for proper functioning of the heart, liver, kidneys, stomach, and bowels. They were recommended also as a remedy for constipation, as a stimulant for the lungs, for eradication of poisons from the system, improving the complexion by cleaning the skin of impurities, and stimulating the growth and appearance of hair. From our present knowledge of nutrition it is not unreasonable to believe that because of the poor diet of the people at that early period, increasing the protein in their diet would have had a beneficial effect on their health.

The Buddhist religion, because of the exclusion of meat from the diet of its people, without doubt, was a major influence in the development of soybeans for food in China, Korea, Japan, and other Oriental countries. It is commonly reported by the food processors of the Orient that the development of the rather sophisticated soybean food products was the result of the Buddhist priests working in their monasteries. This influence is emphasized also by Brandemuhl

(1963) in his history of soybeans where he points out that the introduction of soybeans into Japan "coincides quite closely with the spread of Buddhism in 500–600 A.D."

Oriental Fermented Foods

We do not know how soybeans were first prepared for eating; we can speculate that roasting or boiling in water were the methods used first. The green or immature soybean is easily prepared for the table by cooking in water for a short period (Woodruff and Klaas 1938). However, the mature dry bean requires a long period of cooking for developing a soft texture, and at that stage it is rather bland. Normally, a vegetarian diet is quite bland in comparison with a meat diet where an abundance of flavor is contributed by the fat. The flavor problem of the Chinese people was resolved by the development of fermented soybean products. Because of poor and unavailable records on the development of the various soybean foods the time of their origin is somewhat uncertain and the dates are considered as approximations.

Shoyu—Shoyu or soy sauce, a widely used condiment, is a dark brown liquid made by fermentation of a combination of soybeans and cereal, usually wheat; it has a pleasant, aromatic odor and a salty taste suggesting a meat extract. It is made by fermenting cooked soybeans with a koji or mold preparation *(Aspergillus oryzae)* which is grown on parched or cooked wheat. Komiya (1964) has reported that shoyu originated during the Chau dynasty in 1134–246 B.C. Some historical accounts seem to indicate that the fermentation of shoyu may have been patterned after a fermented fish product used along the coast of Southeast Asia and developed prior to shoyu. Shoyu is one of the oriental foods that is being accepted by western people.

Miso.—The preparation of miso (Japanese) or soy paste (Chinese) is by a fermentation similar to that used for shoyu *(Aspergillus oryzae)* except that the koji is made from rice or barley and the residue which remains after fermentation is not removed. The color and flavor of miso varies widely, from nearly white to nearly black with a corresponding change in flavor depending on the ratio of soybeans to rice; the higher the ratio of rice, the lighter the color and the sweeter the flavor. Miso is used primarily as a soup base although in some areas it is used also for pickling vegetables.

Miso originated in China, probably in the same period as shoyu and spread later to Korea, Japan, and other countries. Until quite recently, the production of miso as well as shoyu were family operations, especially in China. The fermentation of the mixture of soybeans and koji was carried out in large earthen jars and required several months to a year or more for completion. The jars were located in the open and were covered only during rainy periods (Smith 1949). The longer the fermentation period the better the flavor of the product: it is required that the minimum fermentation time shall include one summer period. *Fig.* 1.3 shows the fermentation jars in a large shoyu plant in Shanghai in

FIG. 1.3. SHOYU OR SOY SAUCE PLANT IN SHANGHAI, CHINA, 1948

1948; miso fermentation was carried out in the same type of equipment except that a small operator would use fewer than a dozen jars.

It was very common for rural people to make their own miso. However by 1948, the production of shoyu and miso in Japan had reached a higher stage of technical development. In recent years the Japanese have developed a dry miso soup which has a good flavor; however, it has a limited sale probably because miso itself can be preserved for a long period.

Tempeh.—Tempeh kedele, usually called tempeh (tempe) is a very popular Indonesian food product prepared by a short time fermentation of soybeans using a special species of *Rhizopus.* Tempeh has been used in Indonesia for hundreds of years but the history of its development is now known (Djien and Hessletine 1961). Since Indonesia is located near the equator, it has a limited temperature variation, between $20°$ and $30°$ C, which is optimum for the growth of most microorganisms and favors the development of fermented food products without the need of technical knowledge. The method of making tempeh is given in Chap. 11.

Freshly made tempeh has a very delectable flavor but efforts at preserving its flavor for a period long enough for large scale production and marketing have not yet been successfully accomplished. However, using present processing methods, many people in Indonesia make a living in the preparation and marketing of freshly prepared tempeh, and Djien and Hesseltine (1961) have estimated that half or more of the 17 million bushels of soybeans produced in 1959 in Indonesia were used for making tempeh.

Ontjom.—Ontjom is another Indonesian food product made by short time fermentation of peanuts, coconut presscake, or the residue from soybean milk production. Using a mold belonging to the genus *Neurospora*, Steinkraus *et al.*

(1965) have reported a procedure similar to that used in the preparation for making ontjom, except they used soybean cotyledons as the substrate and reported that the product had an acceptable texture and nutlike flavor.

Natto.—Natto is made by a short period fermentation from cooked soybeans and is reported (Piper and Morse 1943) to have been used first by the Buddhist monks but is now used in a number of localities in Japan. Natto has a grey color, does not have much odor, but has a strong and rather persistent musty flavor. Its flavor would not be acceptable to most occidental tastes and, in fact, many of the Japanese people do not eat natto. It is produced and consumed only in localities where its taste is favored. In the ancient method of preparation the microorganisms used in the fermentation were derived from rice straw used in the process.

To make natto, small portions of well cooked beans were wrapped in straw and placed in a room which was maintained at a temperature of 35–40° C to ferment for about 24 hr. The fermentation process supplies part of the heat to maintain the temperature of the room. During the fermentation period the beans become covered with a viscous, sticky fluid or mucus material which seems to determine to a large extent the quality of the natto. The fluid has the property of forming long threads when touched with the finger; the longer the strings, the better the quality of the natto. Natto is a low cost but highly nutritious food.

Hamanatto.—Hamanatto is made by fermentation of whole soybeans and is produced in a limited area in Japan, primarily in the vicinity of Hamanatsu. Hamanatto should not be confused in any way with natto since different microorganisms are used in fermenting the two products and the flavor of the two products is quite different. Hamanatto has a pleasant flavor resembling that of miso or shoyu but is sweeter. Factors unfavorable to hamanatto are its dark color (it is nearly black) and its high cost. Methods of producing it have never been modernized; thus, its production costs are comparatively high. Hamanatto is reported to have come to Japan from Korea more than 350 yr ago. The ancestors of the people owning the Yamaya Brewery and the Saito Miso Plant of Hamamatsu in Japan are said to have inherited the process from the Buddhist monks who developed it in their temples (Smith 1958).

To make hamanatto, the beans are soaked in water for 4 hr and then steamed lightly for 10 hr. The beans are cooled and their surface is covered with a koji prepared from roasted wheat or barley and placed on trays in a fermenting room for 20 hr. The beans, which now are covered with a green mold, are dried in the sun to 12% moisture. The dry beans are placed in wooden buckets of about 15 gal. capacity. Strips of ginger are placed in the bottom of the buckets before adding the beans and then they are covered with salt water. A cover which fits inside the bucket is placed over the beans and a very heavy weight, nearly 100 lb, is placed on the cover. Fig. 1.4 shows the buckets with the stone weights used during fermentation which requires 6–12 months and must include 1 full

FIG. 1.4. FERMENTATION OF HAMANATTO, JAPAN,
1957

summer. During fermentation the beans acquire a dark reddish color. After fermentation the beans are dried again in the sun which turns them black.

Tao Tjo.—Tao tjo is another fermented product in small scale production in Indonesia and Thailand. It is made by mixing cooked beans with roasted rice flour with some leaves added; the leaves probably supply the fermenting organism, which appears to be mostly *Aspergillus sp.* Within a few days the mass is covered with fungus. It is dried in the sun and then soaked in brine. In a few days sugar and tapeh (rice yeast preparation) are added and it is again placed in the sun. It is served with vegetables, fish, or meat. Tao tjo appears to be a modified (short period fermentation) miso type product.

Kochu Chang.—Kochu chang (Smith 1949) is an interesting fermented product used on a small scale in Korea. Its history is unknown, but it probably originated a long time ago. To make kochu chang, the beans are boiled, mashed, and inoculated from a previous batch of kochu chang, and then hung in sacks to cure for 2–3 months. The fermented material is broken up, dried in the sun, and ground to a fine powder. Finely ground red peppers, salt, and water are added and the combination allowed to age in earthen jars for 2–3 months. To prepare for the table, the kochu chang is cooked with meat and sugar added if available.

Ketjap.—Ketjap is a soy sauce made in Indonesia by a short period fermentation of soybeans using *Aspergillus oryzae.* Black soybeans are soaked and cooked in water and fermented for 2–3 days (Djien and Hesseltine 1961) and then put in a salt brine for about 8 days. The bean mass is filtered and the residue cooked several times with fresh water which is added to the filter to extract all of the solubles. Sugar and various flavors are added, according to the taste of the manufacturer, and the solution concentrated by slow evaporation, and bottled for the market. There is no available history of the development of this product but it appears to be a shortened procedure of the Chinese method of making shoyu.

Oriental Nonfermented Foods

The principal nonfermented soybean foods include soybean milk (Smith and Beckel 1946), tofu, yuba, and bean sprouts. The traditional soy milk is popular only in China and Hong Kong; in the latter it is a rather recent development and is processed and sold as a soft drink.

Soybean Milk.—Soybean or vegetable milk or Fu Chang in Chinese is reported to have been developed and used in China before the Christian era (Piper and Morse 1943) by the philosopher Whi Nain Tze, who is credited also with the development of tofu. The tofu and yuba are closely associated with soy milk since the preparation of the milk is the first step in the processing of tofu and yuba. The traditional milk is made by soaking the beans in water overnight, wet grinding the beans, heating the wet mash to improve flavor and nutritional value, and filtering. Figure 1.5 shows the marketing of soy milk on the streets of Canton, China in 1948.

Tofu.—Tofu (Japanese), Dan fu (Vietnamese), Teou fu or Tou fu ho (Chinese) or bean curd is a cottage cheese-like product formed into a cake, which is precipitated from soy milk by a calcium salt or, in some instances, by concentrated sea water. Shih (1918) states Tou fu ho originated in the Han dynasty during the reign of Huai Nan Wang, A.D. 22 at Luian. Figure 1.6 shows tofu on the Tokyo market in 1957. Tofu can be prepared for the table in many different ways; the most important are in soup and by deep fat frying, the latter called aburage in Japanese and Yu Tou Fu in Chinese. Figure 1.7 shows the preparation of aburage from fresh tofu. The modern methods of processing these products are described in chap. 10.

Wedge Press

The early history of the production and food uses of soybeans in China does

FIG. 1.5. MARKETING SOY MILK, CANTON, CHINA, 1948

FIG. 1.6. FRESH TOFU IN JAPANESE MAR-
KET, 1957

FIG. 1.7. PREPARATION OF ABURAGE FROM TOFU,
JAPAN, 1957

not make reference to the processing of soybeans for oil or to the food uses of
the oil; thus it must be concluded that the processing of beans for oil and meal is
a comparatively recent operation. Their early equipment, which was still in use
in 1948 (Smith, 1949), was ingenious but quite primitive and probably represents the earliest type used for expressing oil from soybeans as well as other
oilseeds. It was called the "wedge press." There were several different designs of
the wedge press but they all operated on the same principle. The beans were
cracked into grits by crushing or coarse grinding between millstones; then
steamed for a short period by placing in gunny bags which were laid on wooden
gratings and placed over tubs of boiling water. The steamed beans were then
formed into a flat disc about 4-5 in. thick by use of a circular steel mold and the
cakes thus formed were placed in the wedge press. (see Fig. 1.8). When the press
was full, wooden wedges were driven by a sledge hammer or other hand-powered

FIG. 1.8. WEDGE PRESS FOR EXPELLING SOYBEAN OIL, CANTON, CHINA, 1948

device to compress the grits until the oil was expelled. In China at that period, most of the soybean oil was used as unrefined or as water washed oil.

In 1948 in the larger cities and Manchuria, the wedge press was being replaced by the hydraulic press. Figure 1.9 shows cakes of hydraulic pressed meal being loaded on a truck on the Bund in Shanghai.

SOYBEANS AND WORLD FOOD PROBLEMS

Green Revolution

With the recent development of high yielding varieties of rice and wheat, high lysine corn, and the general increase in the production of other crops throughout the world (Anon. 1970A), it is now recognized that by providing adequate technology to the production of these crops the world food energy requirements for the next few years can be met. These new developments are referred to as the "Green Revolution." However, because of the low protein content of the cereals and the imbalance of their essential amino acids, they do not supply adequate protein for satisfactory growth of babies and children, and for bodily maintenance of laboring people. Thus, protein supplementation of the cereals is required and soybeans have become heavily involved in this part of the world food program (Gould 1966).

Protein Supplements

The proteins of meat, poultry, milk, and eggs are very expensive compared with vegetable proteins and there is no possibility that they can be produced in quantities adequate for supplementing the cereal proteins (Desrosier 1961). The

FIG. 1.9. CAKES OF HYDRAULIC PRESS MEAL ON THE
BUND, SHANGHAI, CHINA, 1948

other sources of protein which are expected to help supplement the cereals
(USDA 1969) are cottonseed, peanuts, sesame, sunflower, fish protein concentrate, and single cell microorganisms such as algae, fungi, yeast, and bacteria
derived from the fermentation of hydrocarbons. Also, Pirie has developed a
process for the extraction of nutritious protein from green plants (reviewed by
Hartman *et al.* 1967). However, because of its unfavorable color, texture, and
collection problems it has not yet gained any significant acceptance.

At present, soybean protein outranks all of the other proposed supplements
(USDA 1967) in the worldwide nutrition program. This lead over other proteins
is attributed to the abundance of soybeans, their low cost, and an active research
program as well as their historical use for food in the Orient. New methods for
processing the glanded varieties of cottonseed, which eliminate most of the
gossypol, should soon contribute substantially to the food program. Also, the
new varieties of glandless cottonseed, now being introduced in many countries,
from which protein products with a bland flavor and a nearly white color can be
processed, should readily gain an important place in the world food program.

Peanuts are cultivated in many countries, but the United States is the only
one where they are grown primarily for food use. It seems that peanuts have a
low social status in many developing countries, which restricts acceptance of the
whole peanut or peanut protein for food. With a rather minimum anount of
additional research, peanut protein products can be produced to make a very
substantial contribution to the world food problems (Woodroof 1966).

Single cell proteins (Snyder 1970), which present the problem of reducing
their high nucleic acid content before they can be generally accepted for food
use, and fish protein concentrates, which are rather expensive to collect and

process into acceptable food products, will require a great deal more research and development before they can contribute substantially to the food program. However, the single cell proteins may contribute to our food program by their use in animal feeds.

A list of 15 fortified food products, their names, the names of the companies responsible for their production and their general composition are given in Table 1.2 (Anon. 1970B).

The products described in Table 1.2 have been produced for varying periods

TABLE 1.2

HIGH PROTEIN FOOD FORMULATIONS (AID FUNDED)

Product	Producer	Composition	Marketing Status
Bal ahar (blend, like farina)	Government of India	Bulgar wheat, peanut flour, NFDM, vitamins, minerals; 22% protein	Well accepted for past 6 months; each unit produces 50 to 100 tons daily; 8¢ per lb
Cerealina (weaning food)	CPC International (Brazil)	Full-fat soy flour, cornstarch, milk powder, vitamins, minerals; 20% protein	Good acceptance in consumer test market for 21 months
Fortified atta (flour; after cooking, resembles unleavened bread)	Flour mills in Bombay, India	Wheat, peanut flour, vitamins, iron, calcium; 13.5% protein	Incremental cost met by 40% government subsidy; program expansion planned to New Delhi and Calcutta
Golden elbow macaroni	General Foods (Brazil)	Corn flour, soy flour, wheat flour, calcium carbonate, calcium phosphate, iron, vitamin B; 20% protein	Good acceptance; 15¢ per lb
Incaparina (cereal blend with water)	Central Distribuidora, Cia Ltda., Guatemala City, Guatemala	Cottonseed flour, corn flour, vitamins, minerals, torula yeast	Satisfactory volume after 3 yr; 4 ton-per-day production; 20¢ per lb
Kupagani biscuits	Pyott, Ltd., Union of South Africa	Wheat meal, soy meal, sugar, fat, glucose, milk powder, vitamins, minerals; 7.9% protein	Well received
Leche alim (cereal)	Pediatrics Laboratory, University of Chile, Santiago	Toasted wheat flour, fish protein concentrate, sunflower meal, skim milk powder; 27% protein	Institutional product; produced for 1½ yr; 80 ton-per-day production; 13.6¢ per lb

TABLE 1.2 (cobtinued)
HIGH PROTEIN FOOD FORMULATIONS (AID FUNDED)

Product	Producer	Composition	Marketing Status
Modern bread (sandwich, fruit, sweet, masala, milk, brown types)	Commercial bakeries in India	Wheat flour, lysine, vitamins, minerals	Well accepted for 2 yr; 10,000 ton-per-year production
PL (reconstitutable beverage)	Instituto Nacional de Nutricion, Caracas, Venezuela	NFDM, sugar, vitamins, minerals, methionine, 12 different flavors	Rural distribution by government for 14 yr to children referred by physicians
ProNutro cereal and ProNutro soup	Food Corp. (Pty.), Ltd., Durban, Natal, Republic of South Africa	Soy flour, corn flour, peanut flour, NFDM, wheat germ; 22% protein	Full commercial production for 6 yr; 250 tons sold weekly; 2.5¢ per lb (mfg. cost)
Puma (beverage)	Dih, Ltd., Guyana; Monsanto	Vegetable protein, sugar, vitamins, flavor	Good acceptance after 1 yr; marketed through soft drink system
Superamine (reconstitutable powder)	FAO/WHO/ UNICEF with Algerian government	Wheat flour, chickpea flour, lentil flour, dried skim milk, sugar, vitamins, minerals, flavor; 20.9% protein	Production: 140 tons in 1968; 400 tons in 1969; 8000 tons projected by 1974; 23.6¢ per lb
Vita bean (soybean milk)	Yeo Hiap Seng, Ltd., Singapore	Soybeans, sugar, vitamins	Good acceptance for 2 yr; 4500 carton-per-hour production; 6¢ per 10 oz
Vitalia (macaroni products)	Instituto de Investigaciones Tecnoligicas, Bogota, Colombia	Semolina, wheat, soya, corn, rice derivatives; 17.8% protein	Test acceptance good; Proposed production: 6000 tons per year; 10¢ per lb
Yoo Hoo (milklike beverage)	Yoo Hoo Beverage Co., Carlstadt, N.J.	Blend of animal and vegetable protein products; animal protein from nonfat milk solids	Marketed in U.S. and abroad; produced in 14 overseas countries; 3¢ per bottle (mfg. cost)

Source: Anon. (1970B).

and it is not possible to evaluate their relative acceptability or usefulness in the world food program, although some show much promise. ProNutro, one of the older products, has been in full production for 6 yr and is reported to have production of 250 tons weekly. Incaparina, also one of the older products, is rated at 4 tons per day. Leche alim, an institutional product, is produced in Santiago, Chile; 18 months after introduction it had a daily production of 80 tons per day. Vitasoy, a soy milk-like product not mentioned in the table, is produced in Hong Kong and sold as a soft drink; it is reputed to be the most profitable of the protein beverages.

Amino Acids.—Recent developments which have greatly reduced the cost of production of synthetic lysine and methionine have centered much attention in their use for supplementing the amino acids in vegetable proteins. Lysine is presently being used especially for supplementing wheat flour protein in bread. Initial trials in India, Japan, Thailand, and Tunisia indicate that this method of increasing the nutritive value of wheat protein will have wide acceptance (Iwan 1968). It is anticipated that the cost of other essential amino acids, such as tryptophan, may be reduced and become available for protein supplementation. However, the fortification of proteins with lysine must be compared for efficiency with supplementation by soy protein. This has a surplus of lysine, and, when used to supplement cereal-based foods, it not only balances their amino acid profile but also increases their total protein quantity.

CSM.—A new product, sponsored by the USDA Foreign Agricultural Service (USDA 1970 A,B) which is not included in Table 1.2, is made up of gelatinized corn meal, soy flour, and NFDM with added minerals and vitamins, and is referred to as CSM. The toasted soy flour may be either the full-fat or defatted type. The ingredients are listed (USDA 1970 A,B) as follows:

	Lb per 2000 Lb CSM
Corn meal, gelatinized	1276
Soy flour, defatted	484
NFDM, spray-dried	100
Calcium phosphate, dibasic FCC grade, $CaHPO_4$-$2H_2O$	12
Mineral mix	26
Vitamin antioxidant premix	2
Soy oil, refined, stabilized	100

The mineral premix, for each 2000 lb is made up of precipitated calcium carbonate, 12 lb; zinc sulfate .$7H_2O$, 36 gm; ferrous fumarate, 418 gm; and iodized salt (0.007% I_2), 13 lb.

The vitamin-antioxidant premix (USDA 1970B) for each 2000 lb is:

	Grams in Final Food
Thiamine mononitrate	2.5
Riboflavin	3.5
Pyridoxine hydrochloride	1.5
Niacin	45.0
Ca D-pantothenate	25.0
Folacin	1.8
Vitamin B_{12}	36.0 (mg)

Vitamin A (stabilized retinyl palmitate)	15.0 (mil USP units)
Vitamin D (stabilized)	1.8 (mil USP units)
Alpha tocopherol acetate	68,000.0 (IU)
Butylated hydroxyanisole	20.0
Butylated hydroxytoluene	20.0
Ascorbic acid	364.0
Total	2.0 (lb)

Soy flour, defatted (toasted) or starch is added to reach a total weight; additional soy flour may be added as a carrier, if desired.

The CSM, which is used as a gruel or for supplementing other foods, is a highly nutritious, low cost food, with a modified flavor of corn meal, containing 20% protein (Vojnovich and Pfeifer 1970). Shipments totaling 1.2 billion pounds had been made to a number of countries around the world by June 1970; however its degree of success is yet to be determined.

A product similar to CSM is made by replacing the corn meal with wheat flour. Because of the gluten in the wheat the physical properties are somewhat different; this product is designated WSB. Vojnovich and Pfeifer (1970) investigated the stability of coated and uncoated ascorbic acid in CSM and WSB under various storage conditions of moisture and temperatures. In CSM at a temperature of 26° C and moisture level of 11.8% there was a loss of about 30% of uncoated ascorbic acid in 7 months, whereas in WSB about the same loss occured at 45° C in the same period, at 13% moisture. The ascorbic acid coated with ethyl cellulose had substantially greater stability than the uncoated product. (WSB signifies "Wheat-Soy Blend.")

Cottage Industries

An approach to solving the food problems of developing countries, which has received a minimum of effort, is the development and use of "cottage" or "family operated" industries. This approach uses a minimum of capital and technology. An historical example of a successful cottage industry is the production of tofu in Japan and China; tofu has made an important contribution to the food requirements of these countries since it is the largest food use of soybeans. While the processing of tofu is presently being converted to modern, sophisticated operation, there are yet thousands of family operations still in use. Unfortunately, there has been little or no effort to extend the use of tofu outside of its original area, and with the spreading of soybean production to India, Brazil, and other countries, a serious effort should be made to extend the use of tofu in its present or perhaps a modified form to these countries.

Recent work by Mustakas et al. (1967) at the USDA Northern Regional Research Laboratory (NRRL) with AID assistance is an attempt to modify the processing of full-fat soy flour to a cottage industry. In their process a small quantity of soybeans is cleaned, washed, soaked in water, and then cooked in a steel drum. The cooked beans are dried in the sun, cracked and dehulled with simple equipment, and the meats ground into grits or soy flour with small-scale equipment operated by hand, or if available, an electric motor.

Albrecht *et al.* (1966) compared the rate of cooking of soybeans by atmospheric steaming with immersion cooking, with variations in initial moisture, particle size, and hull removal. The rate of cooking was correlated, also, with inactivation of urease and trypsin inhibitor.

They found that the initial moisture in the bean is a major factor influencing the rate of cooking as well as enzyme inactivation. By steaming soybean grits (under 20 mesh) at low moisture (8%) it is possible to inactivate the urease and trypsin inhibitor and retain high protein dispersibility (NSI). They found that immersion cooking of soaked whole soybeans for 5-7 min retained a high NSI, at the same time inactivating the urease. The urease and trypsin inhibitor were inactivated at about the same rate. The removal of the seed coat did not increase the cooking rate.

The Japanese have a product similar to full-fat soy flour, made from whole beans, which they have produced for a number of years and is not one of their traditional soybean products. This product, kinako, can be readily modified for processing on a family or village scale. In this process the clean beans are roasted in an oven to remove the bitter taste and to develop nutritional value and ground to a fine particle size. If desired, a simplified method for dehulling the beans can be added to the process.

Since a family type operation has low labor and distribution costs and requires a minimum of equipment and technology, it has many advantages over the other approach of first building a costly, sophisticated processing operation which may require technically trained people for its operation. As the country develops the sophisticated operations will follow.

BIBLIOGRAPHY

ALBRECHT, W. J., MUSTAKAS, G. C., MCGHEE, J. E., and GRIFFIN, E. L., JR. 1966. Rate studies on atmospheric steaming and immersion cooking of soybeans. Cereal Chem. *43*, 400–407.

ANON. 1959. Soybeans curiosity 40 years ago in U.S. Decatur Herald Rev. Jan. 25, Decatur, Ill.

ANON. 1970A. Green Revolution. Proceedings Subcommittee on National Security Policy and Scientific Developments. Ninety First Congress. U.S. Govt. Printing Office, Washington, D.C.

ANON. 1970B. Fortified foods: the next revolution. Chem. Eng. News *48*, No. 33, 35–43.

ANON. 1971. Blue Book Issue, Soybean Dig., Am. Soybean Assoc., Hudson, Iowa

BABCOCK, G. and SMITH, A. K. 1947. Extending phenolic resin plywood glue with proteinaceous materials. Ind. Eng. Chem. *39*, 85–88.

BAIN, W. M., CIRCLE, S. J. and OLSON, R. A. 1961. Isolated soy protein for paper coating. Tappi Monograph Ser. *22*, 206–241.

BOYER, R. A., ATKINSON, W. T., and ROBINETTE, C. E., 1945. Artificial fibers and manufacture thereof. U.S. Pat. 2,377,854. June 12.

BRANDEMUHL, W. 1963. Soybean History. Aspects of Buddhist Influence. Anthropology Dept., Univ. Wisconsin.

BROTHER, G. H., SMITH, A. K., and CIRCLE, S. J. 1940. Soybean protein—resume and bibliography. USDA, *ACE-62.*

CARTTER, J. L., and HOPPER, T. H. 1942. Influence of variety, environment and fertility level on the chemical composition of soybean seed. USDA Tech. Bull. *787.*

CHEN, P. S., and CHEN, H. D. 1956. Soybeans for Health, Longevity and Economy. The Chemical Elements, South Lancaster, Mass.

COOPER, R. L. 1970. Early lodging: a major barrier to higher yields. Soybean Dig. *30,* No. 3, 12–13.

DESROSIER, N. W. 1961. Attack on Starvation. Avi Publishing Co. Westport, Conn.

DIES, E. J. 1942. Soybeans—Gold From the Soil. Macmillan Co., New York.

DJIEN, K. S., and HESSELTINE, C. W. 1961. Indonesian fermented foods. Soybean Dig. *21,* No. 1, 14–15.

DROWN, M. J. 1943. Soybeans and soybean products as food. USDA Misc. Publ. *534.*

GOSS, W. H. 1944. Processing soybeans. Soybean Dig. *5,* No. 1, 6–9.

GOULD, R. F. (Editor) 1966. World Protein Resources. Am. Chem. Soc. Advan. Chem. Ser. *57.*

HAM, W. E., and SANDSTEDT, R. M. 1944. A proteolytic inhibiting substance in the extract of unheated soybean meal. J. Biol. Chem. *154,* 505–506.

HARTMAN, G. H., AKESON, W. R., and STAHMANN, M. A. 1967. Leaf protein concentrates prepared by spray drying. J. Agr. Food Chem. *15,* 74–79.

HAYWARD, J. W., STEENBOCK, H. and BOHSTEDT, G. 1936A. The effect of heat as used in the extraction of soybean oil upon the nutritive value of the protein of the soybean meal. J. Nutr. *11,* 219–234.

HAYWARD, J. W., STEENBOCK, H. and BOHSTEDT, G. 1936B. The effect of cystine and casein supplements upon the nutritive value of the protein of raw and heated soybeans. J. Nutr. *12,* 275–283.

HIERONYMUS, T. H. 1969. Soybeans: End of an Era. J. Illinois Agr. Econ. *9,* July 2.

IWAN, J. L. 1968. Fortified bread takes hold in India. Cereal Sci. Today *13,* 202, 206.

KOMIYA, A. 1964. Japanese soy sauce offers market for U.S. soybeans. Soybean Dig. *24,* 43.

KROMER, G. W. 1970. U.S. Soybean processing capacity expanding USDA, *FOS-255.*

LIENER, I. E. 1969. Toxic Constituents of Plant Foodstuffs. Academic Press, New York.

MARKLEY, K. S. 1950. Soybeans and Soybean Products. John Wiley & Sons, New York.

MCKINNEY, L. L., DEANIN, R., BABCOCK, G., and SMITH, A. K. 1943. Soybean—modified phenolic plastics. Ind. Eng. Chem. *35,* 905–908.

MUSTAKAS, G., ALBRECHT, W. J., BOOKWALTER, G. N., and GRIFFIN, E. L. 1967. Full fat soy flour by a simple process for villagers. USDA—ARS *71-34.*

NAKAMURA, H., and HIERONYMUS, T. A. 1965. Structure of the soybean processing industry. Univ. Illinois Agr. Expt. Sta. Bull. *706.*

OSBORNE, T. B., and MENDEL, L. B. 1917. The use of soybeans as food. J. Biol. Chem. *32,* 369.

PIPER, C. V., and MORSE, W. J. 1943. The Soybean. Peter Smith, New York.

ROCKEFELLER FOUNDATION. 1968. Strategy for the Conquest of Hunger. Rept. Proc. Rockefeller Found.

SCHAIBLE, P. J. 1970. Poultry: Feeds and Nutrition. Avi Publishing Co., Westport, Conn.

SHIH, C. Y. 1918. Beans and soybean products. Biol. Dept., Soochow Univ., Shanghai, China.

SMITH, A. K., and BECKEL, A. C. 1946. Soybean or vegetable milk. Chem. Eng. News. *24,* No. 1, 54–56.

SMITH, A. K. 1958. Use of United States soybeans in Japan. USDA—ARS *71-12.*

SMITH, A. K., and BECKEL, A. C. 1946. Soybean or vegetable milk. Chem. Eng. News. *24,* No. 1, 54–56.

SMITH, A. K., and MAX, H. J. 1942. Soybean protein: adhesive strength and color. Ind. Eng. Chem. *34,* 817–820.

SNYDER, H. 1970. Microbial sources of protein. Advan. Food Res. *18,* 85–138.

SPILSBURY, C. C. 1969. U.S. soybean market in the Republic of China (Taiwan). USDA—FAS *209.*

STEINKRAUS, K., Lee, C. Y., and BUCK, P. A. 1965. Soybean fermentation by the ontjom mold *neurospora*. Food Technol. *19,* 119–120.

USDA. 1951. Marketing potential for oilseed protein materials in industrial uses. USDA Tech. Bull. *1043.*

USDA. 1967. Proc. Intern. Conf. Soybean Protein Foods. ARS *71–35.* Peoria, Ill., May 17–19. USDA–ARS *71–35.*

USDA. 1969. Conference on Protein-Rich Foods from Oil Seeds. New Orleans, May 15–16. USDA–ARS *72–71.*

USDA. 1970A. Announcement for the purchase of corn-soy milk for use in foreign donation programs. Announcement *CSM–1,* Febr. 7. USDA–FAS.

USDA. 1970B. Announcement *CSM–2,* July. USDA–FAS.

VOJNOVICH, C., and PFEIFER, V. F. 1970. Stability of ascorbic acid in blends with wheat flour, CSM and infant cereals. Cereal Sci. Today. *15*, 317–322.

WOODROOF, J. G. 1966. Peanuts: Production, Processing, Products. Avi Publishing Co., Westport, Conn.

WOODRUFF, S. and KLAAS, H. 1938. Study of soybean varieties with reference to their use as food. Univ. Illinois Agr. Expt. Sta. Bull. *443.*

CHAPTER 2

R. W. Howell
B. E. Caldwell

Genetic and Other Biological Characteristics

NOMENCLATURE AND DESCRIPTION

The soybean, *Glycine max* (L.) Merr., is a member of the family Leguminosae, subfamily Papilionoideae (Hermann 1962). Other nomenclatures which have been used include *Phaseolus max* (L.), *Soja max* (L.) Piper, and *Soja hispida* Moench.

Hermann (1962) has published a revision of the genus *Glycine*. Several hundred species have been assigned to the genus in the past, but Hermann now assigns only ten, including a few subspecies. The species are grouped in three subgenera. *G. max* and *G. ussuriensis* Regel and Maack comprise the subgenus *Soja* (Moench) F. J. Herm.*G. max* is said by Hermann to be "a derivative of *G. ussuriensis* or some Asiatic ancestor closely related to it."

Morphology

The soybean plant has alternate, trifoliolate leaves except at the first two nodes. A pair of opposite simple leaves occur at the second node. The cotyledonary node is considered the first. The leaves, stems, and pods are normally covered with a gray or brown (tawny) pubescence, which is very noticeable at maturity and is useful in variety identification. Hairs on U.S. varieties consist of a long (1-3 mm) cylindrical cell and 1, 2, or 3 basal cells (Singh *et al.* 1971). Five aberrant types of pubescence expressions, each different from the normal by a single gene, have been described by Bernard and Singh (1969). These types are glabrous, curly, dense (3-4 times the normal number of hairs), sparse (1/3-1/4 the normal number of hairs), and puberulent (minute, stubby hairs). Singh *et al.* (1971) reported that yields of normal, dense, and sparse pubescence lines were similar to each other and superior to yields of the curly and glabrous lines. Differences in plant vigor among different pubescence types were attributed to differences in infestation by the potato leafhopper (*Empoasca fabae* Harris). The number of leafhoppers per plant was governed by pubescence type and was the greatest by far in plants of the curly and glabrous types.

Soybeans display an indeterminate or determinate growth habit. Varieties adapted to the northern part of the United States are mostly indeterminate. Those adapted to the southern area are determinate and have a pronounced terminal raceme with as many as 20 flowers. Woodworth (1932, 1933) described the difference between determinateness and indeterminateness as due to a single gene pair. Bernard (1964) has extended the description to include additional growth types.

Flowers are borne in the axillary position and are 6–7 mm in length. A dozen or more flowers may be borne at each node, but many of these will not result in pods and seeds. Counting the number of flowers is not a reliable means of predicting yield, as the numbers of pods and seeds and weight of seeds are strongly influenced by environmental factors.

Soybeans are self-pollinated. Weber and Hanson (1961) estimated that out-crossing under natural conditions is from 0.5 to 1%.

Flowers are usually either purple or white, with purple being dominant. There is some variation in intensity of color and other minor aspects which was discussed by Johnson and Bernard (1963).

Soybean pods may be black, brown, or tan at maturity, two gene pairs being involved (Bernard 1967).

The soybean plant is capable of profuse branching, but under usual cultural conditions, branching is limited by plant-to-plant competitive factors. At recommended planting rates of about 50 kg per hectare in 1-meter rows, branching is limited. However, when individual plants are spaced some distance from other plants, much more abundant branching occurs.

SEED

A detailed description of the soybean seed was given by Williams (1950). Whole soybeans and seed parts are shown in Fig. 2.1. Seeds of most current soybean varieties are nearly spherical. Average seed weight is 120–180 mg. of which the seed coat contributes about 10%. The large-seeded varieties, Kanrich, Disoy, Magna, and Prize, produce seed weighing about 260 mg. Genotypes in germplasm collections vary in seed weight from 15 to more than 500 mg.

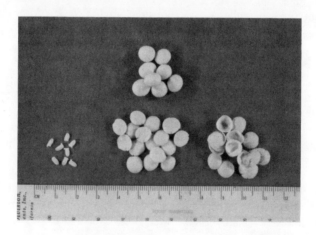

FIG. 2.1. WHOLE SOYBEANS (ABOVE) AND SEED PARTS: LEFT, HYPOCOTYLS; CENTER, COTYLEDONS; RIGHT, HULLS (PRIZE VAR.).

Cotyledons are green before maturity, becoming yellow as the seeds mature. Some genotypes retain green cotyledons at maturity, but this character does not occur in commercial cultivars. Green cotyledons are conditioned by two recessive complementary genes, yellow being dominant. A cytoplasmic factor for green cotyledons has also been reported (Johnson and Bernard 1963).

Many seed coat colors occur in the soybean germplasm collection. Yellow, green, black, and several shades of brown are common. There are also a few genetic types with bicolored seed coats. Seed coats of American commercial cultivars at maturity are yellow except for the hilum. Yellow color has been a major objective in breeding programs for many years. The Ogden cultivar formerly grown in the south has a light green seed coat and is no longer recommended since it is discriminated against by buyers and processors. Green seed coat is determined by a single dominant gene (Johnson and Bernard 1963).

The inheritance of seed coat colors was discussed in detail by Williams (1950) and Johnson and Bernard (1963). Most genotypic differences are related to two loci, *Tt* and *Rr,* with modification possible by combination with other genes. The pattern of color development, including hilum color, is controlled by a series of at least four alleles at the *I* locus.

Hilum colors include black, imperfect black, brown, buff, gray, and yellow. Hilum color is a major factor in cultivar identification. Krober (1962) has studied the chemistry of the pigments in hila and seed coats. The principal pigments are anthocyanins, with characteristic absorbance peaks in the 530–541 nm range (Fig. 2.2). Imperfect black pigment of selections from Richland had much higher absorbance below 500 nm than did black pigment. The extract from Harosoy, which has a yellow hilum, gave a shoulder only in the 500–540 nm region, and no peaks.

Seed coat mottling, as on some of the genotypes illustrated by Williams (1950), may occur in seeds which are predominantly yellow. The mottling appears to be an extension of the pigments of the hilum, but that it is not is evident from the heavy mottling often observed on yellow hilum seed. Johnson and Bernard (1963) summarized the evidence for the inheritance of mottling. It has now been shown that the appearance of mottling may be associated with soybean mosaic virus (Koshimizu and Iizuka 1963; Ross 1963, 1968; Kennedy and Cooper 1967; Wilcox and Laviolette 1968). A single gene for resistance to mottling, showing partial to complete dominance was described by Cooper (1966). The gene was identified in the cultivar Merit.

AREAS OF ADAPTABILITY

The soybean is Asiatic in origin, but includes germ plasm adapted to latitudes from 0° to more than 50°. The United States now has about 60% of the world acreage and accounts for about 75% of world production (Anon. 1969). Mainland China is estimated to have about 20 million acres (half the U.S. total).

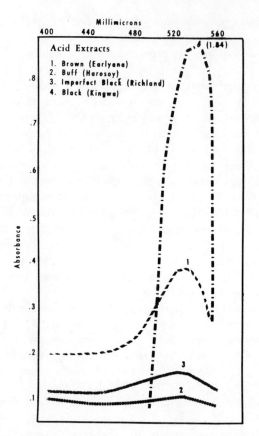

From Krober (1962)

FIG. 2.2 ABSORBANCE OF ACID EXTRACTS
OF SOYBEAN HILA

Other countries with more than 1 million acres are U.S.S.R. 2 million, Indonesia 1.6 million, and Brazil 1.4 million. Indonesian production is at latitudes of 0° to 10°

In the United States and Canada cultivars and experimental lines are grouped into ten maturity groups for convenience in comparing performance (Fig. 2.3). Group 00 includes cultivars adapted to southern Manitoba, northern Minnesota, and North Dakota; Group VIII to the Gulf Coast region.

Production in the United States is almost entirely in the North Central States, the lower Mississippi Valley, and the South Atlantic States (Table 2.1). Acreage has more than doubled since the mid-1950's. In 1969, there were 9 States with more than 1.3 million acres of harvested soybeans. Small acreages or test plantings have been grown in the Northeastern States and in the West, but the crop remains of minor or no commercial significance in those areas.

Maturity Groups

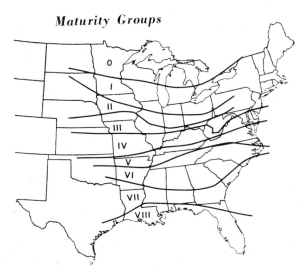

FIG. 2.3. GENERALIZED AREAS OF ADAPTABILITY
OF SOYBEAN CULTIVARS OF VARIOUS
MATURITY GROUPS IN THE UNITED
STATES

Environmental Effects on Adaptability

Day-length has been known as the primary environmental factor controlling change of the soybean from the vegetative to reproductive (flowering) state since the classical work of Garner and Allard (1920). Even earlier, Mooers (1908) noted an effect of planting date on the flowering period. The soybean is a "short day" plant; it will not flower if the day is too long. Borthwick and Parker (1938) and numerous collaborators, beginning in 1938, refined our understanding of the mechanism of the day-length effect. They originated the term "critical day-length" to denote the maximum day-length under which short day plants would initiate flowers. Hamner and Bonner (1938) proposed that the length of the dark period rather than of the light period is significant in floral control, a relationship which has been generally accpted. Hamner has recently denied this hypothesis, citing evidence that Biloxi cultivar fails to flower regardless of the length of the dark period if the photoperiods are 16 hr or longer. He now concludes (Hamner 1969) that "under natural conditions, long days prevent flowering because the photoperiod extends beyond 12 hr (an 'innocuous' day-length) and not because the extension of the photoperiod shortens the dark period to a critical level."

Processes essential to floral induction do occur in the dark, however, and these can be reversed by red light. The light-sensitive pigment, a protein, was isolated in 1959 and named "phytochrome" (Butler et al. 1959).

TABLE 2.1

U.S. SOYBEAN PRODUCTION

State	Acreage Harvested			Yield Per Acre			Production		
	1968	1969	1970	1968	1969	1970	1968	1969	1970
	1000 Acres			Bushels			1000 Bushels		
N. Y.	6	5	6	22.0	21.0	20.0	132	105	120
N. J.	45	46	50	24.0	28.0	24.5	1,080	1,288	1,225
Pa.	22	25	28	24.0	30.0	32.0	528	750	896
Ohio	2,276	2,344	2,438	30.5	29.0	28.5	69,418	67,976	69,483
Ind.	3,246	3,311	3,311	32.0	32.5	31.5	103,872	107,608	104,297
Ill.	6,663	6,730	6,865	31.5	33.5	31.0	209,884	225,455	212,815
Mich.	463	514	524	26.0	23.0	26.0	12,038	11,822	13,624
Wis.	161	174	153	22.0	19.0	21.0	3,542	3,306	3,213
Minn.	3,232	3,068	3,129	22.0	24.5	26.5	71,104	75,166	82,919
Iowa	5,561	5,450	5,832	32.0	33.0	32.0	177,952	179,850	186,624
Mo.	3,663	3,150	3,496	28.0	26.0	26.0	102,564	81,900	90,896
N. Dak.	215	185	176	15.5	16.0	15.0	3,332	2,960	2,640
S. Dak.	300	243	255	17.5	24.5	17.5	5,250	5,954	4,463
Nebr.	782	766	812	23.5	33.5	22.0	18,377	25,661	17,864
Kans.	957	852	1,005	25.0	23.0	15.0	23,925	19,596	15,075
Del.	156	162	162	19.0	29.0	21.0	2,964	4,698	3,402
Md.	209	205	213	25.0	33.0	24.0	5,225	6,765	5,112
Va.	372	361	339	19.0	25.0	19.0	7,068	9,025	6,441
N. C.	972	885	876	17.5	26.5	24.0	17,010	23,453	21,024
S. C.	931	959	997	12.5	22.5	20.5	11,638	21,578	20,439
Ga.	472	467	528	15.0	24.0	22.5	7,080	11,208	11,880
Fla.	143	169	184	24.0	27.0	28.0	3,432	4,563	5,152
Ky.	466	485	558	26.5	28.0	27.0	12,349	13,580	15,066
Tenn.	1,193	1,193	1,229	21.0	24.0	23.0	25,053	28,632	28,267
Ala.	557	641	609	22.0	23.0	23.5	12,254	14,743	14,312
Miss.	2,120	2,290	2,336	27.0	22.0	24.0	57,240	50,380	56,064
Ark.	3,989	4,228	4,313	22.0	20.5	22.5	87,758	86,674	97,043
La.	1,436	1,608	1,688	27.0	19.0	22.5	38,772	30,552	37,980
Okla.	184	204	177	21.0	17.0	17.0	3,864	3,468	3,009
Texas	312	262	158	27.0	29.0	28.0	8,424	7,598	4,424
U. S.	41,104	40,982	42,447	26.8	27.5	26.8	1,103,129	1,126,314	1,135,769

Source: Crop Reporting Board, USDA—ARS, Dec., 1970.

Borthwick and Parker (1938) used the cultivar Biloxi, of Group VIII maturity, almost exclusively in their soybean work. The critical day-length of Biloxi is between 13 and 14 hr. Cultivars of northern adaptation obviously have longer critical day-lengths, since locally adapted cultivars flower readily in the field in northern Minnesota (maximum summer day-length, 16 hr). Some entries in the germ plasm collection of the USDA flower very late or not at all in the southern United States, indicating a critical day-length less than that of Biloxi.

Day-length, in the classical work of Garner and Allard and of Borthwick and Parker, specifically controls the shift from vegetative to reproductive growth. Johnson *et al.* (1960) and Fisher (1963) have extended our knowledge of day-length effects to include other aspects of growth, such as fruit set, length of the pod-filling period, and date of maturity. Effects of day-length on growth subsequent to flowering contribute to cultivar differences in adaptation; effects on growth responses in different cultivars may not be parallel. Benedict *et al* (1964) reported evidence of an interaction of iron and day-length in chlorophyll content and soybean growth.

Byth (1968) observed day-length effects and cultivar x day-length interactions on height at flowering, number of nodes at the time of flowering, and leaf area at flowering. He also studied the inheritance of these differences in an F_2 population derived from a cross of Mamloxi and Avoyelles cultivars. The F_2 mean for days to flowering was consistently smaller than the mid-parent value in all photoperiods. These results suggested that the tendency to low number of days is dominant to that for a longer period to flowering. Differences, however, were small and most of them did not exceed the standard error.

Uniform frequency distribution for most characters in F_2 populations indicated a relatively complex form of inheritance, according to Byth. Segregation beyond the limits of the parents occurred for most traits, particularly in the longer day-lengths, and evidence of substantial genetic control of differences in day-length response between varieties was observed.

Day-length is a major factor determining cultivar adaptability, acting as a "triggering" mechanism to shift the plant from vegetative to reproductive growth. Several weeks elapse between floral induction and the appearance of flowers. The time requirements of floral development and pod-filling, as well as gross vegetative growth, are functions of basic metabolic processes such as photosynthesis, respiration, cell division and elongation, translocation, protein synthesis, etc. These processes are affected directly or indirectly by temperature, moisture, soil fertility, and other elements of the environment. Thus, the accelerating effect of a short day on flowering may be partially offset by a retarding effect of other environmental conditions.

Locally, cultivars may show a difference in adaptibility on different soil types, especially if soil type influences the development of pathogens. A good example involves resistance to *Phytophthora megasperma* Drech. var. *sojae* Hildeb., the organism causing phytophthora rot. Phythophthora rot is more likely to be a problem on poorly-drained or heavy-textured, tight soils, than on well-drained soils. Thus, the cultivar Semmes is successful in the lower Mississippi Valley on such soils when susceptible varieties are not. In the absence of *P. megasperma,* Semmes offers no advantage.

Some anomalies in day-length response are unexplained. The cultivar Hill of maturity group V is well adapted to northern Arkansas, northern Tennessee, and the Bootheel area of southeast Missouri. However, when grown in the tropics,

Hill makes better growth than several varieties of more southern adaptation in the United States. In its area of adaptation, Hill flowers later in relation to its maturity than do most other cultivars. Arksoy is nonresponsive to fluorescent light for control of flowering, whereas most varieties are responsive to fluorescent light.

GROWTH OF THE SOYBEAN PLANT

Germination

A good seed lot should germinate 85–90% under standard conditions used in seed testing laboratories. Grabe (1965) reported that soybean seeds could be safely stored up to 10 yr if their moisture content was maintained below 10%. It is customary for plant breeders to maintain their germ plasm stocks at 5°–10° C and 20–30% RH, which results in seed moisture of 6% or less.

Seeds are usually planted at depths of from 2 to 5 cm, depending on soil type and moisture conditions. Rate of emergence will depend on moisture and temperature, but may also vary among varieties. Differences in speed of emergence are related to seed size and to elongation of the hypocotyl. Grabe and Metzer (1969) reported that hypocotyl elongation of Ford cultivar is inhibited at 25° C. but is similar to that of Hawkeye at 15°, 20°, and 30° C. Hypocotyl elongation at 25° C may be controlled by a single major locus and a modifying gene system (Fehr 1969). Moisture requirements for soybeans are higher than for some other crops. Hunter and Erickson (1952) found that a moisture content of about 50% is necessary for soybean germination. This compared in their tests with a requirement of about 30% for corn and 26% for rice.

Excessive moisture is unfavorable for germination, probably due, in part, to restriction of the oxygen supply. There is little information on the quantitative effects of oxygen level on germination, but it has been shown (Ohmura and Howell 1960) that a thin film of water on the seed significantly interferes with movement of oxygen. Grable (1964) showed that CO_2 may affect the germination of soybeans under different levels of oxygen and moisture.

Temperature effects on germination have been studied extensively in the laboratory, but to only a limited extent under field conditions. Delouche (1953) observed maximum germination in the shortest time with a constant temperature of 30° C. It took about twice as long to reach a given percentage germination at 20° C as at 30° C. It is probable that soil temperatures rarely exceed 20° C during the normal period of soybean germination. A combination of low temperature and high moisture may favor development of certain fungi and bacteria detrimental to seedling growth.

Soybean seeds do not require special environmental treatments to break dormancy, which is negligible in soybeans. Some chemical fungicides are registered for use as seed protectants on soybeans.

Germination requires mobilization of the food materials stored in the seed.

McAlister and Krober (1951) found that reserve proteins were mobilized at a steady rate, but that carbohydrates were quickly depleted in the early phases of germination (Fig. 2.4). Mobilization of fats began as the carbohydrate supply was depleted. After about 2 weeks, protein decreased more slowly than lipids in the cotyledon, each cotyledon retaining about 2 mg of protein after 5 weeks. At that time the cotyledon would be functioning as a leaf.

Abrahamsen and Sudia (1966) reported results similar to those of McAlister and Krober (1951). In their experiments total soluble carbohydrates fell from about 12% to about 3% during the first 4 days of germination. Reducing sugars increased steadily, ultimately accounting for about half of the total soluble carbohydrates in the cotyledon.

Oligosaccharides comprise about 15% of the air-dried weight of soybeans according to Pazur *et al.* (1962). Sucrose, stachyose, and raffinose are present in that order of abundance, and are rapidly metabolized during germination. D-fructose and D-glucose were readily detectable in extracts from germinating seeds, but only traces of D-galactose were found. Pazur *et al.* interpreted this to indicate a rapid utilization of the galactose moiety from the oligosaccharides.

Kasai *et al.* (1966) studied the changes in free and total amino acids and other

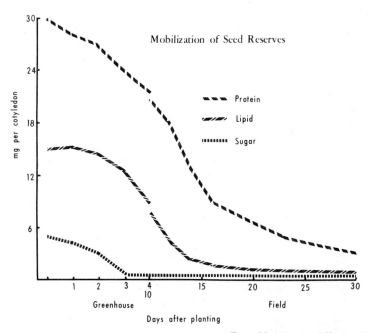

Mobilization of Seed Reserves

■ ■ ■ Protein
▰▰▰ Lipid
▥▥▥ Sugar

Greenhouse
Field

Days after planting

From McAlister and Krober (1951)

FIG. 2.4. PROTEIN, LIPID, AND SUGAR CONTENT OF SOYBEAN COTYLE-DONS DURING GERMINATION

major nitrogen fractions during germination of T201, a non-nodulating line, and T202, a closely related nodulating line. All amino acids except aspartic acid declined (μmole per plant) during nine days' germination in the dark. The authors did not recognize any significant differences between the two lines in amino acid metabolism. Dry weight data were not reported.

Kasai et al. (1966) also observed that γ-glutamyltyrosine and γ-glutamylphenylalanine, previously discovered by Morris and Thompson (1962), declined beginning after about 20 hr of germination and disappeared by the end of the third day. The same group (Ishikawa et al. 1967) have reported ϵ- and γ-glutamylaspartic acid in seedlings. No time course for changes was reported.

Respiratory activity during germination undergoes changes correlated with the pattern of utilizing food reserves. The uptake of oxygen and the amount of metabolically active mitochondrial nitrogen both reach a peak about five days after the start of germination (Howell 1961). The "glyoxylate shunt," by which fatty acids can be converted to carbohydrates, also appears to be active by the fifth day (Carpenter and Beevers 1959; Howell 1961), coinciding with the requirements for lipids as an energy source. Changes in amino acids, fatty acids, and nucleic acids during germination have been the subjects of several recent papers (Miksche 1966; Wilson and Shannon 1963; Dorworth 1967).

Vegetative Growth Habit

The term "growth habit" will be restricted in this section to include only those factors which relate to the appearance of the individual plant or the community of plants constituting the crop in the field. The expression of growth habit reflects both genetic and environmental factors. The interaction of genetic and environmental factors is being studied by several groups. Many questions are still unresolved with respect to the importance of variation in growth habit to productivity and the extent to which environmental effects may be limited genetically. The growth habit is the result primarily of leaflet shape, petiole length and angle, branching, stem growth, and determinateness. Growth habit is often referred to inexactly but usefully as the "canopy." "Canopy" refers not just to the "cover" or upper layer of the crop, but to vertical as well as horizontal properties influencing light penetration and absorption and gas exchange.

Many southern cultivars have a long inflorescence stalk. Height and maturity characteristics also affect growth habit. Johnson and Bernard (1963) reviewed the inheritance of these characters.

An important distinguishing characteristic is leaflet shape. Clark is an example of a cultivar with broad leaflets; SRF300 and Hark illustrate those with narrow leaflets. The genetic symbol Na is assigned to the broad leaflet-narrow leaflet character. An oval leaflet character has also been reported. The number of seeds per pod was associated with leaflet shape (high number with narrow leaf, low number with oval), presumably due to pleiotropic action of the same gene

(Johnson and Bernard 1963). Leaflet shape is quite plastic and environmental factors can exercise a major influence.

Leaflet number is also variable. Three leaflets are normal, but genotypes are known with 5, 7, or more leaflets.

The effects of environment on growth habit are strikingly expressed when a cultivar is grown north or south of its area of adaptation, or when there is a substantial change in plant population. Plants tend to be shorter and earlier when a cultivar is grown nearer the equator than its normal area. This is primarily a response to early induction of flowering by the shorter days in the southern (lower latitude) locations. Flowering retards subsequent vegetative growth, especially in the determinate growth type. The shorter day-lengths may also have direct effects on pod development and maturation, as was discussed by Johnson et al. (1960). When a cultivar is grown at a higher latitude than its area of adaptation, it will continue in the vegetative condition longer than normal because of late flowering associated with longer day-length, and may fail to ripen before frost.

Plant-to-plant competition strikingly affects branching. Plants in a population of 20–30 plants per meter in 1-meter rows will have very few branches. Arrangements with more space between plants, usually referred to as "spaced planting," will have many more branches. Plant breeders "space plant" promising breeding lines in order to obtain the maximum seed increase. This ability of soybean plants to compensate by branching and in other ways for differences in available space, has long been recognized. A practical effect is that gaps of several centimeters in the row or variations of several percentage points in planting rate may have little effect on yield per hectare.

The date of planting has major effects on many aspects of growth habit in southern latitudes, as extensively reviewed by Cartter and Hartwig (1963). Soybeans are versatile, so by choice of cultivar it is possible to utilize a considerable range in planting dates. The optimum time to plant in most areas of the United States is usually in May. Delays in planting shorten each phase of growth (Leffel 1961). There may be some delay in maturity from late planting but because of the compression of each phase of growth, the delay in maturity is less than the delay in planting.

Plant height is influenced by the genotype and planting date. Tallest plants are usually expected from May plantings. The degree of lodging may also be influenced by planting date and height. However, studies summarized by Cartter and Hartwig (1963) illustrate that the degree of lodging is dependent on weather conditions. Lodging resistance has been an objective of soybean breeders for many years and modern cultivars are superior in this respect.

Growth habit may also be modified by chemical treatment. The chemical which has received most attention for this purpose is 2, 3, 5-triiodobenzoic acid (TIBA), which accelerates the change from vegetative to reproductive growth (Greer and Anderson 1965). TIBA treatment results in shorter internodes and changes in leaf size and orientation and in canopy shape to permit light to reach more leaf surface. In some circumstances TIBA gives a yield increase, but the conditions under which TIBA will be beneficial are still being defined.

Gibberellic acid (GA) was extensively tested about a decade ago (Howell *et al.* 1960). GA stimulates intermode elongation. It did not increase yield. This compound has been shown to reduce iron chlorosis (Mitchell and Anderson 1966) and possibly to participate in the control of apical dominance (Ruddat and Pharis 1966).

Reports of stimulation of various aspects of growth have appeared, but chemical control of growth as a soybean commercial practice is still uncertain. Research on chemical growth control is very active and is expected to continue.

Use of chemical defoliants has been explored but appears to offer no advantage in soybean production.

Root Growth

Root growth is probably the least studied of any significant aspect of soybean growth and development. A pioneer study by Borst and Thatcher (1931) reported that soybean roots may penetrate as deeply as 1.6 m. That study has been a classic and for many years has been virtually the only information available on root growth under natural conditions.

More recently (1968), Barber and his students at Purdue, Mitchell at Iowa State, and Lewis at Virginia State, have initiated studies of various aspects of root growth.

Barber's group studied the rooting patterns of 48 genotypes, selected to represent a good sampling of genetic diversity in maturity groups II, III, and IV. From these studies, they chose Harosoy 63 and Aoda cultivars for more detailed study. These cultivars produced similar top dry weights, but Harosoy 63 had a more spreading type of root system than Aoda. Harosoy 63 had a consistently greater root surface throughout the soil profile, especially in the upper 0.5 m (Raper 1968). Both cultivars developed a profusion of lateral root branches about 15 cm below the soil surface. Neither developed an extensive taproot system.

Mitchell (1968) reported similar results in a study of rooting patterns of 12 cultivars. He observed the major root dry matter concentration in the upper 7-15 cm of soil. Roots extended laterally to the middle of 0.7-m rows within 5-6 weeks after planting and to depths of 1.6-2 m in a loam soil by the end of the season. The appearance of roots in Mitchell's study after about 3 weeks' and 2 months' growth, respectively, is shown in Figs. 2.5 and 2.6 (Mitchell 1969).

Peters and Johnson (1960) reported that soybeans extracted water from soil below 1.3 m. Nevertheless, root distribution seems to be primarily in the upper part of the profile. Recent studies tend to confirm earlier evidence, as interpreted by Cartter and Hartwig (1963), that soybean roots are usually concentrated above the 0.6 m depth.

Courtesy of Dr. R. L. Mitchell

FIG. 2.5. ROOT GROWTH OF A
SOYBEAN PLANT AFTER
THREE WEEKS' GROWTH
IN A LOAM SOIL

Lateral roots are just beginning to develop.

Courtesy of Dr. R. L. Mitchell

FIG. 2.6. ROOT GROWTH OF A SOYBEAN PLANT
AFTER TWO MONTHS' GROWTH

Note extensive root development in upper 10–15 cm.

Cobb (1962), by forcing soybean root meristems to grow through minute openings in glass sleeves or plastic blocks, showed that pressure causes a number of growth abnormalities. Precocious initiation of lateral roots was induced and normal tissue differentiation was suppressed. Although there was no differentiation into xylem and phloem, conduction occurred through the compressed zone and a normal stele developed below the compressed zone. Nuclear disorganization also occurred in the compressed zone, but there was an abundance of a substance staining as DNA. Miksche (1966) concluded that DNA synthesis may not be closely related to the beginning of mitosis and later (Miksche and Greenwood 1966) reported the presence of a "quiescent zone" of no DNA synthesis.

Nodulation and Nitrogen Fixation

The soybean, like many other legumes, can obtain nitrogen from the air through a symbiotic relationship with the bacteria, *Rhizobium japonicum* (Kirchner) Buchanan in root nodules. The symbiotic nitrogen fixation system becomes active by about three weeks after planting.

Soils where soybeans have been grown previously usually contain an adequate population of *R. japonicum* to serve as an inoculum. Many different genetic strains of *R. japonicum* exist, as identified by serological techniques and specificity of soybean varieties. Strains may occur at random in a field, or the distribution may reflect differences in soil properties. Usually one strain will predominate, either in the field or in a part of the field where soil conditions are relatively homogeneous (Damirgi *et al.* 1967). Strain 123 predominates in the North Central States, strains 110 and 122 on the East Coast, and strains of serogroup C-2 in the lower Mississippi Valley.

The USDA collection at Beltsville includes about 200 strains of *R. japonicum*. Varying degrees of affinity between bacteria and soybean genotypes have been observed (Abel and Erdman 1964; Caldwell *et al.* 1966). A strain of bacteria may be effective or ineffective in forming nodules. It may be efficient or inefficient in nitrogen fixation even if nodules are formed. Caldwell (1966) and Vest (1970) have shown that the inheritance of two strain-specific ineffective reactions are controlled by two separate dominant genes. These genes have been designated Rj_2 and Rj_3.

It is very difficult to substitute a new strain for one which is indigenous in a soil. This is probably due, in part, to ecological factors favoring one strain. But strains display different degrees of competitiveness for nodulation sites (Johnson and Means 1963, 1964; Johnson *et al.* 1965). Thus, one strain may have a competitive advantage for establishment in the rhizosphere, but may not be able to form nodules. Caldwell and Weber (1970) showed that the strain distribution in the soil may change throughout the growing season. Also, plants sampled early in the season exhibited a different serological pattern from those sampled later. Each nodule contains only one strain, implying infection by a single bacterium. Different strains may be observed from different nodules on a single plant. Caldwell and Vest (1970) observed that when a number of cultivars were exposed to the same soil population, the resultant serological patterns of the genotypes were significantly different. This indicates the selective role of the host genotype.

There has been much work in the past on the pathway of nitrogen fixation in the nodule. It now appears to be agreed that ammonia is probably the earliest stable product of nitrogen fixation by soybean root nodules (Bergersen 1965, 1966).

Studies on nitrogen fixation received an important stimulus with the discovery that the nitrogen-fixing enzyme, nitrogenase, could reduce acetylene to ethylene (Dilworth 1966). The estimation of N_2 fixation in soybean nodules by the acetylene reduction assay was applied by Koch and Evans (1966). This procedure makes it possible to collect nodules in the field, expose them to acetylene immediately in very simple apparatus, and obtain a sample which may be held several hours before analysis by gas liquid chromatography. The method is simple, rapid, and requires less expensive instrumentation than the older technique, which required feeding the heavy isotope, ^{15}N, and analysis by mass spectrometry. The new technique has greatly increased the number of samples which can be analyzed and has made it possible to perform field studies that were impossible with older methods.

Nitrogen fixation by excised nodules was first achieved by Aprison *et al.* (1954). Bergersen (1962) showed that maximum nitrogen fixation by either intact or excised nodules required a partial oxygen pressure of 30 to 60%. At higher levels, oxygen was a competitive inhibitor of nitrogen fixation. Nitrogen fixation by cell-free extracts of the free-living bacteria, *Clostridium pasteurianum,* was first demonstrated by Carnahan *et al.* (1960). Evans and his associates (Evans 1969) and Bergersen and Turner (1967) have established that nitrogenase in symbiotic systems is located in the nodule bacteroids. Koch *et al.* (1967A, B) demonstrated that the reduction of acetylene to ethylene by soybean nodule bacteroids required anaerobic conditions, sodium hydrosulfite, and an ATP generating system. These developments, coupled with the acetylene reduction assay method, have vastly accelerated progress in nitrogen fixation studies during the last few years.

Studies with the acetylene reduction method have led to reconsideration of older opinions as to the duration of nitrogen fixation. The traditional view held that the fixation system probably becomes inoperative about the time seeds begin to develop in the pod, or several weeks prior to maturity. Thus, Bergersen (1962) reported that fixation ceased when nodules are about six weeks old. Such an interpretation may have been due, in part, to confusion of age of plant with age of nodules. New crops of nodules are formed, however, so that active nodules may be present until the plant is nearly mature. The traditional view (no fixation after about the time seed development begins) has been unsatisfactory for a long time in view of evidence such as that recently summarized by Hanway and Thompson (1967) (Fig. 2.7), which showed that the plant obtains a substantial amount of nitrogen subsequent to the start of seed development. Hardy *et al.* (1968), using the acetylene reduction method, reported that nitrogen fixation continued until the leaves began to turn yellow, which in general coincides with the attainment of maximum dry weight in the seeds. Similar results have been obtained by our colleague, C. Sloger, at Beltsville. N_2 fixation capacity of soybean cultivars increases sharply during flowering and remains highly active until near the mature seed stage. Sloger (1969) has also shown that the fixation rates of different soybean genotypes inoculated with the same strain of *R. japonicum* may differ during the first 60 days of growth in the greenhouse.

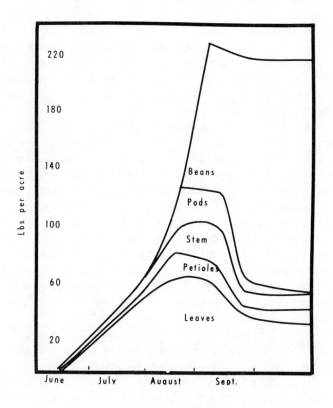

From Hanway and Thompson (1967)

FIG. 2.7. UPTAKE OF NITROGEN AND ITS DISTRI-
BUTION WITHIN THE SOYBEAN PLANT
DURING THE GROWTH CYCLE

The ability to nodulate soybeans is dependent on a single gene (Williams and Lynch 1954) which now has the symbol Rj_1 (Caldwell 1966). C. R. Weber has developed near-isogenic lines differing only in the nodulation character and used these lines for studies on the importance of nodulation to crop productivity (Weber 1966A, B). The non-nodulating lines approach the productivity of their normal counterparts if sufficient nitrogen is applied. Weber (1966B) was unable to achieve higher yields from the non-nodulating than the normal lines regardless of nitrogen supply level.

R. japonicum strains 76, 94, and others induce a chlorosis in soybeans (Erdman *et al.* 1957). The toxic material has been identified as a nonspecific antimetabolite and named "rhizobitoxine" (Owens 1969). It can be isolated from nodules, chlorotic leaves, or the culture medium in which the bacteria have grown. A structure, which is similar to that of cystathionine, suggested by Owens, is shown in Fig. 2.8.

Owens suggests that the toxic reaction results from interference with the metabolism of a sulphur-containing amino acid. Later work with *Salmonella* and spinach indicated that rhizobitoxine inhibits β-cystathionase, thereby inducing a deficiency of methionine and decreased rate of protein synthesis (Owens 1969).

The effect of rhizobitoxine is very nonspecific. Johnson *et al.* (1959) showed that seedlings of many plant species when placed with their roots in a solution containing what is now called rhizobitoxine developed chlorosis.

This condition does not seem to have a major effect on soybean production. Chlorosis in the field is transitory, affected plants growing out of the chlorosis in a few days. So many environmental factors affect yield after the plants recover from chlorosis that it has not been possible so far to get good evidence indicating how much *Rhizobium*-induced chlorosis reduces yield under field conditions.

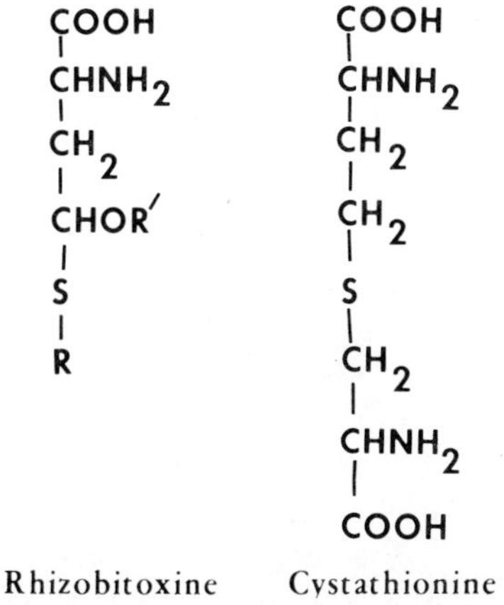

$$
\begin{array}{cc}
\text{COOH} & \text{COOH} \\
| & | \\
\text{CHNH}_2 & \text{CHNH}_2 \\
| & | \\
\text{CH}_2 & \text{CH}_2 \\
| & | \\
\text{CHOR}' & \text{CH}_2 \\
| & | \\
\text{S} & \text{S} \\
| & \backslash \\
\text{R} & \text{CH}_2 \\
& | \\
& \text{CHNH}_2 \\
& | \\
& \text{COOH} \\
\text{Rhizobitoxine} & \text{Cystathionine}
\end{array}
$$

FIG. 2.8. PROPOSED STRUCTURE OF RHIZOBI-
TOXINE AS COMPARED WITH CYSTATHI-
ONINE

Molybdenum and cobalt are required in some of the nodule enzymes (Evans 1954; Ahmed and Evans 1961). While there has been some further experimentation with the use of these minor elements as constitutents of inoculum (Burton and Curley 1966), they are usually present in adequate amounts in the soil. Beneficial results have been obtained from use of molybdenum on soils having a low pH (Harris *et al.* 1965). With the addition of lime no response was obtained from molybdenum. Lavy and Barber (1963) obtained significant yield increase by treating seed containing less than 1.6 ppm with molybdenum.

Fixation is reduced by addition of exogenous nitrogen. Tanner and Anderson (1963) showed that this effect might involve production of nitrite and interference with the auxin system. Small and Leonard (1969) reported that NO_3-N reduced the translocation of ^{14}C-labeled photosynthate to nodules.

Mineral Nutrition and Fertilization

Soybeans have long had a reputation for being unresponsive to direct fertilization. The reputation is derived, in part, from comparisons of the effect of added fertilization on soybeans with that on corn and other nonlegumes. Ohlrogge (1963) pointed out that absorption of mineral nutrients by the soybean crop is apparently not a serious limitation on productivity. Indeed, a 3.36 metric tons per hectare (50-bu per acre) crop of soybeans contains in the seed from 1 ha about 67 kg of K, 24 kg of P, and 200 kg of N at maturity. The work of Hanway and Thompson (1967) shows that the soybean plant retains the ability to take up nutrients until very late in its life cycle.

The symbiotic nitrogen fixation system of soybeans in general excludes the type of beneficial response that nonlegumes give to added nitrogen. An adjustment for the role of nitrogen should be, but frequently is not, made in comparing soybeans with nonlegumes.

One should also recognize that much of the historical data on fertilizer responsiveness of soybeans was obtained at yield levels which are very low compared to current record yields. In the 1968 soybean yield contest 4 growers averaged more than 6.7 metric tons per hectare (100 bu per acre) on 2-ha test plots. L. M. Phillips of Holly Bluff, Mississippi, has averaged over 3 metric tons per hectare on 200 ha or more for several years, and in 1968 averaged 3.3 metric tons per hectare on 300 ha. It is clear that such yields impose major demands on the soil for nutrients and that nutrients withdrawn will have to be replenished.

Most recent work on mineral nutrition has concerned phosphorus, potassium, or lime and in special cases such minor elements as boron, molybdenum, and zinc. Strontium, calcium, and aluminum have also been studied. Cartter and Hartwig (1963), in discussing earlier data of Bray (1961), concluded that soybeans are more responsive than corn, wheat, alfalfa, oats, and clover to low levels of phosphorus as indicated by soil tests. Maximum soybean yields were obtained with smaller additions of phosphorus to soybeans. Wheat and oats were more responsive to potassium than soybeans, but corn, alfalfa, and clover were less so.

It seems clear that soybeans will respond to fertilization on soils with definite deficiencies of certain elements. But it is not possible in most situations to write a prescription for fertilization at high yield levels with the same confidence as it can be done for other crops.

Genetic differences in response to nutrients have been known in soybeans since the work of Weiss (1943) on iron. The iron-inefficient line which he identified, P.I. 54619, has been studied extensively by J. C. Brown and a number of his collaborators. Chlorosis is related to the genotype of the rootstock (Brown 1968). Citrate metabolism is probably involved (Brown and Tiffin 1965; Brown 1966).

Genetic differences in phosphorus nutrition were reported by Howell (1954) and were shown by Bernard and Howell (1964) to be inherited in a simple manner. Fletcher and Kurtz (1964) reported differential effects of phosphorus fertility on soybean cultivars. The controlling system, as in the case of iron, was shown by Foote and Howell (1964) to be associated with the root. By contrast, Kleese (1968) showed that strontium and calcium accumulation in grafted plants under field conditions was controlled by the genotype of the scion. The difference between Kleese's results with Sr and Ca and those of Brown's and Howell's groups with Fe and P are thought to reflect a difference between long-term (Kleese) accumulation and short-term (Brown and Howell) uptake. We do not think a basic difference in mechanisms among the four elements should be inferred.

Genotypic differences in response to aluminum have been reported by Armiger *et al.* (1968).

Photosynthesis

The difference in performance of soybeans and corn crops has puzzled farmers and agronomists since soybeans emerged as a major crop. Major advances in understanding differences between soybeans and corn as representative of major groups of plants have come in recent years in the area of photosynthesis.

It was shown by Hesketh (1963) that photosynthesis in corn and some other grasses in not light-saturated even at light intensities typical of mid-day. By contrast, photosynthesis in soybeans and many other broadleaf plants is saturated at about 1/3 of mid-day light intensity (Bohning and Burnside 1956).

The saturation of photosynthesis as light intensity is increased marks the point where CO_2 becomes limiting. Brun and Cooper (1967) measured soybean photosynthesis in an atmosphere containing 1670 ppm CO_2. Photosynthesis was not saturated at the highest light intensity used, 75,350 lux, or about 3/4 of full sunlight. Cooper and Brun (1967) observed a substantial increase in number of pods and yield of plants grown in a 1350 ppm CO_2 atmosphere. In that case, the higher photosynthesis resulting from higher CO_2 concentration was credited for formation or retention of more pods. Recent field studies suggest that earlier concepts on light saturation, based on studies in the greenhouse or with excised leaves, may need to be revised (Egli *et al.* 1970).

Later it was learned that a component of the respiration of many plant species, including soybeans, requires light. This component, named "photorespiration," does not occur in corn and some other species, mostly grasses (Moss 1966). These facts led to the suggestion that there might be genetic variability within the species so that these characteristics of soybeans might be improved through plant breeding. Curtis *et al.* (1969) examined 36 soybean cultivars selected to represent a wide range of maturities and genetic backgrounds. A 2-fold difference in photosynthetic rate was observed ranging from a high of 24 mg CO_2 dm^{-2} hr^{-1} down to 12. The average photosynthetic rates, however, were poorly correlated with average yields in the field. Although there were some cultivar differences in photosynthetic rate there appeared to be no difference in the light intensity at which photosynthesis was saturated. All cultivars appeared to have similar photorespiration, the compensation point being approximately 40 ppm CO_2.

Moss *et al.* (1969) surveyed about 400 species to determine whether there was variability within species for CO_2 compensation point. They found no differences within species and only three cases of differences within a genus. Widholm and Ogren (1969) grew soybeans in a closed system with corn, in which corn depleted CO_2 to a level below the compensation point for soybeans, to study senescence of soybeans when CO_2 is restricted. When corn plants were in the same closed system or when CO_2 was removed chemically to a level below the compensation point soybean senescence was hastened by increasing oxygen concentration, light intensity, and temperature. Cannell *et al.* (1971) have screened the U.S. collection of soybeans of northern maturity (Groups 00 through IV) for low compensation point, and have found no genotypic variability in compensation point.

It has not been established that differences in photosynthesis between species are due to the presence or absence of photorespiration. Curtis *et al.* (1969) reviewed some of the evidence pertaining to this point.

Moss *et al.* (1969) enumerated several aspects in which species with low CO_2 compensation points differ from those with high compensation points. Species with low compensation points are characterized by an absence of photorespiration, absence of suppression of photosynthesis by oxygen, and ability to fix CO_2 by carboxylation of phosphoenolpyruvate (a 4-carbon pathway). Low compensation point species also have vascular bundle sheath cells which are particularly active in starch formation. Moss *et al.* consider unresolved the question of whether all of these traits are required for a higher photosynthetic potential.

Photosynthetic rate or total photosynthesis may be regulated by the rate of movement of photosynthate out of the leaf. This is the so-called "sink" concept. Hew *et al.* (1967) suggest a hormonal control of photosynthate translocation. Both indoleacetic acid and gibberellic acid stimulate sucrose translocation.

Seed Development

Seed development is initiated with cell divisions to form the embryo very soon after pollination (Pamplin 1962). Cell divisions continue for about two weeks, during which the cotyledons and growing points are differentiated. Bils (1960) observed no mitotic figures in an electron microscope study of cotyledons during the period from 15 days after flowering to maturity. Galitz (1961), however, observed mitotic figures in the radicles throughout the period of seed development.

There is little increase in weight of the seeds until about the tenth day after pollination. Shortly after that, seed weight begins to increase and averages a gain of 6-7 mg dry weight per day for about 3 weeks. The rapid increase begins about 30 days after full bloom (Hanway and Thompson 1967). A small loss in dry matter seems to be associated with ripening (Cartter and Hopper 1942).

There is an endosperm during the early stages of seed development but it is almost completely absorbed and is visible only as the inner layer of the seed coat at maturity.

Oil synthesis lags somewhat behind the initial increases in dry weight (Garner et al. 1914). The oil percentage increases very rapidly after the seed weight reaches about 30 mg and reaches its final percentage content when the seed is about half developed.

Fatty acid composition of soybean oil changes during seed development and is very sensitive to temperature during this period (Howell and Collins 1957). Linolenic acid is at its highest percentage in the youngest seed (Fig. 2.9). Linolenic acid percentage declines during the initial phases of seed development, reaching its final plateau 3-4 weeks before maturity. Linoleic acid percentage is initially low and increases beginning at about the time linolenic reaches its plateau. The final linoleic acid percentage is not reached until shortly before maturity.

Linolenic and linoleic acid percentages are inversely related to temperature within the range of temperature normally encountered in the field (Howell and Collins 1957). This effect of temperature has been confirmed in growth chamber studies.

Rinne (1969) has shown that temperature affects the incorporation of labeled acetate into fatty acids by a soluble preparation from developing soybean cotyledons in a manner that is consistent with temperature effects on whole plants. At 17°C, 13% of the ^{14}C went into oleic, 6% into linoleic, and 13% into linolenic acids. At 37°C, 5% went to oleic and none into linoleic and linolenic. The proportion of ^{14}C incorporated into stearic acid increased from 51 to 79% over the 20° range, accounting for all of the loss from the unsaturated acids. Palmitic acid acquired 15% of the ^{14}C at both temperatures.

The high level of linolenic acid in very young seed seems inconsistent with an

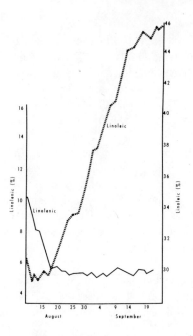

FIG. 2.9. LINOLENIC   AND   LINOLEIC
ACIDS   IN   DEVELOPING   SOY-
BEAN SEEDS

oleic→linoleic→linolenic metabolic pathway of fatty acid synthesis. However, recent studies of Rinne using very short exposure time show the greatest proportion of radioactivity to be in oleic acid when the shortest exposure times are used. The ^{14}C in oleic acid can be readily extrapolated to 100% at zero time (Rinne 1967). Rinne interprets these results as indicating that oleic acid is the first fatty acid formed from acetate in the malonyl pathway and that oleic acid in this system is a precursor of linoleic and linolenic.

Protein content has been studied less extensively than oil content. Protein percentage is quite high in very young seed and increases only slightly as the seeds develop. We are not aware of any data concerning variability in specific amino acids as related to seed development.

Bils and Howell (1963) showed that protein accumulates in conspicuous "protein bodies" in developing cotyledons. These bodies are not homogeneous but do have a very high concentration of protein. They increase in number during seed development and at maturity virtually fill the cell. Bils and Howell reported the presence of numerous conspicuous starch grains throughout most of the period of seed development, but these structures disappeared in the final stages of maturation.

Several plant breeders have been interested for many years in the possibility of breeding cultivars with higher protein percentages. The cultivars Provar and Protana, released in 1969, are the first two cultivars which were developed for higher protein content. Their protein contents in many tests have averaged about 42% dry weight or about 3 percentage points higher than most commercial varieties. Cultivars with substantially higher protein contents than Provar and Protana may be available later.

Information on amino acid content or nutritional status of high protein cultivars is very limited. Such data as is available gives no evidence that methionine will decrease with an increase in protein (Krober and Cartter 1966). The nutritional value of high protein lines has appeared to be equal on a protein unit basis to normal cultivars in very limited feeding studies.

Larsen and Caldwell (1968, 1969) observed two protein components from soybean seed, designated A and B, which were cultivar-specific. Each cultivar contained one or the other component, but not both. The cultivars were crossed and their progeny studied through the F_2 generation. Larsen and Caldwell concluded that the two proteins are controlled by a pair of co-dominant genes at a single locus. Buttery and Buzzell (1968) and Brim et al. (1969) have observed genotypic differences in peroxidases and isoperoxidases.

DISEASES AND WEEDS

Diseases

Soybeans are attacked by about 30 bacterial, fungal, nematode, and viral pathogens. About half of these are of some economic importance. A comprehensive manual on soybean diseases was recently published by Dunleavy et al. (1966).

The magnitude of losses caused by soybean diseases is uncertain but has been estimated to be about $250 million per year. Nearly all soybean fields have some disease varying from a negligible amount to complete loss of all or part of the field.

Some diseases are effectively controlled by use of resistant cultivars. Perhaps the most outstanding example is phytophthora rot caused by *Phytophthora megasperma* Drechs. var. *sojae* Hildeb. This disease constituted a major threat in Illinois, Indiana, and Ohio, and in the heavier-textured soils of the lower Mississippi Valley. Highly resistant cultivars are now available to growers and the disease is under good control. Resistance is conditioned by a single gene, originally designated *Ps,* and later redesignated *Rps* (Hartwig et al. 1968). The pathogen has been shown to vary in pathogenicity and two races were described (Morgan and Hartwig 1965). Hartwig et al. (1968) studied crosses involving Hood (susceptible to both races), D60-9647 (resistant to race 1 and susceptible to race 2), and Semmes (resistant to both races). They concluded that there is an allelo-morphic series of at least three genes, *Rps* (Semmes), Rps^2 (D60-9647), and *Rps* (Hood), with dominance in that order. The frequency of Rps^2 in the U.S. germ plasm collection appears to be very low.

Brown stem rot, caused by *Cephalosporium gregatum* Allington and Chamberlain, is a disease for which no resistant cultivars are available. A substantial research effort is currently directed toward identifying some usable degree of genetic resistance. Nearly complete control was achieved in Iowa with a 5 years corn-1 year soybean crop rotation (Dunleavy and Weber 1967).

Some pathogens affect the seed and therefore seed quality. Purple stain is caused by *Cercospora kikuchii* (T. Matsu and Tomoyasu) Gardner. The cultivars Hill and Lee are moderately resistant to purple stain. Pod and stem blight, caused by *Diaporthe phaseolorum* (Cke. & Ell.) Sacc. var. *sojae* (Lehman) Wehm., although most noticeably a disease of pods and stems, may cause its greatest damage on the seed. Yeast spot is caused by the yeast *Nematospora coryli* Pegl. The spores are carried in the mouth of the green stinkbug and are forced into the developing seeds during feeding. Seeds attacked early may not mature or may remain small and shriveled. Seeds infected later may show a slightly depressed spot varying from yellow to light cream to brown in color. There may be a considerable volume of dead tissue extending well into the infected seeds (Dunleavy *et al.* 1966).

Seed mottling may result from soybean mosaic virus (Ross 1963, 1968). Several species of aphids spread the virus from plant to plant.

The soybean cyst nematode (SCN), *Heterodera glycines* Ichinohe, occurs in the Southeastern States, the lower Mississippi Valley, and in the southern part of Illinois, Indiana, and Missouri. Four races can be distinguished based on anatomical features or virulence. The cultivars Pickett, Custer, and Dyer are resistant to races 1 and 3, the most widely known races, and provide control on most infested fields. PI 90763 is resistant to race 2, which has been identified in Virginia and North Carolina. Resistance to race 4, which was first reported in 1969 in limited areas of Arkansas, Missouri, and Tennessee, has not been identified. The area in which SCN is known to exist is gradually expanding.

The rootknot nematodes *Meloidogyne incognita incognita,* (Kofoid and White, 1919) Chitwood 1949 and *M. hapla* Chitwood 1949 attack soybeans, especially in the south. The cultivars Delmar, Hill, Dyer, Hardee, and Bragg have fairly good resistance to *M. incognita incognita* and *M. incognita acrita* Chitwood 1949.

Many other nematode species have been identified in soybean fields and may reduce yields, but their significance to soybean production is not well-defined. However, the combined damage caused by other nematodes may equal that caused by cyst and rootknot nematodes.

Weeds

Weeds constitute the greatest hazard to soybean production in terms of magnitude of losses which they can cause. Several workers have estimated potential losses to be 1 kg of crop plant for each kilogram of weeds. Actual costs of weeds in U.S. soybean fields, including production losses and cost of control measures, are estimated to be of the order of $400 million per year. About 20% of harvested acreage is treated with herbicides at a cost of about $35 million per year (Anon. 1968).

The most important weeds are cocklebur (*Xanthium pensylvanicum* Wallr.), crabgrass (*Digitaria* sp.), foxtails (*Setaria* sp.), johnsongrass (*Sorghum halepense* (L.) Pers.), lambsquarters (*Chenopodium album* L.), morningglory (*Ipomoea* sp.), nutsedges (*Cyperus* sp.), pigweeds (*Amaranthus* sp.), velvetleaf (*Abutilon theophrasti* Medic.), jimsonweed (*Datura stramonium* L.), ragweed (*Ambrosia* sp.), and smartweed (*Polygonum* sp.) (Anon. 1968). Volunteer corn and sorghum may be weeds in fields where these crops have preceded soybeans in the rotation.

Great advances have been made in chemical weed control in soybeans in recent years. More than 25 herbicides are now registered for use on soybeans. Most of these are for preplanting or preemergence use but some postemergence uses have also been registered. Postemergence materials offer considerable promise for controlling many of the broadleafed weeds. At least one mechanical cultivation is usually required in connection with chemical treatments.

Good cultural practices and seedbed preparation, combined with timely mechanical cultivation, often provide an effective grass-control program. In addition, several available herbicides control most annual grass weeds. Trifluralin (a, a, a-trifluoro-2, 6-dinitro-*N*,*N*-dipropyl-*p*-toluidine) or nitralin [4-(methylsulfonyl)-2,6-dinitro-*N*,*N*-dipropylaniline] , incorporated in the soil before planting, effectively controls the major annual grasses and seedlings of perennial grasses such as johnsongrass in soybeans.

Other herbicides which are useful in controlling annual grasses include alachlor [2-chloro-2′,6′-diethyl-*N*(methoxymethyl) acetanilide] , vernolate (*S*-propyl dipropylthiocarbamate), chloramben (3-amino-2,5-dichlorobenzoic acid), CDAA (*N*,*N*-diallyl-2-chloroacetamide), and linuron [3-(3,4 dichlorophenyl)-1-methoxy-1-methylurea] .

Chloramben or linuron, applied to the soil surface before emergence of the weeds or soybeans, will control many small-seeded broadleaf weeds satisfactorily if rain occurs shortly after application. Some States recommend a mixture of either chlorpropham (isopropyl *m*-chlorocarbanilate) or dinoseb (2-*sec*-butyl-4,6-dinitrophenol) plus naptalam (*N*-1-naphthylphthalamic acid) for control of small-seeded broadleaf weeds and annual grasses.

Fair to good control of jimsonweed, annual morningglory, and cocklebur, and fair control of velvetleaf can be obtained by postemergence application of chloroxuron 3[p-(p-chlorophenoxy) phenyl-1-1-dimethylurea] plus surfactant.

Soybeans can be quite sensitive to herbicides applied after emergence. Often, specialized application techniques are required to avoid excessive injury to the crop. For example, only the lowest 2 cm of the soybean stem will safely tolerate naphtha. Linuron applications are restricted to the lower 7 cm of soybean plants at least 40 cm tall to avoid excessive injury. Chloroxuron, although it can be applied as an overall treatment, is frequently applied so as to contact only the lower third of the soybean plants. Fortunately, basally directed sprays often control weeds more effectively than nondirected treatments. Treatment with 2,4-DB [4-(2,4-dichlorophenoxy) butyric acid] is usually applied only as an overall spray because the target weed (cocklebur) is often taller than the crop when 2,4-DB is used.

Some weeds such as cocklebur or hemp sesbania [*Sesbania exaltata* (Raf.) Cory] sometimes grow taller than the soybeans. To control these weeds, some ingenious methods have been devised for applying herbicides with little or no exposure of the crop. These include a wax bar, impregnated with 2,4-D (2,4 dichlorophenoxyacetic acid). The bars are mounted on a boom several centimeters above the crop so that the wax containing 2,4-D is rubbed off by tall weeds and the shorter soybeans do not contact the bar at all (McWhorter 1966). Another device, the use of which is not yet registered, can apply a spray over the top of the crop and recover up to 95% of the spray not intercepted by tall weeds (McWhorter 1970).

Johnsongrass reduces yield extensively in soybeans in the southern section of the North Central States and in the lower Mississippi Valley. It can be controlled with a preplanting application of dalapon (2,2-dichloropropionic acid) to control growth from rhizomes, plus an application of trifluralin or nitralin at planting time to control seedlings. Crop rotation and timely tillage practices help complete the program (Wax and McWhorter 1968).

In the North Central States soybeans in 50–75-cm rows outyield those in traditional 90–100-cm rows (Weber and Weiss 1948). Lack of suitable weed control measures has retarded the shift to narrow rows. Interest in narrow rows is increasing as better herbicides become available and more corn is grown in narrow rows. If rows become more narrow than 50 cm, mechanical cultivation to supplement chemical treatments will not be feasible. Current research using a combination of preplant tillage and herbicide mixtures appears promising for weed control in close-drilled soybeans (Wax and McWhorter 1968). Currently, however, row widths of at least 50 cm which will permit at least one cultivation appear more feasible than more narrow widths.

Insects and Mites

Insects cause an estimated $75 million loss to soybeans each year and about $12 million is spent on control. About 25 insecticides are registered for use on soybeans.

Insects may damage the crop directly by feeding or may act as vectors, as in the case of virus diseases carried by aphids or bean leaf beetles. Several species of stink bugs attack soybeans. Jackson (1967) considers the corn earworm (*Heliothis zea* (Boddie)) to be the most important pest in Missouri. The green stink bug

(*Acrosternum hilare* (Say)) is also a serious pest in Missouri. Other major insect pests attacking soybeans include the bean leaf beetle (*Cerotoma trifurcata* (Forster)), Mexican bean beetle (*Epilachna varivestis* (Mulsant)), velvetbean caterpillar (*Anticarsia gemmatilis* Häbner)), green cloverworm (*Pathypena scabra* (Fabricius)), garden webworm (*Loxostege similalis* (Guen'ec)), fall armyworm (*Spodoptera frugiperda* (J. E. Smith)), beet armyworm (*Spodoptera exigua* (Hübner)), and grasshoppers (Jackson 1967).

Spider mites (*Tetranychus sp.*) can be serious. They have most often caused problems in the middle Atlantic area. Mites appear to be one of the reasons soybeans have not been grown extensively in California (Jackson 1967). Recent research in the San Joaquin Valley of California placed major emphasis on mites.

Insecticides and miticides have been the principal means of control. Many of these materials are now in disfavor because of residue problems or other deleterious side effects. Oilseed crops such as soybeans concentrate chlorinated hydrocarbons in the seed, from where they may be extracted with the oil.

QUANTITATIVE GENETICS

Many of the agronomically important attributes of the soybean cannot be associated with the action of a few genes. Yield, lodging, height, seed size, protein, and oil are controlled by a number of genes. These genes have varying effects. A detailed review was conducted by Johnson and Bernard (1963).

The effectiveness of a breeding program designed to produce superior varieties is dependent upon the inheritance of traits, the interrelationships among the traits, and the effect of environment.

For quantitative traits, the effectiveness of selection can be generally determined by heritability estimates. Heritability estimates can be computed in several ways, each giving rise to a different value. The most commonly accepted definition of heritability is the ratio of genotypic to phenotypic variance. The variance data is obtained by evaluating segregating populations. These estimates of heritability provide a quantitative estimate as to how these characters will respond to selection. Care must be exercised when using heritability estimates to know the procedure used in determining the estimates.

Reported heritabilities of yield are generally rather low. Johnson and Bernard (1963) report an average heritability for yield of 52%. Estimates for maturity, height, lodging, and seed size are higher (65–90%). Heritability estimates for protein and oil are about 75%. Therefore, selection for yield is more difficult than selection for the other attributes.

Selection of promising lines by the plant breeder is based on several attributes, and the genes for these attributes are distributed throughout the chromosomes and may have related physiological effects. Consequently, the characters are interrelated and their interrelation must be considered in soybean improvement. The average genotypic and phenotypic correlations for eight soybean populations are presented in Table 2.2. A positive correlation existed between yield and

TABLE 2.2

MEAN PHENOTYPIC AND GENOTYPIC CORRELATION FROM EIGHT SOYBEAN POPULATIONS

Character	Yield		Maturity		Lodging		Height		Seed Size		Protein		Oil	
Correlated	Pheno	Geno	Pheno	Geno	Pheno	Geno	Pheno	Geno	Pheno	Geno	Pheno	Geno	Pheno	Geno
Yield	1.000	1.000												
Maturity	0.370	0.594	1.000	1.000										
Lodging	0.256	0.112	0.186	0.226	1.000	1.000								
Height	0.256	0.321	0.410	0.578	0.289	0.393	1.000	1.000						
Seed Size	0.212	0.255	0.175	0.201	-0.120	-0.137	-0.030	-0.057	1.000	1.000				
Protein	-0.139	-0.226	0.010	-0.010	-0.016	0.030	-0.045	-0.040	0.124	0.125	1.000	1.000		
Oil	0.069	0.107	-0.288	0.291	-0.053	-0.016	-0.101	-0.063	0.085	0.082	-0.545	-0.560	1.000	1.000

maturity, height, seed size, lodging, and oil; a negative correlation between yield and protein. A strong negative correlation existed between oil and protein percentage. Genetic improvement in both traits at the same time is, therefore, very difficult. A general approximation has been that a 2-point change in protein percentage will result in a 1-point change for oil in the opposite direction. The small negative correlation between yield and protein reflects the difficulty plant breeders have in developing high yielding cultivars with increased protein. The recently released cultivars Provar and Protana and other high protein lines in advanced stages of development illustrate that breeders can successfully increase protein percentage without much sacrifice of yield.

BIBLIOGRAPHY

ABEL, G. H., and ERDMAN, L. W. 1964. Response of Lee soybeans to different strains of *Rhizobium japonicum*. Agron. J. *56*, 423–424.

ABRAHAMSEN, M., and SUDIA, T. W. 1966. Studies on the soluble carbohydrates and carbohydrate precursors in germinating soybean seed. Am. J. Botany *53*, 108–114.

AHMED, S., and EVANS, H. J. 1961. The essentiality of cobalt for soybean plants grown under symbiotic conditions. Proc. Natl. Acad. Sci. *47*, 24–36.

ANON. 1968. Extent and cost of weed control with herbicides and an evaluation of important weeds, 1965. USDA–ARS *34–102.*

ANON. 1969. Agricultural statistics. USDA.

APRISON, M. H., MAGEE, W. E., and BURRIS, R. H. 1954. Nitrogen fixation by excised soybean root nodules. J. Biol. Chem. *208*, 29–39.

ARMIGER, W. H., FOY, C. D., FLEMING, A. L., and CALDWELL, B. E. 1968. Differential tolerance of soybean varieties to an acid soil high in exchangeable aluminum. Agron. J. *60*, 67–70.

BENEDICT, H. M., SWIDLER, R., and SIMONS, J. N. 1964. Chlorophyll content and growth of soybean plants: possible interaction of iron availability and day length. Science *144*, 1134–1135.

BERGERSEN, F. J. 1962. The effects of partial pressure of oxygen upon respiration and nitrogen fixation by soybean root nodules. J. Gen. Microbiol. *29*, 113–125.

BERGERSEN, F. J. 1965. Ammonia–an early stable product of nitrogen fixation by soybean root nodules. Australian J. Biol. Sci. *18*, 1–9.

BERGERSEN, F. J. 1966. Some properties of nitrogen-fixing breis prepared from soybean root nodules. Biochim. Biophys. Acta *130*, 304–312.

BERGERSEN, F. J., and TURNER, G. L. 1967. Nitrogen fixation by the bacteroid fraction of breis of soybean root nodules. Biochim. Biophys. Acta *141*, 507–515.

BERNARD, R. L. 1964. Soybean work planning conference. (Unpublished Rept.) U.S. Regional Soybean Lab., U.S.D.A. Urbana, Ill.

BERNARD, R. L. 1967. The inheritance of pod color in soybeans. J. Heredity *58*, 165–168.

BERNARD, R. L., and HOWELL, R. W. 1964. Inheritance of phosphorus sensitivity in soybeans. Crop Sci. *4*, 298–299.

BERNARD, R. L., and SINGH, B. B. 1969. Inheritance of pubescence type in soybeans: glabrous, curly, dense, sparse and puberulent. Crop Sci. *9*, 192–197.

BILS, R. F. 1960. Biochemical and cytological changes in the developing soybean cotyledon. Ph.D. Thesis, Univ. Illinois.

BILS, R. F., and HOWELL, R. W. 1963. Biochemical and cytological changes in developing soybean cotyledons. Crop Sci. *3*, 304–308.

BOHNING, R. H., and BURNSIDE, C. A. 1956. The effect of light intensity on rate of apparent photosynthesis in leaves of sun and shade plants. Am. J. Botany *43*, 557–561.

BORST, H. L., and THATCHER, L. E. 1931. Life history and composition of the soybean plant. Ohio Agr. Expt. Sta. Bull. *494*, 1–96.

BORTHWICK, H. A., and PARKER, M. W. 1938. Influence of photoperiods upon the differentiation of meristems and the blossoming of Biloxi soybeans. Botan. Gaz. *99*, 825–839.

BRAY, R. H. 1961. You can predict fertilizer needs with soil tests. Better Crops with Plant Food *45*, 18–27.

BRIM, C. A., USANIS, S. A., and TESTER, C. F. 1969. Organ specificity and genotypic differences in isoperoxidases of soybeans. Crop Sci. *9*, 843–845.

BROWN, J. C. 1966. Fe and Ca uptake as related to root-sap and stem-exudate citrate in soybeans. Physiol. Plantarum *19*, 968–976.

BROWN, J. C. 1968. Iron chlorosis in soybean as related to the genotype of rootstock. 5. Differential distribution of photosynthetic C^{14} as affected by phosphate and iron. Soil Sci. *105*, 159–165.

BROWN, J. C., and TIFFIN, L. O. 1965. Iron stress as related to the iron and citrate occurring in stem exudate. Plant Physiol. *40*, 395–400.

BRUN, W. A., and COOPER, R. L. 1967. Effects of light intensity and carbon dioxide concentration on photosynthetic rate of soybean. Crop Sci. *7*, 451–454.

BURTON, J. C., and CURLEY, R. L. 1966. Compatibility of *Rhizobium japonicum* and sodium molybdate when combined in a peat carrier medium. Agron. J. *58*, 327–330.

BUTLER, W. L., NORRIS, K. H., SIEGELMAN, H. W., and HENDRICKS, S. B. 1959. Detection, assay, and preliminary purification of the pigment controlling photoresponsive development of plants. Proc. Natl. Acad. Sci. (U.S.) *45*, 1703–1708.

BUTTERY, B. R., and BUZZELL, R. I. 1968. Peroxidase activity in seeds of soybean varieties. Crop Sci. *8*, 722–725

BYTH, D. E. 1968. Comparative photoperiodic responses for several soya bean varieties of tropical and temperate origin. Australian J. Agr. Res. *19*, 879–890.

CALDWELL, B. E. 1966. Inheritance of a strain-specific ineffective nodulation in soybeans. Crop Sci. *6,* 427–428.

CALDWELL, B. E., HINSON, K. and JOHNSON, H. W. 1966. A strain-specific ineffective nodulation reaction in the soybean, *Glycine max* L. Merrill. Crop Sci. *6*, 495–496.

CALDWELL, B. E., and WEBER, D. F. 1970. Distribution of *Rhizobium japonicum* serogroups in soybean nodules as affected by planting dates. Agron. J. *62*, 12–14.

CALDWELL, B. E., and VEST, G. 1970. Effects of *Rhizobium japonicum* strains on soybean yields. Crop Sci. *10*, 19–21.

CANNELL, R. Q., BRUN, W. A., and MOSS, D. N. 1971. A search for high net photosynethetic rate among soybean genotypes. Crop Sci. (in press).

CARNAHAN, J. E., MORTENSON, L. E., MOWER, H. F., and CASTLE, J. E. 1960. Nitrogen fixation in cell-free extracts of *Clostridium pasteurianum* Biochim. Biophys. Acta *38*, 188–189.

CARPENTER, W. D., and BEEVERS, H. 1959. Distribution and properties of isocitritase in plants. Plant Physiol. *34*, 403–409.

CARTTER, J. L., and HARTWIG, E. E. 1963. The management of soybeans. *In* The Soybean, A. G. Norman (Editor). Academic Press, New York.

CARTTER, J. L., and HOPPER, T. M. 1942. Influence of variety, environment, and fertility level on the chemical composition of soybean seed. USDA Tech. Bull. *787.*

COBB, G. W. 1962. Some effects of pressure on growth and differentiation of the roots of *Glycine max*. Dissertation Abstr. *24*, 3604.

COOPER, R. L. 1966. A major gene for resistance to seed coat mottling in soybean. Crop Sci. *6*, 290–292.

COOPER, R. L., and BRUN, W. A. 1967. Response of soybeans to a carbon dioxide-enriched atmosphere. Crop Sci. *7*, 455–457.

CURTIS, P. E., OGREN, W. L., and HAGEMAN, R. H. 1969. Varietal effects in soybean photosynthesis and photorespiration. Crop Sci. *9*, 323–327.

DAMIRGI, S. M., FREDERICK, L. R., and ANDERSON, I. C. 1967. Serogroups of *Rhizobium japonicum* in soybean nodules as affected by soil types. Agron. J. *59*, 10–12.

DELOUCHE, J. C. 1953. Influence of moisture and temperature levels on the germination of corn, soybeans, and watermelons. Assoc. Offic. Seed Analysts Proc. *43*, 117–126.

DILWORTH, M. J. 1966. Acetylene reduction by nitrogen-fixing preparations from *Clostridium pasteurianum*. Biochim. Biophys. Acta *127*, 285–294.

DORWORTH, C. D. 1967. Variations in germination, microflora, and fatty acids of soybeans stored under different combinations of moisture content, temperature, and time. Dissertation Abstr. *27*, 2569B.

DUNLEAVY, J. M., CHAMBERLAIN, D. W., and ROSS, J. P. 1966. Soybean diseases. USDA Revised Agr. Handbook *302*.

DUNLEAVY, J. M., and WEBER, C. R. 1967. Control of brown stem rot of soybeans with corn-soybean rotations. Phytopathology *57*, 114–117.

EGLI, D. B., PENDLETON, J. W., and PETERS, D. B. 1970. Photosynthetic rate of three soybean communities as related to carbon dioxide levels and solar radiation. Agron. J. *62*, 411–414.

ERDMAN, L. W., JOHNSON, H. W., and CLARK, F. E. 1957. Varietal responses of soybeans to a bacterial-induced chlorosis. Agron. J. *49*, 267–271.

EVANS, H. J. 1954. Diphosphopyridine nucleotide-nitrate reductase from soybean nodules. Plant Physiol. *29*, 298–301.

EVANS, H. J. 1969. How legumes fix nitrogen. In How Crops Grow, J. G. Horsfall (Editor). Centennial Lecture Ser. Conn. Agr. Expt. Sta.

FEHR, W. R. 1969. Inheritance of hypocotyl length at 25° C in soybeans, *Glycine max* (L.) Merr. Agronomy Abstr. 6.

FISHER, J. E. 1963. The effects of short days on fruit set as distinct from flower formation in soybeans. Can. J. Botan. *41*, 871–873.

FLETCHER, H. F., and KURTZ, L. T. 1964. Differential effects of phosphorus fertility on soybean varieties. Soil Sci. Soc. Am. Proc. *28*, 225–228.

FOOTE, B. D., and HOWELL, R. W. 1964. Phosphorus tolerance and sensitivity of soybeans as related to uptake and translocation. Plant Physiol. *39*, 610–613.

GALITZ, D. S. 1961. A physiological study of the developing soybean seed radicle. Ph.D. Thesis, Univ. Illinois.

GALITZ, D. S., and HOWELL, R. W. 1965. Measurement of ribonucleic acids and total free nucleotides of developing soybean seeds. Physiol. Plantarum *18*, 1018–1021.

GARNER, W. W., and ALLARD, H. A. 1920. Effect of the relative length of day and night and other factors of the environment on growth and reproduction in plants. J. Agr. Res. *18*, 553–606.

GARNER, W. W., ALLARD, H. A., and FOUBERT, C. L. 1914. Oil content of seeds as affected by the nutrition of the plant. J. Agr. Res. *III*, 227–249.

GRABE, D. F. 1965. Storage of soybeans for seed. Soybean Dig. *26*, No. 1, 14–16.

GRABE, D. F., and METZER, R. B. 1969. Temperature-induced inhibition of soybean in hypocotyl elongation and seedling emergence. Crop Sci. *9*, 331–333.

GRABLE, A. R. 1964. The effects of carbon dioxide on germination and growth of corn and soybeans in soil containing different levels of oxygen and moisture. Dissertation Abstr. *24*, 2644.

GREER, H. A. L., and ANDERSON, I. C. 1965. Response of soybeans to Triiodobenzoic acid under field conditions. Crop Sci. *5*, 229–232.

HAMNER, K. C. 1969. *Glycine max* (L.) Merrill. *In* The Induction of Flowering, Some Case Histories, L. T. Evans (Editor). Cornell Univ. Press, Ithaca, N. Y.

HAMNER, K. C., and BONNER, J. 1938. Photoperiodism in relation to hormones as factors in floral initiation and development. Botan. Gaz. *100*, 388–431.

HANWAY, J. J., and THOMPSON, L. E. 1967. How a soybean plant develops. Iowa Agr. Expt. Sta. Spec. Rep. *53*.

HARDY, R. W. F., HOLSTEN, R. D., JACKSON, E. K., and BURNS, R. C. 1968. The acetylene-ethylene assay for N_2 fixation: laboratory and field evaluation. Plant Physiol. *43*, 1185–1207.

HARRIS, H. B., PARKER, M. B., and JOHNSON, R. J. 1965. Influence of molybdenum content of soybean seed and other factors associated with seed source on progeny response to applied molybdenum. Agron. J. *57*, 397–399.

HARTWIG, E. E., KEELING, B. L., and EDWARDS, C. J., Jr. 1968. Inheritance of reaction to phytophthora rot in the soybean. Crop Sci. *8*, 634–635.

HERMANN, F. J. 1962. A revision of the genus *Glycine* and its immediate allies. USDA Tech. Bull. *1268.*

HESKETH, J. D. 1963. Limitations to photosynthesis responsible for differences among species. Crop Sci. *3*, 493–496.

HEW, C. S., NELSON, C. D., and KROTKOV, G. 1967. Hormonal control of translocation of photosynthetically assimilated [14] C in young soybean plants. Am. J. Botany *54*, 252–256.

HOWELL, R. W. 1954. Phosphorus nutrition of soybeans. Plant Physiol. *29*, 477–483.

HOWELL, R. W. 1961. Changes in metabolic characteristics of mitochondria from soybean cotyledons during germination. Physiol. Plantarum *14*, 89–97.

HOWELL, R. W., and COLLINS, F. I. 1957. Factors affecting linolenic and linoleic acid content of soybean oil. Agron. J. *49*, 593–597.

HOWELL, R. W. *et al.* 1960. Response of soybeans to seed-treatment with gibberellin under simulated commercial conditions. Agron. J. *52*, 144–146.

HUNTER, J. R., and ERICKSON, A. E. 1952. Relation of seed germination to soil moisture tension. Agron. J. *44*, 107–109.

ISHIKAWA, Y., HASEGAWA, S., and KASAI, T. 1967. Changes in amino acid composition during germination of soybean. IV. Identification of a- and b-glutamylaspartic acid. Agr. Biol. Chem. (Tokyo) *31*, 490–493.

JACKSON, R. D. 1967. Soybean insect problems. Soybean Dig. *27*, No. 11, 16–18.

JOHNSON, H. W., and BERNARD, R. L. 1963. Soybean genetics and breeding. *In* The Soybean, A. G. Norman (Editor). Academic Press, New York.

JOHNSON, H. W., BORTHWICK, H. A., and LEFFEL, R. C. 1960. Effects of photoperiod and time of planting on rates of development of the soybean in various stages of the life cycle. Botan. Gaz. *122*, 77–95.

JOHNSON, H. W., and MEANS, U. M. 1963. Serological groups of *Rhizobium japonicum* recovered from nodules of soybeans (*Glycine max*) in field soils. Agron. J. *55*, 269–271.

JOHNSON, H. W., and MEANS, U. M. 1964. Selection of competitive strains of soybean nodulating bacteria. Agron. J. *56*, 60–62.

JOHNSON, H. W., MEANS, U. M., and CLARK, F. E. 1959. Responses of seedlings to extracts of soybean nodules bearing selected strains of *Rhizobium japonicum.* Nature *183*, 308–309.

JOHNSON, H. W., MEANS, U. M., and WEBER, C. R. 1965. Competition for nodule sites between strains of *Rhizobium japonicum* applied as inoculum and strains in the soil. Agron. J. *57*, 179–185.

KASAI, T., ISHIKAWA, Y., OBATA, Y., and TSUKAMOTO, T. 1966. Changes in amino acid composition during germination of soybean. I and II. Agr. Biol. Chem. (Tokyo) *30*, 973–981.

KENNEDY, B.W., and COOPER, R. L. 1967. Association of virus infection with mottling of soybean seed coats. Phytopathology *57*, 35–37.

KLEESE, R. A. 1968. Scion control of genotypic differences in Sr and Ca accumulation in soybeans under field conditions. Crop Sci. *8*, 128–129.

KOCH, B., and EVANS, H. J. 1966. Reduction of acetylene to ethylene by soybean root nodules. Plant Physiol. *41*, 1748–1750.

KOCH, B., EVANS, H. J., and RUSSELL, S. 1967A. Reduction of acetylene and nitrogen gas by Breis and cell-free extracts of soybean root nodules. Plant Physiol. *42*, 466–468.

KOCH, B., EVANS, H. J., and RUSSELL, S. 1967B. Properties of the nitrogenase system in cell-free extracts of bacteroids from soybean root nodules. Proc. Natl. Acad. Sci. (U.S.) *58*, 1343–1350.

KOSHIMIZU, Y., and IIZUKA, N. 1963. Studies on soybean virus diseases in Japan. (English Summary). Tohoku Natl. Agr. Expt. Sta. Bull. *27*, 1–103.

KROBER, O. A. 1962. Annual Report. U.S. Regional Soybean Laboratory, Urbana, Ill. (Unpublished).

KROBER, O. A., CARTTER, J. L. 1966. Relation of methionine content to protein levels in soybeans. Cereal Chem. *43*, 321–325.

LARSEN, A. L., and CALDWELL, B. E. 1968. Inheritance of certain proteins in soybean seed. Crop Sci. *8*, 474–476.

LARSEN, A. L., and CALDWELL, B. E. 1969. Source of protein variants in soybean seed. Crop Sci. *9*, 385–387.

LAVY, T. L., and BARBER, S. A. 1963. A relationship between the yield response of soybeans to molybdenum applications and the molybdenum content of the seed produced. Agron. J. *55*, 154–155.

LEFFEL, R. C. 1961. Planting date and varietal effects on agronomic and seed compositional characters in soybeans. Maryland Univ. Agr. Expt. Sta. Bull. *A*–117.

MCALISTER, D. F., and KROBER, O. A. 1951. Translocation of food reserves from soybean cotyledons and their influence on the development of the plant. Plant Physiol. *26*, 525–538.

MCWHORTER, C. G. 1966. Sesbania control in soybeans with 2,4-D wax bars. Weeds *14*, 152–155.

MCWHORTER, C. G. 1970. A recirculating spray system for directed postemergence weed control in row crops. Weed Sci. *18*, 285–287.

MENZ, K. M., MOSS, D. N., CANNELL, R. O., and BRUN, W. A. 1971. Screening for photosynthetic efficiency. Crop Sci. (in press).

MIKSCHE, J. P. 1966. NA synthesis in primary root of *Glycine max* during germination. Can. J. Botan. *44*, 789–794.

MIKSCHE, J. P., and GREENWOOD, M. 1966. Quiescent center of the primary root of *Glycine max*. New Phytologist *65*, 1–4.

MITCHELL, R. L. 1968. Rooting patterns of soybean varieties. Agronomy Abstr., p. 48.

MITCHELL, R. L. 1969. Root development of soybeans. Iowa Farm Sci. *23*, 9–12.

MITCHELL, R. L., and ANDERSON, I. C. 1966. Effect of gibberellic acid in reducing Fe chlorosis in soybeans. Crop Sci. *6*, 111–112.

MOOERS, C. A. 1908. Soy beans vs. cowpeas. Tenn. Agr. Expt. Sta. Bull. *82*, 75–104.

MORGAN, F. L., and HARTWIG, E. E. 1965. Physiologic specialization in *Phytophthora megasperma* var. *sojae*. Phytopathology *55*, 1277–1279.

MORRIS, C. J., and THOMPSON, J. F. 1962. The isolation and characterization of γ-L-glutamyl-L-tyrosine and γ-L-glutamyl-L-phenylalanine from soybeans. Biochemistry. *1*, 706–709

MOSS, D. N. 1966. Respiration of leaves in light and darkness. Crop Sci. *6*, 351–354.

MOSS, D. N., KRENZER, E. G., and BRUN, W. A. 1969. Carbon dioxide compensation points in related plant species. Science *164*, 187–188.

OHLROGGE, A. J. 1963. Mineral nutrition of soybeans. *In* the Soybean, A. G. Norman (Editor). Academic Press, New York.

OHMURA, T., and HOWELL, R. W. 1960. Inhibitory effect of water on oxygen consumption by plant materials. Plant Physiol. *35*, 184–188.

OWENS, L. D. 1969. Toxins in plant diseases: structure and mode of action. Science *165*, 18–25.

PAMPLIN, R. A. 1962. Ph.D. Thesis, Univ. Illinois.

PAZUR, J. H., SHADAKSHARASWAMY, M., and MEIDELL, G. E. 1962. The metabolism of oligosaccharides in germinating soybeans, *Glycine max*. Arch. Biochem. Biophys. *99*, 78–85.

PETERS, D. B., and JOHNSON, L. C. 1960. Soil moisture use by soybeans. Agron. J. *52*, 687–689.

RAPER, C. D., Jr. 1968. Differences in rooting morphology among soybean varieties. Agronomy Abstr., 49.

RINNE, R. W. 1967. Unpublished annual report, U.S. Regional Soybean Laboratory, Urbana, Ill. (Quoted with permission.)

RINNE, R. W. 1969. Biosynthesis of fatty acids by a soluble extract from developing soybean cotyledons. Plant Physiol. *44*, 89–94.

ROSS, J. P. 1963. Interaction of the soybean mosaic and bean pod mottle viruses infecting soybeans. Phytopathology *53*, 887.

ROSS, J. P. 1968. Effect of single and double infections of soybean mosaic and bean pod mottle viruses on soybean yields and seed characters. Plant Disease Reptr. *52*, 344–348.

RUDDAT, M., and PHARIS, R. P. 1966. Participation of gibberellin in the control of apical dominance in soybean and redwood. Planta *71*, 222–228.

SINGH, B. B., HADLEY, H. H., and BERNARD, R. L. 1971. Morphology of pubescence in soybeans and its relationship to plant vigor. Crop Sci. *11*, 13–16.

SLOGER, C. 1969. Symbiotic effectiveness and N_2 fixation in nodulated soybean. Plant Physiol. *44*, 1666–1668.

SMALL, J. G. C., and LEONARD, O. A. 1969. Translocation of C[14] labeled photosynthate in nodulated legumes as influenced by nitrate nitrogen. Am. J. Botan. *56*, 187–194.

TANNER, J. W., and ANDERSON, I. C. 1963. Investigations on non-nodulating and nodulating soybean strains. Can. J. Plant Sci. *43*, 542–546.

VEST, G. 1970. Rj$_3$ – a gene conditioning ineffective nodulation in soybean. Crop Sci. *10*, 34–35.

WAX, L. M., and MCWHORTER, C. G. 1968. Prescription for weed control. Soybean Dig. *28*, No. 4, 12–15.

WEBER, C. R. 1966A. Nodulating and non-nodulating soybean isolines: I. Agronomic and chemical attributes. Agron. J. *58*, 43–46.

WEBER, C. R. 1966B. Nodulating and non-nodulating soybean isolines: II. Response to applied nitrogen and modified soil conditions. Agron. J. *58*, 46–49.

WEBER, C. R., and HANSON, W. D. 1961. Natural hybridization with and without ionizing radiation in soybeans. Crop Sci. *1*, 389–392.

WEBER, C. R., and WEISS, M. G. 1948. Let's push up soybean yields. Soybean Dig. *8*, 17–19.

WEISS, M. G. 1943. Inheritance and physiology of efficiency in iron utilization in soybeans. Genetics *28*, 253–268.

WIDHOLM, J. M., and OGREN, W. L. 1969. Photorespiratory-induced senescence of soybeans. Plant Physiol. *44* (suppl.)

WILCOX, J. R., and LAVIOLETTE, F. A. 1968. Seedcoat mottling response of soybean genotypes to infection with soybean mosaic virus. Phytopathology *58*, 1446–1447.

WILLIAMS, L. F. 1950. Structure and genetic characteristics of the soybean. *In* Soybeans and Soybean Products, K. S. Markley (Editor). John Wiley & Sons, New York.

WILLIAMS, L. F., and LYNCH, D. L. 1954. Inheritance of a non-nodulating character in the soybean. Agron. J. *46*, 28–29.

WILSON, C. M., and SHANNON, J. C. 1963. The distribution of ribonucleases in corn, cucumber, and soybean seedlings. Biochim. Biophys. Acta *68*, 311–313.

WOODWORTH, C. M. 1932. Genetics and breeding in the improvement of the soybean. Illinois Agr. Expt. Sta. Bull. *384*

WOODWORTH, C. M. 1933. Genetics of the soybean. J. Am. Soc. Agron. *25*, 36–51.

CHAPTER 3

A. K. Smith
S. J. Circle Chemical Composition of the Seed

INTRODUCTION

Soybeans are well known for variations in color, size, and shape of the seed and other physical properties as well as their chemical composition. The physical and chemical differences are considerably modified by the heredity of the variety and the influences of the climatic conditions in which they are grown. Their photoperiodic characteristics or length of the day during the growing season is a major factor in controlling the onset of bloom and thus the yield of the plant. Accordingly, each variety produces best within a definite range of latitude.

Piper and Morse (1923) reviewed the early literature on soybeans and reported on the variations in oil and protein and on the iodine number of the oil as well as other factors resulting from varietal and environmental influences. Their report on 500 samples shows a range in protein content of 30–46% and in oil, of 12–24%. Another series of experiments by the USDA, reported by Dies (1942), on the characteristics of 128 varieties of soybeans shows the number of seeds per pound varied from 1232 to 9950, the oil content, from 13.9 to 23.2%, and the protein content from 32.4 to 50.2%. These data were from soybeans introduced from China, Manchuria, Korea, Japan, Siberia, France, and Italy during the years of 1880 to 1940 and grown in different locations in the United States.

During the long period of development of soybeans in China, Korea, Japan, and other oriental countries, the lack of travel and communication between farmers limited the exchange of information on methods of production and exchange of seed. This isolation led to the development of a great number of varieties and strains of soybeans. Piper and Morse (1923), who were responsible for a great many of the introductions from the Orient, stated "among the many varieties introduced from China and Manchuria, it is a very interesting fact that the same variety has rarely been secured a second time unless from the same place. It appears that practically every locality in these countries has its own local variety." To obtain the best varieties for the American farmer, more than 10,000 introductions were made for studies by the USDA and the state experiment stations in plant selection and breeding research.

In preliminary studies on the influence of fertilizer on the composition of soybeans, Cartter (1940) found that, while fertilizer level affected the yield of soybeans and varieties responded differently to different levels of application,

61

the fertilizer application had no noticeable effect on composition. He concluded that the importance of this work seems to be that composition is an inherited characteristic and thus makes possible breeding research on composition and yield as well as other agronomic characteristics.

NITROGENOUS CONSTITUENTS

Nitrogen Conversion Factor

In the analytical determination of protein in soybeans, soybean meal, and other protein products, a nitrogen to protein conversion factor is used. A true conversion factor for soybeans has not been determined. However, U.S. agronomists and commercial handlers and processors have used the value 6.25 since soybeans were introduced into the United States.

Jones (1931) published a list of proposed conversion factors for converting percentage of nitrogen in foods, feeds, seeds, and derived protein concentrates into percentage of protein in which he recommended a conversion factor for soybeans and soybean protein of 5.71. This factor was taken from the work of Osborne and Campbell (1898) in which they isolated and purified several protein fractions from the soybean. The fraction from which the above factor was taken contained 17.5% nitrogen and was called glycinin; it represented less than 10% of the total protein of the bean. Also, they isolated two smaller fractions, which they called legumelin and proteose, for which they reported nitrogen values of 16.12 each. The factor 5.71 originated from this early work and has been quoted frequently in the literature. However, the Osborne and Campbell (1898) report did not claim that their values represent the nitrogen content of the whole bean or any fractions other than those indicated; and, thus, their work does not justify the factor 5.71 for general use.

Agronomists and other analysts in the United States customarily use the factor 6.25 which is legally recognized by the National Soybean Processors Association, the Association of Official Analytical Chemists, the American Soybean Association, and other technical associations engaged in trade in soybeans and soybean products.

Research on soybean protein isolates by Smiley and Smith (1946) indicates that the factor 6.25 is probably higher than justified. Smith (1966) reviewed this problem and stated that "the use of two or more conversion factors in the literature will lead to confusion and difficulty in interpretation and comparison of nutritional and other data obtained in soybean protein investigations; and the use of a factor other than 6.25 will only introduce confusion and should be avoided."

Recently, Tkachuk and Irvine (1969) and Tkachuk (1969) made a different approach to the problem. They have determined nitrogen-conversion factors for a number of cereals and oilseed meals by using the amino acid composition of the meal protein as the basis for their calculations. They determined factors by dividing the weight of the amino acid residues, corrected for water, by the weight of the nitrogen they contain.

The use of the amino acid assay method of determining protein-conversion factors for protein isolates and concentrates is a decided improvement over the earlier methods, provided suitable corrections are made for nonprotein nitrogen (Becker *et al.* 1940) and other variables. To determine a protein factor for the whole seed, a further correction for the seed coat is necessary; the seed coat will vary with the size of the seed and its thickness and the correction will probably be a compromise. For the defatted soybean meal, containing the seed coat as well as the soluble nonprotein nitrogen, Tkachuk (1969) reported a conversion factor of 5.69. Perhaps one of the more important effects of the use of this smaller factor on soybean meal utilization will be in nutritional investigations which use the protein efficiency ratio (PER) for evaluating the protein. However, when the biological value (BV) or the ratio of the retained nitrogen to absorbed nitrogen is determined, a nitrogen to protein factor is irrelevant. In the nutritional evaluation (PER) of food preparations as well as animal feed formulations which are composites of several proteins, a factor for their particular combination of products would have to be determined.

Protein Composition of the Seed

One of the major investigations in the United States on the influence of variety, environment, and fertility level on the chemical composition of soybeans was reported by Cartter and Hopper (1942). A summary of their average results for the principal components of 10 varieties of soybeans grown at 5 locations for a 5-yr period is presented in Table 3.1.

Their data, on a moisture free basis, show average protein values for the 10 varieties range from 40.58% for the Peking to 46.42 for the Mandarin. Cartter and Hopper (1942) concluded from their investigation that the variation in protein content is the result of two factors: (1) locality where the beans are grown that is, soil and other environmental conditions; and (2) variety of the bean. Their report states that for well-composited soybeans the varietal factor is the greater and more consistent of the two. The soybeans in this investigation were not all commercial varieties and the average values for protein do not represent protein values for present-day commercial soybeans.

Notwithstanding the large number of soybean varieties from which U.S. growers may choose, only a few varieties are found in a given region. Their selection from the list of recommended varieties for the region are based largely on their adaptation to latitude, yield, and disease resistance. The recommended varieties for 1968 for the soybean growing areas from north to south are shown in Table 3.2. This table shows location numbers as well as protein and oil composition and seed weight of strains and varieties adapted to similar length of day and other location conditions. The varieties presently in production are

TABLE 3.1

AVERAGE COMPOSITION OF 10 VARIETIES OF SOYBEANS GROWN AT 5
LOCATIONS FOR A 5-YEAR PERIOD (TOTAL SUGARS AS SUCROSE)

Variety or Strain	Protein N × 6.25 (%)	Oil (%)	Total Sugar (%)	Crude Fiber (%)	Total Ash (%)	Iodine Number
	Moisture Free Basis					
Mandarin	46.42	18.16	6.76	5.39	5.37	127.6
Mukden	45.76	19.26	6.83	5.33	5.00	124.6
Dunfield A	41.38	20.97	8.24	5.34	4.65	124.9
Dunfield B	41.42	20.91	8.40	5.45	4.61	124.4
Illini	42.59	19.99	8.83	5.26	4.81	130.5
Manchu	44.06	19.40	7.78	5.42	5.12	130.2
Scioto	42.47	20.29	8.22	5.23	5.17	133.0
T-117	41.86	20.37	8.61	5.51	5.02	123.9
Peking	40.58	17.07	7.50	6.48	5.21	137.7
P.I. 54563–3	42.18	19.91	8.58	5.80	4.97	129.8

Source: Cartter and Hopper (1942).

subject to change as new varieties are developed in the agronomic program of breeding and selecting for improved composition, yield, and disease resistance.

High Protein Soybeans

When the soybean breeding program was initiated in the United States, the high price of oil in relation to the meal encouraged the plant breeders to develop high oil-yielding soybeans. As a result of this program, U.S. soybeans are higher in oil than beans from other countries, a characteristic which sometimes brings a premium over beans from other countries in the export markets. However, with increasing production of soybeans in the United States a large surplus of oil has developed which has caused a marked decrease in its price. It seems unlikely that in the United States the price of oil will ever regain its former position. Now, the dollar value of the meal is greater than that of the oil.

This change in price relationship has encouraged some of the plant breeders to revise their program to include the breeding of high protein soybeans. However, when the level of protein is raised other modifications in composition are brought into play such as a decrease in oil and carbohydrate and a loss in yield. Thus, the economics of the farmer and of the processor in choosing between high oil and high protein involves interrelated factors.

Krober and Cartter (1962) studied the interrelationship of the protein and the nonprotein constituents of the soybean such as oil, sugars, holocellulose, pentosans, crude fiber, ash and seed size. They found that the oil, sugars, and holocellulose were affected most by changes in protein content. When the protein was increased, they found that about 1/3 of the decrease in nonprotein constituents was in sugars, 1/3 in oil, and the remainder in holocellulose and

ANALYTICAL DATA FOR CURRENT SOYBEAN VARIETIES FROM UNIFORM
TEST REPORT FOR 1968, NORTHERN GROUPS 00–IV AND SOUTHERN
GROUPS IVS–VIII SUPPLIED BY USDA REGIONAL SOYBEAN LABORATORY
URBANA, ILLINOIS
MOISTURE FREE BASIS

Uniform Test No.	Variety[1]	Hilum Color	Seed Weight	Protein N × 6.25 (%)	Oil (%)
00	Altona	Black	18.0	39.6	19.7
	Flambeau	Black	16.1	40.7	18.3
	Portage	Yellow	17.6	38.6	19.6
0	Clay	Yellow	17.0	40.2	21.1
	Grant	Black	16.7	40.0	19.7
	Merit	Buff	14.6	39.4	20.6
	Traverse	Yellow	18.0	40.7	19.8
I	Chippewa	Black	16.0	41.0	20.5
	Hark	Yellow	16.6	41.8	20.4
II	Amsoy	Yellow	17.2	38.7	22.0
	Corsoy	Yellow	15.9	39.6	21.5
	Harosoy 63	Yellow	18.0	40.3	21.1
III	Adelphia	Buff	15.8	39.2	21.7
	Calland	Black	17.8	38.7	21.3
	Wayne	Black	16.4	40.4	21.5
IV	Clark 63	Black	15.9	40.0	21.7
	Cutler	Black	18.0	40.4	21.6
	Kent	Black	17.7	40.0	22.0
IVS	Kent	Black	15.5	40.1	22.0
	Delmar	Yellow	14.3	39.2	22.2
	Custer	Black	13.7	36.9	22.0
V	Hill	Brown	11.8	38.5	21.2
	Dare	Buff	12.9	38.6	22.1
	York	Buff	17.3	38.6	20.9
	Dyer	Black	15.1	39.2	20.7
VI	Hood	Buff	13.4	39.3	21.7
	Lee	Black	12.4	40.6	21.1
	Lee 68	Black	12.5	40.5	20.9
	Pickett	Black	12.5	40.1	21.3
	Davis	Buff	13.4	38.8	21.6
VII	Bragg	Black	14.4	40.5	21.5
	Semmes	Black	14.3	41.4	20.9
VIII	Hampton	Buff	14.7	38.7	22.5
	Hardee	Buff	13.8	41.4	21.5

[1] Seed color of all varieties was yellow.

pentosans. However, the changes do not always follow the same pattern. In another lot of low protein beans, more than 1/2 of the increase was in oil with lesser changes in sugars and other constituents. The overall results indicate that the increase in protein was 2–3 times the decrease in oil and about 1/2 of the loss in carbohydrates was in a form which is not readily digestible. They pointed out that in developing high protein beans it may be difficult to gain 3% protein with only 1% loss in oil while maintaining normal yields. These preliminary studies give indications there will be a compromise in the level of protein that can be bred profitably into soybeans.

Krober and Cartter (1966) devised experiments to separate, as far as possible, the genetic and environmental factors that affect protein content of the seed, especially the methionine, and to determine if in developing high protein beans there might be a change in the methionine which would affect nutritional value. They conducted four experiments with as wide a range of protein content and environmental conditions as can ordinarily be found and reported no significant tendency for methionine in the protein to decrease with increasing protein in the seed. They stated "in fact, there was more of a tendency toward a positive relationship, which is favorable to the development of high protein strains of good nutritional quality."

In calculating the value of the oil and meal derived from solvent extraction of 1 bu of soybeans, the yield of oil is approximately 10.7 lb and meal (44% protein) is 47.7 lb (Anon. 1971). The quotations on oil are usually in cents per pound and the meal in dollars per short ton. Using these data, the value of the oil and meal derived from 1 bu of beans can be estimated by multiplying the price of oil by 10.7 and the price of meal in dollars per ton by 0.0238 as the following example assuming the price of oil is 9c per lb and the meal $72.00 per ton:

$$10.7 \times 0.09 = \$0.963 \text{ (oil)}$$
$$72.00 \times 0.0238 = 1.713 \text{ (meal)}$$
$$\text{Total} \quad \$2.676 \text{ per bu}$$

These data can be used along with the yield to estimate the value of the oil and meal derived from an acre of soybeans. With these and other cost data, an estimate can be made at what price of oil and meal an economic advantage can be gained by changing to production of high protein soybeans.

However, in estimating the future trends, the world price of oil must be given consideration. In many countries the price of oil still remains at a high level; thus the production of high oil soybeans for the export market will need to be considered. With a possible need for high oil beans for the export market and high protein beans for the domestic specialty products market, soybean production might be divided between the two types of soybeans.

Garden Type Soybeans

Garden types are soybeans which the Chinese and other oriental people use during the summer as green beans for the table. They were introduced into the U.S. program and tested as a potential garden crop by Lloyd and Burlison (1939), Woodruff and Klaas (1938), and Weiss et al. (1942). The garden type soybeans are sometimes referred to as vegetable or edible soybeans; however, at present the most popular designation is "garden type." The garden varieties can be preserved by freezing and canning much like other vegetables.

Garden type soybeans are not basically different from field varieties but are reported generally to be larger in size, higher in protein, lower in oil, lower in yield, and on reaching maturity they have a tendency to shatter from the pod, resulting in substantial loss if harvested with a combine. Garden varieties are reported to have a better flavor and texture than the regular field beans and have been compared in these qualities to lima beans.

Table 3.3 gives the protein and oil content of several varieties of garden type beans and Table 3.4 compares garden type beans with other common beans and with peas. The garden type contains about twice as much protein as the other beans and peas and 11 times as much oil. Thus, they are much higher in nutritive and caloric value than other garden beans and peas.

Nonprotein Nitrogen

Soybeans contain small quantities of peptides and amino acids having variable molecular dimensions which may occur as the residue of incomplete protein synthesis or possibly the result of protein degradation. Muller and Armbrust

TABLE 3.3

PROTEIN AND FAT CONTENT AND SEED SIZE OF 12 VARIETIES OF MATURE GARDEN TYPE SOYBEANS TESTED AT URBANA, ILLINOIS
(MOISTURE FREE BASIS)

Variety	Protein N X 6.25 (%)	Oil (%)	Weight Avg 3 Yr (Gm/100 Seed)
Very early			
Giant green	39.3	22.4	29.4
Early			
Bansei	36.4	21.6	21.2
Fuji	39.6	21.4	25.9
Midseason			
Illini	38.7	21.4	13.9
Hokkaido	40.4	20.8	31.9
Jogun	40.7	19.9	29.9
Willomi	42.3	19.4	31.1
Late			
Illington	42.9	18.6	25.9
Imperial	41.0	20.5	28.4
Funk Delicious	42.3	20.0	31.7
Emperor	42.2	19.9	29.7
Higan	41.0	18.1	23.4

Source: Lloyd and Burlison (1939).

TABLE 3.4

COMPOSITION OF GARDEN TYPE SOYBEANS AND OTHER
VEGETABLE BEANS AND PEAS

Food	Mois- ture (%)	Protein N × 6.25 (%)	Fat (%)	CHO Total (%)	Ash (%)	Calcium (%)	Iron (%)	Calories (per/Lb)
Green shelled								
Soybeans	70.0	12.2	5.2	11.1	1.52	0.072	0.0029	636
Lima beans	66.5	7.5	0.8	23.5	1.71	0.028	0.0024	595
Broad beans	74.1	8.1	0.6	15.8	1.40			460
Peas	74.3	6.7	0.4	17.7	0.92	0.028	0.00207	460
Mature dry								
Soybeans	7.0	40.6	16.5	30.9	5.0	0.212	0.0103	1973
Lima beans	10.4	18.1	1.5	65.9	4.1	0.071	0.0086	1586
Navy Beans	12.6	22.5	1.8	59.6	3.5	0.158	0.0079	1564
Peas	9.5	24.6	1.0	62.0	2.9	0.084	0.0057	1612
Flour								
Soybean	5.1	42.5	19.9	24.3	4.5	. . .	. . .	2026
Wheat	12.4	11.2	1.0	74.9	0.5	0.021	0.0008	1603

Source: Woodruff and Klaas (1938).

(1940) reported that a protein-free extract of mature soybeans contained adenine, arginine, choline, glycine, betaine, trigonelline, guanidine, tryptophan, and probably canavanine. Glutathione, quaternary amines and other organic nitrogen-containing compounds have been qualitatively identified. These minor nitrogen bearing compounds are classified as nonprotein nitrogen (NPN) of the soybean.

The peptides have a wide variation in molecular dimension, and there is no sharp line which distinguishes a large peptide from a low molecular weight protein molecule: thus, the level of the nonprotein nitrogen constituents in soybeans is a somewhat arbitrary value and is influenced by the method used in its determination.

Becker *et al.* (1940) developed a simple method for the determination of NPN in soybeans which is based on the extraction of either the defatted or full-fat soybeans with 0.8 N trichloroacetic acid. The result from meal ground in a Wiley mill to pass a ½ mm screen was the same as for meal ground in a hammer mill to pass through a 100 mesh screen. They found also that results with 0.8 N trichloroacetic acid were essentially the same as those obtained by using a method which separated the NPN from the protein and other constituents by dialysis through a selected cellophane membrane. Their data show that the concentration of trichloroacetic acid is not critical since TCA values ranging from 0.65 N to 1.0 N give essentially the same results. The value of 0.8 N trichloroacetic acid was taken as the midpoint of this range.

Becker *et al.* (1940) determined the NPN in fat free meal from 12 varieties and strains of soybeans ranging in protein content from a high of 56.4% to a low of 41.4%. Also, they compared NPN values for germinated and ungerminated Dunfield beans. These results are shown in Table 3.5. Their data show a range of NPN from a high of 7.80% to a low of 2.88% based on the weight of the fat free meal. It is apparent from these data there is no correlation between the amount of total nitrogen in the meal and the NPN. The germinated seed contains double the amount of NPN as the ungerminated seed.

Krober and Gibbons (1962) investigated the relationship of the environmental factors which might be expected to influence the level of NPN as well as the relationship of NPN to the level of protein in the seed. In general, they found that the slight tendency of the NPN to increase with increasing protein was too small to be significant and most of their data showed very little effect of location or variety on NPN. The major influence on mature beans appeared to be weather conditions. Unfavorable weather conditions, whether too cold and wet or too hot and dry, were associated with a high percentage of NPN. These

TABLE 3.5

NONPROTEIN NITROGEN
EXTRACTION OF FAT FREE MEAL WITH 0.8 N TRICHLOROACETIC ACID

		Fat-free Meal		Whole Meal		
	Total N/gm of Meal (Mg)	0.8 N N/gm Meal (Mg)	Extract N Extracted (%)	N (%)	Oil (%)	Moisture (%)
Mukden 1936 (Iowa)	90.3	7.04	7.80	7.68	16.48	6.45
Mukden 1937 (Ind.)	84.9	3.10	3.65	7.07	17.30	5.75
Dunfield 1937 (Ohio)	73.2	2.11	2.88	6.00	19.50	5.48
Mukden 1937 (Ohio)	79.8	2.30	2.88	6.77	18.08	5.78
Mukden 1937 (Iowa)	83.1	3.92	4.72	7.00	18.39	5.83
Dunfield 1937 (Iowa)	76.0	2.78	3.66	6.13	19.92	5.74
Illini 1937 (Ark.)	66.2	3.07	4.64	5.03	23.79	5.53
86518 1937 (Iowa)	85.7	6.38	7.44	7.50	15.15	6.10
Dunfield 1937 (Iowa)	70.3	2.61	3.71	5.72	21.24	5.85
Mukden 1937 (Iowa)	78.0	2.74	3.51	6.54	18.60	5.77
Dunfield 1937 (Ind.)	79.3	5.02	6.33	6.89	14.47	6.00
Dunfield 1937 (Ind.)	73.0	2.50	3.42	6.07	20.64	5.67
Ungerminated Dunfield 1937	77.8	2.81	3.61			
Germinated Dunfield 1937	77.8	5.63	7.24			

Source: Becker *et al.* (1940).

conditions may have influenced the proper maturing of the seed. Krober and Collins (1948) reported, also, that weather-damaged beans often are higher in NPN than undamaged beans.

Nitrogen Distribution in Meal Fractions

The defatted soybean meal and its fractions are the basic raw material for preparing or supplementing many food products and mixed feeds and for the preparation of protein isolates, concentrates, and less important residues. Rackis *et al.* (1961) processed defatted meal (obtained from 1958 Hawkeye soybeans) on a laboratory scale into various fractions and determined their relative percentages as well as their nitrogen and protein content; their fractionation procedure is outlined in Fig. 3.1, and the analytical data are given in Table 3.6. The data account for 96.3% of the nitrogen in the defatted meal; however, the nonprotein nitrogen was lost during dialysis and recovery of the whey.

Because soybeans vary in composition, and also because the work was performed on a laboratory scale, the results will be somewhat different from similar data obtained in large-scale processing of the meal. Nevertheless, the results give an approximation of the nitrogen distribution when the seed is fractionated for product development. For example, when the process is carried out on a commercial scale to produce acid precipitated protein, the yield will be 4–6% lower than the 36.9% obtained for the laboratory procedure; and most of the protein which does not appear in the protein isolate will be found in the residue and whey fractions. However, according to the results of Krober and Cartter

TABLE 3.6

YIELD, NITROGEN, AND PROTEIN CONTENT OF SOYBEAN MEAL FRACTIONS
(DRY BASIS)

Fraction	Yield Gm/100 Gm Meal	Nitrogen (%)	Protein (%)	Percentage Total N
Soybean meal	100.0	9.83	61.4	—
Acid-precipitated protein	36.9	16.29	101.9	61.1
Residue	30.3	8.31	52.0	25.6
Total whey solids	31.9	2.86	17.9	9.3
Isolated whey protein	3.9	16.23	101.4	6.4
Phytate-protein complex	0.93	2.98	18.6	0.3
Seed coat	8.0	1.53	9.56	—
Hypocotyl, whole seed basis	2.0	7.90	49.40	—
Acid-precipitated protein of hypocotyl		15.19	95.15	—

Source: Rackis *et al.* (1961).
Total nitrogen recovered 96.3%

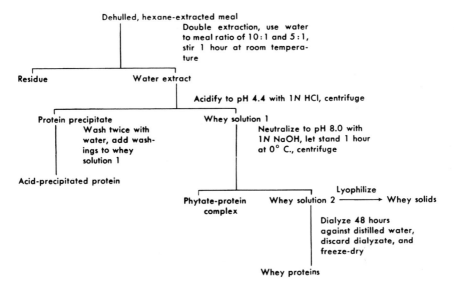

FIG. 3.1. PREPARATION OF SOYBEAN MEAL FRACTIONS

(1962), the variations of nitrogen in the fractions will be influenced more by the mechanics of fractionation than changes in the original level of protein in the soybean.

Amino Acid Distribution in Meal Fractions

In processing defatted meal for producing specialty products as illustrated in Fig. 3.1 there occurs a nonuniform distribution of amino acids in the various fractions which is given in Table 3.7. For example, the data show that the essential amino acids in the acid-precipitated protein are lower than in the meal. This is especially true for the lysine, tryptophan, threonine, methionine, and cystine; the values for the tryphophan, cystine, and cystine plus methionine are 20% or more lower in the protein isolate than in the meal. In the whey protein the histidine, lysine, tyrosine, threonine, cystine, and methionine values are 20% higher than in the meal. This amino acid fractionation is an important nutritional consideration in formulating the meal fractions into food products.

SOYBEAN OIL

It has been demonstrated that the oil content of soybeans and the composition of their fatty acids are influenced by the genetic characteristics of the variety and the climatic environment during the period in which the oil is elaborated. The extremes in fatty acid composition and iodine values of soybean oils are illustrated in Table 3.8 by data taken from Dollear *et al.* (1940).

TABLE 3.7

AMINO ACID COMPOSITION OF SOYBEAN MEAL FRACTIONS

Amino Acid	Whole Meal	Residue	Acid-precipitated Protein	Whey Protein	Hulls	Hypocotyl Meal	Acid-precipitated Protein of Hypocotyl
			(Grams of Amino Acid/16 Gm N)				
Arginine	8.42	7.44	9.00	6.64	4.38	8.32	6.38
Histidine	2.55	2.70	2.83	3.25	2.54	2.60	2.65
Lysine	6.86	6.14	5.72	8.66	7.13	7.45	7.80
Tyrosine	3.90	3.30	4.64	4.67	4.66	3.48	3.78
Tryptophan	1.28	—	1.01	1.28	—	—	—
Phenylalanine	5.01	5.24	5.94	4.46	3.21	3.88	4.22
Cystine	1.58	0.71	1.00	1.82	1.66	1.24	—
Methionine	1.56	1.63	1.33	1.92	0.82	1.72	1.79
Serine	5.57[1]	5.97[1]	5.77[1]	7.62[1]	7.02[1]	4.90[1]	4.50[1]
Threonine	4.31	4.67	3.76	6.18	3.66	4.00	3.82
Leucine	7.72	8.91	7.91	7.74	5.93	6.62	7.22
Isoleucine	5.10	6.02	5.03	5.06	3.80	4.11	4.53
Valine	5.38	6.37	5.18	6.19	4.55	4.82	5.28
Glutamic acid	21.00[1]	17.76[1]	23.40[1]	15.64[1]	8.66	13.78	14.12
Aspartic acid	12.01	12.39	12.87[1]	14.08[1]	10.05[1]	9.74	9.84
Glycine	4.52	5.21	4.56	5.74	11.05	4.25	4.93
Alanine	4.51	5.73	4.48	6.16	3.98	4.69	4.47
Proline	6.28	5.35	6.55	6.66	5.76[1]	4.23	4.38
Hydroxyproline	0	0	—	—	7.57[1]	Trace	0
Ammonia	2.05	2.61	2.20	1.53	1.55[1]	1.40	1.20

[1] Values obtained by extrapolation to zero-hydrolysis time.

TABLE 3.8

IODINE NUMBERS AND FATTY ACIDS OF SOYBEAN OIL

Variety and Location	Year	Iodine No.	Saturated Acids (%)	Unsaturated Fatty Acids			
				Total (%)	Oleic (%)	Lino-leic (%)	Linolenic (%)
Dunfield (Mo.)	1936	102.9	12.0	88.0	60.0	25.0	2.9
Dunfield (Mo.)	1937	124.0	13.2	86.8	34.0	49.1	3.6
Dunfield (Ind.)	1937	127.3	13.1	86.9	34.8	46.0	6.0
Illini (Ill.)	1936	131.6	12.7	87.3	27.7	53.7	5.9
Peking (Ill.)	1937	137.8	12.4	87.6	24.4	56.2	7.3
Seneca (N.Y.)	1938	139.4	11.9	88.1	27.4	55.4	8.0
Wild beans	1938	151.4	13.5	86.5	11.5	63.1	12.1

Source: Dollear et al. (1940).

They reported that the ratio of saturated to unsaturated fatty acids is fairly constant, irrespective of the total amount of oil present in the seed or of the iodine number of the extracted oil.

The average oil content of 10 varieties of soybeans grown at 5 locations during a 5-yr period in the midwest area were reported by Cartter and Hopper (1942) (Table 3.1). The average oil content for the 10 varieties on a moisture free basis is 19.63%. If the noncommercial varieties are eliminated the average is increased slightly to 19.98%. Their results show that the effect of location is as important as the effect of variety on oil content. The oil content of presently recommended soybean varieties is given in Table 3.2.

Table 3.8 shows the principal fatty acid and iodine number of the wild soybean, which is generally recognized as the precursor of present-day beans. Other compositional date of the wild soybean, taken from Dollear et al. (1940), are tabulated below:

	%		%
Nitrogen	7.92	Potassium	1.77
Protein N × 6.25	49.50	Phosphorus	0.91
Ash	6.88	Calcium	0.38
Lipids	5.4	Polysaccharides	
Crude fiber	10.35	as sucrose	5.62

The most notable differences between present-day soybeans and the wild bean are the high protein and fiber and very low oil of the wild bean.

Crawford and Gillingham (1967) made a 2-yr study on the effect of variety and location on the oil and protein content of soybeans in South Carolina. They found that location within the area was a major factor in affecting oil content and they also concluded that oil content may be determined predominantly by variety. They recommended that highest yields will be obtained by planting a given variety in a selected location. However, their report states that location was without effect on protein content.

Howell and Collins (1957) made a study on the effect of environment on the variability of the linolenic and linoleic acid. They found that differences in temperature caused major differences in linolenic and linoleic acid content of soybean oil; they stated that differences associated with temperature are of greater magnitude than associated with variety.

Collins and Sedgwick (1959) studied the fatty acid composition of several varieties of soybeans. The soybean seed samples used in their study were taken from the Uniform Test Groups 0 through VIII of the U.S. Regional Soybean Laboratory program, and grown in 6 to 14 locations. Their results are summarized in Table 3.9.

In the 2 crop years of the study, the soybean oil for the whole series ranged from 5 to 11% in linolenic, 43 to 56% in linoleic, 15 to 33% in oleic, and 11 to 26% in saturated acids. In general, when soybean varieties were grown near the northern range of their area of adaptation, they produced oil which was 1–2 percentage points higher in linolenic and 3–6 percentage points higher in linoleic

TABLE 3.9

FATTY ACID COMPOSITION AND IODINE VALUES OF SOYBEAN OIL FOR YEARS 1956–1957 in FOUR UNIFORM TEST GROUPS

Uniform Test Group	Linolenic Acid (%)	Linoleic Acid (%)	Oleic Acid (%)	Saturated Acids (%)	Iodine Value (%)
I					
Avg 5 varieties	8.59–8.76	48.0–48.4	23.7–23.4	19.5–19.4	131.9–132.7
High-low	9.45–8.19	51.2–45.9	26.4–21.0	20.0–18.9	135.2–129.7
III					
Avg 6 varieties	7.87–8.14	49.3–49.3	23.6–22.8	19.2–19.7	132.0–131.9
High-low	8.70–7.25	51.4–46.0	26.7–21.1	20.7–18.8	134.2–128.9
IV					
Avg 5 varieties	8.14–8.06	49.1–47.8	23.9–26.5	18.9–17.7	132.8–132.2
High-low	8.72–7.46	51.2–45.0	29.3–23.1	19.7–15.1	137.0–136.3
VII					
Avg 3 varieties	7.24–8.02	51.2–54.0	22.9–19.4	18.6–18.6	133.1–136.9
High-low	8.49–6.73	54.9–49.5	26.6–18.2	19.8–17.3	138.0–131.9

Source: Collins and Sedgwick (1959).

acid than when grown at the southern range of their adaptation. Within each group, varieties tended to maintain the same relative order of fatty acid composition of oil at all locations in the 2-yr study.

ASH AND MINERAL CONSTITUENTS

The most extensive investigation of the ash content of soybeans, including the minerals, potassium, phosphorus, and calcium, is that reported by Cartter and Hopper (1942). They investigated 10 varieties, grown at 5 locations, for a 5-yr period; a summary of their average results for ash, potassium, phosphorus, and calcium is given in Table 3.10. The original data show that the ash of 1 of the varieties, Mandarin, was consistently high for the 5-yr period; its highest value for a single sample and for the series was 5.90% and its average for the series was 5.37%. Another variety, Dunfield, was consistently low and had a minimum of 3.67%, which was also low for the series; its average was 4.65%. The average for the 10 varieties for the total period was 4.99%.

The principal mineral components of the ash as reported by Beeson (1941) are shown in Table 3.11. The average values are potassium, 1.83%; phosphorus, 0.78%; magnesium, 0.31%; and sodium, calcium and sulfur, each 0.24%. Since the composition of the ash is made up of the residue of many components of the seed, and since the portion of each component may be influenced in a different degree by variety, climatic, and soil conditions, it is to be expected that the identification of the specific factors which control their concentration would be

TABLE 3.10

AVERAGE ASH, PHOSPHORUS, POTASSIUM, AND
CALCIUM FOR 10 VARIETIES OF SOYBEANS GROWN
AT 5 LOCATIONS FOR 5 YEARS

Variety or Strain	Moisture-free Basis			
	Ash (%)	Phosphorus (%)	Potassium (%)	Calcium (%)
Mandarin	5.37	0.696	1.64	0.386
Mukden	5.00	0.660	1.74	0.240
Dunfield A	4.65	0.626	1.62	0.226
Dunfield B	4.61	0.627	1.58	0.221
Illini	4.81	0.623	1.67	0.252
Manchu	5.12	0.670	1.67	0.313
Scioto	5.17	0.658	1.68	0.343
T-117	5.02	0.654	1.67	0.248
Peking	5.21	0.727	1.75	0.272
Boone	4.97	0.653	1.71	0.253
Average	4.99	0.659	1.67	0.275

Source: Cartter and Hopper (1942).

TABLE 3.11

MINERAL CONTENT OF SOYBEANS (MOISTURE FREE BASIS)

Mineral	No. of Analyses	Range Maximum (%)	Minimum (%)	Mean (%)
Ash	. . .	6.35	3.30	4.60
Potassium	29	2.39	0.81	1.83
Calcium	9	0.30	0.19	0.24
Magnesium	7	0.34	0.24	0.31
Phosphorus	37	1.08	0.50	0.78
Sulfur	6	0.45	0.10	0.24
Chlorine	2	0.04	0.03	0.03
Sodium	6	0.61	0.14	0.24
Boron	5	0.0029	0.0006	0.0019
Manganese	11	0.0041	0.0021	0.0028
Iron	13	0.0133	0.0057	0.0080
Copper	1			0.0012
Barium	. . .			0.0008
Zinc	1			0.0018

Source: Beeson (1941).

difficult to identify and to substantiate. However, Cartter and Hopper (1942) did find consistent differences in ash content based on location, which they concluded were mostly the effect of soil and climate. They concluded, also, that the factors of environment which influenced oil and protein metabolism do not influence total ash accumulation in the seed in the same manner.

Smirnova and Lavrova (1934), reporting earlier on the composition of soybeans, stated that the composition of the ash varies and at the same time the level of the phosphorus varies in direct production to the total amount of ash.

Cartter and Hopper (1942) show a range of potassium for the 10 varieties extending from a high of 2.17% to a low of 1.29% with an average value of 1.67%. This is a narrower range than the values of 2.39 and 0.81% reported by Beeson (1941). When Cartter and Hopper compared their results for potassium with that for phosphorus, they found that seasonal variations had more effect on the metabolism of potassium than on phosphorus.

Cartter and Hopper, in their investigations on calcium in 10 varieties, found a low of 0.163%, a high of 0.470%, and a mean of 0.275%. They found a tendency for the calcium in soybeans to be above average during warm seasons and concluded that temperature plays an important part in determining the amount of calcium stored in the soybean.

PHOSPHORUS CONSTITUENTS

The compounds which contribute phosphorus to the soybean are inorganic phosophorus, phytin, several different phospholipids, and nucleic acids. Analytical investigations show a wide variation in the phosphorus content of soybeans as reported in the ash. Cartter and Hopper (1942) determined the phosphorus content of 10 varieties of soybeans grown over a 5-yr period. Their results (Table 3.10) range from an average low of 0.623% for the Illini variety to an average high of 0.727% for the Peking, with a mean value of 0.659%. In a single year the average low was 0.419% for the Mukden and the high 0.830% for the Peking. For this latter series, the high is approximately double that of the low value.

Phytin and Inorganic Phosphorus

The principal source of phosphorus in soybeans, as in most seeds, is phytin, the calcium-magnesium-potassium salt of inositol hexaphosphoric or phytic acid (Staley Mfg. Co. 1952). The phytates are especially important because of their effect on protein solubility and calcium nutrition. Averill and King (1926) found the phytin phosphorus content of soybeans to vary from 0.505 to 0.727%.

Pons and Guthrie (1946) determined the total and inorganic phosphorus in a number of plant materials, including defatted soybeans, cottonseed, and peanut meals. Their results are in Table 3.12. Also, they show phytin values for

TABLE 3.12

INORGANIC PHOSPHORUS CONTENT OF VARIOUS PLANT MATERIALS
(DRY BASIS)

		Inorganic Phosphorus		
	Total Phos- phorus (%)	1-hr Extrac- tion (%)	24-hr Extrac- tion at 25° C. (%)	24-hr Extrac- tion at 5° C. (%)
Cottonseed meal				
diethyl ether extracted	1.722	0.085	0.085	0.085
Skellysolve B extracted	1.711	0.071	0.073	0.071
Peanut kernels				
Skellysolve F extracted	0.849	0.081	0.082	0.079
Raw cotton fiber	0.028	0.015	0.015	0.015
Sweet potatoes, L-5	0.135	0.075	0.076	0.076
Jerusalem artichokes	0.385	0.083	0.085	0.085
U.S. 13 corn	0.282	0.016	0.016	0.016
Milo	0.274	0.016	0.018	0.017
Federation wheat	0.377	0.018	0.019	0.019
Kharkov wheat	0.445	0.017	. . .	. . .
Wheat straw	0.165	0.118	0.122	0.119
Soybean meal				
Skellysolve F extracted	0.750	0.036	0.037	0.036
Phytin, crude				
from peanuts	14.31	0.070	. . .	. . .
from cottonseed	13.84	0.061	. . .	. . .
Dialyzed peanut protein	0.650	0.014	. . .	. . .
Dialyzed cottonseed protein	1.164	0.013	. . .	. . .

Source: Pons and Guthrie (1946).

cottonseed and peanut meals and total phosphorus for dialysed, isolated cottonseed and peanut proteins.

The cottonseed meal with 1.72% of total phosphorus has more than double
the value of the soybean meal; however, the value of 0.849% for the peanut meal
is nearly the same as for the soybean. The crude phytin values of 13.84% for
cottonseed and 14.31% for peanut are more than double the values reported for
soybeans by Averill and King (1926). For the dialysed, isolated cottonseed
protein, Pons and Guthrie (1946) reported 1.164% phosphorus which is much
higher than the 0.8–1.0% found in isolated soybean protein by Smith and Rackis
(1957).

Earle and Milner (1938) investigated the effectiveness of various solvents for
extracting the phosphorus-containing compounds from soybeans as shown in
Tables 3.13 and 3.14. Of the total phosphorus in the Dunfield variety, which
they studied, they found that phytin phosphorus accounts for approximately
75%; the phosphatide phosphorus, 12%; the inorganic phosphorus, 4.5%; and the
residual phosphorus, 6%. It is possible that the residual phosphorus is nucleic
acid phosphorus, which they did not report. When they extracted with petroleum ether, most or perhaps all of the phytin phosphorus remained in the defatted
meal. They found that only about 0.5% of the total phosphorus, which was

TABLE 3.13

PHOSPHORUS REMOVED FROM SOYBEANS BY SUCCESSIVE EXTRACTION
USING VARIOUS SOLVENTS

| Solvent | Type Phosphorus Removed | Phosphorus[1] Removed from Samples Numbered | | | | | |
		1 (Mg)	2 (Mg)	3 (Mg)	4 (Mg)	5 (Mg)	6 (Mg)
Petroleum ether	Phosphatide	—	—	0.02	0.02	0.03	0.03
Alcohol, 95%	Phosphatide	0.80	0.77	—	—	0.71	0.71
Filter paper from above		0.11	0.11	0.13	0.13	0.14	0.15
Alcohol-HC1, 1st extraction	Inorganic	0.20	0.19	0.76	0.75	0.17	0.19
Alcohol-HC1, 2nd and 3rd extractions	Inorganic	0.06	0.07	0.11	0.12	0.08	0.08
Filter paper from above		0.12	0.12	0.16	0.13	0.09	0.10
HC1, 1.8% in water, 1st extraction	Phytin	4.09	3.86	3.86	3.92	3.90	3.86
HC1, 1.8% in water, 2nd and 3rd extractions	Phytin	0.15	0.18	0.33	0.15	0.23	0.19
Phosphorus in residue		0.23	0.31	0.34	0.35	0.31	0.30

Source: Earle and Milner (1938).

[1] Results expressed in milligrams of phosphorus per gram of whole bean. Total phosphorus in beans 6.02 mg per gram.

TABLE 3.14

DISTRIBUTION OF PHOSPHORUS IN DUNFIELD SOYBEANS CONTAINING
6.02 MG PHOSPHORUS PER GRAM OF WHOLE BEAN

Sample No.	Phosphatide Phosphorus (Mg)	Inorganic Phosphorus (Mg)	Phytin Phosphorus (Mg)	Phosphorus in Residue (Mg)	Total Accounted for (Mg)
1	0.80	0.27	4.45	0.24	5.76
2	0.77	0.26	4.29	0.33	5.61
3	0.91[1]	—	4.44	0.36	5.71
4	0.91[1]	—	5.29	0.37	5.57
5	0.74	0.26	4.33	0.33	5.66
6	0.74	0.28	4.26	0.32	5.60

Source: Earle and Milner (1938).

[1] Also includes inorganic phosphorus.

designated as phosphatide phosphorus, was removed with petroleum ether. This value will probably vary somewhat with the amount of moisture in the meal at the time of extraction, since with increasing moisture level there is an increase in the amount of phosphatides removed with the hydrocarbon solvent.

Phospholipids

The phosphatides or phospholipids are fat-like substances which contain nitrogen and phosphorus. The phosphorus usually occurs as phosphoric acid or inositol in the molecule and the nitrogen as lecithin or cephalin. While the phospholipids are found in most oilseeds they are especially abundant in the soybean. The phosphatides are good emulsifying agents, soluble in alcohol and insoluble in acetone.

The soybean processors refer to the mixture of phospholipids, which are partly removed with the oil in solvent extraction processing, as "soybean lecithin." The lecithin is removed from the oil with a centrifuge after it has been hydrated at an elevated temperature with a small amount of water or steam. Prior to refining, the crude lecithin contains about 30% oil. After removing the oil with acetone it has been further characterized by Scholfield *et al.* (1948) in the Craig apparatus. They found that the defatted "soybean lecithin" consisted of approximately 29% lecithin, 31% cephalin, and 40% inositol-containing phosphilipids. Since the hexane used for extracting the oil removes only part of the phospholipids, we can assume that the fraction left in the meal is approximately the same composition as that which was removed.

Carter *et al.* (1958A,B) isolated and partially characterized, from the inositol-containing phospholipid fraction, a phytoglycolipid, which is described also as a phytosphingosine containing phosphoglycolipid. On hydrolysis they reported that this complex lipid contained phytosphingosine, fatty acids, phosphate, inositol, glucosamine, hexuronic acid, galactose, arabinose, and mannose.

Nucleic Acids

According to Di Carlo *et al.* (1955), whole soybeans contain 1.05% nucleic acid and the defatted soybean meal 1.30% in comparison to yeast in which nucleic acid ranges between 2.0% and 7.7%. They found that soybeans contain very little if any deoxyribonucleic acid (DNA). The composition of soy protein nucleic acid (PNA), terminology used by DiCarlo et al. (1955), as determined by chromatography and spectrophotometric techniques, was found to differ from the PNA isolated from yeasts and other sources. Table 3.15 shows the proportion of bases in yeast and soybean nucleic acids.

Mori *et al.* (1969) have reported that H-RNA in soybeans does not exist in the free state but as some particular component such as ribonucleoprotein.

MINOR ORGANIC CONSTITUENTS

Phenolic Acids

Phenolic acids have a wide distribution in the plant kingdom and although they usually occur in low concentration they have a significant role in soybean foods because of their possible effect on flavor of the soy flour and other products with which they are combined. Arai *et al.* (1966) extracted and identified a number of phenolic acids from hexane-defatted soy flour. They extracted 500 gm of defatted soyflour with 2 liters of 50% ethanol for 5 hrs at 80° C and found this solution to have a strong phenolic-like odor and flavor. On fractionation of the extract and use of paper chromatography, they found the solution to contain nine or more phenolic acids which were considered to have some influence on the flavor of the soy flour. They identified the following acids: syringic, vanillic, ferulic, gentisic, salicylic, *p*-coumaric and *p*-hydroxy-

TABLE 3.15

MOLAR PROPORTIONS OF BASES IN YEAST AND SOYBEAN
RIBONUCLEIC ACID

Base	Yeast PNA		Soy PNA	
	Prepn A[1]	Prepn B[2]	Crude[3]	Purified[4]
Adenine	1.05	1.09	1.03	1.00
Guanine	1.19	1.19	1.29	1.00
Cytosine	0.83	0.77	0.88	0.92
Uracil	0.93	0.95	0.80	0.74

Source: Di Carlo *et al.* (1955).
Molar ratio of purines/pyrimidines:
[1] 1.27
[2] 1.32
[3] 1.37
[4] 1.41

benzoic acids. Of this group the principal component was syringic acid. They identified, also, two isomers of chlorogenic acid, presumably isochlorogenic and chlorogenic acid. They reported that the chlorogenic acids have sour, bitter, astringent, and phenolic-like flavors.

Rackis *et al.* (1970), in studies on soybean factors which cause gas formation in the intestines by the intestinal bacteria, reported that the phenolic acids, especially syringic and ferulic, are effective inhibitors *in vitro* and in the intestinal segments of dogs.

Other Organic Components

In addition to the isolation of phenolic compounds from dehulled soybeans, Fujimaki *et al.* (1969) made an extensive investigation of the minor organic components of the soybean with the objective of identifying the components responsible for the flavor of the bean and derived products. However, many of the products they identified occurred as the result of processing the beans for oil and protein or of the action of natural enzymes, such as lipoxidase, and they found it difficult to distinguish such components. They found, also, that many of the minor components were combined with the protein or other major components of the seed and thus were difficult to isolate and identify. These minor components will be discussed further in Chapter 10 under flavor problems.

SOLUBLE CARBOHYDRATES

The carbohydrates of the soybean have not been studied seriously as a potential source of food or feed. Their principal utilization is in animal feeds where they contribute some calories to the diet, especially for ruminants, since the latter make better use of the polysaccharides than monogastric animals.

The carbohydrates, like other components, have been shown to vary widely in the soybean; for example, O'Kelly and Gieger (1937) reported that the nitrogen free extract varied between 17.93% and 30.18%. The reported variations in the specific carbohydrates may arise partly in the difficulty of separation, purification, and analysis, and partly because of the influence of varietal and environmental factors in production. For example, when there is a change in the ratio of protein to oil, there are also changes in other components of the bean; usually, there is an inverse relationship between the oil and carbohydrates with the protein.

A crude extract of the soluble sugars can be made by refluxing the defatted soybean meal with 60–95% ethanol and filtering. For identifying specific sugars further purification of the extract is necessary.

The principal sugars of the soybean are the disaccharide sucrose, $C_{12} H_{22} O_{11}$, the trisaccharide raffinose, $C_{18} H_{32} O_{16}$, and the tetrasaccharide stachyose, $C_{24} H_{42} O_{21}$. A pentasaccharide verbascose, $C_{30} H_{52} O_{26}$, has been found in very minor amounts. Glucose or other reducing sugars are present in green or immature beans in substantial amounts, but they disappear as the beans approach maturity and the occurrence of glucose in the mature soybean is questionable. The alcohol extract of defatted soybean meal has been reported to contain sucrose, raffinose, stachyose, fructose, galactose, rhamnose, arabinose, and glucuronic acid.

MacMasters *et al.* (1941) reviewed the early work on the carbohydrates of the soybean, and extended these investigations by determining the amount of total sugars, pentosans, and galactans and calculated total carbohydrates by difference for nine varieties of garden type soybeans at different stages of maturity. The total sugars, galactans, and pentosans were selected for study because preliminary data indicated they might be interrelated during growth. The stages investigated were precooking, cooking, postcooking, and maturity. A summary of their results is given in Table 3.16.

These results show that the average values for total carbohydrates (by difference) decreased from the immature precooking stage to maturity from 44.6 to 35.4%, the total sugars decreased from an average of 23.4 to 9.4%, and the reducing sugars decreased from an average of 7.4% to a value too low for analysis. The pentosans and galactans increased from 2.6 to 3.6% and 1.3 to 2.3%, respectively. While other workers have reported small amounts of starch in soybeans, MacMasters *et al.* (1941) were unable to identify starch by microscopic examination. Starch is frequently reported in immature beans but is seldom found in mature beans.

Cartter and Hopper (1942) determined the total sugars which was reported as sucrose, in field type beans for 10 varieties of soybeans, over a 5-yr period at 5 locations; the averages for the 10 varieties are shown in Table 3.1. The Illini variety had the highest average of 8.83% and the Mandarin a minimum average of 6.76%. The individual sugar values, however, range from a low of 2.76% for the Dunfield A variety to a high of 11.97% for the P.I. 54563-3 variety. The

TABLE 3.16

CARBOHYDRATES IN GARDEN TYPE SOYBEANS AT DIFFERENT
STAGES OF MATURITY (VACUUM DRIED BASIS)

Stage of Maturity	Range of Total Carbohydrates (by Difference) (%)	Range of Total Sugars (%)	Range of Reducing Sugars (%)	Range of Pentosans (%)	Range of Galactans (%)
Precooking	38.4–51.3	16.5–31.7	5.8–9.7	2.4–3.1	1.2–1.6
Cooking	31.7–49.7	11.7–22.8	4.8–7.4	2.4–3.1	1.7–2.7
Postcooking	31.6–42.2	5.9–16.2	1.8–5.6	2.8–3.8	1.5–3.1
Mature	31.1–43.9	7.6–10.4	–	3.4–3.8	2.0–2.72

Source: MacMasters *et al.* (1941).

overall average for the 10 varieties was 7.97%. They stated that the greater number of analytical values varied less than 1% from the mean.

A relationship between the oil and carbohydrate content of oilseeds was pointed out by Garner *et al.* (1914) when they stated that "as a consequence of the physiological relationship of oil to carbohydrates, it appears that maximum oil production in the plants requires conditions of nutrition favorable to the accumulation of carbohydrates during the vegetative period and the transformation of carbohydrates into oil during the reproductive period." The results of Cartter and Hopper (1942) agree with this generalization in that the percentage of sugars increases and decreases with the percentage of oil and that when total sugar and oil changes in one direction, the protein content changes in the other.

Sucrose

Kraybill *et al.* (1937) isolated and identified sucrose from soybeans by two methods. Fat free flakes were extracted with 99% ethanol (temperature was not mentioned) and the alcohol extract was concentrated on a steam bath until sugar crystallized on the sides of the beaker. In a second method, an 80% ethanol extract of fat free flakes was concentrated before a fan to a thick syrup. Two liters of water were added and the solution treated with lead acetate; the precipitate was removed and the filtrate treated with barium hydroxide, and the excess lead and barium removed with sulfuric acid. The solution was shaken with ether to remove the acetic acid and then concentrated to a thick syrup. An equal volume of 99% alcohol was added causing the syrup to settle to the bottom. After stirring the syrup in the alcohol on the steam bath, the alcohol was decanted and the solution was left standing for several days for the formation of crystals of sucrose.

In another experiment, Kraybill *et al.* (1937) crystallized sucrose from an acetone extract of fat free flakes. The acetone solution was evaporated and the residue was extracted with ether to remove small amounts of fat; the residue was taken up in water and the sucrose precipitated by the addition of ethanol.

Raffinose

Raffinose is a nonreducing sugar without food value unless it has been hydrolyzed by strong acids into its components of galactose, glucose, and fructose. The raffinose can be hydrolyzed by enzymes two ways. Invertase will hydrolyze the sucrose part of the molecule to give melibiose and D-fructose. The enzyme emulsin, which contains an α-D-galactosidase as well as a β-glucosidase, can hydrolyze the melibiose residue to yield galactose and sucrose. Bottom yeasts, which contain both enzymes, can completely hydrolyze raffinose.

Stachyose and Verbascose

French *et al.* (1953) investigated the hydrolytic products of stachyose as a means of determining the type of linkage between the D-galactose and D-glucose units in the molecule. They found that partial hydrolysis of stachyose by an almond emulsin preparation gave D-galactose and raffinose, sucrose and galactobiose. An acid hydrolysis of stachyose to D-fructose and manninotriose, followed by reduction of the manninotriose to manninotriitol and partial acid hydrolysis gives D-galactose, melibiitol, D-sorbitol, and galactobiose. A periodate oxidation of manninotriitol and manninotriose-1-phenylflavazole confirm the presence of a 1,6 linkage between the D-galactose and D-glucose units in stachyose. They report that stachyose is O-α-D-galactopyranosyl-($1\rightarrow6$)-O-α- D-galactopyranosyl-($1\rightarrow6$)-O-α-D-glucopyranosyl-β-D-fructofuranoside.

Kasai and Kawamura (1966) used dextran gel filtration for the isolation and purification of sucrose, raffinose, stachyose, and verbascose. They found that Sephadex G-15, which may be applied to compounds with molecular weights below 1500, was superior to carbon column chromatography, especially for isolating and purifying of stachyose. The Sephadex G-15 elutes first the highest molecular weight whereas the carbon column elutes the lowest molecular weight.

The sugars were extracted from both the defatted raw and autoclaved flakes. The treatment in the autoclave was at 120° C for 10 min at a moisture level of 20%. The melting points and specific rotations were determined on the sugars and their acetate derivatives, and reported in Table 3.17.

Kawamura and associates (1967) examined the sugars of the whole soybean and of the defatted seed parts for 6 U.S. and 3 Japanese varieties by quantitative paper chromatography. He estimated the maximum variation by his method as ± 10%, but did not have a standard for verbascose. The total sugars were determined by anthrone colorimetry.

The sugars were extracted from the defatted parts of the bean by refluxing with 80% ethanol for 1 hr, filtering, and washing the residue with cold water until the washings gave a negative sugar reaction with anthrone. The combined extracts and washings were concentrated under reduced pressure below 40° C. The average results for the 6 U.S. and 3 Japanese varieties are in Table 3.18.

The author states that the arabinose reported in the hull may not be free but could have come from the water-soluble polysaccharides which were liberated in the course of the isolation of the other components. These data indicate that the Japanese soybeans may be somewhat higher in sugars than U.S. soybeans.

Kawamura's results on a dry basis for U.S. whole soybeans are sucrose 4.5%, raffinose 1.1%, stachyose 3.7%, and traces of arabinose and glucose for a total of 9.3% sugars. For the Japanese beans the values are sucrose 5.7%, raffinose 1.1%, stachyose 4.1%, and traces of arabinose and glucose for a total of 10.9% sugars.

If we assume 20% oil for the whole bean, then the defatted soybean flakes will contain approximately 11.6% total soluble sugars and the Japanese beans 13.6% total soluble sugars. In earlier work and by a different method, Cartter and Hopper (1942), working on 10 varieties of well composited U.S. soybeans determined total sugars as sucrose and found 7.97% for whole soybeans. Assum-

TABLE 3.17

MELTING POINTS AND SPECIFIC ROTATIONS OF SUGARS AND ACETATES OF SUGAR FROM RAW AND AUTOCLAVED SOYBEAN FLAKES

| | Melting Point, °C | | $[\alpha_D]$ | |
	Found	Literature	Found	Literature
Sugars				
Sucrose	176–180	184–185	+ 66.7/24° (H_2O)	+ 66.5° (H_2O)
Raffinose	118–120	118–120	+123.4/20° (H_2O)	+123.1° (H_2O)
Stachyose	170–172	170	+146.2/20° (H_2O)	+146.3° (H_2O)
Acetates of Sugars				
Sucrose octaacetate	87	87	+ 59.0/20° ($CHCl_3$)	+ 59.6° ($CHCl_3$)
Raffinose hendecaacetate	98–101	99–101	+ 97.4/16° (EtOH)	+ 92° (EtOH)
Stachyose tetradecaacetate	94–95	95–96	+120.1/16° (EtOH)	+120.2° (EtOH)

Source: Kasai and Kawamura (1966).

TABLE 3.18

SUGARS IN SEED PARTS OF U.S. AND JAPANESE SOYBEANS

	Whole Soybeans (%)	Defatted Cotyledons (%)	Defatted Hypocotyl (%)	Hull (%)
	U.S. Soybeans: Avg 6 Varieties			
Sucrose	4.5	6.2	6.0	0.58
Raffinose	1.1	1.4	1.7	0.11
Stachyose	3.7	5.2	8.4	0.39
Verbascose				
Arabinose	0.002			0.023
Glucose	0.005			0.06
	Japanese Soybeans: Avg 3 Varieties			
Sucrose	5.7	7.4	9.6	0.64
Raffinose	1.1	1.4	2.1	0.16
Stachyose	4.1	5.4	6.7	0.45
Verbascose				
Arabinose	0.001			0.015
Glucose	0.007			0.04

Source: Kawamura (1967).

ing 20% oil in the soybeans this would be equivalent to 9.96% of total sugars in the defatted flakes, a value which agrees quite well with the 9.3% found by Kawamura.

INSOLUBLE CARBOHYDRATES OF COTYLEDONS

A limited amount of work on the polysaccharides of the cotyledons has been reported by Aspinall et al. (1967). Starting with defatted meal (they did not mention the presence or absence of the seed coat) the soluble sugars were extracted with boiling ethanol-water (4:1) and the protein removed with dilute alkali. However, they emphasized that they experienced difficulty in eliminating all of the protein.

The major part of the polysaccharides was isolated from the deproteinized meal by extraction with ammonium oxalate or, preferably, with ethylenediamin-etetraacetic acid, disodium salt. Some additional polysaccharides were extracted with alkali although the latter fraction was not further examined. The fraction obtained by extraction with the oxalate contained a mixture of acid polysaccharides and arabinogalactan. Further extraction with water gave a mixture of arabinan and arabinogalactan.

In another experiment, they extracted defatted meal with phenol-acetic acid-water (1:1:1) and then with EDTA to give the same components as extracting the deproteinized meal with ammonium oxalate. The acidic polysaccharides extracted from the cotyledons were said to possess structures of extreme complexity. Their results indicated that these polysaccharides may be regarded as belonging to the pectic group of substances.

SEED COAT

The total amount of seed coat or hull which enters the U.S. soybean processing plants annually, at the present level of processing, is about 1.6 million short tons. The hulls, which are the least valuable part of the seed, account for about 1/2 of the 6% of fiber which is present in the undehulled, defatted meal. Since fiber is usually held to a minimum in poultry feeds, most of the hulls are diverted into ruminant feeds and some are used as a carrier for vitamins. Although the nutritional value (Chap. 7) of the hulls is low, there has not been a serious effort made to upgrade this factor.

The percentage of hull varies somewhat with the size of the seed, the larger the seed the lower the proportion of hull. The percentage of hulls and the seed size, the size being measured by the weight in grams per 100 seeds, were determined for several varieties of soybeans by Cartter and Hopper (1942) and are reported in Table 3.19. The smallest seed in the group is the Peking variety, which has 12.98% hull and a seed weight of 6.29 gm, whereas the T-117 has 7.4% hull and a seed weight of 15.37 gm. The average hull for the 10 varieties is 8.28%. However, when the Peking, which is a black-seeded variety, is eliminated the average is 7.75%.

Kawamura and associates (1967) measured the relative amounts of hull, hypocotyl and cotyledon for 6 U.S. and 3 Japanese varieties and also determined

TABLE 3.19

PERCENTAGE OF HULLS FOR 10 VARIETIES AND STRAINS GROWN
AT ONE LOCATION AND WEIGHT OF 100 SOYBEANS
(MOISTURE EQUILIBRATED AT 70° F. AND 18% RELATIVE HUMIDITY)

Variety	Hull (%)	Seed Size Wt/100 (Gm)
Mandarin	8.49	14.88
Mukden	7.51	14.43
Dunfield A	7.71	14.47
Dunfield B	7.61	15.21
Illini	8.14	13.52
Manchu	7.36	14.79
Scioto	7.32	14.79
T-117	7.40	15.37
Pekin	12.98	6.92
P.I. 54563-3	8.25	12.82
Avg	8.28	13.6
Avg without Pekin	7.75	14.4

Source: Cartter and Hopper (1942).

the crude protein, fat, nitrogen free extract, plus fiber and ash in each part; his results are in Table 3.20.

The hull of the mature bean is hard, water resistant, and protects the cotyledons and hypocotyl or germ from damage which may be caused by weathering, harvesting, insects, and transportation. The dark scar on the seed is called the hilum, and at one end of the hilum is the micropyle, or small opening in the seed coat, which under favorable conditions will permit the entrance of moisture. The hull is loosely attached to the cotyledons and when the dry mature seed is cracked, as in oil mill processing, the hull is detached fairly readily and separated from the cotyledons by aspiration.

The color of the hull may be different from that of the cotyledons; it may be black, brown, blue, mottled, green, or various shades of yellow. The present trend in plant breeding is to develop varieties with a light yellow seed coat. A white coated seed would be the most desirable. In processing soybeans for oil it is not practical to attempt a complete removal of the hulls; thus, soybeans having a dark colored seed coat will leave dark specks in the meal, an undesirable factor when used in food products.

Chemical Composition

The composition of soybean hulls is difficult to determine and analytical results reported from different laboratories often do not agree. Nelson *et al.* (1950) determined the composition of a large number of crop seed hulls

TABLE 3.20

AVERAGE COMPOSITION OF SOYBEAN SEED PARTS OF 6 U.S. AND
3 JAPANESE VARIETIES

	Whole Soybeans (%)	Full-fat Cotyledons (%)	Full-fat Hypocotyl (%)	Hull (%)
	U.S. Soybeans: Avg 6 Varieties			
Crude protein	40.4	43.4	40.8	9.0
Crude fat	22.3	24.3	12.0	0.9
N-free extract				
+ fiber	31.9	27.4	42.7	86.2
Ash	4.9	5.0	4.5	4.0
	Japanese Soybeans: Avg 3 Varieties			
Crude protein	39.2	41.7	40.7	8.6
Crude fat	18.4	20.0	11.2	1.3
N-free extract				
+ fiber	37.4	33.3	44.0	85.6
Ash	5.0	5.0	4.2	4.5

Source: Kawamura (1967).

Average yield for the 3 parts was cotyledons 90.3%, hypocotyl 2.42%, and hull 7.25%.

(including soybean hulls), nut shells, and fruit pits. Table 3.21 gives the analysis of raw and cooked soybean hulls as reported by Nelson *et al.* The principal components of the hull are cellulosic type materials 49.3%, pentosans 22.6%, lignin 4.5%, ash 5.7%, and nitrogen 1.6%.

Whistler and Saarnio (1957) analyzed soybean hulls and found that acetone-extracted hulls contained 64% alpha cellulose, 16% hemicellulose extractable with an alkaline solution, and 8% lignin. One of the celluloses was a galactomannan which was extractable with water at 40° C in a 2% yield. They found that the galactomannan has a D-galactose unit to a D-mannose unit in the ratio of 2:3 which they state compares with guaran which has an approximate ratio of 1:2, reported for the fractionated guaran. The soybean galactomannan yielded on acid hydrolysis only D-galactose and D-mannose which were separated chromatographically and obtained in crystalline form.

In another investigation, Sanella and Whistler (1962) reported a partial analysis of soybean hulls which gave them alpha cellulose 49.8%, lignin 7.8%, hemicellulose A 10.6%, hemicellulose B 6.0%, crude protein 13.6%, and ash 4.9%. The hemicellulose B was an alkali-soluble hemicellulose which was not precipitated with acid. This fraction contained an acid polysaccharide as a major component and on purification and hydrolysis they identified D-xylose, L-arabinose, D-glucose and a galactomannan occurring in molar ratios of 14:1:3:3.

Aspinall *et al.* (1967) also reported on the polysaccharides of the seed coat. They found four general types of polysaccharides, namely galactomannans, a group of related polysaccharides of the pectic type, xylan, and cellulose. In a further study of the above fractions an estimation was made of the various hydrolytic products. However, a quantitative evaluation of the hydrolytic prod-

TABLE 3.21

PARTIAL CHEMICAL ANALYSIS[1] OF SOYBEAN HULLS
(OVEN DRY BASIS)

	Raw Inventory No. CD 12701 (%)	Cooked Inventory No. CD 12702 (%)
Ash content	5.7	5.6
Solubility in:		
Alcohol-benzene	4.6	4.9
Hot water	17.1	16.9
1% NaOH solution	45.6	46.0
Lignin	4.5	5.7
Pentosans	22.6	22.9

[1] Analysis supplied by E. G. Helman, Northern Regional Research Laboratory, USDA-ARS, Peoria, Ill.

ucts is complicated by the fact that the same sugars may be derived on hydrolysis of more than one type of polysaccharide. For example D-galactose and D-glucose are important constituents of the galactomannans and hemicellulose, respectively; also, both sugars are formed as hydrolysis products from acidic polysaccharides. Digestion of the hulls with cold 72% sulfuric acid followed by hydrolysis with dilute acid indicated the formation of 72% reducing sugars.

From their work, Aspinall *et al.* (1967) estimated the hulls contain 9-11% glactomannans, 10-12% acidic polysaccharides, 9-10% xylan hemicellulose, and about 40% cellulose. They found the hulls to contain about 11% protein and peptides and they state that the remaining material, not specifically accounted for, is probably lignin.

Amino Acids

The soybean seed coat contains 1.53-2.0% nitrogen. Although the source of the nitrogen has not been completely identified it is the equivalent of 10-12% protein. Rackis *et al.* (1961) determined the amino acids in the seed coat, along with the amino acids in the meal and protein fraction of the meal. The results reported in Table 3.7 show that most of the essential and related amino acids in the seed coat are in much lower concentration than in the meal, particularly phenylalanine, arginine, methionine, isoleucine, and leucine. However, lysine is slightly higher and tyrosine is 20% higher than in the meal. The seed coat contains more than double the amount of glycine found in the meal and 7.56% hydroxyproline, whereas the dehulled meal and residue from protein isolation do not contain hydroxyproline. At periods of 72 and 90 hrs of hydrolysis there was extensive destruction of hydroxyproline and a moderate destruction of proline; thus, the values were obtained by extrapolating to zero.

BIBLIOGRAPHY

ANON. 1971. Blue Book Issue. Am. Soybean Assoc., Hudson, Iowa.
ARAI, S., SUZUKI, H., FUJIMAKI, M., and SAKURAI, Y. 1966. Studies on flavor components of soybeans. II. Phenolic acids in defatted soybean flour. Agr. Biol. Chem. (Tokyo) *30*, 364-369.
ASPINALL, G. O., BEGBIE, R., and MCKAY, J. E. 1967. Polysaccharide components of soybeans. Cereal Sci. Today June, 223-228, 260.
AVERILL, H. P., and KING, C. G. 1926. Phosphorus content of soybeans. J. Am. Chem. Soc. *49*, 724-728.
BECKER, H. C., MILNER, R. T., and NAGEL, R. A. 1940. A method for determination of nonprotein nitrogen in soybean meal. Cereal Chem. *17*, 447-457.
BEESON, K. C. 1941. The mineral composition of crops with special reference to the soils in which they were grown. USDA Misc. Publ. *369*.
CARTER, H. E. *et al.* 1958A. Biochemistry of the sphingolipids. J. Biol. Chem. *233*, 1309-1314.
CARTER, H. E. *et al.* 1958B. Biochemistry of the sphingolipids X. J. Am. Oil Chemists' Soc. *35*, 335-343.

CARTTER, J. L. 1940. Effect of environment on composition of soybean seed. Soil Sci. Soc. Am. Proc. 5, 125–130.

CARTTER, J. L., and HOPPER, T. H. 1942. Influence of variety, environment, and fertility level on the chemical composition of soybean seed. USDA Tech. Bull. 787.

COLLINS, F. I., and SEDGWICK, V. E. 1959. Fatty acid composition of several varieties of soybeans. J. Am. Oil Chemists' Soc. 36, 641–644.

CRAWFORD, D. E., and GILLINGHAM, J. T. 1967. Oil and protein content of soybeans grown in South Carolina. S. Carolina Agr. Expt. Sta. Bull. AE–303.

DI CARLO, F. J., SCHULTZ, A. S., and KENT, A. M. 1955. Soybean nucleic acids. Arch. Biochem. Biophys. 55, 253–256.

DIES, E. J. 1942. Soybeans–Gold from the Soil. Macmillan Co., New York.

DOLLEAR, F. G., KRAUCZUNAS, P., and MARKLEY, K. S. 1940. The chemical composition of some high iodine number soybean oils. Oil Soap 17, 120–121.

EARLE, F. R., and MILNER, R. T. 1938. The occurrence of phosphorus in soybeans. Oil Soap 15, 41–42.

FRENCH, D., WILD, G. M., and JAMES, W. J. 1953. Constitution of stachyose. J. Am. Chem. Soc. 75, 3664–3666.

FUJIMAKI, M. et al. 1969. Proteolytic enzyme application to soybean protein in relation to flavor. Univ. Tokyo, Japan, Dept. Agr. Chem., U.S. Dept. Agr., Project No. UR-A11-(40)-8.

GARNER, W. W., ALLARD, H. A., and FOUBERT, C. L. 1914. Oil content of seeds as affected by nutrition of the plant. J. Agr. Res. 3, 227–249.

HEATHCOTE, J. G. 1950. The protein quality of oats. Brit. J. Nutr. 4, 145.

HOWELL, R. W., and COLLINS, F. I. 1957. Factors affecting linolenic and linoleic acid content of soybeans. Agron. J. 49, 593–597.

JONES, D. B. 1931. Factors for converting percentage of nitrogen in foods and feeds into percentage protein. USDA Circ. 113.

KASAI, T., and KAWAMURA, S. 1966. Soybean oligosaccharides. Isolation by gel filtration and identification by acetylation. Kagawa Univ. Fac. Tech. Bull. 18, No. 1.

KAWAMURA, S. 1967. Quantitative paper chromatography of sugars of the cotyledon, hull and hypocotyl of soybeans of selected varieties. Kagawa Univ. Fac. Tech. Bull. 15, 117–131.

KAWAMURA, S., and TADA, M. 1967. Isolation and determination of sugars from the cotyledon, hull and hypocotyl of soybeans by carbon column chromatography. Kagawa Univ. Fac. Tech. Bull. 15, 138–141.

KAWAMURA, S., TADA, M., and IRIE, N. 1967. Effect of autoclaving on sugars of defatted soybean flakes from selected varieties. Kagawa Univ. Fac. Tech. Bull. 15, 147–153.

KRAYBILL, H. R., SMITH, R. L., and WALTER, E. D. 1937. The isolation of sucrose from soybeans. J. Am. Chem. Soc. 59, 2470–2471.

KROBER, O. A., and CARTTER, J. L. 1962. Quantitative interrelationships of protein and nonprotein constituents of soybeans. Crop Sci. 2, 171–172.

KROBER, O. A., and CARTTER, J. L. 1966. Relation of methionine content to protein levels in soybeans. Cereal Chem. 43, 320–325.

KROBER, O. A., and COLLINS, F. I. 1948. Effect of weather damage on the chemical composition of soybeans. J. Am. Oil Chemists' Soc. 25, 296–298.

KROBER, O. A., and GIBBONS, S. J. 1962. Nonprotein nitrogen in soybeans. J. Agr. Food Chem. 10, 57–59.

LLOYD, J. W., and BURLISON, W. L. 1939. Eighteen varieties of edible soybeans. Univ. Illinois Bull. 453.

MACMASTERS, M. M., WOODRUFF, S., and KLAAS, H. 1941. Studies on soybean carbohydrates. Ind. Eng. Chem. 13, 471–474.

MORI, T., IBUKI, F., MATSUSHITA, S., and HATA, T. 1969. Characteristics of ribonucleic acid contained in soluble fraction from soybeans. J. Agr. Biol. Chem. 33, 1229–1235.

MULLER, E., and ARMBRUST, K. 1940. Non-protein constituents in soybeans. Hoppe-Seylers Z. Physiol. Chem. 263, 41–46.

NELSON, G. A., TALLEY, L. E., and ARONOVSKY, S. I. 1950. Chemical composition of grain and seed hulls, nuts, shells and fruit pits. Cereal Chem. *8,* 59–68.

O'KELLY, J. F., and GIEGER, M. 1937. Effect of variety, maturity and soundness on certain soybean seed and oil characteristics. Mississippi Agr. Expt. Sta. Bull. *24.*

OSBORNE, T. B., and CAMPBELL, G. F. 1898. Proteids of the soybean. J. Am. Chem. Soc. *20*, 419–428.

PIPER, C. V., and MORSE, W. J. 1923. The Soybean. McGraw-Hill Book Co., New York.

PONS, W. A., Jr., and GUTHRIE, J. D. 1946. Determination of inorganic phosphorus in plant materials. Ind. Eng. Chem. *18,* 184–186.

RACKIS, J. J. *et al.* 1961. Amino acids in soybean hulls and oil meal fractions. J. Agr. Food Chem. *9,* 406–412.

RACKIS, J. J. *et al.* 1970. Soybean factors relating to gas production by intestinal bacteria. J. Food Sci. *35,* 634–639.

SANELLA, J. L., and WHISTLER, R. L. 1962. Isolation and characterization of soybean hull hemicellulose B. Arch. Biochem. Biophys. *98,* 116–119.

SCHOLFIELD, C. R., DUTTON, H. J., and TANNER, F. W., JR. 1948. Components of soybean lecithin. J. Am. Oil Chemists' Soc. *25,* 368–372.

SMILEY, W. G., and SMITH, A. K. 1946. Preparation and nitrogen content of soybean protein. Cereal Chem. *23,* 288–296.

SMITH, A. K. 1966. Basis of nitrogen to protein conversion factor for soy protein in relation to nutritive value. Fleischwirtschaft *10,* 1106–1110.

SMITH, A. K., and RACKIS, J. J. 1957. Phytin elimination in soybean protein isolation. J. Am. Chem. Soc. *79,* 633–637.

SMIRNOVA, M. I., nd LAVROVA, M. N. 1934. Variation of the chemical composition in different soybean varieties. Bull. Appl. Botany Genet. Plant Breeding 1–103. (Russian with English Summary.)

STALEY, A. E., Mfg. Co. 1952. Phytin and Inositol. A bibliography. A. E. Staley Mfg. Co., Decatur, Ill.

TKACHUK, R. 1969. Nitrogen to protein conversion factor for cereals and oilseed meals. Cereal Chem. *46*, 419–423.

TKACHUK, R., and IRVINE G. N. 1969. Amino acid composition of cereals and oilseeds meals. Cereal Chem. *46*, 206–218.

WEISS, M., WILSIE, C. P., LOWE, B., and NELSON, P. M. 1942. Vegetable soybeans. Iowa State Coll. Agr. Expt. Sta. Bull. *P39.*

WHISTLER, R. L., and SAARNIO, J. 1957. Galactomannan from soybean hulls. J. Am. Chem. Soc. *79,* 6055–6057.

WOODRUFF, S., and KLAAS, H. 1938. A study of soybean varieties with reference to their use as food. Univ. Illinois Agr. Expt. Sta. Bull. *443.*

CHAPTER 4

W. J. Wolf | # Purification and Properties of the Proteins

INTRODUCTION

Work done in the past 20 yr shows that proteins prepared by earlier workers were mixtures and that several of the proteins undergo complex reactions. At least seven soybean proteins now appear to be made up of subunits, which may be disrupted under a variety of conditions. Because of their subunit structure the major soybean proteins have molecular weights ranging from about 200,000 to 600,000. In the native state, these large molecules can form still higher particle sizes either through association-dissociation reactions or by forming disulfide-linked polymers. Because of this complexity it is necessary to fractionate soybean proteins before detailed studies are made on them.

This chapter describes progress made in the isolation and characterization of soybean proteins in the past two decades and points out where further work is needed. Coverage is mainly on "storage proteins," but biologically active proteins, such as trypsin inhibitors and hemagglutinins that are also discussed in Chap. 6, are included for completeness. A detailed summary of the chemistry of soybean proteins up to 1948-1949 was compiled by Circle (1950). Brief reviews since 1948-1949 are also available (Wolf and Smith 1961; Bain *et al.* 1961; Wolf 1969A, 1970A).

NOMENCLATURE

At present there is no nomenclature system generally accepted for soybean proteins. However, this problem is currently under study and a detailed discussion of past terminology, as well as proposals under consideration, can be found elsewhere (Wolf 1969B). Some of the names being considered have recently been introduced (Catsimpoolas 1969A; Catsimpoolas and Ekenstam 1969), but a final decision on terminology has not been reached.

The nomenclature system based on approximate sedimentation coefficients as introduced by Naismith (1955) has been used extensively in the past decade. Because of its adoption by many workers this terminology, as exemplified in Fig. 4.1 is used here. Figure 4.1 also illustrates a shortcoming of the ultracentrifuge terminology: Sedimentation properties of soybean proteins depend on conditions of buffer composition, pH, and other factors. For example, a portion of the 7S fraction observed at pH 7.6, 0.5 ionic strength, dimerizes at 0.1 ionic strength to form a 9S peak, as indicated at the bottom of Fig. 4.1.

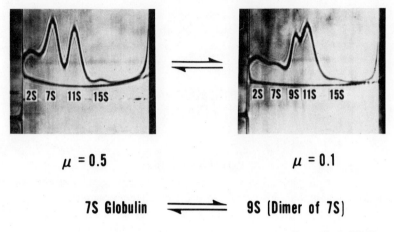

$\mu = 0.5$ $\mu = 0.1$

7S Globulin $\rightleftharpoons$ 9S (Dimer of 7S)

From Wolf (1969B)

FIG. 4.1. EFFECTS OF IONIC STRENGTH ON THE ULTRACENTRI-
FUGE PATTERN FOR WATER-EXTRACTABLE SOYBEAN
PROTEINS AT pH 7.6

SUBCELLULAR STRUCTURE

Many seeds, particularly those rich in protein and oil, contain numerous subcellular inclusions that are the storage sites for proteins, lipids, and other constituents. The protein storage particles are called protein bodies or aleurone grains while lipid deposits are called spherosomes. Protein bodies and spherosomes have been identified in soybean cotyledons by electron microscopy (Bils and Howell 1963; Saio and Watanabe 1966, 1968; Tombs 1967). Figure 4.2 shows an electron micrograph of a soybean cotyledon in which these structural elements are identified. The protein bodies vary from 2 to 20 μ in diameter (Tombs 1967), but many fall into the narrower range of about 5–8 μ (Fig. 4.2). The spherosomes are interspersed between the protein bodies and are about 0.2–0.5 μ in diameter.

Soybean protein bodies have been isolated by three procedures. Saio and Watanabe (1966) homogenized soybeans in cottonseed oil and then separated the protein bodies by differential centrifugation in cottonseed oil-carbon tetrachloride mixtures. Composition of their protein bodies is given in Table 4.1.

Tombs isolated protein bodies from hexane-extracted soybean flour (350 mesh) by sucrose density gradient centrifugation at pH 5, the pH of minimum solubility of the major proteins, to prevent disruption of the protein bodies. The protein bodies often sediment as 2 bands; a light fraction (density less than 1.30) and a heavy fraction (density less than 1.32). Analyses of the total protein bodies and the two fractions are given in Table 4.1, plus a partial analysis of the starting soybean flour for comparative purposes. Figure 4.3 shows scanning electron micrographs of protein bodies in soybean flour and after isolation by Tombs' procedure.

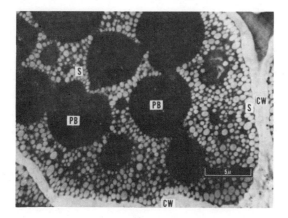

From Saio and Watanabe (1968)

FIG. 4.2. ELECTRON MICROGRAPH OF A SEC-
TION OF MATURE SOYBEAN COTYLE-
DON

Seed was soaked in water overnight, fixed with osmium
tetroxide, and strained with uranyl acetate and lead
citrate. Protein bodies (PB), spherosomes (S), and cell
wall (CW) are identified.

TABLE 4.1

COMPOSITION OF SOYBEAN PROTEIN BODIES[1]

Constituent	Total Preparation[3] (%)	Preparation[4] Total (%)	Light (%)	Heavy (%)	Defatted Soybean[4] (%)	Preparation No.[5] 1 (%)	2 (%)	3 (%)	4 (%)
Protein (N × 5.8)[2]	65.0	82.5	97.5	78.5	50.0	35	52	39	34
Total phosphorus	0.94	0.48	0.84	0.90	—	—	—	—	—
Ribonucleic acid	0.53	1.29	0.43	2.04	1.66	—	—	—	—
Phospholipid	—	1.0	—	—	2.25	—	—	—	—
Total lipid	—	11.3	1.5	5.6	8.6	29	44	23	60
Phytic acid	—	1.35	2.6	2.3	2.24	—	—	—	—
Carbohydrate	8.5	3.0	—	—	—	—	—	—	—
Ash	7.7	—	—	—	—	—	—	—	—
Total	82.7	99.3	100.8	87.4	64.8	—	—	—	—

[1] Calculated on a moisture-free basis.
[2] Nitrogen-to-protein conversion factor based on amino acid composition data (Tombs 1967).
[3] Saio and Watanabe (1966).
[4] Tombs (1967).
[5] Komoda *et al.* (1968).

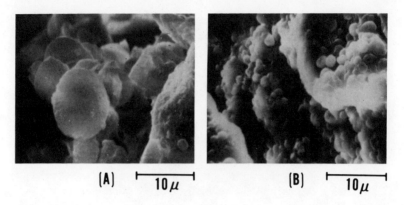

(A) ⊢‾‾10μ‾‾⊣   (B) ⊢‾‾10μ‾‾⊣

From Wolf (1970B)

FIG. 4.3. SCANNING ELECTRON MICROGRAPHS OF PROTEIN BODIES;
(A) IN UNDEFATTED SOYBEAN FLOUR AND (B) ISOLATED
BY SUCROSE DENSITY GRADIENT CENTRIFUGATION

The third method for preparation of protein bodies was reported by Komoda *et al.* (1968). They soaked soybeans overnight in water, homogenized them in 0.5 *M* sucrose, and centrifuged to obtain a pellet fraction believed to be protein bodies.

The protein bodies isolated by the three different procedures differ widely in protein content (Table 4.1). The preparations of Saio and Watanabe approach the heavy fraction of Tombs in protein content. The pellet fractions of Komoda *et al.*, however, are considerably lower in protein content than the protein bodies obtained by the other two methods. Tombs' heavy fraction contained cell wall fragments and cytoplasmic attachments while his light fraction appeared to be almost completely protein. The preparations reported by Komoda *et al.* appear low in protein because of the high lipid contents (23–60%). Since no electron microscopic data are given, it is not clear whether the lipids represent contamination by spherosomes or are associated with the protein bodies *in vivo*. The latter might be removed from the preparations of Saio and Watanabe during the carbon tetrachloride centrifugation, whereas Tombs used hexane-defatted meal for his studies. Attempts to avoid defatting are reported, but intact soybean cotyledons are difficult to homogenize and intractable emulsions of the protein bodies and spherosomes form (Tombs 1967).

Proteins from the protein bodies and the water-extractable proteins from defatted meal differ little by gel filtration and polyacrylamide gel electrophoresis (Saio and Watanabe 1966). Apparently, isolation of the protein bodies does not result in a pronounced fractionation on the basis of differences in distribution of proteins in the protein bodies and the cytoplasm. This conclusion is supported by Tombs' data, which indicate that at least 60–70% of the total protein is stored in the protein bodies. Although polyacrylamide gel electrophoresis suggested that only the 11S ultracentrifugal component was located in the protein bodies (Tombs 1967), Catsimpoolas *et al.* (1968A) observed at least six components in protein bodies by disc immunoelectrophoresis. Disc electrophoretic patterns for the cotyledon proteins and the proteins of isolated protein bodies differed slightly.

Ultracentrifuge patterns of the protein body proteins likewise are similar to patterns of water-extractable proteins from defatted meal; 2S, 7S, 11S, and 15S fractions are present, but there is a reduced concentration of 2S fraction in the protein bodies (Wolf 1970B). The latter result supports Tombs' studies showing that trypsin inhibitor (a protein of the 2S fraction of the water-extractable proteins) is located in the cytoplasm rather than in the protein bodies.

Properties of soybean proteins known at present are largely based on preparations isolated by extracting defatted soybean meal or flakes with aqueous solvents. For example, globulins prepared by acid-precipitation contain low molecular weight compounds that may not be associated with the proteins *in vivo* but that may have interacted with the proteins during the initial water extraction of the meal (Nash *et al.* 1967). Such interactions might be prevented by isolating the protein bodies before extracting the proteins. Information about the merits of isolating the proteins by extraction of protein bodies versus the classical procedure of extracting defatted meal is needed, and also knowledge of the subcellular location of nonprotein constituents would be helpful.

The "prepackaging" of the majority of the proteins in soybeans suggests the possibility of developing methods of milling and separation whereby the protein bodies could be isolated in relatively pure form on an industrial scale. Such a process preferably would not use water so as to eliminate a waste disposal problem (soybean whey) associated with present methods for processing soybean protein isolates and certain protein concentrates.

PROTEIN EXTRACTION

Extractability of soybean proteins is influenced by a variety of factors including moist heat treatment (toasting) of the meal, method of oil extraction, particle size, meal age, temperature, solvent-to-meal ratio, pH, and salt concentration. Because these factors are reviewed in detail by Circle (1950), emphasis here will be only on extraction conditions used in basic studies. Work reported since 1950 is included where appropriate.

Preparation of Flakes and Meal

Soybean hulls contain water-soluble pigments, including anthocyanins, which may interact with proteins during their extraction from the meal (Smiley and Smith 1946).[1] Soybeans should therefore be cracked and dehulled if the

---

[1] Also see previously unpublished data by O. A. Krober cited in Chap. 2.

required equipment is available. After dehulling, cracked beans are ground or preferably flaked and extracted with hexane or diethyl ether at or near room temperature. Moist heat and solvents like alcohols and acetone should be avoided if maximum extractability of the proteins is desired (Belter and Smith 1952; Smith *et al.* 1951). The defatted flakes may be used directly or may be ground in a hammer mill before aqueous extraction of the proteins.

Recent studies by Smith *et al.* (1966) should be consulted for details about preparation of undenatured meal. They extracted 90-95% of the total nitrogenous constituents with water (pH 6.5-6.8) or dilute alkali (pH 7.2) from carefully prepared meals. Defatted meals can be stored at room temperature, but protein extractability slowly decreases as the meals age (Smith and Circle 1938; Nash *et al.* 1971). Methods for preventing or minimizing the decrease in protein extractability on aging are still unknown.

Extraction of Meal

Extraction Solvents.—A large variety of aqueous solvents has been used to extract proteins from defatted soybean meal (Circle 1950). In many studies only the percentage of the total meal protein extracted was reported; few characterization studies are available on the proteins extracted with different solvents. Of all the solvents tried, water, water plus dilute alkali (pH 7-9), and aqueous solutions of sodium chloride (0.5-2 M) are among the most efficient for extracting proteins (Smith and Circle 1938; Smith *et al.* 1938, 1966). Because these solvents are also mild, they should yield the proteins in an undenatured state. Ultracentrifugal studies failed to show significant differences between water and 1 M sodium chloride extracts of defatted meal (Wolf and Briggs 1956). Likewise, tris-citrate buffer and water extracts of defatted meal were similar by starch gel electrophoresis (Shibasaki and Okubo 1966).

Meal-to-Solvent Ratio.—For many laboratory studies a 1:10 meal:solvent extraction is adequate with or without a second extraction at a 1:5 ratio. A 1:20 or 1:40 ratio may yield a greater amount of the total protein but also results in more dilute solutions for subsequent fractionation steps. An exception to the 1:10 meal:solvent ratio is the 1:5 meal:water ratio recommended for isolation of crude 11S component (cold-insoluble fraction) by cryoprecipitation (Briggs and Wolf 1957; Wolf and Sly 1967).

Extraction Temperature.—Aqueous extracts of defatted meal are generally prepared at room temperature. In a few studies extractions were made at refrigerator temperatures (Danielsson 1949), but no advantages are known for working at low temperatures except for enzyme isolation (Wang and Anderson 1969). In cryoprecipitation of crude 11S protein, a 1:5 meal:water extract at room temperature (25° C) is saturated with respect to 11S component. If the extract is prepared at 40° C, more of the 11S component dissolves; hence, greater yields of 11S protein are obtained when the extract is cooled to 0-4° C (Wolf and Sly 1967).

Effect of pH.—Distilled water extracts of defatted soybean meal have a pH of about 6.4–6.6. Extraction at higher pH by adding alkali increases the amount of protein extracted by 5–10%, but lowering the pH drastically reduces the amount of extractable protein (Fig. 4.4). A minimum in protein solubility exists between pH 4 and 5, the isoelectric region for the major proteins. Extracts can also be made at pH values below the isoelectric region, but irreversible changes in the 11S protein, the major protein, have been noted at pH 2–3 (Wolf and Briggs 1958; Catsimpoolas *et al.* 1969A). Accordingly, extractions in the pH range of 6.5–9.0 are recommended.

FRACTIONATION METHODS

Quaternary structures of the major soybean proteins limit the rigor of fractionation procedures that can be applied. If conditions are used that dissociate the quaternary structures into subunits, one is faced with separating a larger number of different molecules whose relationship to the parent proteins (subunit assemblies) is unknown. Many of the conditions required to disrupt quaternary structures also cause irreversible conformation changes in the subunits. Fractionation methods, therefore, must be sufficiently mild to maintain the quaternary structures intact. A variety of fractionation techniques have been applied to soybean proteins, but most give only partial separations. Consequently, purification of a given protein requires a combination of two or more methods based on different properties of the protein.

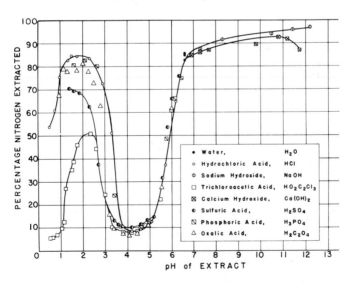

From Smith and Circle (1938)

FIG. 4.4. EXTRACTABILITY OF SOYBEAN MEAL PROTEINS AS A FUNCTION OF pH

Fractional Precipitation

Isoelectric Precipitation.—Precipitation of soybean proteins from aqueous or alkaline extracts by acidification was placed on a firm basis by Smith and Circle (1938) in their studies on the relationship between pH and solubility of the proteins (Fig. 4.4). The minimum in solubility at about pH 4.2 corresponds to the apparent isolectric point of the major proteins. Adjustment of water or dilute sodium hydroxide extracts of defatted meal to pH 4.0–4.2 precipitates about 90% of the extracted protein. Isoelectric precipitation is therefore useful for concentrating globulins and for separating them from such minor constituents as sugars and salts, which are extracted from the meal together with the proteins. Acid precipitation also separates globulins from minor proteins found in the pH 4.2-soluble fraction or whey (Smith *et al.* 1955). Disadvantages of isoelectric precipitation of the proteins particularly from an aqueous extract of defatted meal include: (a) little, if any, fractionation of the globulins; (b) combination of phytate and possibly other low molecular-weight compounds with the globulins (McKinney *et al.* 1949; Smith and Rackis 1957); and (c) insolubilization of some of the proteins (Wolf *et al.* 1964; Nash and Wolf 1967; Nash *et al.* 1971). Because soybeans contain proteolytic enzymes with pH optima of 5.0–5.4 (Pinsky and Grossman 1969), exposure of the proteins to this pH region may also result in their modification through proteolysis.

Moving boundary electrophoresis showed that isoelectrically precipitated protein is a mixture of at least 4 to 5 components (Briggs and Mann 1950). In the ultracentrifuge, acid-precipitated proteins separate into 4 distinct fractions with sedimentation coefficients of 2, 7, 11, and 15S; the sedimentation pattern differs only slightly from the pattern for the unfractionated proteins obtained by extraction of meal with 10% sodium chloride (Naismith 1955). Whey proteins consist of a portion of the 2S and 7S fractions observed in a water extract of defatted meal (Wolf and Briggs 1959; Eldridge *et al.* 1966). Acid-precipitated proteins are also heterogeneous by hydroxylapatite chromatography (Wolf and Sly 1965) and by starch gel electrophoresis (Puski and Melnychyn 1968; Shibasaki and Okubo 1966).

Use of Metal Cations.—For centuries metal cations, particularly calcium and magnesium, have been used as precipitants for heated soybean proteins in making tofu in the Orient. Smith *et al.* (1938) found that 0.0175 N calcium chloride precipitated about 80% of the proteins in a water extract of defatted meal but they did not characterize the proteins which precipitated. Briggs and Mann (1950) found that addition of calcium chloride to aqueous meal extracts after removal of the cold-precipitable protein (also referred to as cold-insoluble fraction and now known to be mainly 11S ultracentrifugal component) caused additional protein to precipitate. Wolf and Briggs (1959) showed that the protein precipitated by calcium ion is mainly 11S component. A more detailed study revealed that the 11S and 15S fractions are quantitatively precipitated by adding calcium chloride to 0.1 N and then cooling (Wolf and Sly 1967). However under these conditions, about 1/3 of the 2S and 1/2 of the 7S fraction are also precipitated. This method is therefore attractive if the remaining 2S and 7S fractions are desired. Following this technique Koshiyama (1965) purified a 7S globulin. Closely related to precipitation of soybean proteins by calcium ions is the use of calcium chloride to fractionally extract the proteins from defatted meal (see Fractional Extraction).

Cryoprecipitation.—One of the simplest and mildest methods for partially purifying the 11S protein consists of making a concentrated water extract (with a high meal:water extraction ratio) of defatted soybean meal at 25–40° C and then cooling the extract to near 0° C. The extract clouds up and a precipitate forms that can be removed by centrifuging in the cold (Briggs and Mann 1950; Briggs and Wolf 1957). This process, called cryoprecipitation, yields a protein fraction (cold-precipitable protein or cold-insoluble fraction) that appears homogeneous by moving boundary electrophoresis but comprises up to four fractions by ultracentrifugation (Naismith 1955; Wolf and Briggs 1959). The 11S ultracentrifuge component, however, may make up 69–88% of the cryoprecipitate (Wolf and Sly 1967). Cryoprecipitation is the method of choice for initial purification of the 11S protein.

High concentrations of sodium chloride or sucrose inhibit cryoprecipitation. The 11S component can be precipitated almost completely from a water extract by adjusting the extract to 0.1 N with calcium chloride or to pH 5.4 before cooling. Under these conditions the cryoprecipitate includes increased amounts of the 2S, 7S, and 15S fractions as compared to the precipitate obtained in the absence of calcium chloride at pH 6.4–6.6 (Wolf and Sly 1967).

Although the 15S fraction is a minor component in the water-extractable proteins, it invariably occurs with the 11S protein in the cryoprecipitate. This behavior indicates that the 15S fraction is also a cryoprotein. If the cold-insoluble fraction is dissolved at 25° C in pH 4.6 acetate buffer, 0.5 ionic strength, and then cooled to 0°–2° C, the 15S fraction precipitates quantitatively along with a part of the 11S protein. This procedure, with some additional steps, can be used to purify the 11S protein as obtained by cryoprecipitation from a water extract. Preparations of 11S with purities ranging from 89 to 94% result (Eldridge and Wolf 1967; Wolf and Tamura 1969).

Ammonium Sulfate Precipitation.—Naismith (1955) fractionated a 10% sodium chloride extract of defatted soybean meal by ammonium sulfate precipitation. He obtained a small fraction at 50% saturation and 5 additional fractions between 50 and 85% saturation. All six fractions were mixtures when analyzed by ultracentrifugation. Part of the heterogeneity may be caused by disulfide polymerization of the 7S and 11S components since the analysis buffer did not contain mercaptoethanol. Naismith studied the sedimentation behavior of partially purified 7S and 11S fractions at different ionic strengths and pH values.

Fractionation of crude 11S protein (cold-insoluble fraction) between 51 and 66% saturation with ammonium sulfate at pH 7.6, followed by precipitation at pH 4.0 between 26 and 40% saturation, yields 11S preparations of about 90% purity (Wolf *et al.* 1962).

Shvarts and Vaintraub (1967) purified 11S protein by zone precipitation on a Sephadex G-100 column with an ammonium sulfate gradient from 40 to 73% saturation.

Acid-precipitated globulins can be fractionated by ammonium sulfate precipitation to give a 7S globulin of 85% purity with 2S and 11S contaminants. The 2S impurity is removed by gel filtration (Roberts and Briggs 1965).

A number of minor proteins from soybean whey (fraction soluble at pH 4.5) that have been purified by ammonium sulfate precipitation are listed in Table 4.2.

Fractionation With Organic Solvents.—Descriptions of soybean proteins fractionated by precipitation with organic solvents, such as alcohols and acetone, appear in only a few instances. These include a crude trypsin inhibitor (Bowman 1946; Birk *et al.* 1963), lipoxygenase (Mitsuda *et al.* 1967) and hemagglutinin (Liener 1953). Because soybean proteins, in common with many other proteins, are sensitive to organic solvents, such precipitants must be used with caution (Smith *et al.* 1951; Roberts and Briggs 1963; Wolf *et al.* 1964; Fukushima 1969A).

Fractional Extraction

Neutral Salts.—Soybean meal proteins exhibit unusual solubilities in neutral salt solutions (Smith *et al.* 1938). Such solutions extract less protein from meal than does water, and in dilute salt solutions the solubility curve shows a sharp minimum at a salt concentration that varies with the salt used. Ultracentrifugal

TABLE 4.2

SOYBEAN WHEY PROTEINS PURIFIED BY AMMONIUM SULFATE PRECIPITATION

Protein	Saturation Limits[1]	Reference
Acid phosphatase	0.40–0.65	Mayer *et al.* (1961)
β-Amylase	33–50	Gertler and Birk (1965)
Hemagglutinin	0.40–0.70	Lis *et al.* (1966B)
Lipoxygenase	40–60	Mitsuda *et al.* (1967)
Chalcone-flavanone isomerase	45–60	Moustafa and Wong (1967)
Cytochrome c	0.40–0.76	Fridman *et al.* (1968)
Allantoinase	35–70	Wang and Anderson (1969)

[1] Numbers expressed as decimals are fractional saturation values, whereas others are percentage saturations.

analyses of the proteins extracted by different concentrations of sodium chloride and calcium chloride showed that lowered extractability of protein results primarily from decreased solubility of the 11S and 15S fractions (Wolf and Briggs 1956). Reduction in extractability of these fractions was more pronounced with calcium chloride than with sodium chloride. It is therefore possible to fractionate soybean meal proteins by selective extraction with salt solutions, but this approach has not been studied further.

Salts at pH 4.5.—When soybean meal is dispersed in water with sufficient hydrochloric acid to reach pH 4.5, only about 10% of the nitrogenous compounds will dissolve. If either sodium chloride or calcium chloride is added to the dilute acid, the amount of nitrogen extracted increases linearly with increasing salt concentration. Extraction levels off at 65% of the meal nitrogen with either 0.7 N sodium chloride or 0.3 N calcium chloride. Acid without salt extracts only 2S and 7S components, but the amounts of these fractions extracted increase and reach maxima as the salt concentration is raised. At higher salt concentrations the 11S and 15S fractions also begin to dissolve but do not reach maximum extractabilities until after extraction of the 2S and 7S fractions is complete (Anderson and Wolf 1967).

Acid-precipitated globulins can be extracted at pH 4.8 with increasing concentrations of sodium chloride to yield 11S preparations approaching 90% purity (Wolf and Briggs 1959). Extraction of the acid curd with 0.2 N salt removes most of the 2S and 7S globulins, and subsequent extractions with 0.35 N salt solubilize the 11S component. Approximately 20% of the total protein in the acid curd failed to dissolve in phosphate buffer (pH 7.6, ionic strength 0.5, 0.01 M mercaptoethanol) as noted in other studies (Wolf *et al* 1964; Nash and Wolf 1967).

Chromatography

Hydroxylapatite.—Chromatography of water-extractable proteins, acid-precipitated globulins, cold-insoluble fraction, and several other soybean proteins on hydroxylapatite is described by Wolf and Sly (1965). This procedure yields a clean separation of a portion of the 2S fraction from the other ultracentrifuge fractions and reveals that the 7S fraction consists of at least two different proteins. Both 7S and 11S globulins of 80–85% purity are obtainable by chromatography on this adsorbent, but the 11S and 15S fractions do not separate from each other. Other soybean proteins that have been purified by hydroxylapatite chromatography include β-amylase (Gertler and Birk 1965), cytochrome c (Fridman *et al*. 1968), hemagglutinin (Lis *et al*. 1966B), and 2S globulins (Vaintraub 1965; Vaintraub and Shutov 1969).

Elution from hydroxylapatite can be effected by a gradient (Wolf and Sly 1965) or a stepwise (Wolf and Sly 1967) procedure. Chromatographic resolution is influenced by buffer cations (Wolf and Sly 1964) and slow flow rates are recommended (Vaintraub 1965; Koshiyama 1968A).

Modified Polysaccharides.—Ion-exchange chromatography of soybean proteins on modified polysaccharides is described in numerous papers, but most applications deal with separations of only minor proteins like those of soybean whey. Selected examples are cited to illustrate applications of the methods.

Diethylaminoethyl (DEAE)-cellulose and DEAE-Sephadex.—Chromatography of whey proteins on DEAE-cellulose revealed more than 13 components in this mixture. Proteins identified include hemagglutinin, β-amylase, phosphatase, and trypsin inhibitors (Rackis *et al.* 1959). Subsequently 4 trypsin inhibitor fractions (Rackis and Anderson 1964) and 4 hemagglutinins were isolated (Lis *et al.* 1966A) by DEAE-cellulose chromatography. Separation of the four hemagglutinins illustrates the resolving power of DEAE-cellulose when proper conditions are known since these proteins were not separated by carboxymethylcellulose or hydroxylapatite chromatography or by polyacrylamide gel electrophoresis. Four hemagglutinins with different isoelectric points have also been isolated by isoelectric focusing, but they were immunochemically identical (Catsimpoolas and Meyer 1969).

A 2.3S and a 2.8S protein were purified by chromatography on DEAE-cellulose after preliminary fractionation of the globulins by ammonium sulfate precipitation and hydroxylapatite chromatography (Vaintraub and Shutov 1969).

Chromatography of cold-insoluble fraction on DEAE-cellulose at pH 7.6 yielded 11S preparations of 85–90% purity with 7S and 15S fractions still present (Wolf *et al.* 1962). Differences in charge between the various proteins apparently are too small to effect separations under the conditions used.

Chromatography of urea-treated 11S protein on DEAE-cellulose in 4 M urea was used by Vaintraub (1967), Okubo and Shibasaki (1967), and Okubo *et al.* (1969) to separate protein subunits. Despite differences in pH, buffer concentration, and column size employed in the two laboratories, the results appear in good agreement with each other. Okubo and Shibasaki also used this technique to separate subunits of a 7S protein into two major fractions.

Catsimpoolas *et al.* (1967) utilized DEAE-Sephadex chromatography to purify 11S protein obtained by ammonium sulfate fractionation of cold-insoluble fraction (Wolf *et al.* 1962). Since the resulting 11S protein was homogeneous by disc electrophoresis, DEAE-Sephadex was presumed to remove the 15S contaminant normally present in the ammonium sulfate-fractionated preparation. Chromatography on DEAE-cellulose, however, failed to remove the 15S contaminant (Wolf *et al.* 1962).

Catsimpoolas and Ekenstam (1969) chose DEAE-Sephadex to isolate two proteins, designated β-conglycinin and γ-conglycinin on the basis of immunoelectrophoresis. Beta-conglycinin is proposed to be the major component of a 7S protein isolated by Roberts and Briggs (1965), whereas γ-conglycinin makes up the 7S protein prepared by Koshiyama (1965).

Carboxymethyl (CM)-cellulose and CM-Sephadex. — Birk *et al.* (1963) isolated the acetone-insoluble trypsin inhibitor (Bowman 1946) by CM-cellulose chromatography slightly below the isoelectric point of the inhibitor. Gertler and Birk (1965) found CM-celluslose useful for removing trypsin inhibitor and lipoxygenase from β-amylase, whereas Lis *et al.* (1966A) separated inactive proteins from hemagglutinins with this adsorbent. However, the multiple forms of hemagglutinin were not resolved from each other on CM-cellulose.

CM-Sephadex chromatography is the last purification step in the preparation of crystalline lipoxygenase (Mitsuda *et al.* 1967).

Ion-exchange Resins. — Ion-exchange resins have been used to only a limited extent for chromatography of soybean proteins. Cytochrome c from soybeans was purified on Amberlite-50 by the classical procedure for isolating this enzyme from other sources (Fridman *et al.* 1968). A glycopeptide with a molecular weight of about 4600 was obtained by Pronase digestion of hemagglutinin, gel filtration, and ion-exchange chromatography on Dowex-50 (Lis *et al.* 1966B). A glycopeptide of about 9870 in molecular weight was isolated in a similar manner from a 7S globulin (Koshiyama 1969A).

Gel Filtration. — The broad range of molecular sizes for soybean proteins is best demonstrated by gel filtration of the water-extractable proteins on Sephadex G-200. Hasegawa *et al.* (1963) observed 7 protein peaks in the effluent when water-extractable proteins were placed on a 200-cm column (Fig. 4.5A) as compared to only 4 fractions that are resolved by the ultracentrifuge (Fig. 4.5B). Peak 1 was attributed primarily to turbidity, but this fraction has an absorption maximum at 260 mμ and gives a positive test for ribose with orcinol (Okubo and Shibasaki 1967; Obara and Kimura 1967). These results indicate that peak 1 contains ribonucleic acid. Ultracentrifugal compositions for other peaks are shown in Fig. 4.5B. The 11S protein eluted almost completely in peak 3 but was accompanied by a 7S fraction. Peaks 4, 5, and 6 also contained proteins of the 7S group while peaks 7 and 8 consisted of 2S proteins. Peaks 9–11 did not appear to contain protein.

Gel filtration of the high-molecular-weight portion of soybean globulins (corresponding to peaks 1 to 4 of Fig. 4.5A) showed that 7S protein occurred in peaks 2 to 4 (Koshiyama 1969B). The 7S protein was of 2 types: 1 dimerized when the ionic strength was changed from 0.5 to 0.1 while the other 7S protein did not. The dimerizing variety of 7S occurred in peaks 2 to 4 but was concentrated in peak 3; the leading half of peak 3 contained only this type of 7S protein. The nondimerizing 7S protein occurred in peak 2, the trailing half of peak 3 and peak 4. The dimerizing 7S protein represented about 75% of the total 7S protein and about 25% of the total protein in the globulin subfraction. These estimates are comparable with values for the dimerizing 7S in the water-extractable proteins: 60% of the total 7S and 20% of the total protein (Wolf and Sly 1967; Wolf 1969B). Heterogeneity of the 7S fraction is also demonstrated

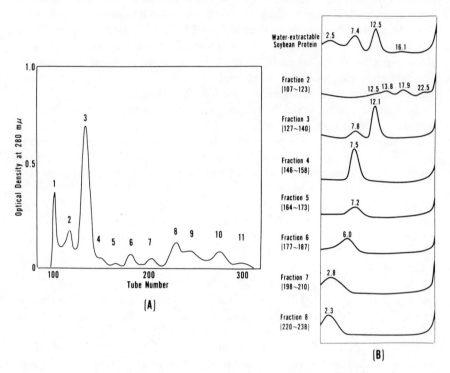

From Hasegawa et al. (1963)

FIG. 4.5. GEL FILTRATION OF WATER-EXTRACTABLE SOYBEAN PROTEINS
ON SEPHADEX G-200

(A) Elution diagram for 150 mg of protein with 5 ml fractions; (B) ultracentrifuge patterns
for fractions obtained in A. Numbers in parentheses indicate tube numbers for each
fraction, and sedimentation is from left to right. Sedimentation coefficients are given above
peaks.

by hydroxylapatite chromatography (Wolf and Sly 1965). From these results it
is apparent that the 11S component is the major protein of soybeans.

Elution of 7S proteins with the 15S and 11S fractions (peaks 2 and 3 of Fig.
4.5A) is surprising because of the large differences in sedimentation rates for the
3 proteins. Further work is needed to explain this behavior of the 7S proteins.

Koshiyama (1965, 1968A) isolated the dimerizing 7S globulin and demonstrated that it corresponded to the protein that elutes with the 11S protein (peak 3,
Fig. 4.5A) from Sephadex G-200. A synthetic mixture of 7S and 11S proteins
tended to separate only slightly. Recycling gel filtration will probably separate
the two proteins, but the process is slow, and yields of 11S are low (Eldridge and
Wolf 1967). Column length is an important factor for resolution of the high-
molecular-weight fractions.

Other uses of Sephadex G-200 include characterization of protein bodies (Saio and Watanabe 1966), heat denaturation of soybean proteins (Saio *et al.* 1968B), and purification of chalcone-flavanone isomerase (Moustafa and Wong 1967). Gel filtration of soybean whey proteins on Sephadex G-100 yielded six major fractions, which were heterogeneous by disc electrophoresis (Catsimpoolas and Leuthner 1969A); this technique should be a useful adjunct to other methods for fractionating these proteins.

Sephadex G-25 and G-50 easily separate glycopeptides from pronase digests of soybean hemagglutinin (Lis *et al.* 1966B) and of 7S globulin (Koshiyama 1969A).

CHEMICAL PROPERTIES OF SOYBEAN PROTEINS

Nitrogen Content

Table 4.3 lists nitrogen contents for unfractionated proteins, several fractions, and various purified proteins. Since the purified proteins differ in nitrogen

TABLE 4.3

NITROGEN CONTENT OF SOYBEAN PROTEINS

Protein Preparation	Nitrogen Content (%)	Reference
Sodium chloride-extractable proteins	16.42	Wolf et al. (1966)
Water-extractable proteins	16.51[1]	Wolf et al. (1966)
Acid-precipitated globulins	16.20[1]	Wolf et al. (1966)
Acid-precipitated globulins	16.24	Nash et al. (1967)
Alcohol-extracted, acid-precipitated globulins	17.04	Nash et al. (1967)
Glycinin	17.45	Osborne and Campbell (1898)
Whey proteins	15.22[1]	Wolf et al. (1966)
Cold-insoluble fraction	17.46[1]	Wolf et al. (1966)
Cold-soluble fraction	16.12[1]	Wolf et al. (1966)
11S	17.62[1]	Wolf et al. (1966)
7S	15.5	Roberts and Briggs (1965)
7S	15.91	Koshiyama (1968B)
Hemagglutinin	13.2	Lis et al. (1966B)
β-Amylase	16.3	Gertler and Birk (1965)
Trypsin inhibitor (Kunitz)	16.74	Kunitz (1947)
Trypsin inhibitor (Bowman-Birk)	15.36	Frattali (1969)

[1] Mean of three or more preparations.

content, mixtures of the proteins will vary in nitrogen contents depending on the relative amounts of each protein present. Nitrogen contents of the proteins are also influenced by nonprotein impurities, such as phytate and alcohol-extractable materials. For example, extraction of acid-precipitated globulins with aqueous ethanol increased nitrogen content from 16.24 to 17.04% (Table 4.3).

Comparison of the nitrogen content of glycinin reported by Osborne and Campbell (1898) with other values in Table 4.3 shows that glycinin must consist mainly of 11S protein. Indeed, the nitrogen content of cold-insoluble fraction (crude 11S) most closely agrees with the value for glycinin, and the two protein preparations are similar in composition by ultracentrifugation (Naismith 1955). Inability of Smiley and Smith (1946) to obtain preparations with nitrogen values similar to those of glycinin probably resulted from their use of acid to adjust the proteins to pH 4.1–4.5 after dialysis. Under these conditions the 2S and 7S globulins are precipitated with the 11S protein; whereas during dialysis of a neutral solution, pH drops slowly toward the isoelectric region, and the 11S component is precipitated preferentially (Wolf and Briggs 1959).

Nonprotein Constituents in Soybean Proteins

Carbohydrates are present in varying amounts in many preparations of soybean proteins (Wolf et al. 1966), but only hemagglutinin (Lis et al. 1966B) and 7S globulin (Koshiyama 1969A) have been shown to be glycoproteins. Hemagglutinin contains 3–5 glucosamine and 25 mannose residues per mole as compared to 12 glucosamine and 39 mannose residues per mole in the 7S globulin. These two proteins are among those with the lowest nitrogen contents in Table 4.3.

Noncovalently bound carbohydrates have also been isolated from soybean proteins (Smiley and Smith 1946). Nash et al. (1967) removed 3.6% of the weight of soybean globulins by alcohol extraction and identified saponins and sitosterol glycoside in the extract. Other components in the alcoholic extract were phosphatidyl choline, phosphatidyl ethanolamine, genistein, triglycerides, and several unidentified compounds. The significance of these nonprotein materials in soybean globulins is still unknown. The phosphatides may originate from the membranes surrounding the protein bodies (Tombs 1967). Alcohol extraction does not remove all the nonprotein compounds from globulins since additional saponins were isolated from alcohol-extracted protein (Eldridge and Wolf 1969A).

In addition to phosphatides, soybean proteins may contain other phosphorus compounds. Phytate is a major contaminant of globulins prepared by isoelectric precipitation (McKinney et al. 1949; Smith and Rackis 1957).

Interaction of phytate with soybean proteins is influenced by pH and such cations as calcium. Calcium phytate, which is insoluble above pH 6, remains soluble above pH 10 when protein is present. When a mixture of calcium ion, phytate, and protein are filtered through Sephadex G-75, calcium and phytate elute with the protein (Saio *et al.* 1967). Equilibrium dialysis studies with crude 11S protein confirm an interrelationship between calcium ion and phytate in the binding of these ions by the protein (Saio *et al.* 1968A).

Puski and Melnychyn (1968) used 3 different methods to remove phytate from acid-precipitated globulins and found 70% of the phosphorus removed by each method. The treated globulins contained about 0.2% phosphorus. This residual phosphorus is not removed by alcohol extraction although phosphatides are removed by this process (Nash *et al.* 1967). The residual phosphorus may be mainly ribonucleic acid. Koshiyama and Iguchi (1965) reported 3.1% ribonucleic acid for acid-precipitated globulins, a percentage which corresponds to a phosphorus content of 0.26% for the protein assuming 8.5% phosphorus in the ribonucleic acid (DiCarlo *et al.* 1955). Shutov and Vaintraub (1967), however, report only 0.27% ribonucleic acid in acid-precipitated soybean globulins. Further studies are needed to resolve this discrepancy.

Bai and Pin (1964) isolated from soybeans proteins which yielded serine-*O*—phosphate on hydrolysis. Based on isolated yields these phosphoproteins occur as minor constituents.

Amino Acid Composition

Amino Acid Analyses.—Data for purified soybean proteins are assembled in Tables 4.4 and 4.5. Included are results for defatted soybean meal, the source material for the proteins. Analyses for meal are expressed as grams of amino acid residues per 16 gm N since the nitrogen-to-protein conversion factor is unknown and since no correction was made for nonprotein nitrogen in the meal. Cytochrome c, trypsin inhibitors, β-amylase, hemagglutinin, and lipoxygenase are minor proteins but are of interest because of their biological activities.

The Bowman-Birk trypsin inhibitor [2] is unusually high in cystine but devoid of glycine and tryptophan. In contrast to the Bowman-Birk inhibitor, hemagglutinin is free of cystine. The 7S globulins isolated by Koshiyama (1968B) and Roberts and Briggs (1965) differ significantly in overall amino acid contents and molecular weights. The two 7S proteins were, however, remarkably low in methionine content, and both underwent a characteristic monomer-dimer reaction with change in ionic strength. Gel filtration of 7S globulins on Sephadex G-200 suggests that the protein capable of monomer-dimer formation is eluted over a broader range than is expected for a single protein (Koshiyama 1969B). Possibly there are two or more 7S proteins with the ability to dimerize at low ionic strength. Catsimpoolas and Ekenstam (1969) report that the 7S globulins prepared by the methods of Koshiyama and of Roberts and Briggs are different

---

[2] See Frattali (1969) and Steiner and Frattali (1969) for a description of this protein; it appears to be the same as the 1.9S trypsin inhibitor isolated by Yamamoto and Ikenaka (1967).

TABLE 4.4

AMINO ACID COMPOSITION OF PURIFIED SOYBEAN PROTEINS

Amino Acid	Defatted Meal,[1] Gm Amino Acid Residue/16 Gm N	Trypsin Inhibitor[2] (Bowman-Birk)		Cytochrome c[3]		Trypsin Inhibitor[4] (Kunitz)		2.8S Globulin[5]		β-Amylase[6]	
		Gm Amino Acid Residue/100 Gm Protein	Residues/7,975 Gm	Gm Amino Acid Residue/100 Gm Protein	Residues/12,000 Gm	Gm Amino Acid Residue/100 Gm Protein	Residues/21,500 Gm	Gm Amino Acid Residue/100 Gm Protein	Residues/32,600 Gm	Gm Amino Acid Residue/100 Gm Protein	Residues/61,700 Gm
Arginine	7.55	3.84	2	2.73	2	7.14	10	6.46	13	5.12	20
Histidine	2.25	1.56	1	2.26	2	1.46	2	0.64	2	2.58	12
Lysine	6.01	7.65	5	12.88	12	7.38	12	5.55	14	6.59	32
Tyrosine	3.51	3.91	2	6.28	7	3.17	4	2.21	4	7.96	30
Tryptophan	1.17	0	0	2.05	1	1.66	2	2.34	4	3.59	12
Phenylalanine	4.46	3.58	2	4.42	4	6.62	10	6.53	15	5.18	22
1/2 Cystine	1.34	17.68	14	0.61	1	2.07	4	1.86	6	0.88	5
Cysteine	—	0	0[4]	0.62	1[7]	0	0[8]	—	—	0.89	9
Methionine	1.37	1.47	1	1.57	2	1.56	3	0.63	2	1.98	9
Serine	4.61	9.58	9	5.09	7	4.76	12	4.11	15	4.66	33
Threonine	3.66	2.41	2	6.32	7	3.63	8	3.37	11	3.72	23
Leucine	6.66	2.71	2	6.96	7	7.85	15	6.36	18	11.30	62
Isoleucine	4.40	2.62	2	3.06	3	7.84	15	8.06	24	5.63	31
Valine	4.55	1.12	1	2.23	3	7.11	15	6.77	22	6.17	38

Glutamic acid	18.42	11.12	7	11.56	10	11.43	19	9.43	24	14.69	70
Aspartic acid	10.38	16.53	12	9.78	10	15.57	29	13.45	38	15.52	83
Glycine	3.44	0.05	0	4.88	10	4.45	17	3.94	23	4.88	53
Alanine	3.60	3.63	4	5.05	8	2.89	9	2.53	12	4.88	42
Proline	5.30	6.90	6	5.78	7	4.99	11	4.54	15	6.32	40
Total[9]	92.68	96.36	72	93.51	103	101.58	197	88.78	262	111.65	617
Ammonia	1.93	1.20	6	—	—	0.95	13	—	—	1.60	62

[1] Rackis et al. (1961).
[2] Frattali (1969).
[3] Fridman et al. (1968).
[4] Yamamoto and Ikenaka (1967).
[5] Vaintraub and Shutov (1969). Residues/mole of protein are uncorrected for hydrolysis and chromatographic losses.
[6] Gertler and Birk (1965).
[7] Based on presence of only one 1/2-cystine/mole.
[8] Wu and Scheraga (1962).
[9] Does not include cysteine values.

TABLE 4.5

AMINO ACID COMPOSITION OF PURIFIED SOYBEAN PROTEINS

Amino Acid	Defatted Meal[1] Gm Amino Acid Residue/16 Gm N	Hemagglutinin[2] Gm Amino Acid Residue/100 Gm Protein	Residues/100,000 Gm	Lipoxygenase[3] Gm Amino Acid Residue/100 Gm Protein	Residues/108,000 Gm	7S Globulin[4] Gm Amino Acid Residue/100 Gm Protein	Residues/180,000 Gm	7S Globulin[5] Gm Amino Acid Residue/100 Gm Protein	Residues/330,000 Gm	11S Globulin[6] Gm Amino Acid Residue/100 Gm Protein	Residues/360,000 Gm
Arginine	7.55	2.81	18	6.10	42	7.91	91	7.1	176	7.04	162
Histidine	2.25	1.92	14	3.46	27	1.48	19	1.7	51	2.28	60
Lysine	6.01	5.13	40	6.39	54	6.15	86	4.8	147	4.24	119
Tyrosine	3.51	2.61	16	6.17	41	3.25	36	3.2	76	3.87	85
Tryptophan	1.17	3.35	18	3.48	20	0.29	3	–	–	1.36	26
Phenylalanine	4.46	5.45	37	5.09	37	6.58	81	5.1	135	4.93	121
1/2 Cystine	1.34	Traces	Traces	0.73	8	0.22	4	–	–	1.36	48
Cysteine	–	0	0	0.38	4	0	0	–	–	–	–
Methionine	1.37	0.79	6	2.07	17	0.22	3	0.17	5	1.26	34
Serine	4.61	6.62	76	6.02	75	5.61	116	3.2	154	6.52	268
Threonine	3.66	5.05	50	5.26	56	2.38	43	1.4	55	4.49	160
Leucine	6.66	7.13	63	10.71	102	8.84	141	5.9	209	6.28	200
Isoleucine	4.40	4.30	38	6.25	60	5.52	88	3.7	131	5.05	161
Valine	4.55	4.46	45	5.96	65	4.30	78	3.1	119	4.30	156

Glutamic acid	18.42	6.71	52	12.31	103	17.99	251	15.3	475	21.81	603
Aspartic acid	10.38	11.16	97	11.74	110	12.22	191	10.3	363	12.18	381
Glycine	3.44	2.28	40	3.63	69	2.17	68	1.4	110	3.64	230
Alanine	3.60	4.34	61	4.76	72	2.95	75	1.8	110	3.36	169
Proline	5.30	4.56	47	5.35	60	3.65	71	3.3	139	4.89	181
Total[7]	92.68	78.67	718	105.48	1018	91.73	1445	71.5	2455	98.86	3,164
Ammonia	1.93	1.28	80	1.52	102	1.61	181	—	—	—	—
Mannose	—	4.50	25	—	—	3.75	38	—	—	—	—
Glucosamine	—	5.37–8.96	3–5	—	—	1.19	12	—	—	—	—

[1] Rackis et al. (1961).
[2] Lis et al. (1966B).
[3] Stevens et al. (1970).
[4] Koshiyama (1968B).
[5] Roberts and Briggs (1965).
[6] Shvarts and Vaintraub (1967).
[7] Does not include cysteine values.

by disc electrophoresis and disc immunoelectrophoresis. However, Catsimpoolas and Ekenstam did not examine their 7S samples for the ability to dimerize at low strength in order to demonstrate identity with the preparations of the other workers.

The most significant differences between the 7S globulins and the 11S globulins are the 5- to 6-fold higher contents of tryptophan, methionine, and 1/2-cystine in the latter.

Other amino acid analyses on soybean proteins include: acid-precipitated globulins and whey proteins (Rackis *et al.* 1961), 11S protein and acid-precipitated globulins (Catsimpoolas *et al.* 1967), partially purified 7S and 11S globulins (Fukushima 1968), 11S protein and its basic subunits (Okubo *et al.* 1969), and a partially purified 2.3S protein (Vaintraub and Shutov 1969).

Sulfhydryl Content.—Available data for cysteine contents of purified proteins are included in Tables 4.4 and 4.5. Lipoxygenase contains four sulfhydryl groups per mole, but their reaction with 5,5'-dithiobis-(2-nitro-benzoic acid) requires that the enzyme be denatured (Stevens *et al.* 1970). Sulfhydryl contents of the major globulins have received little attention. The 7S globulin isolated by Koshiyama (1965; 1968B) appears to contain two disulfide bonds but no sulfhydryl groups although there is a 7S protein in the globulin fraction that forms polymers presumably linked by disulfide bonds (Nash and Wolf 1967). This protein in the depolymerized state contains at least two sulfhydryl groups per mole if polymers larger than the dimer are formed.

No data are available concerning the sulfhydryl content of the 11S protein. Kelley and Pressey (1966) report about 0.6 sulfhydryl equivalents per 10^5 gm of acid-precipitated globulins as compared to 1.0–1.2 sulfhydryl equivalents per 10^5 gm of glycinin (mixture of 7S and 11S globulins) obtained by Smirnova *et al.* (1959).

Primary Structures

Little is known about the primary structures of soybean proteins with one exception. Ikenaka *et al.* (1963) reported the sequence of the first five amino acids in the N-terminal position of Kunitz' trypsin inhibitor. Subsequently, Brown *et al.* (1966) described the sequence of amino acids around the two disulfide bonds of the inhibitor, while Ozawa and Laskowski (1966) reported an arginyl-isoleucine sequence at the active site. Thirty-nine amino acid residues out of a total of about 200 are accounted for by the sequences known at present. Ikenaka and coworkers in Japan are doing further work on the primary sequence of this protein (Sealock and Laskowski 1969). Hopefully, the entire sequence will be determined in the near future.

For other soybean proteins only terminal amino acids are known (Table 4.6). On the basis of N-terminal amino acids the 7S and 11S molecules contain a minimum of 9 and 12 polypeptide chains, respectively. Determination of the primary structure of these multichained proteins must, therefore, be preceded by isolation and purification of the subunits.

TABLE 4.6

TERMINAL AMINO ACID RESIDUES OF PURIFIED SOYBEAN PROTEINS

Protein	Number and Kind of Groups/Mole		Reference
	N-Terminal	C-Terminal	
Kunitz trypsin inhibitor	1 Aspartic acid	1 Leucine	Yamamoto and Ikenaka (1967)
Bowman-Birk trypsin inhibitor	1 Aspartic acid[1]	–	Yamamoto and Ikenaka (1967)
2.3S Globulin	0.2 Lysine[2]		Vaintraub and Shutov (1969)
2S Globulin	1 Aspartic acid	–	Vaintraub and Shutov (1969)
β-Amylase	1 Glutamic acid	1 Glycine	Gertler and Birk (1965)
Hemagglutinin	2 Alanine	1 Serine Aspartic acid[3] Glutamic acid[3] Alanine[3]	Wada et al. (1958)
2.8S Globulin	1 Aspartic acid 1 Alanine 1 Glycine 1 Valine 2 Serine 1 Tyrosine 1 Glutamic acid 1 Leucine	–	Koshiyama (1968B)
11S Globulin	8 Glycine 2 Leucine (Isoleucine) 2 Phenylalanine	–	Catsimpoolas et al. (1967)

[1] Assuming 1.9S inhibitor is same as the Bowman-Birk inhibitor.
[2] Protein was not homogeneous as indicated by yields of other N-terminal residues.
[3] May be penultimate rather than terminal amino acid.

Primary structures of hemagglutinin and 7S globulin are more complex than those of other soybean proteins because they are glycoproteins. In hemagglutinin the carbohydrate units are attached to a single aspartic acid residue, probably through amido linkage with the β-carboxyl carbon of aspartic acid (Lis et al. 1969). Although the mode of attachment of the sugar residues to 7S globulin is not known, the carbohydrates appear to be linked as a single unit (Koshiyama 1969A).

Disulfide Polymerization

Polymers in Defatted Meal.—When water-extractable proteins are ultracentrifuged with and without 0.01 M mercaptoethanol in the analysis buffer, the mercaptan causes the fast-sedimenting materials to disappear and increases the amounts of 7S and 11S proteins (Wolf and Sly 1967). The fast-sedimenting proteins apparently are disulfide polymers of the 7S and 11S proteins. Approximately 30–35% of the total 7S and 10–12% of the total 11S protein are polymerized in this manner. Since the polymers are observed in the extracts which have received a minimum of manipulation (dialysis against buffer), they probably pre-exist in defatted meal. It is not known whether disulfide polymerization occurs during deposition of the proteins in the seed or during subsequent aging of the seed or meal.

Polymerization During Protein Precipitation.—Crude 11S protein isolated by cryoprecipitation and dialysis contains soluble and insoluble polymers that are depolymerized by 0.01 M mercaptoethanol and other disulfide-cleaving reagents (Briggs and Wolf 1957). Although disulfide polymers occur in the water-extractable proteins before cryoprecipitation, formation of insoluble polymers is favored by precipitation. Cryoprecipitation causes up to 40% of the 11S protein to polymerize as compared to only about 10% polymerization with N-ethyl maleimide present during precipitation. More extensive polymerization of 11S occurs during precipitation by dialysis against distilled water; some of the polymers are insoluble in buffer after this treatment.

Insolubilization of 7S and 11S proteins as a result of disulfide polymerization occurs during isoelectric precipitation of the globulins (Nash and Wolf 1967). Insoluble disulfide polymers of the 7S and 11S globulins are also noted in glycinin (Kretovich et al. 1956; Kretovich and Smirnova 1957).

PHYSICAL PROPERTIES OF SOYBEAN PROTEINS

Solubility

Solubility of soybean proteins is sensitive to pH and salts as discussed earlier (Protein Extraction; Fig. 4.4). The globulins are insoluble in the isoelectric region at low salt concentrations, but they are appreciably soluble if salts are added (Smith and Circle 1938). For example, the 11S protein is soluble at pH 4.0 in 1 M sodium chloride but precipitates at a lower concentration of ammonium sulfate than when precipitation is carried out at pH 7.6 (Wolf et al. 1962).

Solubility of the globulins is also influenced by phytate (Smith and Rackis 1957); phytate should be removed before attempting solubility studies. Another factor that affects solubility of some soybean proteins is disulfide polymerization, previously discussed. Solubilities of acid-precipitated globulins in buffer with and without 0.01 M mercaptoethanol have been described, but results were expressed as percentages of the total globulins rather than as absolute solubilities (Wolf et al. 1963; Kelley and Pressey 1966; Nash and Wolf 1967). Phosphate buffer of pH 7.6 and 0.5 ionic strength has been used extensively in solubility studies. This buffer containing 0.01 M mercaptoethanol is often referred to as standard buffer and tends to separate native (soluble) from denatured (insoluble) forms of the proteins as noted in alcohol denaturation studies (Roberts and Briggs 1963; Wolf et al. 1964). However, solubilization of denatured proteins by standard buffer has been noted (Nash and Wolf 1967), and about 1/2 of heat-denatured 11S protein remains soluble in the buffer (Wolf and Tamura 1969).

Kunitz (1947) applied the phase-rule solubility test to his crystalline trypsin inhibitor; this test indicated a high degree of purity. Nonetheless, commercial preparations of crystalline inhibitor may be impure (Eldridge et al. 1966). Briggs and Mann (1950) applied the phase-rule solubility test to crude 11S protein (cold-insoluble fraction) and obtained evidence of heterogeneity. Heterogeneity is now known to arise from non-11S contaminants and disulfide polymers of 11S component (Briggs and Wolf 1957). Solubilities of 11S protein at $0°-2°$ C as a function of ionic strength at pH 4.6 (in 0.01 M mercaptoethanol) are reported by Eldridge and Wolf (1967).

Molecular Size

Molecular weights and other physical properties of purified soybean proteins are listed in Table 4.7. Undoubtedly there are still many unidentified proteins in soybean meal, but it is already evident that the proteins cover a wide range of molecular sizes. At least five proteins have been isolated and shown to be constituents of the 2S fraction of water-extractable proteins: Bowman-Birk trypsin inhibitor, cytochrome c, Kunitz trypsin inhibitor, and two 2S globulins. A sixth protein, chalcone-flavanone isomerase, reportedly has an $s_{20,w}$ value of 1.6–2S but has not been characterized further (Moustafa and Wong 1967).

Molecular weight of the Bowman-Birk trypsin inhibitor is concentration-dependent and various values are recorded in Table 4.7. A monomer-dimer-trimer equilibrium with a monomer molecular weight of 8000 is proposed to explain this behavior (Millar et al. 1969). Allantoinase, β-amylase, hemagglutinin, lipoxygenase, and 7S globulin(s) are present in the 7S fraction of water-extractable proteins (Table 4.7). The two 7S globulin preparations agree in their sedimentation properties but differ in their molecular weights. Further work is necessary to clearly establish whether the 2 isolation procedures yield different proteins or whether the molecular weight reported by Roberts and Briggs is too high by a factor of 2.

Only one protein corresponding to the 11S fraction of the water-extractable proteins has been isolated; the 11S globulin can thus be considered the major protein of soybeans. The 11S protein has a molecular weight of about 350,000, which is typical of major globulins of other seeds.

On the basis of its sedimentation rate, the 15S fraction is estimated to have a molecular weight of 1/2 million or more. Catsimpoolas et al. (1969A) suggest that the 15S fraction is a polymer of glycinin (11S half-molecules[3]); a trimer and tetramer would correspond to respective molecular weights of 525,000 and 700,000. Polymers of 11S resembling the 15S fraction have been observed (Wolf et al. 1962; Shvarts and Vaintraub 1967), but the 15S fraction observed in the

---

[3] In this terminology the 11S protein would be a dimer of glycinin.

TABLE 4.7

PHYSICAL PROPERTIES OF PURIFIED SOYBEAN PROTEINS

Protein	V_{20} Ml/Gm	$s_{20,w}^{\circ}$ × 10¹³	$D_{20,w}$ × 10⁷	Mol Wt	Mol Wt Method[1]	Isoelectric Point	$E_{1\,Cm}^{1\%}$ 280 Mμ	$[\eta]$, Dl/Gm	References
Bowman-Birk trypsin inhibitor	0.69	2.3[2]	9.03	24,000	SD	4.2	4.8		Birk et al. (1963)
		1.9[2]		16,400	AE	4.0	4.4		Frattali (1969); Yamamoto and Ikenaka (1967)
				7,975	AA				Frattali (1969)
				8,000	SE				Millar et al. (1969)
				8,250	GF				Kakade et al. (1970)
				12,000	GF				Fridman et al. (1968)
Cytochrome c		1.8[3]							
Kunitz trypsin inhibitor	0.698	2.29		21,500	AE	4.5	9.44	0.028	Wu and Scheraga (1962); Kunitz (1947); Edelhoch and Steiner (1963); Jirgensons et al. (1969)
							10.6	0.034	
2.3S Globulin		2.28		18,200	SE				Vaintraub and Shutov (1969)
2.8S Globulin		2.80		32,600	SE	4.4	9.06		Vaintraub and Shutov (1969)
Allantoinase				50,000	SE		5.48		Yuan (1969)
β-Amylase		4.67	7.47	61,700	SD	5.85	17.3		Gertler and Birk (1965)
				69,000	AA				
Hemagglutinin		6.4, 6.1[4]	5.72	105,000	SD	6.1	15.7		Pallansch and Liener (1953); Wada et al. (1958)
				89,000	LS				
				110,000	SD				
		6.0	5.0			5.85, 6.00, 6.10, 6.20[5]			Lis et al. (1966B); Catsimpoolas and Meyer (1969)

Protein				Mol. wt.	Method				References
Lipoxygenase	0.750	5.62	5.59	102,000	SD	5.4			Theorell et al. (1947)
		5.65		102,000	GF	5.65			Mitsuda et al. (1967); Catsimpoolas (1969B)
7S Globulin	0.734	7.95	3.85	108,000	SE		17.4		Stevens et al. (1970)
	0.729	7.92		330,000	AE		6.4		Roberts and Briggs (1965)
	0.725			186,000	AE	4.9	5.47	0.0638	Koshiyama (1968B)
				180,000	SE				
				193,000	SV				
				210,000	GF				
11S Globulin	0.719	12.20	2.91	333,000	LS	~5.0[6]	9.2		Wolf and Briggs (1959); Wolf (1956)
				356,000	LS				

[1] AA—Amino acid analysis; AE—approach to equilibrium; GF—gel filtration; LS—light scattering; SD—sedimentation-diffusion; SE—sedimentation equilibrium; SV—sedimentation-viscosity.
[2] Value for 1% solution.
[3] Value for solution with optical density of 0.2 at 267 mμ.
[4] Value for 1.5% solution.
[5] Values for four hemagglutinins.
[6] Based on solubility measurements.

water-extractable proteins has not been isolated and characterized to establish its chemical identity with 11S protein.

The water-extractable proteins often contain a small amount of material sedimenting ahead of the 15S fraction. Designated as the >15S fraction, this diffuse boundary probably represents a mixture of proteins with molecular weights approaching 1 million. Since soybean urease has an $s_{20,w}$ value of 18S at pH 7.0 in a sucrose density gradient (Tanis and Naylor 1968), it is one of the proteins in the >15S fraction. The >15S fraction may also contain the nucleic acid or nucleoprotein that elutes near the exclusion volume of Sephadex G-200 (Okubo and Shibasaki 1967).

Molecular Structure and Conformation

Information about the secondary, tertiary, and quaternary structures of most soybean proteins is limited. Kunitz trypsin inhibitor was crystallized over 20 yr ago and has received more attention than any other soybean protein, but its detailed structure is still unknown. From the standpoint of size, trypsin inhibitor is a relatively simple molecule; consequently, the task of unraveling the structures of more complex molecules, such as the 7S and 11S globulins, is even greater.

Bowman-Birk (1.9S) Inhibitor.—Ultraviolet difference spectral studies indicate that the two tyrosine residues are accessible to solvent in the native state; high concentrations of urea or guanidine hydrochloride cause no changes attributable to exposure of buried phenolic groups (Steiner and Frattali 1969). Circular dichroism measurements revealed little, if any, α-helical structure (Ikeda *et al.* 1968; Steiner and Frattali 1969). Its high cystine content (Table 4.4), however, suggests that the molecule is extensively crosslinked and, therefore, has a definite three-dimensional structure. At concentrations of 0.1% and higher the Bowman-Birk inhibitor also possesses quaternary structure since it associates into dimer and possibly trimer. A 1% solution contains approximately equal mole fractions of monomer and dimer and 7% trimer as estimated by sedimentation equilibrium studies (Millar *et al.* 1969). Further study of this system by osmometry confirmed a monomer-dimer equilibrium although no trimer was indicated at concentrations up to 0.3% (Harry and Steiner 1969).

Kunitz Trypsin Inhibitor.—This protein possesses a single polypeptide chain crosslinked by two disulfide bonds. Intrinsic viscosity (Table 4.7), fluoresence polarization, and other physical properties indicate that the inhibitor is compact, low in asymmetry, and rigid in structure (Steiner and Frattali 1969). Circular dichroism measurements failed to detect either α-helical or β-structures in the native protein (Jirgensons *et al.* 1969). Nonetheless, large changes in the circular dichroism spectra occurred after cleavage of the disulfide bonds. Reduction of both disulfide crosslinks abolishes trypsin inhibitor activity, and activity is restored by reoxidation (Steiner 1965). However, one disulfide can be reduced selectively without loss of activity (DiBella and Liener 1969). Reducing both disulfides also increases the intrinsic viscosity of the inhibitor in buffer as a result of increased unfolding of the molecule. Further unfolding occurs when reduced or reduced-alkylated inhibitor is treated with 8 M urea; the unfolded protein has an intrinsic viscosity of 0.25–0.27 dl per gm and an electrophoretic mobility (in polyacrylamide gel) 0.4–0.5 that of the native inhibitor (Eldridge and Wolf 1969B).

When Kunitz inhibitor interacts with trypsin, the arginyl-isoleucine peptide bond is cleaved at the reactive site of the inhibitor without loss of activity. However, removal of the resulting C-terminal arginine by treatment with car-boxypeptidase B destroys activity of the inhibitor (Ozawa and Laskowski 1966). The arginyl residue at the reactive center was replaced enzymatically by a lysyl group with nearly complete recovery in activity (Sealock and Laskowski 1969).

Hemagglutinin.–Practically nothing is known about the structure of this glycoprotein. The absence of cystine crosslinks (Table 4.5) suggests that the molecule may be fairly flexible and subject to conformational changes under milder conditions than are required to alter conformation of such molecules as the trypsin inhibitors. The presence of 2 N-terminal residues is indicative of 2 polypeptide chains and, therefore, 2 subunits. Disruption of the quaternary structure with phenol-acetic acid-mercaptoethanol-urea yielded two subunits as detected by disc electrophoresis (Catsimpoolas and Meyer 1969).

Lipoxygenase.–This enzyme appears to consist of two subunits of 58,000 mol wt. Dissociation into subunits occurs with guanidine hydrochloride or sodium dodecyl sulfate (Stevens *et al.* 1970).

7S Globulin.–Fukushima (1968) investigated the secondary and tertiary structures of a partially purified 7S globulin that appears similar to the 7S proteins isolated by Koshiyama (1968B) and Roberts and Briggs (1965) on the basis of carbohydrate and amino acid analysis. Optical rotatory dispersion and infrared measurements failed to detect appreciable α-helical structure but suggested that antiparallel β-structure and disordered regions predominated in the molecule. Only about 40% of the peptide hydrogens exchanged with deuterium. Fukushima concluded that the molecules are folded compactly even though large regions of disordered (nonhelical) structure occur. Ultraviolet difference spectra in urea solutions showed that the tyrosine residues are buried in the interior of the molecule while the tryptophan residues are accessible to solvent. Fukushima proposed that hydrophobic bonding is important in maintaining the tertiary structure.

Koshiyama (1968B) reported an intrinsic viscosity of 0.0638 dl per gm for a purified 7S globulin. This value is consistent with a relatively compact structure [4]. Isoelectrically precipitated globulins have an intrinsic viscosity of 0.052 dl per gm (Wolf *et al.* 1963).

---

[4] See Table III in review by Tanford (1968) for comparison of intrinsic viscosities of native and denatured proteins.

End group analysis (Table 4.6) indicates that there are at least nine polypeptide chains in the 7S globulin isolated by Koshiyama. Based on a molecular weight of 180,000, the 7S globulin has a quaternary structure of 9 subunits with average molecular weight of 20,000. The extent of disulfide crosslinking between polypeptide chains is unknown but must be small since only two cystines occur per mole of 7S globulin (Table 4.5).

Changes in the quaternary structure of 7S globulin(s) as a function of pH and ionic strength are reported by Roberts and Briggs (1965) and Koshiyama (1968C). Figure 4.6 summarizes changes in sedimentation properties of 7S globulin under various conditions as described by Koshiyama. At neutrality the 7S protein undergoes a distinctive reversible reaction with changes in ionic strength. At 0.5 ionic strength the protein has a molecular weight of 180,000–210,000, but at 0.1 ionic strength the protein sediments at a rate of about 9S (at 1% concentration) and has a molecular weight of 370,000. The 9S form is thus a dimer of the 7S protein. Based on formation of the 9S form at 0.1 ionic strength, about 60% of the total 7S fraction in water-extractable proteins can undergo dimerization (Fig. 4.1) (Wolf and Sly 1967). At pH 2 and low ionic strength the 7S globulin is converted into 2S and 5S species presumably as a result of dissociation into subunits. Conversion into the 2S and 5S forms is inhibited by ionic strengths of 0.1 and higher and is reversed by dialysis of the protein against pH 7.6, 0.5 ionic strength buffer. In 0.01 N sodium hydroxide the protein is irreversibly converted to a slowly sedimenting form (0.4S).

The 7S globulin isolated by Roberts and Briggs (1965) exhibited some of the same reactions shown in Fig. 4.6 but they reported respective molecular weights of 330,000 and 660,000 for the monomer and dimer forms. They also observed conversion of their 7S preparation to slow sedimenting species by the detergent, sodium octyl benzene sulfate, and by high concentrations of urea. Dissociation

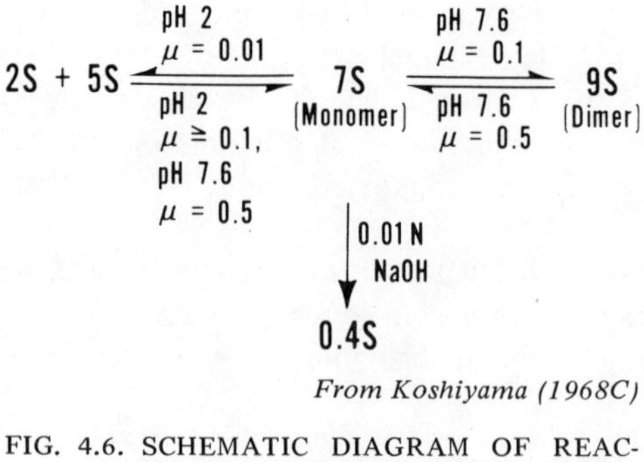

$$2S + 5S \xrightleftharpoons[\substack{pH\ 2 \\ \mu \geq 0.1, \\ pH\ 7.6 \\ \mu = 0.5}]{\substack{pH\ 2 \\ \mu = 0.01}} \underset{\text{(Monomer)}}{7S} \xrightleftharpoons[\substack{pH\ 7.6 \\ \mu = 0.5}]{\substack{pH\ 7.6 \\ \mu = 0.1}} \underset{\text{(Dimer)}}{9S}$$

$$\downarrow \substack{0.01\ N \\ NaOH}$$

$$0.4S$$

From Koshiyama (1968C)

FIG. 4.6. SCHEMATIC DIAGRAM OF REACTIONS OF 7S GLOBULIN ACCORDING TO KOSHIYAMA

of 7S globulin into subunits by treatment with urea or sodium dodecyl sulfate has been confirmed by molecular weight determinations. In 8 M urea the protein has a molecular weight of only 22,500 (Koshiyama 1970).

Disc electrophoresis of 7S globulin (Koshiyama preparation) in phenol-acetic acid-mercaptoethanol-urea yielded 9 major and 5 minor bands. In the absence of mercaptoethanol 6 major and 13 minor bands occurred; some of the minor bands were attributed to aggregates formed by thiol-disulfide interchange (Catsimpoolas et al. 1968B).

11S Globulin.—Optical rotatory dispersion and infrared measurements suggest that the secondary and tertiary structures of 11S protein are similar to the structures of the 7S globulin (Fukushima 1968). The 11S protein contains little, if any, a-helical structure but may consist of antiparallel β-structure and disordered regions. As noted in the 7S globulin, only about 40% of the peptide bond hydrogens in the 11S globulin exchanged with deuterium. Ultraviolet difference spectral studies indicated that the tyrosine and tryptophan residues of the 11S protein are buried in hydrophobic regions of the molecule (Fukushima 1968; Catsimpoolas et al. 1969A). Fukushima further demonstrated that denaturation of the protein by alcohols depends on hydrophobicity of the alcohols; n-butanol was a much more effective denaturant than ethanol or methanol. Native 11S protein is attacked very slowly by the proteinase of Aspergillus sojae but is readily hydrolyzed if the protein is denatured by alkaline treatment (pH 12.6). The 11S molecule, therefore, appears to be a compact structure stabilized by hydrophobic bonds.

The 12 N-terminal groups of 11S protein (Table 4.6) indicate that the molecule has a complex quaternary structure. Changes in the quaternary structure as a function of experimental conditions are shown schematically in Fig. 4.7. At 0.1 ionic strength the 11S protein is partially associated into faster sedimenting forms (reaction A); this reaction is reversed by increasing the ionic strength (Naismith 1955). Breakdown into subunits (reactions B and C) through the formation of a 7S intermediate (half molecules) occurs under a variety of conditions (Wolf and Briggs 1958). Dissociation under the mildest conditions (pH 7.6, 0.01 ionic strength) is reversible, but irreversibility was noted when the protein was dissociated at pH 8.6 (Eldridge and Wolf 1967). Urea in high concentrations, anionic detergents, and extremes of pH are effective dissociating agents. Because dissociation processes generally are irreversible, unfolding of the subunits also must occur. A more detailed discussion of conditions causing irreversible conformation changes is given elsewhere (Denaturation of Soybean Proteins). Although only a single step is shown for conversion from half molecules to unfolded subunits (Fig. 4.7), a multistep process may be involved. For example, in dissociation of 11S at pH 2.2 the slowest sedimenting fraction is 5S at 0.2 ionic strength, 4S at 0.1 ionic strength, and 2S at 0.01 ionic strength (Wolf et al. 1958). The 4S and 5S forms may be aggregates (dimers, trimers, etc.)

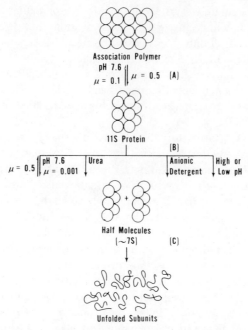

Association Polymer
pH 7.6
$\mu = 0.1$ $\mu = 0.5$ (A)

11S Protein
                              (B)
$\mu = 0.5$ | pH 7.6 | Urea   Anionic | High or
           | $\mu = 0.001$ Detergent | Low pH

+

Half Molecules
(~7S)   (C)

Unfolded Subunits

FIG. 4.7 SCHEMATIC DIAGRAM OF CHANGES
IN QUATERNARY STRUCTURE OF
11S PROTEIN

of the 2S form. The extent of disulfide crosslinking between polypeptide chains
is also unknown.

An interesting result of starch gel electrophoresis of 11S protein in 5-7 M
urea is the observation of basic protein bands (Shibasaki and Okubo 1966; Puski
and Melnychyn 1968). Vaintraub (1967) reported 18 bands on polyacrylamide
gel electrophoresis of 11S protein after treatment with 6 M urea at pH 6.76.
Several of the bands migrated slightly at pH 8.6. Chromatography of urea-dis-
sociated 11S protein on diethylaminoethyl-cellulose yielded 4 fractions, each
consisting of 2 subfractions. The first fraction was not adsorbed on the column
at pH 8.0 and contained one band that migrated toward the cathode plus several
others that moved only slightly toward the anode. These results also suggest that
some of the subunits of the 11S molecule are basic.

On analysis of the chromatographic fractions for N-terminal amino acids,
Vaintraub found that the first (basic) fraction contained predominantly N-terminal glycine. The other (acidic) fractions contained N-terminal glycine, isoleucine (leucine), and phenylalanine with clear evidence of redistribution of N-terminal residues as compared to the parent 11S molecule. Apparently then, there
are four kinds of subunits in the 11S protein based on differences in the

N-terminal analyses. Vaintraub attributed the 18 gel electrophoretic bands to isomerization of the 4 groups of subunits, but heterogeneity within each kind of subunit is a distinct possibility. Because the subunit mixture obtained after urea treatment of the 11S protein had an $s_{20,w}$ value of 2.56S in pH 6.9, ionic strength 0.1 buffer, urea dissociation of the protein is irreversible.

Independently, Okubo and Shibasaki (1967) treated 11S protein with 8 M urea followed by chromatography on diethylaminoethyl-cellulose and also separated 4 major fractions in good agreement with Vaintraub's results. Approximately 20% of the 11S protein was recovered as the basic subunits with isolectric points of pH 7–9, a sedimentation coefficient of 1.21S, molecular weight of about 36,000, and with glycine in the N-terminal position (Okubo et al. 1969). The acidic subunits contained leucine (isoleucine) and phenylalanine N-terminal groups.

Catsimpoolas (1969C) confirmed the existence of basic subunits in the 11S molecule by isoelectric focusing in 6 M urea and 0.2 M mercaptoethanol. Among the 6 fractions were: 3 acidic that had isoelectric points of pH 4.75, 5.15, and 5.40; while 3 were basic with isoelectric points at pH 8.00, 8.25, and 8.50. The isolated fractions agreed with the six bands after disc electrophoresis of the 11S protein in phenol-acetic acid-mercaptoethanol-urea.

Discovery of basic subunits in the 11S protein is an unexpected finding and its significance in terms of the quaternary structure of the protein is still unknown. Electron microscopy of the 11S molecule suggests that it is made up of 2 annular-hexagonal structures, each containing 6 subunits (Catsimpoolas 1969C). Alternation of acidic and basic subunits is proposed as an aid to stabilizing the molecule through ionic bonding. Saio et al. (1970) recently proposed an 11S structure consisting of two oval split-rings facing each other.

Heat will also disrupt the quaternary structure of 11S protein. Changes in 11S protein induced by heat are discussed later (Denaturation of Soybean Proteins).

15S Globulin.—Although this protein has not been isolated and characterized, there is evidence that it also has a quaternary structure. In studies on 11S protein where the 15S fraction occurred as a contaminant, dissociation into subunits apparently took place under conditions similar to those outlined in Fig. 4.7 (Wolf and Briggs 1958; Wolf et al. 1958).

Urease.—Unpurified soybean urease sediments as an 18S form in a pH 7 sucrose density gradient, but at pH 4.8 it is converted into a 13.3S form. Similar changes in jackbean urease are attributed to a splitting of the 18S unit into halves (Tanis and Naylor 1968).

Electrochemical Properties

Electrophoresis.—Use of this technique to determine the electrochemical properties of soybean proteins has been very limited. Reasons for this paucity include low solubility of the globulins in the isoelectric region (pH 4–5), lack of purified proteins until recently, and a greater interest in the use of gel electrophoresis as a tool for assessing homogeneity and for detecting subunits.

Moving Boundary Electrophoresis.—Briggs and Mann (1950) demonstrated heterogeneity of glycinin and other globulin preparations by moving boundary electrophoresis, but their resolution of components was poor. Likewise, Kondo *et al.* (1953) and Smith *et al.* (1955) observed only limited separations of proteins now known to be mixtures. The latter workers, for example, obtained 4–5 peaks on electrophoresis of acid-precipitated globulins; 1 peak accounted for nearly 90% of the total protein. However, this protein mixture separates into 4 peaks in the ultracentrifuge and the major component (11S protein) is less than 50% of the total protein (Wolf and Sly 1965). A synthetic mixture of 7S and 11S globulins did not separate electrophoretically at pH 7.6, thereby showing limitations of this technique (Fukushima 1968).

Moving boundary electrophoresis is, however, useful for determining isolectric points if solubility problems can be overcome. Isoelectric points for 2.8S globulin, hemagglutinin, lipoxygenase, and 7S globulin (Table 4.7) were determined by this method. Electrophoresis has been used to study interaction of phytate with soybean globulins, but results are difficult to interpret because the protein components and protein-phytate complexes did not resolve well (Smith and Rackis 1957). This system needs further study since the presumed protein-phytate complexes migrated slower than the uncomplexed protein, whereas one would predict the opposite behavior on the basis of the high negative charge of phytate at pH 7.6.

Gel Electrophoresis.—Shibasaki and Okubo (1966) observed 13–14 bands on starch gel electrophoresis of soybean proteins in tris-citrate buffer, pH 8.6, containing 7 M urea and 0.02 M mercaptoethanol. Crude 11S component (cold-insoluble fraction) formed 3 major bands plus 3 minor bands that migrated toward the cathode. These three minor fractions were subsequently isolated and shown to be basic polypeptide chains (Okubo *et al.* 1969). In the absence of urea, the crude 11S migrated as a major band plus 4–5 minor bands.

Puski and Melnychyn (1968) analyzed soybean globulins by starch gel electrophoresis using 5 M urea in tris-hydrochloride buffer, pH 8.7, or in 1 N acetic acid, pH 3.5. The globulins separated into 14 bands in alkaline gels and into 15 bands in acid gels. Purified 7S globulin (Roberts and Briggs preparation) migrated as 7 bands in alkaline and as 12 bands in acid gels, whereas purified 11S protein formed 18 bands in alkaline and 10 bands in acid gels. Puski and Melnychyn also observed the basic protein bands detected by Shibasaki and Okubo.

Vaintraub (1967) obtained 18 bands by polyacrylamide gel electrophoresis of 11S protein after treatment with 6 M urea and he isolated basic subunits as described earlier (Molecular Structure and Conformation—11S Globulin). Polyacrylamide gel electrophoresis with 8 M urea has also been used to study soybean whey proteins and trypsin inhibitors (Eldridge *et al.* 1966; Eldridge and Wolf 1969B).

Catsimpoolas *et al.* (1967) introduced polyacrylamide gel disc electrophoresis for characterizing soybean proteins. They purified the 11S protein until a single major band resulted, but on treatment with guanidine hydrochloride, the protein separated into more than 12 bands. The method was later applied to: soybean proteins during germination (Catsimpoolas *et al.* 1968B), soybean whey proteins (Catsimpoolas *et al.* 1969B), Kunitz trypsin inhibitor (Catsimpoolas *et al.* 1969C), hemagglutinin (Catsimpoolas and Meyer 1969), purification of 2S and 7S globulins (Catsimpoolas and Ekenstam 1969), and association-dissociation of 11S protein (Catsimpoolas *et al.* 1969A).

The disc electrophoretic technique was modified by using phenol-acetic acid-mercaptoethanol-urea as a dissociating solvent to detect subunits in proteins with quaternary structures (Catsimpoolas *et al.* 1968B). In this solvent, 11S component formed 12 bands while 7S globulin (Koshiyama preparation) separated into 14 bands. Changes in the proteins during germination were followed with this method.

Genetic variants of an unidentified protein (Larsen and Caldwell 1968) and of trypsin inhibitor (Singh *et al.* 1969) have been detected by polyacrylamide gel electrophoresis.

Immunoelectrophoresis.—Combinations of immunochemical techniques with gel electrophoresis are discussed later (Immunochemical Properties of Soybean Proteins).

Isoelectric Focusing.—Catsimpoolas and coworkers have extensively studied usefulness of this technique as a tool for isolating and characterizing soybean proteins. Isoelectric focusing of whey proteins yields a spectrum of proteins with isoelectric points ranging from pH 3.38 to 10. Isolation of the isoelectrically focused proteins and analysis by disc electrophoresis shows that many of the fractions are heterogeneous. Multiple forms of trypsin inhibitor and hemagglutinin reported by others were confirmed by isoelectric focusing (Catsimpoolas *et al.* 1969B). Kunitz trypsin inhibitor isolated from whey and commercial inhibitor samples by isoelectric focusing is pure by disc electrophoresis and immunoelectrophoresis (Catsimpoolas *et al.* 1969C). Four hemagglutinins isolated by isoelectric focusing of whey proteins had isoelectric points in the range of 5.85–6.20 (Table 4.7) (Catsimpoolas and Meyer 1969).

Isoelectric focusing of whey proteins also yielded lipoxygenase homogeneous by disc electrofocusing, disc electrophoresis, and immunoelectrophoresis (Catsimpoolas 1969B). Multiple forms of lipoxygenase as reported by Guss *et al.* (1967) and Christopher *et al.* (1970) were not observed.

The most significant finding by isoelectric focusing is the separation of 11S protein (referred to as glycinin) into acidic and basic subunits by Catsimpoolas (1969C) as described earlier (Molecular Structure and Conformation—11S Globulin).

Advantages of isoelectric focusing are the ability to test for homogeneity and to obtain isoelectric points with small amounts of protein. Isoelectric focusing in polyacrylamide gels requires only 0.2–0.4 mg of protein (Catsimpoolas 1968D).

The method is limited, however, by the small samples that can be purified and by the insolubility of some proteins at their isoelectric points. The latter limitation prevents fractionation of the major soybean globulins with their quaternary structures intact since solvents, such as 6 M urea, are required to dissolve the proteins.

Proteins with small differences in isoelectric points can be separated. For example, Kunitz and Bowman-Birk trypsin inhibitors whose isoelectric points differ by only 0.2 pH unit were separated by using narrow pH gradients (Catsimpoolas 1969D).

Titration Studies.—Few studies are available on hydrogen ion equilibria of soybean proteins. Wu and Scheraga (1962) measured the titration curve for Kunitz trypsin inhibitor and found the curve reversible over the entire pH range. Total number of ionizable groups including guanidyl groups was 62 per mole of inhibitor. Harry and Steiner (1969) published a titration curve for Bowman-Birk trypsin inhibitor but did not analyze their data in terms of numbers of various ionizable groups involved. The titration curve was reversible below neutrality, but the curves were no longer superimposable after titration to high pH.

Malik and Jindal (1968) titrated a preparation referred to as glycinin and believed to be a single protein. The method of isolation, however, indicates that their material consisted of acid-precipitated globulins, which are a mixture of at least four proteins (Wolf and Sly 1965) and which were probably denatured by the high pH used during extraction of the proteins from defatted meal. Their sample likely was also contaminated by phytate since isoelectric precipitation causes phytate to interact and precipitate with the proteins (Smith and Rackis 1957). Notwithstanding these limitations, the data should be a useful approximation of the ionic groups in soybean globulins.

DENATURATION OF SOYBEAN PROTEINS

Current information indicates that soybean proteins have compact structures as opposed to random coil-like conformations typical of milk caseins (McKenzie 1967). Denaturation studies of soybean proteins, therefore, provide several types of information: (a) the limits of the conditions under which the native structures are stable, (b) the nature of the conformation changes occurring during denaturation, and (c) information about the structures of the native protein molecules. For example, a study of proteins in urea or guanidine hydrochloride may reveal the presence of subunits. If subunits are found, the native protein has a quaternary structure and therefore a higher degree of complexity than a protein that contains only a single polypeptide chain.

For purposes of definition, denaturation is interpreted here as a major change from the native structure without alteration of the amino acid sequence (Tanford 1968). Limitations of this definition are discussed by Tanford.

Heat Denaturation

Denaturation of soybean proteins by moist heat is well known and has long been used to eliminate antinutritional factors (believed to be proteins) in soybean meals and flours used in feeds and foods (Chap. 9). Even though heat is a common physical treatment given to most foods either during processing or cooking, surprisingly little is known about the reactions that soybean proteins undergo when they are heated. Early studies dealt with the effects of heat and moisture on defatted meal and were primarily concerned with measurements of the amounts of soluble protein remaining after heating (Beckel *et al.* 1942; Belter and Smith 1952). Changes in extractability of the different protein components as a function of heating time are reported by Shibasaki *et al.* (1969). Water, buffer, buffer-mercaptoethanol, and buffer-mercaptoethanol-urea were used as extraction solvents. The various extracts were analyzed by starch gel electrophoresis.

When soybean meal with an equal weight of water is autoclaved above 100° C, the water-soluble protein decreases to a minimum and then increases again as heating is continued (Fukushima 1959A). Likewise, heating at 100° C with greater amounts of water solubilizes large amounts of protein even after heating for 30 min. Fukushima (1959B), therefore, developed a proteolytic digestion method for measuring extent of denaturation to overcome difficulties inherent in solubility methods.

Mann and Briggs (1950) observed that heating water extracts of defatted meal precipitated protein which appeared to be primarily nonglobulin (whey proteins) as determined by electrophoresis. Heating the whey proteins in buffer (pH 7.6, 0.1 ionic strength) prevented precipitation, and, instead, a single, nearly symmetrical peak formed. This electrophoretic peak apparently is an aggregate formed by interaction between the different components. When cold-precipitable protein (now known to be primarily 11S protein) was added to the whey proteins, it also was incorporated into the aggregate by heating.

Watanabe and Nakayama (1962) heated water-extractable soybean proteins at pH 7.0 and confirmed formation of aggregates. After the proteins were heated for 10 min at 100° C, only a 5S and a 2S fraction were detectable by ultracentrifugation; after 30 min of heating the 5S fraction disappeared leaving only the 2S fraction. Saio *et al.* (1968B) presented additional evidence for heat aggregation of soybean proteins. Gel filtration of water-extractable, acid-precipitated, and calcium-precipitated proteins, before and after heating, showed that the slowly eluting (low molecular weight) fractions were converted to the fraction that eluted first; i.e., the high-molecular-weight materials.

Circle *et al.* (1964) studied the effect of heat on aqueous dispersions of a commercial preparation of soybean globulins (sodium soy proteinate form).[5]

---

[5] Solubilities and ultracentrifugal compositions of laboratory and commercial preparations of soybean globulins are reported by Nash and Wolf (1967).

Gels formed when protein dispersions of 8% or higher were heated for 10–30 min at 70° C. In the concentration range of 8–12% the gels broke down when heated at 125° C. This behavior may be related to resolubilization of protein observed by Fukushima (1959A) when meal is heated at 100° C for prolonged times with excess water or at temperatures about 100° C.

Circle and co-workers also examined the influence of additives on gelation; included were salts, lipids, starch, gums, and reducing agents. Sodium sulfite and cysteine markedly reduced viscosity of unheated 10% dispersions and inhibited gelation. Evidently, cystine crosslinks contribute to the gel structure, but other interactions undoubtedly also play a part.

Catsimpoolas and Meyer (1970) propose that heating soybean globulins at concentrations greater than 8% converts them to a progel state that gels on cooling. Gel-to-progel conversion is reversible on heating. Excessive heat or addition of chemicals, such as disulfide-cleaving agents, forms a "metasol" state that does not gel. Effects of pH and ionic strength on viscosities of the progels and gels were examined.

Heat gelation of soybean globulins, precipitated by hydrochloric acid or calcium chloride, is reported by Aoki (1965A,B,C). Methods are described for evaluation of the gels and effects of time, temperature, protein concentration, and salt concentration were examined. Most studies were conducted with 20% protein solutions. Sodium bisulfite markedly decreased gel strength in the range of 3–11mM. Effects of low concentrations of urea and guanidine hydrochloride on gelation were complex, but at high concentrations they tended to weaken the gels (Aoki and Sakurai 1968).

Proteins isolated from a defatted meal with a nitrogen solubility index (NSI) of 63 gave stronger gels than proteins from meals with NSI values of 27, 42, or 83. If the extracts from the meals of different NSI values were heated to 95° C for 5 min before precipitating the proteins with acid, gels subsequently prepared (at 90° C for 50 min) were similar in strength and other properties. Although solubilities of the thermal gels in phosphate buffer, 6 M urea, or 0.01 M mercaptoethanol were low, combination of the three solvents dissolved 90% or more of the gels (Aoki and Sakurai 1969). This solubility behavior indicates that gel structure depends upon disulfide, plus hydrogen, and/or hydrophobic bonds.

Only limited data are available on heat denaturation of purified soybean proteins. Kunitz (1948) made a classical study of the kinetics and thermodynamics of heat denaturation of crystalline trypsin inhibitor. This protein, consisting of a single polypeptide chain crosslinked by two disulfide bonds (Wu and Scheraga 1962), is inactivated when heated in solution but regains its activity reversibly when the solution is cooled. Side reactions must occur, however, when soybean meal is heated since there are no reports of reversibility of inhibitor activity in heated meal.

Liener (1958) studied kinetics of heat inactivation of soybean hemagglutinin from pH 4 to 9.5. Maximum stability occurred at pH 6-7. No evidence for reversibility of denaturation was reported.

Catsimpoolas *et al.* (1969A) heated 11S protein solutions and followed the reaction by turbidity measurements. Turbidity increased rapidly above 70° C and protein precipitated at 90° C. Disc electrophoresis showed that dissociation into subunits occurred at 90° C but that undissociated 11S also remained; immunodiffusion likewise indicated incomplete denaturation of the protein.

Wolf and Tamura (1969) heated 11S protein at 100° C and followed changes in ultracentrifugal composition as a function of heating time (Fig. 4.8). In 5 min of heating the 11S component disappeared and was converted into a soluble aggregate and a 3-4S fraction. However, on continued heating the aggregate precipitated and only the 3-4S fraction was left in solution. Heating in 0.01 *M* mercaptoethanol hastened formation of the precipitate and no soluble aggregate was detectable as an intermediate form. Since precipitation still occurred in 0.5 *M* mercaptoethanol, sulfhydryl-disulfide interchange does not appear to contribute to the precipitation reaction. When the protein was heated in 0.01 *M*

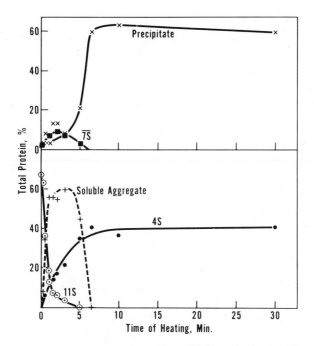

From Wolf and Tamura (1969)

FIG. 4.8. CHANGES IN ULTRACENTRIFUGAL COM-
POSITION OF 11S PROTEIN AS A FUNC-
TION OF TIME AT 100°C

N-ethylmaleimide, the 11S protein again disappeared, but no precipitation of protein took place. Instead, a soluble aggregate of 58–67S and a 3–4S fraction formed. It was concluded that heating disrupts the quaternary structure of the 11S protein and separates the subunits into a soluble and an insoluble fraction by the following three steps:

$$11S \xrightarrow{\quad (a) \quad} \text{A-subunits} + \text{[B-subunits]}$$
$$\downarrow (b)$$
$$\text{Soluble aggregates}$$
$$\downarrow (c)$$
$$\text{Insoluble aggregates}$$

In reaction a, A-subunits represent the 3–4S fraction, which remains soluble, while B-subunits represent that portion of the 11S molecule which is converted into aggregates through reactions b and c. Reaction b apparently is very rapid since B-subunits were not detected in an unaggregated state. Reaction c is catalyzed by sulfhydryl groups but blocked by N-ethylmaleimide. It was proposed that reaction c involves hydrophobic interactions which are promoted by cleaving the disulfide bonds in the molecule.

Additional reactions probably occur when the 11S protein is heated in the presence of other soybean proteins since Watanabe and Nakayama (1962) did not observe that the 3–4S fraction was formed when water-extractable proteins were heated.

Catsimpoolas et al. (1970) followed changes in turbidity of 11S protein solutions as a function of heating time under various conditions. Rate and extent of aggregation were increased by low ionic strength and by mercaptoethanol. Maximum aggregation occurred between pH 4.0 and 6.0.

Denaturation by Extremes of pH

The overall effect of high or low pH on the major proteins (7S and 11S globulins) appears to be dissociation of the molecules into subunits possibly by electrostatic repulsions between the high positive or negative charges on the proteins at extreme pH values. The subunits are often irreversibly altered by conditions necessary for complete dissociation.

Acid pH.—Soybean globulins prepared by acid precipitation were progressively converted from a mixture of 2S, 7S, 11S, and 15S fractions observed at alkaline pH values to 2–3S and 7S fractions as the pH was lowered from 3.8 to 2.0 at ionic strength, 0.06 (Rackis et al. 1957). The 11S fraction was observed at pH 3.0 when the ionic strength was 0.15, but at higher ionic strengths the 11S fraction was not stable and appeared to aggregate. A complex series of association-dissociation reactions influenced by pH and ionic strength is suggested by this behavior in acid solution.

Similar conclusions were reached by Kretovich *et al.* (1958) in studies on glycinin (prepared according to Osborne and Campbell 1898) and a glycinin subfraction in acid solutions. At pH 2.2 the glycinin subfraction, which consisted of 7S and 11S components in 10% sodium chloride, sedimented at a rate of 2.36S that corresponded to a molecular weight of 19,800. Dissociation of the 7S and 11S globulins into subunits in acid is thus established. However, the concept of a system of identical subunits in reversible states of aggregation as proposed by Kretovich and co-workers is now known to be incorrect on the basis of differences in physical, chemical, and immunochemical properties between the various globulins and their subunits (Wolf *et al.* 1962; Okubo and Shibasaki 1967; Puski and Melnychyn 1968; Fukushima 1968; Catsimpoolas and Ekenstam 1969).

The 11S protein undergoes changes in sedimentation rate and optical rotation as pH and ionic strength are lowered below pH 4. Dissociation into 2–3S subunits occurs through a 7S intermediate. Only a 2S species is detectable at pH 2.2, 0.01 ionic strength, and levorotation (sodium D line) is $81-82°$ as compared to $43-47°$ in the undissociated state. The tendency of the 11S to dissociate into subunits is counteracted by increasing ionic strength, but conformation changes still occur if the pH is low enough. For example, at pH 2.2, ionic strength 1.0, no 11S component was detectable and the protein precipitated on standing (Wolf *et al.* 1958). Low pH treatment of 11S protein is irreversible as measured by ultracentrifugation (Wolf and Briggs 1958). Catsimpoolas *et al.* (1969A) observed exposure of tyrosine and tryptophan residues in the 11S molecule at pH 2.0 and found the protein to be irreversibly modified as detected by disc electrophoresis.

A 7S globulin isolated by Koshiyama (1968C) sedimented as 2 peaks of 1.92S and 5.47S in 0.01 N hydrochloric acid (pH 2). On dialysis of the acid solution to pH 7.6 the protein exhibited the 7S$\rightleftharpoons$9S monomer-dimer reaction with change in ionic strength that is typical of the native protein (Fig. 4.6). On prolonged acid treatment part of the protein precipitated when it was brought back to pH 7.6, but the soluble portion still underwent the monomer-dimer reaction. In 0.1 N hydrochloric acid, conversion to the slowly sedimenting form was less complete and the protein was irreversibly modified to a greater extent than by 0.01 N acid. Addition of salts to acid solutions of 7S globulin inhibited the conformation changes observed at low ionic strength.

Alkaline pH.—Solutions of soybean globulins in sodium hydroxide (pH 12) increase in viscosity and form gels if protein concentration is 14.5% or higher (Kelley and Pressey 1966). The globulin mixture dissolved in alkali sediments with an $s_{20,w}$ value of about 3S. Conversion of the globulins to the 3S form is nearly complete in 15 min at pH 12. If the alkaline solution is dialyzed against phosphate buffer (pH 7.6, ionic strength 0.5, 0.01 M mercaptoethanol), only 3S and 7S peaks are observed. When the alkali treatment is stopped by dilution and pH is adjusted to 4.5 to precipitate the protein, large decreases in protein solubility occur. Alkali thus causes irreversible changes in the molecules.

Partial dissociation of 11S protein into subunits occurs even near neutrality if ionic strength is low and divalent cations, such as calcium, are absent. Thus dialyzing the 11S protein at pH 7.6, ionic strength 0.01, dissociates it into 7S and 2-3S units (Wolf and Briggs 1958; Wolf and Tamura 1969). Dissociation is increased by raising the pH from 7.6 to 9.0, but reassociation into 11S is low when the sample is brought back to pH 7.6, ionic strength 0.5. Disc electrophoresis also indicates that alkali-induced (pH 11) changes in the 11S molecule are irreversible (Catsimpoolas et al. 1969A).

Optical rotatory dispersion measurements suggest that alkali disrupts the internal structure of the 11S molecule. Such disorganization of the structure makes the 11S molecule more susceptible to attack by a protease from Aspergillus sojae (Fukushima 1968).

When 7S globulin is dissolved in 0.01 N sodium hydroxide (pH 12) it is converted into a very slow sedimenting form (Fig. 4.6). Adjusting the alkaline solution to pH 7.6, ionic strength 0.5 buffer did not restore the 7S form (Koshiyama 1968C). Evidently dissociation of the protein into subunits under these conditions is also accompanied by irreversible conformational changes as noted in studies on the globulins (Kelley and Pressey 1966).

Denaturation by Organic Solvents

Smith et al. (1951) studied the effects of time, temperature, and concentration of methanol, ethanol, isopropanol, and acetone on extractability of proteins in defatted meal. Aqueous solutions of the organic solvents were more effective than water or the pure solvents in denaturing the proteins. Denaturation of soybean proteins by organic solvents is complete in about 5 min. Fukushima (1969A) measured denaturation of the proteins in defatted meal by a variety of organic solvents. Water-immiscible solvents were weak denaturants but water-miscible solvents in combination with water were stronger denaturants than the pure solvents. Solvents completely miscible with water, such as the lower alcohols, were most effective as denaturants at characteristic water:solvent ratios. The ability of the lower alcohols to denature the proteins increased as the hydrocarbon chain length increased; the order of effectiveness was: methanol< ethanol<propanol<butanol.

Electrophoretic studies by Mann and Briggs (1950) indicated that the globulin fraction is the group of proteins that is most markedly affected by alcohols. Treatment of the globulins with ethanol showed that maximum denaturation occurred at 60% alcohol and that the 7S fraction was the most sensitive while the 2S fraction was most stable to ethanol (Roberts and Briggs 1963). With isopropanol, maximum denaturation of the globulins occurs with 40% alcohol (Wolf et al. 1964). Again, the 7S fraction was most readily denatured (insolubilized) while the 2S fraction was comparatively stable as demonstrated by ultracentrifugation and hydroxylapatite chromatography.

Fukushima (1968) treated 7S and 11S globulins with eight different alcohols and found that denaturation increased as the hydrocarbon chain length of the alcohols increased as noted in studies on meal (Fukushima 1969A).

Fukushima (1969A) proposed that the major proteins have structures with hydrophobic groups, buried in the interior, which help stabilize the molecules to heat in water. Alcohols presumably are able to penetrate to the interior and can disrupt the hydrophobic bonding while water breaks hydrogen bonds in more polar regions near the surface. The combination of water and alcohol is, therefore, more effective than either solvent alone. This conclusion is supported by the fact that soybean meal proteins denatured with aqueous ethanol are more completely hydrolyzed by proteolytic enzymes than are proteins denatured by moist heat (Fukushima 1969B).

Denaturation by Detergents

Data on denaturation of soybean proteins by detergents are scarce. Wolf and Briggs (1958) measured the binding of sodium octyl benzene sulfonate by 11S protein by equilibrium dialysis and analyzed the detergent-protein complexes in the ultracentrifuge. Binding of detergent caused dissociation of the 11S species into a 3S form through a 7S intermediate. Dissociation was complete when approximately 375 moles of detergent were bound per mole of protein. Dissociation of 15S contaminant in the 11S sample also occurred. Removal of the detergent by dialysis did not reform the 11S component; irreversible conformation changes in the subunits caused aggregates to form instead.

Detergents have also been used to precipitate proteins from soybean whey (Smith *et al.* 1962). Enzymes normally associated with whey proteins were inactive in the detergent-protein precipitates. Soybean hemagglutinin, a whey protein, is inactivated by sodium decylbenzene sulfonate (Liener 1958).

Effects of Urea and Guanidine Hydrochloride

A major effect of urea on soybean globulins is dissociation of the high-molecular-weight proteins into subunits. Treatment of 11S protein with 1.5 and 3.0 M urea caused partial dissociation into 7S and 3S forms (Wolf and Briggs 1958). Kelley and Pressey (1966) confirmed these results in studies on the globulin fraction and found that the reactions were reversible on removing the urea. In 6 M urea the globulin fraction was largely converted into 1-2S material plus a small fraction of 4S, but only partial reversal of the protein to its original sedimentation distribution occurred when the urea was removed by dialysis.

Aoki and Sakurai (1968) studied the effects of urea and guanidine hydrochloride denaturation on thermal gelation of soybean globulins prepared by acid or calcium chloride precipitation. High concentrations of the two denaturants weakened the gels.

Starch-gel electrophoresis with 5-7 M urea separates soybean globulins and globulin fractions into a large number of bands presumably as a result of dissociating the 7S and 11S molecules into subunits (Shibasaki and Okubo 1966;

Puski and Melnychyn 1968). Vaintraub (1967) likewise observed a large number of protein bands on polyacrylamide gel electrophoresis of 11S protein after treatment with urea.

Optical rotatory dispersion and ultraviolet difference spectral studies on the 7S and 11S indicate that high concentrations of urea disrupt native structures. In the 7S globulin, tyrosine groups appear buried in the interior of the molecule, whereas in the 11S globulin both tyrosine and tryptophan residues are unexposed to solvent until they are denatured (Fukushima 1968). Catsimpoolas *et al.* (1969A) confirmed these effects of urea on the 11S protein and found that treatment of the protein with 6 M urea is irreversible as measured by disc electrophoresis and immunochemical techniques.

In contrast to the effects of urea on the 7S and 11S globulins, soybean trypsin inhibitor changes little in conformation in 9 M urea unless high temperatures or alkaline pH are included in the treatment (Edelhoch and Steiner 1963). If the two disulfide bonds are reduced, the inhibitor unfolds in 8 M urea as measured by viscosity and gel electrophoresis (Eldridge and Wolf 1969B).

High concentrations of urea are required to inactivate soybean hemagglutinin while guanidine hydrochloride is a much more effective denaturant for this protein (Liener 1958).

IMMUNOCHEMICAL PROPERTIES OF SOYBEAN PROTEINS

Introducing immunodiffusion and immunelectrophoresis to characterize soybean proteins, Catsimpoolas and Meyer (1968) have published extensively on applications of these techniques. Their initial study demonstrated at least 5 components in water-extractable proteins and 3 components in the isoelectrically precipitated globulins. Purified 11S protein formed only one precipitin band when diffused against antibodies for the water-extractable proteins. The 11S protein retained its immunochemical properties on heating up to 80° C; at higher temperatures the protein precipitated. Single diffusion measurements indicated that quantitative estimates of 11S protein are possible by this method.

Single immunodiffusion techniques for estimation of Kunitz trypsin inhibitor are also described (Catsimpoolas *et al.* 1969D; Catsimpoolas and Leuthner 1969B). This approach offers the possibility that the amounts of various proteins present in soybean meal can be determined quantitatively.

Immunochemical techniques were used to demonstrate homogeneity of soybean lipoxygenase (Catsimpoolas 1969B) and hemagglutinins (Catsimpoolas and Meyer 1969). Although four different hemagglutinins were isolated by isolectric focusing, they were immunochemically identical.

Protein bodies contained six antigenic components by disc immunoelectrophoresis. One of the components was 11S protein while another was identical to the 7S globulin isolated by Koshiyama (1965). On germination, the 11S component was detectable up to the 16th day, whereas the 7S component disappeared after the 9th day (Catsimpoolas *et al.* 1968A). Immunochemical techniques also proved useful in following fractionation of soybean globulins (Catsimpoolas and Ekenstam 1969) and to study dissociation of 11S protein into subunits (Catsimpoolas *et al.* 1969A).

BIBLIOGRAPHY

ANDERSON, R. L., and WOLF, W. J. 1967. Fractional extraction of soybean meal proteins with salt solutions at pH 4.5. Abstr. *A 13*, 153rd Meeting Am. Chem. Soc.

AOKI, H. 1965A. Studies on the gelation of soybean protein. I. On the methods determining the physical properties (chewy properties) of gels. J. Agr. Chem. Soc. Japan *39* 262–269.

AOKI, H. 1965B. Studies on the gelation of soybean protein. II. On the fundamental factors affecting the gelation of soybean protein. J. Agr. Chem. Soc. Japan *39*, 270–276.

AOKI, H. 1965C. Studies on the gelation of soybean protein. III. On the effect of alkaline salts. J. Agr. Chem. Soc. Japan *39*, 277–285.

AOKI, H., and SAKURAI, M. 1968. Studies on the gelation of soybean protein. IV. On the effects of reducing agents, oxidizing agents and protein denaturants. J. Agr. Chem. Soc. Japan *42*, 544–552.

AOKI, H., and SAKURAI, M. 1969. Studies on the gelation of soybean protein. V. On the effects of heat denaturation. J. Agr. Chem. Soc. Japan *43*, 448–456.

BAI, N. K., and PIN, P. 1964. Vegetable phosphoproteins. Compt. Rend. Soc. Biol. (French) *158*, 2054–2056.

BAIN, W. M., CIRCLE, S. J., and OLSON, R. A. 1961. Isolated soy proteins for paper coating from a manufacturer's viewpoint. *In* Synthetic and Protein Adhesives for Paper Coating, L. H. Silvernail and W. M. Bain (Editors). Tappi Monograph Series No. 22, 206–241.

BECKEL, A. C., BULL, W. C., and HOPPER, T. H. 1942. Heat denaturation of protein in soybean meal. Ind. Eng. Chem. *34*, 973–976.

BELTER, P. A., and SMITH, A. K. 1952. Protein denaturation in soybean meal during processing. J. Am. Oil Chemists' Soc. *29*, 170–174.

BILS, R. F., and HOWELL, R. W. 1963. Biochemical and cytological changes in developing soybean cotyledons. Crop Sci. *3*, 304–308.

BIRK, Y., GERTLER, A., and KHALEF, S. 1963. A pure trypsin inhibitor from soya beans. Biochem. J. *87*, 281–284.

BOWMAN, D. E. 1946. Differentiation of soybean antitryptic factors. Proc. Soc. Exptl. Biol. Med. *63*, 547–550.

BRIGGS, D. R., and MANN, R. L. 1950. An electrophoretic analysis of soybean protein. Cereal Chem. *27*, 243–257.

BRIGGS, D. R., and WOLF, W. J. 1957. Studies on the cold-insoluble fraction of the water-extractable soybean proteins. I. Polymerization of the 11S component through reactions of sulfhydryl groups to form disulfide bonds. Arch. Biochem. Biophys. *72*, 127–144.

BROWN, J. R., LERMAN, N., and BOHAK, Z. 1966. The amino acid sequences around the disulfide bonds of soybean trypsin inhibitor. Biochem. Biophys. Res. Commun. *23*, 561–565.

CATSIMPOOLAS, N. 1968. Micro isoelectric focusing in polyacrylamide gel columns. Anal. Biochem. *26*, 480–482.

CATSIMPOOLAS, N. 1969A. A note on the proposal of an immunochemical system of reference and nomenclature for the major soybean globulins. Cereal Chem. *46*, 369–372.

CATSIMPOOLAS, N. 1969B. Isolation of soybean lipoxidase by isoelectric focusing. Arch. Biochem. Biophys. *131*, 185–190.

CATSIMPOOLAS, N. 1969C. Isolation of glycinin subunits by isoelectric focusing in urea-mercaptoethanol. Federation European Biochem. Soc. Letters *4*, 259–261.

CATSIMPOOLAS, N. 1969D. Isoelectric focusing in narrow pH gradients of Kunitz and Bowman-Birk soybean trypsin inhibitors. Separ. Sci. 4, 483–492.

CATSIMPOOLAS, N., CAMPBELL, T. G., and MEYER, E. W. 1968A. Immunochemical study of changes in reserve proteins of germinating soybean seeds. Plant Physiol. 43, 799–805.

CATSIMPOOLAS, N., CAMPBELL, T. G., and MEYER, E. W. 1969A. Association-dissociation phenomena in glycinin. Arch. Biochem. Biophys. 131, 577–586.

CATSIMPOOLAS, N., and EKENSTAM, C. 1969. Isolation of alpha, beta, and gamma conglycinins. Arch. Biochem. Biophys. 129, 490–497.

CATSIMPOOLAS, N., EKENSTAM, C., and MEYER, E. W. 1969B. Separation of soybean whey proteins by isoelectric focusing. Cereal Chem. 46, 357–369.

CATSIMPOOLAS, N., EKENSTAM, C., and MEYER, E. W. 1969C. Isolation of the pI 4.5 soybean trypsin inhibitor by isoelectric focusing. Biochim. Biophys. Acta 175, 76–81.

CATSIMPOOLAS, N., EKENSTAM, C., ROGERS, D. A., and MEYER, E. W. 1968B. Protein subunits in dormant and germinating soybean seeds. Biochim. Biophys. Acta 168, 122–131.

CATSIMPOOLAS, N., FUNK, S. K., and MEYER, E. W. 1970. Thermal aggregation of glycinin subunits. Cereal Chem. 47, 331–344.

CATSIMPOOLAS, N., and LEUTHNER, E. 1969A. The major pH 4.5 soluble proteins of soybean cotyledons. I. Separation by gel filtration, disc electrofocusing and immunoelectrophoresis. Biochim. Biophys. Acta 181, 404–409.

CATSIMPOOLAS, N., and LEUTHNER, E. 1969B. Immunochemical methods for detection and quantitation of Kunitz soybean trypsin inhibitor. Anal. Biochem. 31, 437–447.

CATSIMPOOLAS, N., and MEYER, E. W. 1968. Immunochemical properties of the 11S component of soybean proteins. Arch. Biochem. Biophys. 125, 742–750.

CATSIMPOOLAS, N., and MEYER, E. W. 1969. Isolation of soybean hemagglutinin and demonstration of multiple forms by isoelectric focusing. Arch. Biochem. Biophys. 132, 279–285.

CATSIMPOOLAS, N., and MEYER, E. W. 1970. Gelation phenomena of soybean globulins. I. Protein-protein interactions. Cereal Chem. 47, 559–570.

CATSIMPOOLAS, N., ROGERS, D. A., CIRCLE, S. J., and MEYER, E. W. 1967. Purification and structural studies of the 11S component of soybean proteins. Cereal Chem. 44, 631–637.

CATSIMPOOLAS, N., ROGERS, D. A., and MEYER, E. W. 1969D. Immunochemical and disc electrophoresis study of soybean trypsin inhibitor SBTIA-2. Cereal Chem. 46, 136–144.

CHRISTOPHER, J., PISTORIUS, E., and AXELROD, B. 1970. Isolation of an isozyme of soybean lipoxygenase. Biochim. Biophys. Acta 198, 12–19.

CIRCLE, S. J. 1950. Proteins and other nitrogenous constituents. In Soybeans and Soybean Products, K. S. Markley (Editor). John Wiley & Sons, New York.

CIRCLE, S. J., MEYER, E. W., and WHITNEY, R. W. 1964. Rheology of soy protein dispersions. Effect of heat and other factors on gelation. Cereal Chem. 41, 157–172.

DANIELSSON, C. E. 1949. Seed globulins of the Gramineae and Leguminosae. Biochem. J. 44, 387–400.

DIBELLA, F. P., and LIENER, I. E. 1969. Soybean trypsin inhibitor. Cleavage and identification of a disulfide bridge not essential for activity. J. Biol. Chem. 244, 2824–2829.

DICARLO, F. J., SCHULTZ, A. S., and KENT, A. M. 1955. Soybean nucleic acid. Arch. Biochem. Biophys. 55, 253–256.

EDELHOCH, H., and STEINER, R. F. 1963. Structural transitions of soybean trypsin inhibitor. II. The denatured state in urea. J. Biol. Chem. 238, 931–938.

ELDRIDGE, A. C., ANDERSON, R. L., and WOLF, W. J. 1966. Polyacrylamide-gel electrophoresis of soybean whey proteins and trypsin inhibitors. Arch. Biochem. Biophys. 115, 495–504.

ELDRIDGE, A. C., and WOLF, W. J. 1967. Purification of the 11S component of soybean protein. Cereal Chem. 44, 645–652.

ELDRIDGE, A. C., and WOLF, W. J. 1969A. Crystalline saponins from soybean protein. Cereal Chem. *46*, 344–349.

ELDRIDGE, A. C., and WOLF, W. J. 1969B. Polyacrylamide-gel electrophoresis of reduced and alkylated soybean trypsin inhibitors. Cereal Chem. *46*, 470–478.

FRATTALI, V. 1969. Soybean inhibitors. III. Properties of a low molecular weight soybean proteinase inhibitor. J. Biol. Chem. *244*, 274–280.

FRIDMAN, C., LIS, H., SHARON, N., and KATCHALSKI, E. 1968. Isolation and characterization of soybean cytochrome c. Arch. Biochem. Biophys. *126*, 299–304.

FUKUSHIMA, D. 1959A. Studies on soybean proteins. I. Water dispersibility of protein of defatted soybean flour as a criterion for degree of denaturation. Bull. Agr. Chem. Soc. Japan. *23*, 7–14.

FUKUSHIMA, D. 1959B. Studies on soybean proteins. II. A new method for quantitative determination of the degree of denaturation of protein in soybean flour. Bull. Agr. Chem. Soc. Japan. *23*, 15–21.

FUKUSHIMA, D. 1968. Internal structure of 7S and 11S globulin molecules in soybean proteins. Cereal Chem. *45*, 203–224.

FUKUSHIMA, D. 1969A. Denaturation of soybean proteins by organic solvents. Cereal Chem. *46*, 156–163.

FUKUSHIMA, D. 1969B. Enzymatic hydrolysis of alcohol-denatured soybean proteins. Cereal Chem. *46*, 405–418.

GERTLER, A., and BIRK, Y. 1965. Purification and characterization of a β-amylase from soya beans. Biochem. J. *95*, 621–627.

GUSS, P. L., RICHARDSON, T., and STAHMANN, M. A. 1967. The oxidation-reduction enzymes of wheat. III. Isoenzymes of lipoxidase in wheat fractions and soybean. Cereal Chem. *44*, 607–610.

HARRY, J. B., and STEINER, R. F. 1969. Characterization of the self-association of a soybean proteinase inhibitor by membrane osmometry. Biochemistry *8*, 5060–5064.

HASEGAWA, K., KUSANO, T., and MITSUDA, H. 1963. Fractionation of soybean proteins by gel filtration. Agr. Biol. Chem. (Tokyo) *27*, 878–880.

IKEDA, K., HAMAGUCHI, K., YAMAMOTO, M., and IKENAKA, T. 1968. Circular dichroism and optical rotatory dispersion of trypsin inhibitors. J. Biochem. (Tokyo) *63*, 521–531.

IKENAKA, T., SHIMADA, K., and MATSUSHIMA, Y. 1963. The N-terminal amino acid sequence of soybean trypsin inhibitor. J. Biochem, (Tokyo) *54*, 193–195.

JIRGENSONS, B., KAWABATA, M., and CAPETILLO, S. 1969. Circular dichroism of soybean trypsin inhibitor and its derivatives. Makromol. Chem. *125*, 126–135.

KAKADE, M. L., SIMONS, N. R., and LIENER, I. E. 1970. The molecular weight of the Bowman-Birk soybean protease inhibitor. Biochim. Biophys. Acta *200*, 168–169.

KELLEY, J. J., and PRESSEY, R. 1966. Studies with soybean protein and fiber formation. Cereal Chem. *43*, 195–206.

KOMODA, M., MATSUSHITA, S., and HARADA, I. 1968. Intracellular distribution of tocopherol in soybean cotyledons. Cereal Chem. *45*, 581–588.

KONDO, K., MORI, S., and KAJIMA, M. 1953. Studies on proteins (56). On the components of soybean protein. I. Kyoto Univ. Res. Inst. Food Sci. Bull. *11*, 1–23.

KOSHIYAMA, I. 1965. Purification of the 7S component of soybean proteins. Agr. Biol. Chem. (Tokyo) *29*, 885–887.

KOSHIYAMA, I. 1968A. Chromatographic and sedimentation behavior of a purified 7S protein in soybean globulins. Cereal Chem. *45*, 405–412.

KOSHIYAMA, I. 1968B. Chemical and physical properties of a 7S protein in soybean globulins. Cereal Chem. *45*, 394–404.

KOSHIYAMA, I. 1968C. Factors influencing conformation changes in a 7S protein of soybean globulins by ultracentrifugal investigations. Agr. Biol. Chem. (Tokyo) *32*, 879–887.

KOSHIYAMA, I. 1969A. Isolation of a glycopeptide from a 7S protein in soybean globulins. Arch. Biochem. Biophys. *130*, 370–373.

KOSHIYAMA, I. 1969B. Distribution of the 7S proteins in soybean globulins by gel filtration with Sephadex G-200. Agr. Biol. Chem. (Tokyo) *33*, 281–284.

KOSHIYAMA, I. 1970. Dissociation into subunits of a 7S protein in soybean globulins with urea and sodium dodecyl sulfate. Agr. Biol. Chem. (Tokyo) *34*, 1815–1820.

KOSHIYAMA, I., and IGUCHI, N. 1965. Studies on soybean protein. I. A ribonucleoprotein and ribonucleic acids in soybean casein fraction. Agr. Biol. Chem. (Tokyo) *29*, 144–150.

KRETOVICH, V. L., and SMIRNOVA, T. I. 1957. Oxidation-reduction conditions as a factor of enzymatic activity of plant proteins. Biokhimiya *22*, 102–110; Biochemistry (USSR) (English Transl.) *22*, 96–103.

KRETOVICH, V. L., SMIRNOVA, T. I., and FRENKEL, S. Ya. 1956. Investigation of the reserve proteins of the soybean by means of the ultracentrifuge. Biokhimiya *21*, 842–847; Biochemistry (USSR) (English Transl.) *21*, 864–869.

KRETOVICH, V. L., SMIRNOVA, T. I., and FRENKEL, S. Ya. 1958. The submolecular structure of glycinin and the conditions for its reversible association. Biokhimiya *23*, 547–557; Biochemistry (USSR) (English Transl.) *23*, 513–522.

KUNITZ, M. 1947. Crystalline soybean trypsin inhibitor. II. General properties. J. Gen. Physiol. *30*, 291–310.

KUNITZ, M. 1948. The kinetics and thermodynamics of reversible denaturation of crystalline soybean trypsin inhibitor. J. Gen. Physiol. *32*, 241–263.

LARSEN, A. L., and CALDWELL, B. E. 1968. Inheritance of certain proteins in soybean seed. Crop Sci. *8*, 474–476.

LIENER, I. E. 1953. Soyin, a toxic protein from the soybean. I. Inhibition of rat growth. J. Nutr. *49*, 527–539.

LIENER, I. E. 1958. Inactivation studies on the soybean hemagglutinin. J. Biol. Chem. *233*, 401–405.

LIS, H., FRIDMAN, C., SHARON, N., and KATCHALSKI, E. 1966A. Multiple hemagglutinins in soybean. Arch. Biochem. Biophys. *117*, 301–309.

LIS, H., SHARON, N., and KATCHALSKI, E. 1966B. Soybean hemagglutinin, a plant glycoprotein. I. Isolation of a glycopeptide. J. Biol. Chem. *241*, 684–689.

LIS, H., SHARON, N., and KATCHALSKI, E. 1969. Identification of the carbohydrate-protein linking group in soybean hemagglutinin. Biochim. Biophys. Acta *192*, 364–366.

MCKENZIE, H. A. 1967. Milk proteins. Advan. Protein Chem. *22*, 55–234.

MCKINNEY, L. L., SOLLARS, W. F., and SETZKORN, E. A. 1949. Studies on the preparation of soybean protein free from phosphorus. J. Biol. Chem. *178*, 117–132.

MALIK, W. U., and JINDAL, M. R. 1968. Hydrogen ion equilibria of soybean protein. J. Phys. Chem. *72*, 3612–3616.

MANN, R. L., and BRIGGS, D. R. 1950. Effects of solvent and heat treatments on soybean proteins as evidenced by electrophoretic analysis. Cereal Chem. *27*, 258–269.

MAYER, F. C., CAMPBELL, R. E., SMITH, A. K., and MCKINNEY, L. L. 1961. Soybean phosphatase. Purification and properties. Arch. Biochem. Biophys. *94*, 301–307.

MILLAR, D. B. S., WILLICK, G. E., STEINER, R. F., and FRATTALI, V. 1969. Soybean inhibitors. IV. The reversible self-association of a soybean proteinase inhibitor. J. Biol. Chem. *244*, 281–284.

MITSUDA, H., YASUMOTO, K., YAMAMOTO, A., and KUSANO, T. 1967. Study on soybean lipoxygenase. I. Preparation of crystalline enzyme and assay by polarographic method. Agr. Biol. Chem. (Tokyo) *31*, 115–118.

MOUSTAFA, E., and WONG, E. 1967. Purification and properties of chalcone-flavanone isomerase from soya bean seed. Phytochemistry *6*, 625–632.

NAISMITH, W. E. F. 1955. Ultracentrifuge studies on soya bean protein. Biochim. Biophys. Acta *16*, 203–210.

NASH, A. M., ELDRIDGE, A. C., and WOLF, W. J. 1967. Fractionation and characterization of the alcohol extractables associated with soybean proteins. Nonprotein components. J. Agr. Food Chem. *15*, 102–108.

NASH, A. M., KWOLEK, W. F., and WOLF, W. J. 1971. Denaturation of soybean proteins by isoelectric precipitation. Cereal Chem. *48*, 360–368.

NASH, A. M., and WOLF, W. J. 1967. Solubility and ultracentrifugal studies on soybean globulins. Cereal Chem. *44*, 183–192.

OBARA, T., and KIMURA, M. 1967. Gel filtration fractionation of the whole water-extractable soybean proteins. J. Food Sci. *32*, 531–534.

OKUBO, K., ASANO, M., KIMURA, Y., and SHIBASAKI, K. 1969. On basic subunits dissociated from C (11S) component of soybean proteins with urea. Agr. Biol. Chem. (Tokyo) *33*, 463–465.

OKUBO, K., and SHIBASAKI, K. 1967. Fractionation of main components and their subunits of soybean proteins. Agr. Biol. Chem. (Tokyo) *31*, 1276–1282.

OSBORNE, T. B., and CAMPBELL, G. F. 1898. Proteids of the soy bean *(Glycine hispida)*. J. Am. Chem. Soc. *20*, 419–428.

OZAWA, K., and LASKOWSKI, M., Jr. 1966. The reactive site of trypsin inhibitors. J. Biol. Chem. *241*, 3955–3961.

PALLANSCH, M. J., and LIENER, I. E. 1953. Soyin, a toxic protein from the soybean. II. Physical characterization. Arch. Biochem. Biophys. *45*, 366–374.

PINSKY, A., and GROSSMAN, S. 1969. Proteases of the soyabean. II. Specificity of the active fractions. J. Sci. Food Agr. *20*, 374–375.

PUSKI, G., and MELNYCHYN, P. 1968. Starch-gel electrophoresis of soybean globulins. Cereal Chem. *45*, 192–201.

RACKIS, J. J., and ANDERSON, R. L. 1964. Isolation of four soybean trypsin inhibitors by DEAE-cellulose chromatography. Biochem. Biophys. Res. Commun. *15*, 230–235.

RACKIS, J. J., SASAME, H. A., ANDERSON, R. L., and SMITH, A. K. 1959. Chromatography of soybean proteins. I. Fractionation of whey proteins on diethylaminoethylcellulose. J. Am. Chem. Soc. *81*, 6265–6270.

RACKIS, J. J., SMITH, A. K., BABCOCK, G. E., and SASAME, H. A. 1957. An ultracentrifugal study on the association-dissociation of glycinin in acid solution. J. Am. Chem. Soc. *79*, 4655–4658.

RACKIS, J. J. *et al.* 1961. Amino acids in soybean hulls and oil meal fractions. J. Agr. Food Chem. *9*, 409–412.

ROBERTS, R. C., and BRIGGS, D. R. 1963. Characteristics of the various soybean globulin components with respect to denaturation by ethanol. Cereal Chem. *40*, 450–458.

ROBERTS, R. C., and BRIGGS, D. R. 1965. Isolation and characterization of the 7S component of soybean globulins. Cereal Chem. *42*, 71–85.

SAIO, K., KOYAMA, E., and WATANABE, T. 1967. Protein-calcium-phytic acid relationships in soybean. I. Effects of calcium and phosphorus on solubility characteristics of soybean meal protein. Agr. Biol. Chem. (Tokyo) *31*, 1195–1200.

SAIO, K., KOYAMA, E., and WATANABE, T. 1968A. Protein-calcium-phytic acid relationships in soybean. II. Effects of phytic acid on combination of calcium with soybean meal protein. Agr. Biol. Chem. (Tokyo) *32*, 448–452.

SAIO, K., MATSUO, T., and WATANABE, T. 1970. Preliminary electron microscopic investigation on soybean 11S protein. Agr. Biol. Chem. (Tokyo) *34*, 1851–1854.

SAIO, K., WAKABAYASHI, A., and WATANABE, T. 1968B. Effects of heating on soybean meal proteins. J. Agr. Chem. Soc. Japan *42*, 90–96.

SAIO, K., and WATANABE, T. 1966. Preliminary investigation on protein bodies of soybean seeds. Agr. Biol. Chem. (Tokyo) *30*, 1133–1138.

SAIO, K., and WATANABE, T. 1968. Observation of soybean foods under electron microscope. J. Food Sci. Technol. Japan *15*, 290–296.

SEALOCK, R. W., and LASKOWSKI, M., JR. 1969. Enzymatic replacement of the arginyl by a lysyl residue in the reactive site of soybean trypsin inhibitor. Biochemistry *8*, 3703–3710.

SHIBASAKI, K., and OKUBO, K. 1966. Starch gel electrophoresis of soybean proteins in high concentration of urea. Tohoku J. Agr. Res. *16*, 317–329.

SHIBASAKI, K., OKUBO, K., and ONO, T. 1969. Food chemical studies on soybean proteins. V. On the insoluble protein components of defatted soybean heated by steaming. J. Food Sci. Technol. Japan *16*, 22–26.

SHUTOV, A. D., and VAINTRAUB, I. A. 1967. Composition of fraction B of soybean globulins. Biokhimiya *32*, 1220–1226; Biochemistry (USSR) (English Transl.) *32*, 1006–1010.

SHVARTS, V. S., and VAINTRAUB, I. A. 1967. Isolation of the 11S component of soya bean protein and determination of its amino acid composition by an automatic chromato-polarographic method. Biokhimiya *32*, 162–168; Biochemistry (USSR) (English Transl.) *32*, 135–140.

SINGH, L., WILSON, C. M., and HADLEY, H. H. 1969. Genetic differences in soybean trypsin inhibitors separated by disc electrophoresis. Crop Sci. *9*, 489–490.

SMILEY, W. G., and SMITH, A. K. 1946. Preparation and nitrogen content of soybean protein. Cereal Chem. *23*, 288–296.

SMIRNOVA, T. I., POGLAZOV, B. F., and KRETOVICH, V. L. 1959. Amperometric titration of SH-groups of glycinin. Biokhimiya *24*, 758–760; Biochemistry (USSR) (English Transl.) *24*, 695–697.

SMITH, A. K., and CIRCLE, S. J. 1938. Peptization of soybean proteins. Extraction of nitrogeneous constituents from oil-free meal by acids and bases with and without added salts. Ind. Eng. Chem. *30*, 1414–1418.

SMITH, A. K., CIRCLE, S. J., and BROTHER, G. H. 1938. Peptization of soybean proteins. The effect of neutral salts on the quantity of nitrogeneous constituents extracted from oil-free meal. J. Am. Chem. Soc. *60*, 1316–1320.

SMITH, A. K., JOHNSON, V. L., and DERGES, R. E. 1951. Denaturation of soybean protein with alcohols and with acetone. Cereal Chem. *28*, 325–333.

SMITH, A. K., NASH, A. M., ELDRIDGE, A. C., and WOLF, W. J. 1962. Recovery of soybean whey protein with edible gums and detergents. J. Agr. Food Chem. *10*, 302–304.

SMITH, A. K., and RACKIS, J. J. 1957. Phytin elimination in soybean protein isolation. J. Am. Chem. Soc. *79*, 633–637.

SMITH, A. K., SCHUBERT, E. N., and BELTER, P. A. 1955. Soybean protein fractions and their electrophoretic patterns. J. Am. Oil Chemists' Soc. *32*, 274–278.

SMITH, A. K. et al. 1966. Nitrogen solubility index, isolated protein yield, and whey nitrogen content of several soybean strains. Cereal Chem. *43*, 261–270.

STEINER, R. F. 1965. The reduction and reoxidation of the disulfide bonds of soybean trypsin inhibitor. Biochim. Biophys. Acta *100*, 111–121.

STEINER, R. F., and FRATTALI, V. 1969. Purification and properties of soybean protein inhibitors of proteolytic enzymes. J. Agr. Food Chem. *17*, 513–518.

STEVENS, F. C., BROWN, D. M., and SMITH, E. L. 1970. Some properties of soybean lipoxygenase. Arch. Biochem. Biophys. *136*, 413–421.

TANFORD, C. 1968. Protein denaturation. Advan. Protein Chem. *23*, 121–282.

TANIS, R. J., and NAYLOR, A. W. 1968. Physical and chemical studies of a low-molecular-weight form of urease. Biochem. J. *108*, 771–777.

THEORELL, H., HOLMAN, R. T., and AKESON, A. 1947. Crystalline lipoxidase. Acta Chem. Scand. *1*, 571–576.

TOMBS, M. P. 1967. Protein bodies of the soybean. Plant Physiol. *42*, 797–813.

VAINTRAUB, I. A. 1965. Isolation of the 2S-component of soybean globulins. Biokhimiya *30*, 628–633; Biochemistry (USSR) (English Transl.) *30*, 541–545.

VAINTRAUB, I. A. 1967. The heterogeneity of the subunits of the 11S protein of soybean seeds. Mol. Biol. *1*, 807–814; Mol. Biol. (USSR) (English Transl.) *1*, 671–676.

VAINTRAUB, I. A., and SHUTOV, A. D. 1969. Isolation and certain properties of the 2.8S protein of soybean seeds. Biokhimiya *34*, 984–992; Biochemistry (USSR) (English Transl.) *34*, 795–802.

WADA, S., PALLANSCH, M. J., and LIENER, I. E. 1958. Chemical composition and end groups of the soybean hemagglutinin. J. Biol. Chem. *233*, 395–400.

WANG, L. C., and ANDERSON, R. L. 1969. Purification and properties of soybean allantoinase. Cereal Chem. *46*, 656–663.

WATANABE, T., and NAKAYAMA, O. 1962. Study of water-extracted protein of soybean. J. Agr. Chem. Soc. Japan *36*, 890–895.

WOLF, W. J. 1956. Physical and chemical studies on soybean proteins. PhD. Thesis, Univ. Minn.

WOLF, W. J. 1969A. Chemical and physical properties of soybean proteins. Baker's Dig. *43*, No. 5, 30, 31, 34–37.
WOLF, W. J. 1969B. Soybean protein nomenclature: A progress report. Cereal Sci. Today *14*, 75, 76, 78, 129.
WOLF, W. J. 1970A. Soybean proteins: Their functional, chemical, and physical properties. J. Agr. Food Chem. *18*, 969–976.
WOLF, W. J. 1970B. Scanning electron microscopy of soybean protein bodies. J. Am. Oil Chemists' Soc. *47*, 107–108.
WOLF, W. J., BABCOCK, G. E., and SMITH, A. K. 1962. Purification and stability studies of the 11S component of soybean proteins. Arch. Biochem. Biophys. *99*, 265–274.
WOLF, W. J., and BRIGGS, D. R. 1956. Ultracentrifugal investigation of the effect of neutral salts on the extraction of soybean proteins. Arch. Biochem. Biophys. *63*, 40–49.
WOLF, W. J., and BRIGGS, D. R. 1958. Studies on the cold-insoluble fraction of the water-extractable soybean proteins. II. Factors influencing conformation changes in the 11S component. Arch. Biochem. Biophys. *76*, 377–393.
WOLF, W. J. *et al.* 1958. Behavior of the 11S protein of soybeans in acid solutions. I. Effects of PH, ionic strength and time on ultracentrifugal and optical rotatory properties. J. Am. Chem. Soc. *80*, 5730–5735.
WOLF, W. J., and BRIGGS, D. R. 1959. Purification and characterization of the 11S component of soybean proteins. Arch. Biochem. Biophys. *85*, 186–199.
WOLF, W. J., ELDRIDGE, A. C., and BABCOCK, G. E. 1963. Physical properties of alcohol-extracted soybean proteins. Cereal Chem. *40*, 504–514.
WOLF, W. J., and SLY, D. A. 1964. Effects of buffer cations on chromatography of proteins on hydroxylapatite. J. Chromatog. *15*, 247–250.
WOLF, W. J., and SLY, D. A. 1965. Chromatography of soybean proteins on hydroxylapatite. Arch Biochem. Biophys. *110*, 47–56.
WOLF, W. J., and SLY, D. A. 1967. Cryoprecipitation of soybean 11S protein. Cereal Chem. *44*, 653–668.
WOLF, W. J., SLY, D. A., and BABCOCK, G. E. 1964. Denaturation of soybean globulins by aqueous isopropanol. Cereal Chem. *41*, 328–339.
WOLF, W. J., SLY, D. A., and KWOLEK, W. F. 1966. Carbohydrate content of soybean proteins. Cereal Chem. *43*, 80–94.
WOLF, W. J., and SMITH, A. K. 1961. Food uses of soybean protein. II. Physical and chemical properties of soybean protein. Food Technol. *15*, No. 5, 12, 13, 16, 18, 21, 23, 26, 28, 31, 33.
WOLF, W. J., and TAMURA, T. 1969. Heat denaturation of soybean 11S protein. Cereal Chem. *46*, 331–344.
WU, Y. V., and SCHERAGA, H. A. 1962. Studies of soybean trypsin inhibitor. I. Physicochemical properties. Biochemistry *1*, 698–705.
YAMAMOTO, M., and IKENAKA, T. 1967. Studies on soybean trypsin inhibitors. I. Purification and characterization of two soybean trypsin inhibitors. J. Biochem. (Tokyo) *62*, 141–149.
YUAN, L. F. H. 1969. Studies on allantoinase. Dissertation Abstr. *29*, 4524–B–4525–B.

CHAPTER 5

A. C. Eldridge | # Organic Solvent Treatment of Soybeans and Soybean Fractions

INTRODUCTION

The organic lipophilic solvent used throughout the soybean processing industry is a low molecular weight hydrocarbon usually referred to as hexane. It is an excellent solvent for triglycerides and other lipids, low in cost, and so effective that it is not likely to be replaced by another solvent when the defatted meal is intended for use by the animal feed industry.

When the meal from hexane extraction is desolventized at a low temperature it has a bitter-beany flavor and in this state the meal is not acceptable in most food products. When the meal is desolventized with steam most of the undesirable flavors are eliminated and the meal can be processed for use in many foods. However, steaming may be detrimental for certain food applications. Nutty flavors normally are developed when soybean meal is heated; these flavors may be desirable or undesirable depending on the use of the product. Steam also causes denaturation of the protein; thus, meals which have been heat treated for use in feeds are inappropriate for protein isolation.

Because of the increasing interest in high protein foods, attention has been given to finding a new solvent or a combination of solvents which will completely remove the undesirable flavors along with the oil to produce a truly bland soybean meal. This chapter will evaluate the research on organic solvents other than hexane which have been investigated, or used, to improve flavor, color, and to modify the properties of soybean meal, protein concentrates, and isolates.

ALCOHOLS

The use of alcohol(s) for the extraction of soybeans, soybean meal, and soybean proteins has been discussed briefly in earlier treatises by Langhurst (1951) and Kuiken (1958). More recently, the extraction or treatment of soy materials with aqueous alcohols has been reinvestigated and much interest has been shown in the process. Therefore, a thorough review on the use of alcohols for extraction of soybeans, soy meal, and soy protein is needed.

Many alcohols have been investigated on a laboratory scale. The particular alcohol selected for extraction will alter not only the chemical and physical properties of the extracted meals and protein, but also the flavor, functionality, and color properties. Other problems which will have to be considered are residual solvent in the products (Black *et al.* 1961) and economics of the process.

Alcohols Investigated

The principal alcohols used for extraction of soybean meals and protein are methyl, ethyl, and isopropyl. A lesser amount of research has been done with *n*-propyl, *n*-butyl, and isobutyl alcohol. References to work on each of these solvents have been published by Eldridge (1969).

Material Removed

Early studies on the use of aqueous methyl and ethyl alcohols to extract soybeans were made by Oriental researchers. Mashino (1929) published data on the total extractive matter, crude protein extracted, and the ash removed with various concentrations of methanol and ethanol. Okatomo (1937) reported the use of ethanol as a solvent for batch extraction of soybeans in a Manchurian plant. Data was published on the solubility of soybean oil in various ethanol-water ratios. Their data were plotted by Beckel *et al.* (1948A) and are shown in Fig. 5.1.

Since few quantitative data were available showing the solubility of nonprotein nitrogen components, lecithins, and carbohydrates in aqueous alcohols, Nagel *et al.* (1938) studied the solubility of these meal constituents in alcohol-water solutions. These researchers used ethanol and methanol between zero and 100% and consecutive extractions. The soluble nitrogenous constituents increased from 0.47 to 5.22% as the ethanol concentration decreased from 100 to 60%. For methanol the solubility of the nitrogenous constituents was somewhat higher. Below 60% alcohol the solubility curve rose rapidly.

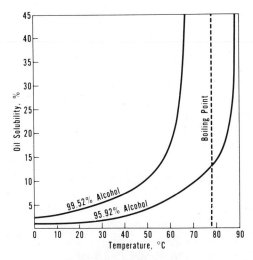

FIG. 5.1. SOLUBILITY OF SOYBEAN OIL IN
ETHYL ALCOHOL

Beckel *et al.* (1948A) surveyed the effects of various solvents, including ethanol, on the color and flavor of soybean meal and on the color of the isolated protein. Six solvents investigated were carbon tetrachloride, trichloroethylene, dichloroethylene, isobutyl alcohol, isopropyl alcohol, and ethanol. Reflectance measurements of extracted flakes and of protein isolated from them showed that all solvents improved the color of the meal and its products. In addition, Beckel and Smith (1944) reported that ethanol served as a debittering agent for soy meal. From these studies it is apparent that alcohol has advantages over other solvents for improving the color and flavor of soy products.

The amount of material removed from hexane-defatted soybean meal by ethanol depends on the concentration of alcohol, the temperature of the extraction, extraction time, and particle size of the meal. To provide a meal with good flavor and color and also with a high nitrogen solubility, 95% ethanol was used and the extraction conducted at ambient temperature (Smith *et al.* 1951). Under these conditions 1.8–2.6% of material (Mashino 1929; Mustakas *et al.* 1961) will be removed. About 2/3 of the alcohol extractives were carbohydrate while the other 1/3 was minor components. The minor components contribute part of the undesirable flavor and were studied by Teeter *et al.* (1955) and Fujimaki *et al.* (1965).

If soybean protein is isolated from meals that have not been treated to remove the minor components, the minor components tend to associate with the isolated protein. Smiley and Smith (1946) found that soybean protein isolated from hexane-defatted meal contained 4% of a brown syrupy material, which could be removed by alcohol washing of the isolated protein. Removal of materials associated with isolated soybean protein was investigated further by Eldridge *et al.* (1963B). The quantity of material removed from isolated soybean protein with aqueous alcohol and several mixed solvents was determined. An example of results obtained is shown in Fig. 5.2. Carefully prepared isolated soybean protein was washed with aqueous ethanol. The quantity of solids and soluble esters removed reached a maximum when the alcohol concentration reached approximately 85%. The nitrogen content of the washed protein also reached a maximum at about the same concentration of alcohol. These results indicate the need for a mixed solvent to remove the impurities associated with acid precipitated protein of the soybean.

In the same paper, Eldridge *et al.* (1963B) investigated the material extracted from isolated soybean protein by several solvent mixtures. Table 5.1 shows the results obtained. Aqueous alcohols remove more material than do alcohols alone. The water-alcohol mixtures are probably removing carbohydrate materials which are not removed when hydrocarbon-alcohol solvents are used. More recent results (Honig *et al.* 1969) indicate hexane-ethanol and 95% ethanol both remove the flavor components from defatted soybean meal but 95% ethanol removes more material.

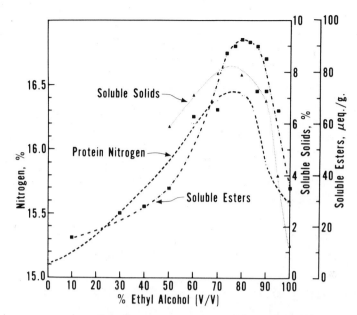

FIG. 5.2. EFFECT OF ETHYL ALCOHOL CONCEN-
TRATION ON EXTRACTION OF NONPRO-
TEINACEOUS COMPONENTS FROM ACID-
PRECIPITATED SOY CURD AND ON THE
STABILITY OF THE PROTEIN FOAM

Very limited information is available on the materials that are removed from soybean products with the solvents shown in Table 5.1. A study by Nash *et al.* (1967) identified phosphatidyl choline, phosphatidyl ethanolamine, saponins, sitosterol glycoside, and genistein in ethanol extracts of isolated soybean proteins. Triglycerides and other neutral compounds were also present in the mixture.

Other minor components reported in alcoholic extracts of soybean meal or protein are listed in Table 5.2.

Chemical-Physical Properties

Extracting or washing soybean meal or protein with alcohol may substantially alter the solubility, color, flavor, and functional properties of the protein. Okano and Ninomiya (1929) studied solubility of the water- and salt-soluble proteins after the oil had been removed by alcoholic extraction. Sato and Sakai (1929) also studied the extractability of the protein after treatment of the meal with alcohol. In more detailed studies, Smith *et al.* (1951) determined the effect of alcohols on the protein by measuring the nitrogen that dispersed in water after solvent treatment. Results (Smith *et al.* 1951) of treating soybean meal with alcohols at 30° C for 30 min are shown in Fig. 5.3. Apparently insolubilization

TABLE 5.1

EFFECT OF EXTRACTING CRUDE, ACID-PRECIPITATED
SOYBEAN PROTEIN WITH VARIOUS SOLVENTS

	Amount of Material Extracted		
Extraction Solvent	Soluble Solids (%)	Chloroform Solubles (%)	Soluble Esters (μeg/gm)
95% Methyl alcohol	7.3	—	81
97% Methyl alcohol	7.4	3.1	103
86% Ethyl alcohol	7.1	3.0	82
80% n-Propyl alcohol	8.4	3.5	113
82% Isopropyl alcohol	8.0	2.8	106
Water-saturated n-butyl alcohol	7.7	6.3	101
n-Butyl alcohol	0.5	0.4	6
Water-saturated isobutyl alcohol	6.0	5.5	101
Isobutyl alcohol	0.6	0.3	7
Water-saturated diethyl ether	1.8	0.3	11
Anhydrous diethyl ether	1.3	0.1	7
80% Acetone	7.0	2.7	83
Acetone	2.5	0.6	9
Hexane-benzene-ethyl alcohol (1:1:1)	2.0	0.6	18
Benzene-ethyl alcohol (4:1)	1.8	0.4	21
Hexane-ethanol (4:1)	1.2	1.3	12
Diethyl ether-ethyl alcohol (1:1)	1.5	1.1	6

of the protein takes place rapidly and then remains fairly constant after a contact time of about 15 min.

Mann and Briggs (1950) examined by electrophoresis the effects of heat and solvent on the peptizability of soybean protein. Hot extraction with alcohols reduces solubility of the proteins. They report that globulins are more easily denatured by aqueous alcohols than are nonglobulin proteins.

In more definitive work on the denaturation of soy proteins by alcohols, Roberts and Briggs (1963) and Wolf et al. (1963) investigated denaturation of the protein utilizing an analytical ultracentrifuge. The four major components of soybean globulins—namely the 2, 7, 11, and 15S components (Chap. 4)—are insolubilized or denatured at different rates. The alcohols were found to act differently on the various proteins; for example, aqueous butanol insolubilized most of the 7, 11, and 15S components whereas aqueous methanol or ethanol had only slight effect on these components.

TABLE 5.2

MINOR CONSTITUENTS EXTRACTED FROM DEFATTED
SOYBEAN MEAL BY AQUEOUS ALCOHOLS

Substance Reported	Reference
Aliphatic carbonyls	Fujimaki *et al.* (1965)
Antiperotic factors	Kratzer *et al.* (1959)
Carbohydrates	Kawamura and Kimura (1965)
Carbonyls	Janicek and Hrdlicka (1964)
Estrogens	Cheng *et al.* (1953)
Flavones and resene	Okano and Beppu (1940)
Genistin	Walz (1931); Walter (1941)
Isoflavones	Okano and Beppu (1939)
Ketones, carboxylic acids, esters, phenols, ether, hydrocarbon	Teeter *et al.* (1955)
Nucleotides	Wang (1969)
Phenolic acids	Arai *et al.* (1966)
Phosphatides	Earle and Milner (1938)
Phospholipids	McKinney *et al.* (1937)
Saponins	Burrell and Walter (1935)
Saponins	Birk *et al.* (1963)
Sitosterols	Kondo and Mori (1936)
Stigmasterol	Kraybill *et al.* (1940)

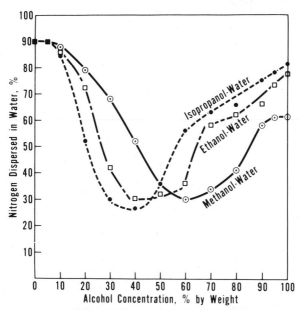

FIG. 5.3. CHANGE IN WATER DISPERSIBILITY OF
NITROGENOUS COMPONENTS OF SOY-
BEAN MEAL AT pH 6.5 AFTER TREAT-
MENT WITH VARIOUS ALCOHOL CON-
CENTRATIONS

Fukushima (1969A) demonstrated the effect of ethanol and heat on the disruption of the soybean protein structure. Apparently very low or very high concentrations of alcohol cause less denaturation than do intermediate concentrations, which is in general agreement with the work of Smith *et al.* (1951).

Alcohol denaturation of the protein sometimes may have definite advantages. For example, Fuskushima (1969B) has shown that alcohol denaturation of the protein in soybean meal permits greater proteinase action in soy sauce fermentation and thus increases the yield by about 9%. The process developed by Fukushima (1965) works on either full-fat or defatted soybean flour and is now being used in Japan. Other instances where alcohol denaturation may have advantages are in the modification of functional properties.

Flavor, Functionality, and Color

Beckel and Smith (1944) state that "the bitter principle, beany taste, and color undoubtedly are the principle obstacles to the use of the soybean as a human food in the United States." Apparently these three problems are still prevalent today since Eley (1968) reports these same problems as the factors limiting the unrestricted use of soybean products in foods. During this 25-yr period, many patents granted and papers published pertain to improving the flavor and color of the isolated protein. Seemingly, none of these procedures or processes have completely solved the problem.

Flavor.—Although many publications discuss the flavor of soy products, few give results of organoleptic evaluations of these materials. For example, Beckel *et al.* (1948A) found that protein isolated from alcohol-extracted soybean flakes had improved flavor, but they conducted no formal taste evaluation of their products. Moser *et al.* (1967) evaluated soybean flours organoleptically and showed that 80% of the taste panel could detect raw full-fat soy flours when the flours were diluted 1:750 with wheat flour. Flavor responses by the taste panel were beany, bitter, and green beany. Both 80% methanol and ethanol treatment greatly improved the flavor scores. Even though steaming of the alcohol-washed flakes further improved the flavor, the flavor was not completely removed.

Certainly, much more research is needed to determine the nature of the flavor constituents of soybean flours. Then, hopefully, solvent systems can be devised to remove these undesirable components.

Functionality.—Alcohol treatment of soybean meals or soybean proteins can alter their functional properties. Major changes occur in the solubility, viscosity, and gelling characteristics of the protein, which are discussed in Chapters 4 and 10. (Eldridge *et al.* 1963A,B; Eldridge and Nash, 1965).

One of the unusual properties of a product named Gelsoy prepared from alcohol-washed flakes (Beckel *et al.* 1949) was its tendency to gel and to form whips at low protein concentrations (Beckel and Belter 1951; Beckel *et al.* 1948C; DeVoss *et al.* 1950). Stable foams were produced with 8–10% protein dispersions after heating and aerating the mixture. Suggested uses for such foams were for meringues, stabilizers, marshmallows, and certain bakery goods (Glabe *et al.* 1956). Gelsoy produced strong gels, which were self-supporting. Paulsen and Horan (1967) changed viscosity, wetting properties, and other functional properties by alcoholic treatment of soybean flours.

If soybean proteins are washed with 80–90% alcohol, the washed protein will form stable foams (Eldridge *et al.* 1963A) at low protein concentrations that are different from foams produced from Gelsoy (Glabe *et al.* 1956).

Color.—Probably, color is a less important factor than either flavor or functionality because many food products are already colored. Again, many articles claim that color has been improved by certain treatments, but supporting data are rarely given. Beckel *et al.* (1948A) used reflectance measurements to determine the color of various samples. Since protein isolated from flakes which had been solvent extracted with ethanol had less color than protein isolated from hexane-extracted flakes, undoubtedly color of the soy products can be improved.

Commercial Applications

The use of alcohol to produce soy products with improved flavor, functionality, and color appears to have potential. Early research on the treatment of soybean meal with alcohol produced the product Gelsoy, mentioned above. Although there was some industrial interest in Gelsoy, this product has not developed to commercial production.

In about 1960, several soybean protein concentrates appeared on the market. One soy protein concentrate (see Chap. 9) is prepared by aqueous alcohol leaching of defatted flakes (Meyer 1966). This concentrate is a commercial success and its manufacture continues. The concentrate is used as a meat extender, in bakery products, and in several other food applications.

Data available in the literature indicate the possibility of preparing soybean products of improved food quality from alcohol-washed flakes. Soybean flakes can be prepared by hexane extraction with a minimum amount of heat in desolventizing the meal (Belter *et al.* 1954; Mustakas *et al.* 1962) followed by an alcohol-washing procedure (Mustakas *et al.* 1961). This procedure has been investigated by commercial firms (Kuramoto 1966; Beaber and Obey 1962).

Another alternative is to extract the full-fat flakes directly with aqueous alcohol (Beckel *et al.* 1946, 1948B). After the extraction process, the meals are processed by normal procedures, and yields of protein isolates would be expected to be about 25%, based on starting flakes (Belter *et al.* 1944). Belter *et al.* (1944) have described a pilot plant for preparing soybean protein isolate from ethanol-extracted flakes and compared the isolate with that obtained from hexane-extracted flakes; they found the isolate from the ethanol-extracted flakes to be superior in color and flavor (Moser *et al.* 1967; Beckel and Smith 1944) and have different functional properties. (Eldridge *et al.* 1963A).

MIXED SOLVENTS

Many reports deal with extraction of soybean meal with mixed solvent systems (Beckel 1944; Eldridge 1969). Much of the literature suggests that a polar-nonpolar solvent mixture will provide a more thorough extraction of the minor lipid-like components in soybean meal than a nonpolar solvent. Since many solvent systems studied are not azeotropic mixtures, they would require changes in present processing procedures. Two of the more promising solvent mixtures, are the azeotropes of hexane-ethyl alcohol (79:21), and hexane-methyl alcohol (72:28).

Ayers and Scott (1952) studied the extraction rates for cottonseed and full-fat soybean flakes using n-hexane and n-hexane-alcohol mixtures. When soybean flakes were extracted with n-hexane for 5 hr, 21.3% of lipids were removed, but when another portion of the starting flakes were extracted with n-hexane containing 5% alcohol, 23.6% of lipids were removed. The total lipid material which could be extracted by any of the alcohol-hexane mixtures exceeded by 2-3% that which could be removed by n-hexane alone.

Nielsen (1960) extracted full-fat soybean flakes with hexane at $60°$ C until the air-dried meal contained 0.7% residual lipid. When the previously hexane-extracted meal was treated with hexane-ethyl alcohol (80:20), the difficultly extractable phosphatides were also removed. The phosphatides were not the same composition as those removed by hexane and were not completely identified by Nielsen.

The use of hexane-alcohol to extract soybean flour has recently been patented (Ferrara and Dalby 1968) and the process reportedly improves not only the color, but also the flavor of soybean products. More detailed studies on the use of hexane-alcohol to improve defatted soy products have been described by Honig *et al.* (1969). After hexane extraction, the azeotrope removed an additional 2.3-3.3% of material, which contained many of the undesirable flavors present in soybean meal. Phospholipids, sterols, triglycerides, carbohydrates, amino acids, and isoflavones were all identified in the hexane-alcohol solubles. The flavor components in the mixture were concentrated by silicic acid column chromatography, but attempts to characterize any of these undesirable flavor components failed.

Recent studies on the organoleptic evaluations (Eldridge *et al.* 1970) of soy flours and protein isolates from hexane-alcohol-extracted flakes showed significant odor and flavor improvement when the treated materials were compared with samples from hexane-extracted soy flours. It is apparent that flavor and odor are improved when soybeans are extracted with the mixed solvent.

CHLORINATED SOLVENTS

Chlorinated hydrocarbons have been investigated as solvents for extracting soybeans both experimentally and commercially primarily because they are nonflammable and are efficient solvents for oil removal (Goss 1952; Arnold and Choudhury 1962).

Trichloroethylene

Of the chlorinated solvents investigated, trichloroethylene (TCE) gained the most widespread use. Commercial production of TCE-extracted soybean meal (TESOM) was achieved in Europe, Japan, and the United States (Pritchard *et al.* 1952). However, TESOM proved to be toxic to cattle and other animals (Pritchard *et al.* 1956). Stockman (1916) first described a toxicity produced in bovines by TESOM. Later, several cases of the toxicity were observed in Duren, Germany; the disorders were called "Duren Disease" by Stang (1927).

Production of TESOM reached a maximum of about 2% of the total meal production in 1951 according to Goss (1952). However, severe and fatal aplastic anemia occurred in Minnesota diary cattle that were fed TESOM as a part of their ration (Pritchard *et al.* 1952). Toxicity of different lots of TESOM varied considerably. Meal prepared from beans recently harvested proved to be much more toxic than meal from beans that had been stored for some time (Pritchard *et al.* 1956).

When TCE was fed to calves for long periods, the animals produced no clinical or metabolic disorders of TESOM toxicity (Pritchard *et al.* 1956). The meal's toxicity could not be ascribed to auto-oxidation products of TCE or to triethylamine, which was used as a stabilizer for TCE (McKinney *et al.* 1955; Seto and Schultze 1954). Chemical analysis of toxic TESOM showed 0.5 mole less sulfhydryl group per 10^6 gm and about 20 ppm more chlorine than did hexane-extracted soybean meal prepared from the same beans (McKinney *et al.* 1957A). Evidence showed that TCE would react when heated with cysteine or reduced glutathione, or both (McKinney *et al.* 1957B).

S-(*trans*-dichlorovinyl)-L-cysteine was synthesized (McKinney *et al.* 1957B) from cysteine and TCE. When the derivative was fed to calves, it produced the complete clinical, hemotologic, and post mortem picture of severe asplastic anemia in the bovine.

As far as known, TCE is no longer used to extract oil from soybeans, and it is not recommended as a solvent for this purpose.

Other Chlorinated Solvents

Chlorinated solvents other than TCE have been investigated for defatting soybeans. These include dichloromethane, 1,2-dichloroethane, 1,2-dichloropropane, 1,2,3-trichloropropane, 1,1,1-trichloroethane, dichlorodifluoromethane, 1,2,2-trifluorotrichloroethane, carbon tetrachloride, and chloroform. Most research on these chlorinated solvents pertains to various engineering problems, such as solvent efficiency, specific gravities of solvent-oil mixtures, boiling points, and the like. In any event, the use of chlorinated solvents should be examined carefully because of toxicity problems that may arise. For example,

Canadian researchers (Morrison and Munro 1965; Munro and Morrison 1967A,B) have shown that fish protein concentrate prepared by extracting cod fillets with 1,2-dichloroethane was toxic to growing rats. Apparently, chlorocholine chloride, a toxic substance which can be produced during extraction, was not responsible for the toxicity.

BIBLIOGRAPHY

ARAI, S., SUZUKI, H., FUJIMAKI, M., and SAKURAI, Y. 1966. Studies on flavor components in soybean. II. Phenolic acids in defatted soybean flour. Agr. Biol. Chem. (Tokyo) *30,* 364–369.

ARNOLD, L. K., and CHOUDHURY, R. B. R. 1962. Extraction of soybeans with four hydrocarbon solvents. J. Am. Oil Chemists' Soc. *39,* 378.

AYERS, A. L., and SCOTT, C. R. 1952. A study of extraction rates for cottonseed and soybean flakes by using *n*-hexane and various alcohol-hexane mixtures. J. Am. Oil Chemists' Soc. *29,* 213–218.

BEABER, N. J., and OBEY, J. H. 1962. Method for producing organoleptically bland protein. U.S. Pat. 3,043,826. July 10.

BECKEL, A. C. 1944. A bibliography on the solvent extraction of vegetable oils from raw materials—with special attention to soybeans. Oil Soap *21,* 264–270.

BECKEL, A. C., and BELTER, P. A. 1951. Reversible vegetable gel. U.S. Pat. 2,561,333. July 24.

BECKEL, A. C., BELTER, P. A., and SMITH, A. K. 1946. Laboratory study of continuous vegetable oil extraction: Countercurrent extractor, rising-film evaporator and oil stripper. Ind. Eng. Chem. Anal. Ed. *18,* 56–58.

BECKEL, A. C., BELTER, P. A., and SMITH, A. K. 1948A. Solvent effects on the products of soybean-oil extraction. J. Am. Oil Chemists' Soc. *25,* 7–9.

BECKEL, A. C., BELTER, P. A., and SMITH, A. K. 1948B. The nondistillation alcohol extraction process for soybean oil. J. Am. Oil Chemists' Soc. *25,* 10–11.

BECKEL, A. C., BELTER, P. A., and SMITH, A. K. 1949. A new soybean product—Gelsoy. Soybean Dig. *10,* No. 1, 17–18, 40.

BECKEL, A. C., DEVOSS, L. I., BELTER, P. A., and SMITH, A. K. 1948C. Soy whip. U.S. Pat. 2,444,241. June 29.

BECKEL, A. C., and SMITH, A. K. 1944. Alcohol extraction improves soya flour flavor and color. Food Ind. *16,* 616, 664.

BELTER, P. A., BECKEL, A. C., and SMITH, A. K. 1944. Soybean protein production: Comparison of the use of alcohol-extracted with petroleum-ether-extracted flakes in a pilot plant. Ind. Eng. Chem. *36,* 799–803.

BELTER, P. A., BREKKE, O. L., WALTHER, G. F., and SMITH, A. K. 1954. Flash desolventizing. J. Am. Oil Chemists' Soc. *31,* 401–403.

BIRK, Y., BONDI, A., GESTETNER, B., and ISHAAYA, I. 1963. A thermostable haemolytic factor in soybeans. Nature (London) *197,* 1089–1090.

BLACK, L. T., KIRK, L. D., and MUSTAKAS, G. C. 1961. The determination of residual alcohol in defatted alcohol-washed soybean flakes. J. Am. Oil Chemists' Soc. *38,* 483–485.

BURRELL, R. C., and WALTER, E. D. 1935. A saponin from soybeans. J. Biol. Chem. *108,* 55–60.

CHENG, E. W.-K. *et al.* 1953. Estrogenic activity of isoflavone derivatives extracted and prepared from soybean oil meal. Science *118,* 164–165.

DEVOSS, L. I., BECKEL, A. C., and BELTER, P. A. 1950. Vegetable gel. U.S. Pat. 2,495,706. Jan. 31.

EARLE, F. R., and MILNER, R. T. 1938. The occurrence of phosphorus in soybeans. Oil Soap *15,* 41–42.

ELDRIDGE, A. C. 1969. A bibliography on the solvent extraction of soybeans and soybean products 1944–1968. J. Am. Oil Chemists' Soc. *46,* 458A, 460A, 462A, 464A, 496A, 498A, 500A, 502A.

ELDRIDGE, A. C., HALL, P. K., and WOLF, W. J. 1963A. Stable foams from unhydrolyzed soybean protein. Food Technol. *17*, 1592–1595.
ELDRIDGE, A. C., and NASH, A. M. 1965. Process of producing soybean proteinate. U.S. Pat. 3,218,307. Nov. 16.
ELDRIDGE, A. C., WOLF, W. J., NASH, A. M., and SMITH, A. K. 1963B. Alcohol washing of soybean protein. J. Agr. Food Chem. *11*, 323–328.
ELDRIDGE, A. C. et al. 1970. Sensory evaluation of soy flours and protein isolates from hexane-alcohol extracted flakes. Abstr. 55th Meeting Am. Assoc. Cereal Chemists. 66.
ELEY, C. P. 1968. Food uses of soy protein. *In* Marketing and Transportation Situation. USDA–ERS *388*.
FERRARA, P. J., and DALBY, G. 1968. Wheat germ and soybean process for extracting glutathione therefrom. U.S. Pat. 3,396,033. Aug. 6.
FUJIMAKI, M., ARAI, S., KIRIGAYA, N., and SAKURAI, Y. 1965. Studies of flavor components in soybean. I. Aliphatic carbonyl compounds. Agr. Biol. Chem. (Tokyo) *29*, 855–863.
FUKUSHIMA, D. 1965. Method for treatment of soybean proteins. U.S. Pat. 3,170,802. Febr. 23.
FUKUSHIMA, D. 1969A. Denaturation of soybean proteins by organic solvents. Cereal Chem. *46*, 156–163.
FUKUSHIMA, D. 1969B. Enzymatic hydrolysis of alcohol-denatured soybean proteins. Cereal Chem. *46*, 405–418.
GLABE, E. F. et al. 1956. Uses of Gelsoy in prepared food products. Food Technol. *10*, 51–56.
GOSS, W. H. 1952. Trends in the oilseed industry. J. Am. Oil Chemists' Soc. *29*, 253–257.
HONIG, D. H., SESSA, D. J., HOFFMAN, R. L., and RACKIS, J. J. 1969. Lipids of defatted soybean flakes: Extraction and characterization. Food Technol. *23*, 803–808.
JANICEK, G., and HRDLICKA, J. 1964. Study of changes during thermic and hydrothermic processes. V. Carbonyl derivatives produced in debittering process of soybeans. Sci. Papers Inst. Chem. Technol., Prague. Food Technol. (Prague) *8*, 107–110.
JOHNSON, D. W. 1969. Oilseed proteins–properties and applications. Food Prod. Develop. *3*, No. 8, 78, 82, 84, 87.
KAWAMURA, S., and KIMURA, T. 1965. A note on the identification of sugars from soybeans through crystalline derivatives. Kagawa Univ. Fac. Agr. Tech. Bull. *17*, 26–28.
KONDO, K., and MORI, S. 1936. Sitosterol-*d*-glucoside of soybeans. J. Chem. Soc. (Japan) *57*, 1128–1131.
KRATZER, F. H. et al. 1959. Fractionation of soybean oil meal for growth and antiperotic factors. Poultry Sci. *38*, 1049–1055.
KRAYBILL, H. R., THORTON, M. H., and ELDRIDGE, K. E. 1940. Sterols from crude soybean oils. Ind. Eng. Chem. *32*, 1138–1139.
KUIKEN, K. A. 1958. Effect of other processing factors on vegetable protein meals. *In* Processed Plant Protein Foodstuffs, A. M. Altschul (Editor). Academic Press, New York.
KURAMOTO, S. 1966. Preparation of yeast-raised bakery products utilizing an isolated soy protein. U.S. Pat. 3,252,807. May 24.
LANGHURST, L. F. 1951. Solvent extraction processes. *In* Soybeans and Soybean Products, Vol. II. K. S. Markley (Editor). John Wiley & Sons, New York.
MCKINNEY, L. L., UHING. E. H., WHITE, J. L., and PICKEN, J. C., Jr. 1955. Autoxidation products of trichloroethylene. J. Agr. Food Chem. *3*, 413–419.
MCKINNEY, L. L. et al. 1957A. Toxic protein from trichloroethylene-extracted soybean oil meal. J. Am. Oil Chemists' Soc. *34*, 461–466.
MCKINNEY, L. L. et al. 1957B. S-(dichlorovinyl)-L-cysteine: An agent causing fatal aplastic anemia in calves. J. Am. Chem. Soc. *79*, 3932–3933.
MCKINNEY R. S., JAMIESON, G. S., and HOLTON, W. B. 1937. Soybean phosphatides. Oil Soap *14*, 126–129.
MANN, R. I., and BRIGGS, D. R. 1950. Effects of solvent and heat treatment on soybean proteins as evidenced by electrophoretic analysis. Cereal Chem. *27*, 258–269.
MASHINO, M. 1929. The purification of soybean protein. II. Influence of water on the purification by lower alcohols. J. Soc. Chem. Ind. (Japan) Suppl. Binding *32*, 312–313.

MEYER, E. W. 1966. Soy protein concentrates and isolates. Proc. Intern. Conf. Soybean Protein Foods, USDA, Peoria, Ill. USDA–ARS 71-35.

MORRISON, A. B., and MUNRO, I. C. 1965. Factors influencing the nutritive value of fish flour. IV. Reaction between 1,2-dichloroethane and protein. Can. J. Biochem. 43, 33–40.

MOSER, H. A. et al. 1967. Sensory evaluation of soy flour. Cereal Sci. Today 12, 296, 298–299, 314.

MUNRO, I. C., and MORRISON, A. B. 1967A. Factors influencing the nutritive value of fish flour. V. Chlorocholine chloride, a toxic material in samples extracted with 1,2-dichloroethane. Can. J. Biochem. 45, 1049–1053.

MUNRO, I. C., and MORRISON, A. B. 1967B. Toxicity of 1,2-dichloroethane-extracted fish protein concentrate. Can. J. Biochem. 45, 1779–1781.

MUSTAKAS, G. C., KIRK, L. D., and GRIFFIN, E. L., Jr. 1961. Bland undenatured soybean flakes by alcohol washing and flash desolventizing. J. Am. Oil Chemists' Soc. 38, 473–478.

MUSTAKAS, G. C., KIRK, L. D., and GRIFFIN, E. L., Jr. 1962. Flash desolventizing defatted soybean meals washed with aqueous alcohols to yield a high-protein product. J. Am. Oil Chemists' Soc. 39, 222–226.

NAGEL, R. H., BECKER, H. C., and MILNER, R. T. 1938. The solubility of some constituents of soybean meal in alcohol-water solutions. Cereal Chem. 15, 766–774.

NASH, A. M., ELDRIDGE, A. C., and WOLF, W. J. 1967. Fractionation and characterization of alcohol extractables associated with soybean proteins. Nonprotein components. J. Agr. Food Chem. 15, 102–108.

NIELSEN, K. 1960. The composition of the difficulty extractable soybean phosphatides. J. Am. Oil Chemists' Soc. 37, 217–219.

OKANO, K., and BEPPU, I. 1939. Coloring matter in soybean. I. Isolation of four kinds of isoflavone from soybean. Agr. Chem. Soc. Japan Bull. 15, 110.

OKANO, K., and BEPPU, I. 1940. Coloring matter in soybean. II. Three kinds of flavones and resene. J. Agr. Chem. Soc. Japan 16, 369–372.

OKANO, K. and NINOMIYA, M. 1929. Denaturation of soybean protein during oil extraction with alcohol. I. Rept. Central Lab. South Manchuria Ry. Co. 14, 7–9. (German)

OKATOMO, S. 1937. Studies on the alcoholic extraction of soybean oil. Contemp. Manchuria 1 (3), 83–101.

PAULSEN, T. M., and HORAN, F. E. 1967. Changes in functional characteristics of soya flours treated with ethanol. Abstr. 52nd Meeting Am. Assoc. Cereal Chemists 40–41.

PRITCHARD, W. R., REHFELD, C. E., and SAUTTER, H. H. 1952. Aplastic anemia of cattle associated with ingestion of trichloroethylene-extracted soybean oil meal. I. Clinical and laboratory investigation of field cases. J. Am. Vet. Med. Assoc. 121, 1–8.

PRITCHARD, W. R. et al. 1956. Studies on trichloroethylene-extracted feeds. I. Experimental production of acute aplastic anemia in young heifers. J. Am. Vet. Res. 17, 425–454.

ROBERTS, R. C., and BRIGGS, D. R. 1963. Characteristics of the various soybean globulin components with respect to denaturation by ethanol. Cereal Chem. 40, 450–458.

SATO, M., and SAKAI, H. 1929. Extraction of soybean oil with alcohol. Rept. Central Lab. South Manchuria Ry. Co. 14, 1–7; Chem. Abstr. 25, 1694.

SETO, T. A., and SCHULTZE, M. O. 1954. Metabolism of trichloroethylene in the bovine. Proc. Soc. Exptl. Biol. Med. 90, 314–316.

SMILEY, W. G., and SMITH, A. K. 1946. Preparation and nitrogen content of soybean protein. Cereal Chem. 23, 288–296.

SMITH A. K., JOHNSEN, V. L., and DERGES, R. E. 1951. Denaturation of soybean protein with alcohols and with acetone. Cereal Chem. 28, 325–333.

STANG, V. 1927. The cause of düren disease. Landwirtsch. Versuchsta., 105, 179–183. (German)

STOCKMAN, S. 1916. Cases of poisoning in cattle by feeding on meal from soybean after extraction of the oil. J. Comp. Pathol. Therap. 29, 95–107.

TEETER, H. M. et al. 1955. Investigations on the bitter and beany components of soybeans. J. Am. Oil Chemists' Soc. 32, 390–397.

WALTER, E. D. 1941. Genistin (an isoflavone glucoside) and its aglucone, genistein, from soybeans. J. Am. Chem. Soc. *63*, 3273–3276.

WALZ, E. 1931. Isoflavone and saponinglucosides in *Soja hispida*. Ann. Chem. *489*, 118–155.

WANG, L. C. 1969. Effect of alcohol washing and autoclaving on nucleotides of soybean meal. J. Agr. Food Chem. *17*, 335–340.

WOLF, W. J., ELDRIDGE, A. C., and BABCOCK, G. E. 1963. Physical properties of alcohol-extracted soybean proteins. Cereal Chem. *40*, 504–514.

CHAPTER 6

J. J. Rackis | **Biologically Active Components**

Technological developments in the soybean industry enable raw soybeans of low food value to be converted into many protein products with greatly increased nutritional qualities. Because of their having high-quality protein and important functional properties, soy proteins find increasing use in foods. An unusually large number of thermostable and heat-labile substances, capable of eliciting diverse nutritional, biological, and physiological responses in man and animals, have been reported in soybeans. As new processes are developed, it will be necessary to obtain more information about these substances and their effects on body functions and, if necessary, re-establish optimum conditions for their inactivation or removal.

Chemical composition and nutritive value of soy products with respect to protein and other essential nutrients are discussed in Chap. 3 and 7, respectively. Also omitted here will be the guides that determine proper processing conditions required to convert raw soybean meal of low nutritive value to toasted soy products of high-protein quality.

"Nutritive value" and "biological-physiological effects" are differentiated by restricting the former to the ability of soy products to supply amino acids and other essential nutrients and the latter to substances that interfere with the utilization of essential nutrients or cause adverse reactions.

ENZYMES

Circle (1950) last summarized the literature covering enzymes in soybeans and the enzymatic changes that occur during germination. Only those enzymes present in the mature seed will be considered here. A compilation of the enzymes found in soybeans is given in Table 6.1. Lipoxygenase, in raw soy flour, is used in bread dough as a bleach for wheat flour and to improve flavor. This is the only commercial application of the enzymes in soybeans.

Amylases

Highly active α- and β-amylases account for most of the amylolytic activity of soybean meal. Starch-liquefying and starch-saccharifying enzymes are also present (Ofelt et al. 1955A; Learmonth and Wood 1960; Pomeranz and Lindner 1960; Pomeranz and Mamaril 1964).

α-Amylases of soybeans and other plants have been purified 150–700 fold (Greenwood et al. 1965A). The soybean Z-enzyme is an α-amylase (Banks et al. 1960; Cunningham et al. 1962; Manners 1962). Kinetics and hydrolysis patterns

158

TABLE 6.1

ENZYMES FOUND IN SOYBEANS

Enzyme	References
Allantoinase	Lee and Roush 1964; Vogels et al. 1966; Wang and Anderson 1969
Amylases	See text
Ascorbicase	Rangnekar et al. 1948; Sakakibara et al. 1965
Chalcone-flavanone isomerase	Moustafa and Wong 1967
Coenzyme Q	Page et al. 1959
Cytochrome C	Fridman et al. 1967
Glucosyl transferases	
Arylamine N-glucosyltransferase	Frear 1968
Glucan synthetase	Frydman et al. 1966
Steryl-glucosyltransferase	Hou et al. 1968
Glycosidases	Nishizawa 1951
Hexokinases	Axelrod et al. 1952; Ito 1960
Lactic dehydrogenase	Bronovitskaya and Kretovich 1970A; King 1970
Lipases	See text
Lipoperoxidase	See text
Lipoxygenase	See text
Malic dehydrogenase	Bronovitskaya and Kretovich 1970B
α-Mannosidase	Catsimpoolas and Meyer 1969
Peroxidase	Sakakibara et al. 1965; Buttery and Buzzell 1968
Phosphatases	Nakamura 1951; Peng and Wang 1954; Kumar et al. 1957; Mayer et al. 1961
Phosphorylase	Nakamura 1951; Yin and Sun 1949
Transaminases	Maekawa et al. 1959; Kretovich et al. 1964
Urease	See text
Uricase	Sen and Smith 1966

of α-amylolysis of starch have been studied by Greenwood et al. (1965B,C). Soybean amylase appears to attack highly branched carbohydrates more readily than other α-amylases (Geddes 1968). Pomeranz (1963) concluded that, unlike α-amylases from other sources, soybean α-amylase does not require -SH groups for activity.

β-Amylase with only traces of Z-enzyme has been isolated from defatted soy flour and its physicochemical properties have been determined (Gertler and Birk 1965). Gertler and Birk (1966) report that the -SH groups of the pure enzyme do not participate in substrate binding but are involved in the catalytic step. Almost complete loss of activity occurs when the 5-SH groups are complexed with p-mercuribenzoate.

β-Amylase activity in soybean extracts is much higher than that of several legumes tested; however, amylase must be inhibited to detect phosphorylase (Nakamura 1951).

No qualitative differences in amylolytic activity were found in three soybean varieties—Adams, Harosoy, and Lincoln (Birk and Waldman 1965). Young immature seeds contained only 34% of the amylase activity of mature soybeans. The biological function of amylases in soybeans is unknown since mature seeds do not contain starch.

The hydrolysis pattern of soybean β-amylase on various substrates has been studied by a number of investigators (Maruo and Kobayasi 1951; Bird and Hopkins 1954; Wild 1954).

Lipases

Perl and Diamant (1963) isolated two distinct lipase fractions from defatted soybean meal. Lipolytic activity of soybeans can be enhanced (Kunert 1962). The lipases hydrolyze soybean oil at a faster rate than other vegetable oils (Gavrichenkov *et al.* 1969).

Lipoperoxidase

Hematin compounds, such as hemoglobin and catalase, can catalyze oxidation of unsaturated fatty acids (Maier and Tappel 1959). Blain and Styles (1959) report that soybean extracts have a lipoperoxidase activity similar to cytochrome c in that preformed linoleate peroxide is used by both to bleach β-carotene. After developing a procedure to differentiate between hematin and lipoxygenase (lipoxidase) activity, Blain and Styles (1959) concluded that crude soybean extracts contain 2 lipoperoxidase factors: 1 predominating in water extracts and not affected by preformed diene; the other existing in buffer extracts and more active in the presence of preformed diene. The lipoperoxidase factor, which bleaches β-carotene in the presence of linoleate hydroperoxide, will also enzymatically destroy linoleate hydroperoxide in the absence of β-carotene (Blain and Barr 1961). A lipohydroperoxidase is present in defatted soy flour (Gini and Koch 1961). Their studies explain the rapid decrease in hydroperoxide when high levels of crude soybean lipoxygenase extracts are reacted with linoleic acid (Dillard *et al.* 1961). The evidence for a reaction sequence between lipoxygenase and peroxidase points to an existence of an unknown mechanism of action of lipoxygenases in legumes (Gini and Koch 1961).

Lipoxygenase (Lipoxidase)

Lipoxygenase catalyzes the oxidation of lipids containing a *cis,cis*-1,4-pentadiene system by molecular oxygen to hydroperoxides. Methods of purification, properties, and mechanism of lipoxygenase reactions have been reviewed by Andre (1964). Primary oxidation products of both lipoxygenase-catalyzed reactions and autoxidation are thought to involve the formation of 9- and 13-hydroperoxide isomers. Vioque and Holman (1962) isolated the isomeric ketodienes as secondary oxidation products. Dolev *et al.* (1967A,B) report exclusive formation of the 13-hydroperoxide isomer with a crystalline soybean lipoxygenase. Other crystalline preparations showed no apparent specificity; both the 9- and 13-

carbons of linoleic acid were oxidized. The data also indicate that the hydroperoxide oxygen came from the gaseous phase. Hamberg and Samuelsson (1965) reported that lipoxygenase-catalyzed hydroperoxidation is relatively specific with regard to the carbon atom attacked. Oxidation of polyunsaturated acids varying in chain length and position of the double bond was studied by Holman et al. (1969). The nature of the catalytic sites of lipoxygenase (Mitsuda et al. 1967A,B), formation of H_2O_2 (Allen 1968) inhibitors (Blain and Shearer 1965), and formation of free radicals (Walker 1963) are described. Effect of lipoxygenase action on the formation of derived flavors in full-fat soy flour has been investigated (Wilkens et al. 1967; Mattick and Hand 1969; Mustakas et al. 1969).

Yasumoto et al. (1970) have shown that nordihydroguaiacetic acid, the most effective phenolic antioxidant, affects lipoxygenase in two ways: as a conventional antioxidant as well as a potent inactivator of the enzyme. The lipoxygenase catalyzes its own destruction in the presence of both substrates, oxygen, and fatty acid (Smith and Lands 1970).

Lipoxygenases are found in many plants and in some microorganisms, but not in animal tissue. Highest lipoxygenase activity occurs in soybeans. Crystallized or highly purified soybean lipoxygenase contains no prosthetic groups and requires no coenzyme or metal activator.

Generally, dehulled defatted soy flour provides the source of the enzyme; however, soybean whey is the most suitable starting material for its isolation (Catsimpoolas 1969). Tofu and soy milk produced by traditional oriental methods contain residual lipoxygenase activity (Andre and Hou 1964). Some lipoxygenase activity may occur in soybeans low in moisture (14.7%) (Kopeikovskii and Kashevatskaya 1966).

Crystalline lipoxygenase, purchased from several supply houses, contains 8,000–20,000 units per milligram of material. Catsimpoolas (1969) developed a procedure for preparing lipoxygenase having 110,000 units per milligram of protein from soy whey. The enzyme was homogeneous by several criteria including immunoelectrophoresis with antisoybean whey protein sera.

Several workers have shown that plants usually possess multiple forms of lipoxygenase, one attacking only the free fatty acid and the other attacking triglycerides in addition to the free acid. Markert and Moeller (1959) proposed the term "isoenzyme" to describe different proteins with similar enzymatic activity. Isoenzyme is now recommended by the Standing Committee on Enzymes of the International Union of Biochemistry (Wilkinson 1965) to describe the multiple enzyme forms occurring in a single species. Four isoenzymes of lipoxygenase are present in soybeans (Guss et al. 1967, 1968). A specific staining procedure was developed to detect lipoxygenase isoenzymes of crude soybean extracts in polyacrylamide gels after electrophoresis. Although the multiplicity of bands having lipoxygenase activity may indicate polymerized lipoxygenase protein, careful control during electrophoresis to prevent formation of artifacts indicates that lipoxygenase most likely exists in multiple molecular forms (Hale *et al.* 1969). Dillard *et al.* (1961) and Dolev *et al.* (1967A) provide additional evidence that soybean lipoxygenase may be a mixture of more than one enzyme. Isoenzyme of soybean lipoxygenase, distinct from the lipoxygenase (E.C. 1.13.1.13) of Theorell has been isolated (Christopher *et al.* 1970; Yamamoto *et al.* 1970).

Older reports indicate that carotene oxidase activity is identical to soybean lipoxygenase activity in partially purified preparations. Kies *et al.* (1969) conclude that lipoxygenase and carotene oxidase are two distinct enzymes. Several methods are available to determine lipoxygenase activity (Koch *et al.* 1958; Surrey 1964; Mitsuda *et al.* 1967C). A direct assay is available of linoleate oxidation by lipoxygenase and hemeproteins in soy extracts (Ben-Aziz *et al.* 1970).

Proteinases

The more acceptable term proteinase will be used in place of proteases to refer to proteolytic enzymes. Early research on soybean proteinases is reviewed by Circle (1950). The name "Soyin" has been proposed for the soybean proteinases (Laufer *et al.* 1944) and for soybean hemagglutinin (Liener 1953). Soyin is no longer used for either nomenclature.

Conditions of storage govern whether proteinase activity (Taguchi and Echiga 1968) of raw, defatted soybean meal may either decrease to zero or increase 120–130%. Ofelt *et al.* (1955B) reported that proteinases of defatted soy flour are associated with water-insoluble components from which they can be extracted by treatment of a slurry or water suspension with sonic vibration or by vigorous treatment in a blender. These treatments yielded extracts having greater proteinase activity than that obtained by aqueous glycerol extraction. Recently, Weil *et al.* (1966) described purification and properties of six proteinases from the whey protein fraction of defatted soy flour. Eighty to ninety percent of the proteinase activity was extracted with water by mixing in a blender. Both Ofelt *et al.* (1955B) and Weil *et al.* (1966) report that soybean proteinases are papain-like. Pinsky and Grossman (1969) found that none of the six proteinase fractions had trypsin-like activity. These data indicate that soybean trypsin inhibitors may not inhibit proteolysis in soybean extracts.

Contrary to previous reports, Ofelt *et al.* (1955B) concluded that proteinases were not responsible for the softening effect of soy flour on bread doughs, since neither oxidizing nor reducing agents had any effect on proteinase activity of soy flour. Also, proteinase activity of commercial defatted soy flours is much lower than that in wheat flour.

Urease

No quantitative differences were found in the urease activity of three soybean varieties—Adams, Harosoy, and Lincoln. Urease activity of young seeds was only 3% of that in mature seeds (Birk and Waldman 1965). Also, no urease was found in leaves, stems, or empty pods. Smith *et al.* (1956) showed that urease activity varied according to variety and location. The hulls have very low urease activity and the hypocotyl, on a weight basis, has nearly twice the activity of cotyledons. Several factors that affect the urease assay were also determined. Papain contains a nondialyzable factor that accelerates urease activity in soybean extracts (Bahaden and Saxena 1964). Many plant seeds have much more urease activity than soybeans, whereas other seeds have almost no urease (Tai 1953). The inactivation of urease by moist heat provides a quality control guide to determine the degree of heat treatment received by various soy products (see Chap. 9).

PROTEINASE INHIBITORS

Significant advances in our understanding of the chemistry and action of proteolytic enzymes have resulted in new concepts in the structure-activity relationships between proteinases and proteinase inhibitors. Central to these advances and new concepts is the latest research on soybean trypsin inhibitors, which for this review are referred to as proteinase inhibitors.

The role of natural proteinase inhibitors in physiological and pathological reactions in man and animals has been reviewed (Weyer 1968; Vogel *et al.* 1968). Several reviews are available on the relationship between animal nutrition and soybean trypsin inhibitors (Rackis 1965; Liener and Kakade 1969); and on chemical and physical properties of plant proteinase inhibitors (Birk 1968; Vogel *et al.* 1968; Liener and Kakade 1969; Steiner and Frattali 1969). A series on trypsin inhibitors appears in a recent volume of *Methods in Enzymology* (Kassell 1970).

Isolation and Characterization

As of 1970, there may be at least 7 to 10 proteinase inhibitors in soybeans. Rackis *et al.* (1959, 1962) isolated two trypsin inhibitors designated $SBTIA_1$ and $SBTIA_2$. $SBTIA_2$ was identical to the Kunitz inhibitor (Kunitz 1946). Later, Rackis and Anderson (1964) separated four trypsin inhibitors in soybeans designated $SBTIB_1$ -B_2, -A_1, and -A_2. Birk *et al.* (1963A) isolated still another acetone-insoluble trypsin inhibitor, originally designated "purified inhibitor AA" but now referred to as the Bowman-Birk inhibitor (Frattali 1969). Yamamoto and Ikenaka (1967) obtained 2 soybean trypsin inhibitors, 1 of which appears to be identical with the Kunitz inhibitor and with $SBTIA_2$. The second inhibitor, designated 1.9S, contains 13 disulfide bonds per molecule compared to 2 disulfide bonds in the Kunitz inhibitor and $SBTIA_2$. Frattali and Steiner (1968) demonstrated three trypsin inhibitors, designated F_1, F_2, and F_3, in commercial preparations of Kunitz inhibitor. The major component, F_2, is apparently identical to the Kunitz inhibitor. Frattali (1969) suggested that the Bowman-Birk inhibitor and the 1.9S inhibitor of Yamamoto and Ikenaka (1967) may be identical. After recalculation of the original values (Eldridge and Wolf 1969), the sedimentation coefficients of 1.7–1.8 of A_1, B_1, and B_2 indicate molecular weights of 14,000–16,000. Although F_1 and F_3 have molecular weights of 18,000 and 23,000, respectively (Frattali and Steiner 1968), amino acid analyses indicate they are distinct from each other and from Kunitz and Bowman-Birk inhibitors. The relationship between the several proteinase inhibitors of the soybean is discussed by Obara *et al.* (1970).

Electrophoresis of native and reduced-alkylated soybean trypsin inhibitors, isolated according to Rackis and Anderson (1964), show them to be different proteins rather than intra- or intermolecular disulfide polymers of a single protein (Eldridge and Wolf 1969). Since the reaction between trypsin and Kunitz inhibitor is almost instantaneous and forms a complex with an extremely low dissociation constant, Lanchantin *et al.* (1969) developed a procedure for forming an enzyme-inhibitor complex on Kunitz inhibitor-equilibrated gel filtration columns. This technique and the immunochemical assay of Catsimpoolas *et al.* (1969B) may help to differentiate between the several trypsin inhibitors in soybeans.

The significance of the presence of multiple trypsin inhibitors in soybeans is unknown, although Frattali and Steiner (1968) regard them to be the result of genetic heterogeneity. That genetic variants of inhibitors may occur in some soybean strains was reported by Singh *et al.* (1969). The purity of commercial preparations continues to be a problem. Frattali and Steiner (1969B) have described a preparative electrophoretic procedure for purifying commercial preparations of two soybean proteinase inhibitors: Kunitz inhibitor (Kunitz 1946) and Bowman-Birk inhibitor (Birk *et al.* 1963A). Frattali (1969) has improved the procedure for isolating identical preparations of the Bowman-Birk inhibitor from two different soybean varieties, Lee and Hawkeye.

Catsimpoolas *et al.* (1969B) showed that commercial preparations of $SBTIA_2$ of Rackis *et al.* (1962) were identical to the classical Kunitz inhibitor by immunochemical and disc electrophoresis techniques. The amount of apparent impurities in the preparations varied from 0 to 43% and appeared to be a function of the acrylamide concentration in the gel and the mode of gel polymerization. If an antisoybean water-extract serum is used, the major band of the $SBTIA_2$ inhibitor is immunochemically identical with a corresponding protein in the total soybean water extract; therefore, the inhibitor is not an artifact of isolation. Immunospecificity of antiserum with $SBTIA_2$ was further ascertained by reaction of the antiserum with soybean whey proteins separated by isoelectric focusing (Catsimpoolas *et al.* 1969A). Although several of the whey protein components exhibited trypsin inhibitor activity, only the component with a peak at pH 4.47 corresponding to the major band in the $SBTIA_2$ preparation gave an immunoprecipitin band with antiserum.

Physicochemical and Enzymatic Properties

Except for the enzymatic aspects, little new information on the physicochemical properties of soybean trypsin inhibitors has appeared since the reviews of Liener and Kakade (1969) and of Steiner and Frattali (1969).

That Kunitz trypsin inhibitor is a globular protein of nonhelical conformation has been substantiated by numerous reports. Rotatory dispersion analyses suggest that the structure of the inhibitor resembles a flexible, fully disordered, random-coil (Jirgensons 1967). However, circular dichroism studies show that Kunitz inhibitor has a peculiar structure with no α-helix or β-structure in its native state (Jirgensons et al. 1969). The Bowman-Birk inhibitor associates reversibly to form a monomer-dimer mixture (Millar et al. 1969; Harry and Steiner 1969). These studies indicate that the true molecular weight is 8000 in agreement with the amino acid data of Frattali (1969). Previously, the inhibitor was reported to have a molecular weight of 20,000–24,000 (Birk et al. 1963A; Birk 1968).

Kunitz inhibitor has two disulfide bonds. The Arg(64)-Ile(65) bond which is cleaved by trypsin lies within one of the disulfide bridges (Ozawa and Laskowski 1966). Only one of the disulfide bonds is essential for activity (DiBella and Liener 1969). Carmel and Bohak (1968) have shown that after complete reduction of both disulfide bonds, the native inhibitor can be reformed and about 95% of its activity regained. Other physicochemical parameters of the reformed inhibitor were also identical to the native form. Kato and Tominaga (1970) have reported that the cleavage of the 2 methionine residues of the native Kunitz inhibitor, with 2 disulfide bonds remaining intact, yielded 2 inactive fragments which can associate to form an active noncovalently bonded derivative.

Liener and Kakade (1969) have summarized the enzymes that are and are not inhibited by plant proteinase inhibitors. Changes in this list for soybean proteinase inhibitors are discussed below:

Of the 5 protein fractions in pronase having proteolytic activity, only 1 had trypsin-like activity and was inhibited by Kunitz and Bowman-Birk inhibitors (Trop and Birk 1970). Other proteinases inhibited by Kunitz inhibitor include: kallikrein (Vogel et al. 1968); factor X_a esterase and thrombin (Lanchantin et al. 1969); cocoonase (Kafatos et al. 1967); and shrimp trypsin (Gates and Travis 1969). Human serum and plasma trypsins reportedly are inhibited by Kunitz inhibitor; however, human trypsin of the pancreas is not (Travis and Roberts 1969). Also Kunitz inhibitor does not inhibit sea pansy trypsins (Coan and Travis 1969), trypsin-like proteinases of starfish (Camacho et al. 1970), or the esterase activity of thrombin (Lanchantin et al. 1969). Evidence now indicates that Kunitz and Bowman-Birk inhibitors have no effect on the in vitro proteolytic activity of Tribolium larvae (Birk 1968).

Mechanism of Interaction

The chemical events leading to the cleavage of Kunitz inhibitor and resynthesis of a "new" trypsin inhibitor are shown in Fig. 6.1. The interaction of

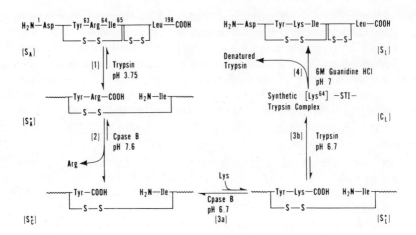

From Sealock and Laskowski (1969)
Courtesy of American Chemical Society

FIG. 6.1. MECHANISM OF INTERACTION OF KUNITZ SOYBEAN TRYPSIN INHIBITOR WITH TRYPSIN AND ENZYMATIC FORMATION OF A "NEW" TRYPSIN INHIBITOR

virgin Kunitz inhibitor (S_A) with catalytic amounts of trypsin leads to the cleavage of a single Arg(64)-Ile(65) peptide bond in the inhibitor to form an active modified inhibitor (S_A*) (Finkenstadt and Laskowski 1967; Ozawa and Laskowski 1966). Circular dichroism and fluorescence analysis show that formation of modified soybean trypsin inhibitor is optimal at pH 4.0 and that tyrosyl or tryptophyl residues are also involved (Ishida et al. 1970). The trypsin-modified [Arg(64)-Ile bond cleaved] inhibitor retained its antitryptic activity; however, removal of the newly formed arginyl terminal residue with carboxypeptidase B produced an inactive protein (des-64-Arg-modified) inhibitor (S_C*). Trypsin can resynthesize the Arg(64)-Ile bond to reform the virgin inhibitor (Finkenstadt and Laskowski 1967). The existence of the same equilibrium mixture of virgin and modified inhibitor when either form was incubated with trypsin provided additional evidence that a specific bond cleavage is the primary step in the interaction of trypsin-Kunitz inhibitor (Niekamp et al. 1969). By enzymatic reactions, Sealock and Laskowski (1969) have succeeded in adding a lysine group at position 64 (previously occupied by arginine) to form a synthetic inactive protein, modified-Lys(64)-inhibitor (S_L*). This synthetic protein, by forming a trypsin-inhibitor complex (C_L) can be converted into a free virgin [64-lysine]-inhibitor (S_L). This new protein is almost fully as active as the original Kunitz inhibitor (S_A).

There is a striking similarity between cocoonase and bovine trypsin (Kafatos et al. 1967). Recently Hixson and Laskowski (1970) provided evidence that both proteinases interact with Kunitz inhibitor at the same reactive site to hydrolyze and resynthesize the Arg(64)-Ile bond.

Thrombin also interacts with Kunitz inhibitor at the same reactive site as trypsin (Lanchantin *et al.* 1969). Both native and trypsin-modified Kunitz inhibitor [Arg(64) bond cleaved] effectively inhibit thrombin which, in turn, blocks the initial activation of prothrombin and the subsequent conversion of prothrombin to thrombin. Both thrombin and the thrombin-Kunitz inhibitor complex can cleave Arg or Lys bonds in fibrinogen. Trypsin and thrombin hydrolyze the same bonds (Arg-Ile and Lys-Ile) during the activation of trypsino-gen-chymotrypsinogen and prothrombin, respectively. As a result, most trypsin-like enzymes are inhibited by inhibitors by what appears to be a competitive inhibition mechanism because of the availability of Arg or Lys residues in the inhibitor. Therefore, the requirement of either an Arg or a Lys residue in the active site of trypsin inhibitors is well established (Haynes and Feeney 1968A; Fritz *et al.* 1969). Because of steric overlap or conformational changes, however, competitive inhibition can also occur when two enzymes have different reactive sites.

Feinstein *et al.* (1966) confirmed the findings of Laskowski's group with chicken ovomucoid, but not with that of turkey and cassowary. The last two retained inhibitory activity after modification with trypsin and subsequent treatment with carboxypeptidase. Although soybean and lima bean inhibitors combine only with catalytically active trypsin, Feinstein and Feeney (1966) demonstrated that inactive derivatives of trypsin and *a*-chymotrypsin reacted strongly with other proteinase inhibitors. Feinstein and Feeney (1966) suggest that conformational changes in modified proteinases may account for their inability to complex with these inhibitors. Modified Kunitz inhibitor lost its activity when the new *a*-amino group generated by cleavage with trypsin was amidinated. Amidination of the trypsin-modified chicken ovomucoid resulted only in a slow loss of activity. Unlike most trypsin inhibitors, however, native soybean trypsin inhibitor and chicken ovomucoid do not require amino groups for activity (Haynes *et al.* 1967).

Based on these findings, Haynes and Feeney (1968B) suggest that some trypsin inhibitors may inhibit proteinases by an entirely different mechanism from that proposed by Laskowski and co-workers. They concluded that limited proteolysis is not essential for inhibition. To act as proteinase inhibitors, a specific amino acid residue (specific for each proteinase) in the inhibitor mole-cule exhibits a high degree of affinity for the enzyme and, in turn, may resist proteolysis. Papaioannou and Liener (1970) have proposed a scheme for the involvement of specific tyrosine and amino groups in the interaction of trypsin and Kunitz inhibitor.

Nearly all the trypsin inhibitory activity of Kunitz inhibitor and chicken ovomucoid was abolished by modification of their arginyl residues with a selective reagent, 1,2-cyclohexanedione (Liu *et al.* 1968). A number of other trypsin inhibitors require lysyl residues for activity. It has been suggested that Kunitz inhibitor can be classified as an arginine-type inhibitor, which is insensitive to modification of its amino groups. Although native trypsin in the free state is readily inactivated by various agents, Pudles and Bachellerie (1968) showed that trypsin is stable as a trypsin-inhibitor complex. Many proteinase inhibitors inhibit several enzymes (Vogel et al. 1968). This inhibition does not always take place at the same reactive site. Such inhibitors have been named "doubleheaded" by Rhodes et al. (1960). Bowman-Birk inhibitor, which inhibits both trypsin and chymotrypsin on separate nonoverlapping reactive sites (Birk 1968), is a doubledheaded inhibitor. This inhibitor can be trypsin modified by cleavage of a single peptide bond (Lys-X or Arg-X), which leads to a change in the rate of reaction with trypsin (Birk et al. 1967; Frattali and Steiner 1969A). Subsequent treatment with carboxypeptidase B eliminates its antitryptic activity, but both the trypsin-modified and trypsin-modified carboxypeptidase B-treated inhibitors retain their chymotrypsin inhibitory activity (Birk et al. 1967). Chymotrypsin also cleaves a single peptide bond (Tyr-X or Phe-X) in the Bowman-Birk inhibitor. The modified inhibitor exhibits greatly reduced chymotrypsin inhibitory activity but maintains its trypsin inhibitory activity. An extensive discussion of enzyme-inhibitor interaction and specificity in molecular terms by Laskowski and Sealock (1971) appears in the third edition of The Enzymes.

Physiological Significance

In plants, most postulated functions for proteinase inhibitors fit into three categories: (a) maintain dormancy by preventing autolysis, (b) regulate protein synthesis and metabolism, and (c) prevent attack by predatory insects.

In mature soybeans, the proteinase inhibitors do not inhibit the proteolytic enzyme system (Ofelt et al. 1955B; Birk 1968; Pinsky and Grossman 1969). No inhibitory activity was detected in leaves, stems, and empty pods (Birk and Waldman 1965); however, trypsin-inhibiting activity amounting to 50% of that in the mature soybean is already present in young (3-week-old) seeds. No loss in trypsin-inhibiting activity occurred in soybeans germinated up to 1 week. Kakade et al. (1969) found that trypsin-inhibiting activity varies widely in several soybean varieties. Singh et al. (1969) pointed out that almost all soybean varieties contain the Kunitz inhibitor, but that in certain soybean lines this inhibitor is missing, and that another trypsin inhibitor is present. Possibly genetic variation of these 2 inhibitors is controlled by a single gene with 2 codominant alleles.

Because of the presence of trypsin-like enzymes in insects and bacteria that invade seeds, some specific metabolic defense mechanism against these invaders has been attributed to plant proteinase inhibitors. However, Kunitz inhibitor does not inhibit all trypsin-like enzymes (Travis and Roberts 1969). That a different proteinase inhibitor could function as a defense mechanism is indicated by isolation of an inhibitor designated C_1,V which inhibits the growth and proteolytic activity of *Tribolium confusum* larvae (Birk *et al.* 1963B; Birk 1968). The purified Kunitz inhibitor and the Bowman-Birk inhibitor had no effect on the *in vitro* proteolytic activity of *Tribolium* larvae. Although soybean trypsin inhibitors do not inhibit germination, Kunitz inhibitor strongly inhibits root growth of several plants including soybeans. Szilagyi (1968) postulates that the inhibitors block the incorporation of amino acids into proteins.

Provided that the sole function of proteinase inhibitors is selective inhibition of distinct proteinases, one must assume that the physiological function of these inhibitors is to protect the soybean against undue activation of proteolytic enzymes in the life of the soybean other than in the mature seed. It may be that apart from their proteinase inhibitory activity the inhibitors may have other enzymic properties; otherwise, their synthesis is rudimentary, reflecting a meaningless genetic cell function.

Assay Procedures

There are several modifications of the casein digestion procedure of Kunitz (1947). Rackis (1966) has indicated that the titration procedure of Wu and Scheraga (1962), with benzoyl-L-arginine ethyl ester as the synthetic substrate, was not suitable for determining antitryptic activity of crude soybean extracts. Kakade *et al.* (1969, 1970B) reported on several modifications for determining trypsin and chymotrypsin inhibitor activity in crude soybean extracts, which determine inhibitor activity in relation to the number of trypsin and chymotrypsin units inhibited, TUI and CUI, respectively.

Often, it is desirable to establish absolute amounts of trypsin inhibitor in soybean samples in order to compare results obtained by several investigators. To do so, a unit of trypsin-inhibiting activity (γ "pure" trypsin inhibited per milligram of protein) should be established. Most commercial preparations of crystalline soybean trypsin inhibitor contain appreciable amounts of impurities so that it becomes necessary to establish the purity of trypsin. Kakade *et al.* (1969) reported that the purity of their trypsin preparation was 56%; on this basis, 1γ of pure trypsin was calculated to have an activity of 1.79 trypsin units (TU) by the casein method and 1.90 TU by the benzoyl-DL-arginine-*p*-nitroanilide (BAPA) method. An international inhibitor unit defined as that amount which inhibits the activity of one international trypsin unit has been proposed by Fritz *et al.* (1968) (see also Vogel *et al.* 1968). The reliability of present assay procedures is much in doubt especially when applied to crude extracts of soy protein products and to toasted products having low trypsin-inhibiting activity.

A procedure for calculating the relative distribution of several trypsin inhibitors in various extracts from an analysis of "crossing diagrams" in electrophoresis has been described by Nakamura and Wakeyama (1961). Catsimpoolas and Leuthner (1969) have developed an immunochemical method for the quantitative determination of the Kunitz inhibitor.

Effect of Processing

Most plant proteinase inhibitors are inactivated by heat, an effect which is generally accompanied by an enhancement of nutritive value of the protein (Liener 1962). At 100° C, only 15 min of steaming is required to achieve maximum protein efficiency and to inactivate the trypsin inhibitors of either full-fat or defatted flakes (Rackis 1966). Trypsin-inhibitor activity of whole soybeans is also readily destroyed by atmospheric steaming for only 20 min, provided the beans are tempered to about 25% moisture before steaming. At lower moisture levels, more time or higher temperatures are required to inactivate the soybean trypsin inhibitors. When moisture of whole soybeans was raised to 60% or more by overnight soaking, boiling for only 5 min was sufficient to inactivate the inhibitor (Albrecht *et al.* 1966). Trypsin-inhibiting activity is destroyed at about the same rate as urease under similar cooking conditions (Albrecht *et al.* 1966). Rapid rate of inactivation of inhibitors in full-fat and defatted meal compared with whole soybeans, cotyledons, and chips was attributed to both particle size and the flaking process (Rackis 1966). Trypsin-inhibiting activity of soy milks, bread, and other products containing raw soybeans is readily destroyed by cooking (Liener and Kakade 1969), particularly at pH above 7 (Badenhop and Hackler 1970).

There was no evidence of pancreatic hypertrophy in rats fed tempeh, a fermented soybean product (Smith *et al.* 1964). In preparing the product, the beans are boiled for 30 min before fermentation, which would be sufficient heat to destroy the trypsin inhibitor. Presumably the trypsin inhibitor is also inactivated when the beans are ground and cooked before making into tofu (see Chap. 10).

Raw soy protein isolates contain sufficient levels of trypsin-inhibiting activity to cause significant pancreatic hypertrophy in rats (Rackis *et al.* 1963). The trypsin-inhibiting activity of the raw protein isolates represents incomplete separation of the whey proteins during the process of precipitating the protein curd. Soybean whey accounts for at least 70% of the trypsin-inhibiting activity of raw soybean meal; and the water-insoluble residue—a by-product in the manufacture of isolates—contains low residual trypsin-inhibiting activity (Rackis 1966).

Trypsin inhibitor content of soy isolate infant formulas vary widely (Theuer and Sarett 1970). An edible grade of soy protein isolate was found to contain about 4 μg STI per 100 μg protein (assuming a stoichiometric relationship between STI and the amount of trypsin inhibited). Additional heat treatment improved the nutritional value of other commercially available soybean concentrates and isolates (Longenecker *et al.* 1964). The level of STI remaining in a soy product can serve as an index of the adequacy of heat treatment of soy milks (Van Buren *et al.* 1964).

Nutritional and Physiological Effects of Proteinase Inhibitors

General Aspects.—By live steam treatment, a process referred to as toasting, the nutritive value of raw soy flour can be improved to that for meat and milk. Only 10-15 min of steaming at atmospheric pressure is required (Rackis 1965; Albrecht *et al.* 1966). The increase in nutritive value parallels the destruction of trypsin inhibitor activity as shown in Fig. 6.2. Pancreatic hypertrophy also decreases as inhibitors are destroyed. Toasted soybean meal is nutritionally superior to raw meal when fed to swine, chicks, rats, mice, turkey poults, and humans.

Raw meal causes growth inhibition, depresses metabolizable energy of the diet, reduces fat absorption, enlarges the pancreas, and stimulates hypersecretion of pancreatic enzymes in chicks, mice, and rats. Kidney transaminase activity is reduced in rats fed raw soybean meal, while other tissue enzymes showed no change (Borchers 1964).

Chicks fed raw soybean meal consumed five times the amount of oxygen and had much lower liver and muscle glycogen contents than chicks fed autoclaved soybean meal (Saxena *et al.* 1962). Amino acid supplements overcame these effects.

No enterokinase is found in the intestinal juice of chickens fed raw soybean diets (Lepkovsky *et al.* 1970B).

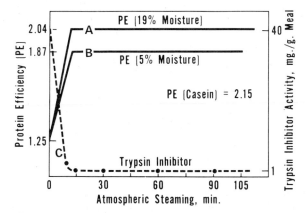

FIG. 6.2. EFFECT OF AUTOCLAVING ON PROTEIN EFFICIENCY (PE) AND TRYPSIN INHIBITOR ACTIVITY OF RAW SOYBEAN MEAL

Conditions: Live steam at atmospheric pressure, 100°C; curve A, PE of meals with 5% moisture before autoclaving; curve B, PE of meals with 19% moisture before autoclaving; curve C, decrease in trypsin inhibitor activity with time of autoclaving. Mean pancreas weight in rats: gm/100 gm body weight; casein control, 0.51; raw meal, 0.73; meal toasted 15 min, 0.49.

Soybean whey proteins, representing 6-8% of total protein in raw defatted meal, account for most of these effects and toasting of the whey destroys the factors (Rackis *et al.* 1963; Garlich and Nesheim 1966). Kunitz inhibitor (Rackis 1965; Garlich and Nesheim 1966; Sambeth *et al.* 1967), Bowman-Birk inhibitor (Gertler *et al.* 1967), and SBTIA$_1$ (Lyman *et al.* 1962), which are all in the whey protein, have similar effects on chicks and rats but not to the same extent as raw meal or whey protein. *p*-Aminobenzamidine, a synthetic trypsin inhibitor, caused growth inhibition, pancreatic enlargement, and exaggerated secretion of pancreatic enzymes in rats similar to that produced by the natural trypsin inhibitor of navy beans (Kakade *et al.* 1970A). In comparative rat feeding tests, Rackis (1965) found that Kunitz inhibitor can account for all the pancreatic hypertrophic effects and for about 30-60% of the growth-inhibiting properties of raw meal. Garlich and Nesheim (1966) reported that Kunitz inhibitor depressed growth rate and metabolizable energy and caused pancreatic hypertrophy in chicks. Other whey protein fractions greatly potentiate the antinutritional effects of the trypsin inhibitor in rats and chicks (Sambeth *et al.* 1967). These studies demonstrate that multiple antinutritional factors are present in raw meal. Of considerable interest in this respect is the recent isolation of a low molecular fraction from soybean whey which inhibits mouse growth without causing pancreatic hypertrophy (Schingoethe *et al.* 1970). The fraction which is free of trypsin inhibitor activity contains several peptides, one of which may be a glycopeptide. Rackis (1965) previously had reported that a whey dialyzate fed at twice the level normally present in raw meal inhibited rat growth and caused pancreatic hypertrophy even though trypsin inhibitor activity was low.

The water-insoluble residue fraction also depresses growth and causes pancreatic hypertrophy in rats and chicks (Rackis *et al.* 1963; Garlich and Nesheim 1966). Bielorai and Bondi (1963) state that the residue fraction, even after extensive extraction, inhibits tryptic digestion of casein and fish meal. Rackis (1966), however, showed that the low level of trypsin-inhibiting activity of the residue was not associated with the known proteinase inhibitors. A growth inhibitor in soybeans may exist in a bound form (Borchers 1963).

The whey protein feeding results of Rackis *et al.* (1963) and Garlich and Nesheim (1966) are in direct contrast to those of Saxena *et al.* (1963B) and Pubols *et al.* (1964), who reported that the residue and not the whey protein depressed growth and caused hypertrophy. The contradictory results of these latter workers can be attributed to their low yields of the crude trypsin inhibitor fraction (whey proteins), since commercial "brew" flakes, a mildly cooked soybean meal, were used. Their reported yield of the whey fraction was 0.5% compared with the usual yield of at least 3% (Rackis *et al.* 1963). Nesheim and Garlich (1966) have observed that some samples of brew flakes have 25-50% of the trypsin-inhibiting activity of raw meal prepared in the laboratory.

Some other conflicting reports in the literature may be the failure to assay for trypsin-inhibitor activity. Commercial trypsin-inhibitor preparations vary widely in purity (Rackis *et al.* 1962; Eldridge *et al.* 1966) and may even be devoid of activity (Garlich and Nesheim 1966). Complicating the situation further is the finding that some of the soybean proteinase inhibitors cause pancreatic hypertrophy but differ in their effect on growth and metabolizable energy (Gertler *et al.* 1967; Garlich and Nesheim 1966). Furthermore, the various experimental techniques applied to many different animals have never been evaluated in relation to one another.

Gorrill and Thomas (1967) and Gorrill *et al.* (1967) have observed poor growth, but reduced trypsin and chymotrypsin secretion, and no pancreatic hypertrophy in young calves fed raw meal. Raw soy flour does not cause pancreatic hypertrophy in pigs, and pancreatic juice secretion is lower than with a heated soybean meal diet (Pekas 1966). Pancreatic hypertrophy occurs in laying hens (Bray 1964). Adult rats fed raw meal maintain body weight in spite of prolonged pancreatic hypertrophy (Booth *et al.* 1964).

Pancreatic Hypertrophy and Enzyme Secretion.—In rats, maximum hypertrophy occurs in nine days of feeding raw meal (Rackis 1965). Increased pancreatic secretion is almost immediate in rats (Lyman *et al.* 1962), whereas in chicks hypertrophy and pancreatic juice secretion is delayed days following the feeding of soybean trypsin inhibitor (Lepkovsky *et al.* 1965; Kakade *et al.* 1967; Nitsan and Alumot 1964; Nitsan and Bondi 1965). Pancreatic hypertrophy and increased enzyme secretion also occur in rats fed a protein-free amino acid diet containing soybean trypsin inhibitor. Hypertrophy occurs also in germ-free chicks fed raw soybean meal (Coates *et al.* 1970).

Some studies indicate enlargement of the pancreas is hyperplasia of the acinar cells in which there is an increase in cell number (Kakade *et al.* 1967; Salman *et al.* 1968; Bredenkamp and Luck 1969) as well as a depletion of zymogen granules (Salman *et al.* 1967). Others report that pancreatic enlargement is caused by an increase in cell size; i.e., hypertrophy (Konijn and Guggenheim 1967; Saxena *et al.* 1963C). Pancreatic islets as well as other organs are normal (Booth *et al.* 1960B; Bielorai 1969). No histopathological damage was observed in rat pancreas hypertrophied for six months, and pancreatic hypertrophy was reversible in rats (Booth *et al.* 1964) and in chicks (Salman and McGinnis 1969). Growth rate of rats fed raw meal is comparable to toasted meal diets when supplemented with certain essential amino acids (Booth *et al.* 1960B; Khayambashi and Lyman 1966), or more protein and calories (Fisher and Shapiro 1963), but hypertrophy still occurred. However, animal growth and protein efficiency of toasted meal diets can be further increased when supplemented with amino acids, particularly methionine.

According to Lyman and Lepkovsky (1957) and Lyman (1957), excessive amounts of pancreatic enzymes are secreted into the intestine, some of which are then excreted in the feces when raw meal and trypsin-inhibitor concentrates are fed to rats. Trypsin inhibitors do not inhibit proteinase activity in the intestinal tract of mice and rats (Haines and Lyman 1961; Nitsan and Bondi

1965). In chicks, the zymogens accumulate in the acinar cells (Saxena *et al.* 1963C). The delay in secretion of the enzymes, together with higher levels of trypsin inhibitor in the intestines of chicks, because of less peptic inactivation of the inhibitors, leads to temporary inhibition of proteolysis. Many workers suggest that decreased protein digestion rather than loss of endogenous protein is responsible for growth inhibition in chicks fed raw meal. Bielorai and Bondi (1963) and Bielorai (1969) found a correlation between low pancreatic digestion of raw soybean meal and content of antitryptic factors in raw soybean and groundnut meals.

Depancreatized chickens fed raw soy meal gained less weight and utilized food less efficiently compared with heated soy diets. As in intact chickens, there is an increase in trichloroacetic acid precipitable nitrogen in the intestinal contents of depancreatized chickens fed raw soy (Lepkovsky *et al.* 1970A). Based on these findings and an analysis of enzymatic activity in the intestines, they suggest that the reduction in protein digestion, because of the presence of inhibitors, rather than the endogenous loss of nitrogen resulting from the excessive pancreatic juice secretion, is primarily responsible for the poor nutritive value of raw soybeans. Growth depression and an increase in intestinal nitrogen as protein also occurs in germ-free chicks fed raw soybean meal (Coates *et al.* 1970). Since growth inhibition in conventional chicks was significantly greater than in germ-free, it was postulated that the intestinal microflora potentiate the antinutritional effects of raw soybeans by formation of additional factors due to microbial action on heat-labile components in raw soy meal.

Saxena *et al.* (1963B) concluded that the lack of specific enzymes was responsible for the incomplete digestion of soybean protein and that part of the undigested protein is absorbed to produce hypertrophy. There is a delay in release of pancreatic enzymes in young chicks fed soybean trypsin inhibitor (Lepkovsky *et al.* 1965; Kakade *et al.* 1967; Nitsan and Alumot 1964; Nitsan and Bondi 1965), which may be responsible for decreased protein digestion and growth inhibition in chicks up to three weeks of age (Bielorai and Bondi 1963). After three weeks of age, proteolytic activity in the intestine increases continuously, and ultimately, there is a loss of nitrogen and sulfur almost entirely of pancreatic origin (Lepkovsky *et al.* 1959).

Trypsin, which contains 8.7% cystine, accounts for 1/2 of the cystine excreted by the rat fed raw soybeans (Barnes *et al.* 1965B). The first limiting amino acids in soybean protein are cystine plus methionine. Barnes *et al.* (1965B) attribute the beneficial influence of antibiotics on overcoming the growth inhibiting effects of raw meal in chicks and rats by preventing sulfur amino acid degradation by the intestinal microflora. Kakade *et al.* (1970A) suggest that antibiotics increase the intestinal absorption of cystine sufficiently to meet the requirement of cystine for synthesis of pancreatic enzymes. Loss of protein may also occur from a sloughing off of intestinal mucosa in rats (DeMuelenaere 1964).

Barnes and Kwong (1965) and Barnes *et al.* (1965A) proposed that Kunitz inhibitor creates a cystine deficiency which then enhances the conversion of methionine to cystine. The increased synthesis of cystine in the pancreas was 7-10 times greater than in the liver. Frost and Mann (1966) found that Kunitz inhibitor interferes with the incorporation of cystine into protein and with the enzyme, cystathionine synthase, whereas Borchers *et al.* (1965) report that a factor in soybeans interferes with metabolism of threonine and valine. These same amino acids, together with methionine, produced maximum growth in rats fed raw soybean meal (Booth *et al.* 1960B; Borchers 1962).

Homogenates of hypertrophied pancreas of chicks fed raw meal have higher amounts of trypsin and chymotrypsin and lower levels of amylase (Salman *et al.* 1967). Nitsan and Gertler (1969) said that pancreas of chicks also contained increased levels of elastase when fed raw soybean meal. Supplementation of trypsin inhibitors to a heated diet increased proteolytic activity as did raw meal, but amylase activity increased instead of decreasing. Dal Borgo *et al.* (1968) found that total secretion of trypsin nearly doubled in chicks fed raw meal, whereas amylase, lipase and chymotrypsin activity was not significantly different from that of autoclaved meal. These effects were much greater when starch rather than glucose was the carbohydrate source.

Dietary conditions can have an appreciable effect on changes in pancreatic enzymes (Haines and Lyman 1961; Lyman *et al.* 1962; Goldberg and Guggenheim 1964; Lepkovsky *et al.* 1965, 1966; Pubols *et al.* 1964). Ma'ayani and Kulka (1969) concluded raw soybeans have two different effects on the pancreas: (a) to increase the secretion of digestive enzymes in general and (b) to increase synthesis of proteolytic enzymes specifically. Therefore, coupled with an increased secretion of all pancreatic enzymes and particularly trypsin into the intestines, there is a higher proteolytic activity and reduced amylase in the pancreas itself. These and other workers postulate that a regulatory mechanism adjusting pancreatic function to proteolytic activity in the intestine is stimulated by the soybean trypsin inhibitor. Since the trypsin inhibitor does not stimulate increased synthesis of amylase, a greater depletion of amylase would result from the increased secretion of pancreatic juice.

Considerable changes occur in the enzyme profile, both in pancreas tissue and its juice secretion, when soybeans are fed to rats and chicks. Konijn *et al.* (1970A,B) regard these changes in rat pancreas as an adaptive mechanism similar to that which occurs with a protein-rich diet: an increase in the synthesis of trypsin and chymotrypsin and a decrease in lipase, elastase, and amylase. These latest results imply that the Bowman-Birk inhibitor is primarily responsible for the pancreatic effect. Konijn *et al.* (1970B) also indicate that the immediate response of the pancreas in rats is a decrease in protein synthesis similar to that observed in chickens. Synthesis of selective enzymes then increases under prolonged stimulation by both the Kunitz and Bowman-Birk inhibitors resulting ultimately in hypertrophy and loss of endogenous protein.

According to Lepkovsky and Furuta (1970), raw soybean meal inhibits the synthesis of lipase in the pancreas and appears to stimulate excessive secretion of intestinal lipase in chickens.

Proposed Mechanisms.—The mechanism whereby unheated soybean meal causes poor fat absorption in young chicks in unknown (Garlich and Nesheim 1966). However, the same protein fractions that cause pancreatic hypertrophy, contract the gallbladder, accelerate bile secretion, and decrease fat absorption (Sambeth et al. 1967). Serafin and Nesheim (1970) have reported that undigested protein in raw soybean meal may bind bile acids and elevate their rate of fecal excretion, thereby depressing fat absorption as much as 50% below normal in young chicks (up to 2 weeks of age).

According to Nesheim and Garlich (1966), who discuss the reasons for divergent results recorded by other workers, depression in protein digestibility may be responsible for the difference in metabolizable energy value between raw and toasted meals in chicks. This proposal would be consistent with impaired proteolysis observed in young chicks (Alumot and Nitsan 1961) and with reduced nitrogen digestibility in chicks with ileostomies (Lepkovsky et al. 1965). And yet, only a slight decrease in protein digestibility occurs in rats (DeMuelenaere 1964). Although metabolizable energy of a raw meal diet remains unchanged during prolonged feeding (Nesheim and Garlich 1966), inhibition of proteolysis in the intestinal tract of chicks disappears with age (Nitsan and Alumot 1964). Since extra calories and protein counteract growth-inhibiting properties of raw meal (Fisher and Shapiro 1963), the utilization of energy and protein is apparently altered. Deviations from normal carbohydrate metabolism observed in rats adapted to raw meal may result from restricted food intake (Goldberg 1970). Since more glucagon and insulin-like activity was secreted by the pancreas, it was suggested that diverting the protein to gluconeogenesis may partly explain growth inhibition of rats fed raw meal.

Two inhibitory, growth-impairing mechanisms resulting from raw soybean meal in the diet have been suggested by Gertler et al. (1967): (a) The trypsin inhibitors stimulate pancreatic hypertrophy and synthesis of proteolytic enzymes, thus increasing the requirements for amino acids and ultimate loss of endogenous nitrogen. For the chick, at least initially, the temporary inhibition of proteolysis by the trypsin inhibitor would further increase the amino acid requirement. (b) Raw meal contains a protein fraction which becomes digestible only after heating, as evidenced by lower digestibility of raw meal or greater growth-inhibiting properties of raw meal compared with toasted meal supplemented with some of the trypsin inhibitors (Gertler et al. 1967; Rackis 1965).

Saxena et al. (1963B) concluded that lack of specific enzymes was responsible for the incomplete digestion of dietary protein and that part of the undigested protein is absorbed to produce pancreatic hypertrophy. Liener and Kakade (1969) have proposed a scheme that would explain the effect which trypsin inhibitors have on the metabolism of certain amino acids, as well as exogenous and endogenous losses of protein, and on the nutritive value of soybean protein. Hormonal influences should be considered more carefully. Khayambashi and Lyman (1969) noted that the effect of soybean trypsin inhibitor on the secretion of pancreatic enzymes may be a hormonal stimulus. Konijn *et al.* (1969, 1970A) suggest that the trypsin inhibitors or other heat-labile substances in soybeans elicit the formation or secretion of pancreozymin, which can also cause pancreatic hypertrophy and increased secretion of enzymes. The pronounced secretion of pancreatic enzymes within 30 min following the ingestion of soybean proteinase inhibitor (Lyman *et al.* 1962), coupled with the observation that the gallbladder contracts and secretes bile, indicated to Melmed and Bouchier (1969) that the pancreozymin-cholecystokinine hormone might be a mediator of the trypsin inhibitor-induced effect on the pancreas. These workers suggested that the primary function of the endogenous pancreatic trypsin inhibitor is to activate hormonal factors that stimulate enzyme synthesis in acinar cells for repletion of digestive enzymes and that dietary trypsin inhibitors have the same effect. The fact that trypsin inhibitors account for only 30–60% of the growth-inhibiting properties of raw meal (Rackis 1965) and that other whey protein fractions potentiate the antinutritional effects of the trypsin inhibitor (Sambeth *et al.* 1967), may indicate that the interaction of trypsin inhibitors with the growth inhibitor factor (Schingoethe *et al.* 1970) is required in order to obtain the complete antinutritional effects of raw meal.

HEMAGGLUTININS

Various terms have been used interchangeably to refer to plant proteins that agglutinate red blood cells: phytohemagglutinins, phytagglutinins, and lectins. Initially, a hemagglutinating protein isolated from soybeans was called soyin, but at present the name "soybean hemagglutinin" is preferred since a proteolytic enzyme had earlier been called soyin. Phytohemagglutinins are reviewed by Jaffe (1969).

Isolation and Characterization

A hemagglutinin from soybeans was first isolated by Liener and Pallansch (1952). The presence of multiple hemagglutinins has been reported by several workers (Rackis *et al.* 1959; Stead *et al.* 1966; Lis *et al.* 1966). Four different forms of hemagglutinins, designated A, B, C, and D, were characterized by isolectric focusing by Catsimpoolas and Meyer (1969). The four hemagglutinins are immunochemically identical. The major hemagglutinin (form B) is identical with that first described by Liener and Pallansch (1952) and is dissociable into two subunits.

All four hemagglutinins are glycoproteins containing mannose and glucosamine. The major hemagglutinin contains 4.5% mannose and 1% glucosamine. Lis *et al.* (1966) showed that the amino acid composition of the four proteins was very similar but that there were differences in carbohydrate content. Catsimpoolas and Meyer (1969) suggest that the multiple forms of hemagglutinin with different isoelectric points may be formed by hydrolysis of labile amide groups of glutamine or asparagine residues and of mannose residues by α-mannosidase during isolation. The hemagglutinins are concentrated in the whey protein fraction.

Chemical modification and inactivation studies have been reviewed by Jaffe (1969). Lis *et al.* (1969) have shown that the glycopeptide isolated from soybean hemagglutinin contains mannose, N-acetylglucosamine, and aspartic acid and that the aspartic acid is linked to N-acetylglucosamine in the form of 1-L-β-aspartamido(2-acetamido)-1,2-dideoxy-β-D-glucose.

Biological Effects and Detection

Toxicity of soybean hemagglutinin when injected into mice is comparable to many other plant lectins, but ricin is a thousand times more toxic. The LD $_{50}$ for soybean hemagglutinin for young rats is about 50 mg per kg, but it has no lethal action when administered to rats by stomach tube at a level of 500 mg per kg (Liener and Rose 1953). According to Liener (1953), hemagglutinin is responsible for about 50% of the growth inhibition of rats fed raw soybean meal. Birk and Gertler (1961) found that hemagglutinins are responsible for only a small part of the inhibition of growth of chicks and rats, since hemagglutinating activity compared to trypsin-inhibiting activity was readily inactivated by gastric digestion.

Raw soy flour contains about 3% hemagglutinin (Liener and Rose 1953). The ability of soybean extracts to agglutinate blood cells of many animals varies widely (Jaffe 1969). Soybean extracts agglutinate human blood cells at low temperature. Detection and tests used for determining hemagglutinating activity are discussed by Jaffe (1969). A photometric method devised by Liener (1955) has proved successful for the quantitative determination of hemagglutinating activity, particularly for soybeans. Using this procedure, DeMuelenaere (1965) and Stead *et al.* (1966) both state that hemagglutination and toxicity appear to be associated with the same factor. Although crude soybean extracts usually contain both trypsin-inhibiting properties and hemagglutinating activity, intraperitoneal injection of crystalline Kunitz inhibitor is nontoxic (DeMuelenaere 1965). Many reports indicate that hemagglutinating activity of raw soybeans is readily destroyed by moist heat treatment.

ALLERGENIC FACTORS

Only a few actual cases of soybean allergenicity have occurred in spite of the great increase in the use of soybeans for food. In fact, soybean milk prepared from soy flour is often used as a hypoallergenic substitute for infants allergic to human milk and cow's milk. Ratner *et al.* (1955), in agreement with Glaser and Johnstone (1953A,B), concluded that soybean milk is an ideal hypoallergenic foodstuff and that soybean oil and soy sauce were devoid of allergenicity. Milks, based on soy protein isolates, effectively alleviate allergenic symptoms while maintaining an adequate nutritional state (Cowan *et al.* 1969).

In 7 yr, only 2 cases of soybean sensitivity were documented among 1500 employees in a soybean meal processing plant (Ratner *et al.* 1955). Asthmatic symptoms have occurred in a few employees of a plywood plant where soy flour was used as a glue (Perlman 1965). Less than 15% of newborn infants developed some clinical manifestations when human milk or cow's milk was replaced completely with a soy milk formula. Soybeans are an innately weak antigen in guinea pigs (Ratner and Crawford 1955) and in humans (Ratner *et al.* 1955). Sensitization of guinea pigs to raw soybean protein extracts was not enhanced by multiple sensitizing injections or inhalation of dust from raw soy flour. A soybean allergen has been isolated by Spies *et al.* (1951). The preparation, however, is quite heterogeneous as indicated by Sephadex chromatography and ultracentrifugation studies (Rackis, unpublished data). Several workers have said that the soybean allergen is quite thermostable (Ratner *et al.* 1955; Crawford *et al.* 1965; Perlman 1965). A strong positive skin reaction can be obtained with soybean preparations heated at $100°$ C for 30 min. Heating to $180°$ C for 30 min was required to almost abolish the skin reaction (Perlman 1965). Only 15 min of steaming at $100°$ C inactivates the antinutritional factors in raw soy flour and achieves a maximum protein efficiency ratio (Rackis 1965).

FLATUS FACTORS

A recent monograph summarizes for the first time what is now known and what is needed for future investigations on gastrointestinal gas (Berk 1968). Flatulence has been attributed to swallowed air, bacterial fermentation, gastrointestinal secretions, and ingestion of many foods, including soybeans. The main types of flatus are nitrogen, carbon dioxide, methane, and hydrogen gases depending upon the individual diet and microflora spectrum in the intestinal tract. Oxygen and malodorous volatile substances including ammonia, volatile amines, hydrogen sulfide, and acids also may occur in small amounts. Although the egestion of rectal gas is the most common complaint, nausea, cramps, diarrhea, and even pain may occur in varying degrees.

Human Studies

Experiments in which four adult male subjects consumed various toasted soybean products commercially manufactured (Steggerda *et al.* 1966) revealed that the gas-producing factors reside mainly in the low-molecular-weight carbohydrate fraction. Data in Table 6.2 show that individuals vary widely in the amount of flatus produced and that whey solids and 80% ethanol extractives have the highest flatus activity. Whey solids and ethanol extractives contain about 60 to 80% water-soluble carbohydrate, respectively. The carbohydrates exist primarily as sucrose, raffinose, and stachyose. Soybean hulls, fat, water-insoluble polysaccharides (residue product), and protein are not associated with

TABLE 6.2

EFFECTS OF SOY PRODUCTS ON FLATUS IN MAN

Product[1]	Daily Intake (Gm)	Flatus Volume (cc/hr) Average	Range
Full-fat soy flour	146	30	0–75
Defatted soy flour	146	71	0–290
Soy protein concentrate	146	36	0–98
Soy proteinate	146	2	0–20
Water-insoluble residues[2]	146	13	0–30
Whey solids[3]	48	300[4]	–
80% Ethanol extractives[3]	27	240	200–260
Navy bean meal	146	179	5–465
Basal diet	146	13	0–28

[1] All products were toasted with live steam at 100° C for 40 min.
[2] Fed at a level three times higher than that present in the defatted soy flour diets.
[3] Amount equal to that present in 146 gm of defatted soy flour.
[4] One subject; otherwise four subjects per test.

flatulence production to any significant degree (Steggerda *et al.* 1966). Flatus production in humans consuming pork and beans can be effectively inhibited by the incorporation of an antimicrobial compound into the diet (Steggerda 1968). High levels of sodium caseinate or soybean proteinate also appear to inhibit flatulence in humans (Steggerda *et al.* 1966).

Calloway and Murphy (1968) claim that analysis of expired air can be used as a measure of intestinal gas formation. These workers found that the water-soluble oligosaccharides, raffinose and stachyose, also evoke the usual flatulent patterns of breath hydrogen which occur with many bean diets. Levitt and Ingelfinger (1968) have shown that hydrogen and methane occur in human flatus.

Flatus formed during flatulent conditions is composed chiefly of gas produced by fermentation of carbohydrates. The types of microorganisms present in the intestinal tract are of primary importance in determining the composition and volume of gas produced (Gall 1968; Steggerda 1968; Levitt and Ingelfinger 1968; Calloway and Murphy 1968). An analysis of orally and rectally expelled hydrogen and methane production in man indicates that the pattern of expelled gases reflect differences in type, site, and abundance of intestinal microorganisms (Calloway and Burroughs 1969).

In Vivo and In Vitro Studies

Richards and Steggerda (1966) and Rackis *et al.* (1970B) have shown that more than 80% of the gas production in surgically prepared intestinal segments of anesthetized dogs incubated with navy bean and soybean meal homogenates occurred in the ileum and colon. Richards and Steggerda (1966) also found that intestinal gas production in segmented intestinal loops was completely inhibited by bacteriostatic agents (Vioform and Mexaform) and antibiotic mixtures which effectively destroy anaerobic spore-forming bacteria *(Clostridia)* in the intestinal tract. Vioform and Mexaform inhibited flatulence in humans also (Steggerda 1968). Other intestinal microflora may be capable of producing gas from various substrates including stachyose in the ileum and colon (Calloway 1966; Calloway *et al.* 1966). Kawamura and Kasai (1968) examined the ability of *E. coli* to decompose soybean oligosaccharides. Antibiotics greatly reduce the utilization of oligosaccharides compared to sucrose (Yoshida *et al.* 1969).

An *in vitro* technique previously developed by Richards *et al.* (1968) was used by Rackis *et al.* (1970B) to show that toasted, dehulled, defatted soybean meal contains a gas-producing and gas-inhibiting factor(s) that can be extracted with aqueous ethyl alcohol. The oligosaccharides—sucrose, raffinose, and stachyose—are associated with the gas-producing factor when incubated in thioglycollate media with anaerobic bacteria taken from the ileal and colonic intestinal segments of dogs. Some of the phenolic acids in soybean meal—syringic and ferulic acid—are effective gas inhibitors *in vitro* and in intestinal segments of dogs (Rackis *et al.* 1970B).

In vitro studies also indicate that the soybean products that cause flatulence in humans (Steggerda *et al.* 1966) were equally effective in producing gas when incubated with anaerobic cultures isolated from dog colon biopsies. Only those complex polysaccharides that can be enzymatically hydrolyzed to simple mono-saccharides act as substrates to produce flatus (Rackis *et al.* 1970B).

Other Considerations

Several reports now indicate that the fermentative degradation of carbohy-drates by the microflora in the ileum and colon is the primary factor causing flatulence in humans consuming certain foods. The apparent increase in anaerob-ic bacteria in the intestine stimulated by the ingestion of some flatulent foods (Rackis *et al.* 1970A; Rockland *et al.* 1970), coupled with an increase in fermentable carbohydrates in the lower intestine, results in a rapid rate of gas production with the formation of high concentrates of CO_2 and H_2. The nature of the carbohydrates can vary depending upon the flatulent food consumed. Gitzelmann and Auricchio (1965) have shown that a-galactosidase activity was not present in human intestinal mucosa, since after ingestion of raffinose and stachyose by a normal and a galactosemic child, there was no absorption of galactose. Hydrolysis products of raffinose and stachyose, however, were found in the feces. Calloway *et al.* (1966) observed that human ileal and colonic microflora are able to utilize stachyose and produce gas. As a result, bacterial hydrolysis in the lower intestine of the raffinose and stachyose in soy products can be one source of flatus.

According to Hellendoorn (1969), the flatulent effect of beans is caused by fermentation of undigested starch, which reaches the lower intestine because of accelerated passage down the gastrointestinal tract. Other factors in flatulent foods that interfere with transportation of gases across the intestinal wall and that inhibit carbonic anhydrase have been postulated as contributing to flatulence (Calloway and Murphy 1968). Omitted from this review is the prevalence of flatulence in certain racial groups with intestinal lactase deficiency and other carbohydrate intolerances.

SAPONINS

An informative review of saponins, including the latest on soybean saponins, has been written by Birk (1969). Briefly, whole soybeans contain about 0.5% saponin. Isolated soybean saponins do not harm chicks, rats, and mice even when fed at a level three times higher than the level in diets containing soy flour. Neither saponins nor sapogenins could be detected in the blood of these animals. Purified soybean saponins have little or no taste and are not bitter (Rackis *et al.* 1970A), contrary to frequent reports in the literature. Soybean saponins are more complex than previously recognized (Wolf and Thomas 1970).

STEROLS AND TRITERPENE ALCOHOLS

Sterols (phytosterols) are widely distributed in the vegetable kingdom. Plant sterols are poorly absorbed through the intestinal mucosa. Plant sterol recovery in the feces was complete in all men when a diet of mixed general foods was consumed; however, in one subject fed several formula diets, only 25–58% of the ingested plant sterols were recovered from the stool (Denbesten *et al.* 1970). The loss of sterols was attributed to bacterial degradation in the lower gut. Nevertheless, what appears to be β-sitosterol is present in small amounts in practically all animal tissue and appears to be derived from the diet (D'Hollander and Chevallier 1969). Miettinen (1967) demonstrated that in the adrenals of rats on an ordinary diet, plant sterols account for about 5% of the total sterols. The significance of the presence of plant sterols in animal tissue is not known. Although an exhaustive literature survey was not made for this chapter, apparently plant sterols do not have a detrimental effect on the physiological and nutritional status of the animal. Soy sterols (1% of the diet) appear to have a nonabsorptive antihypercholesterolemic action in the chicken (Konlande and Fisher 1969).

Kiribuchi *et al.* (1965, 1966, 1967) believe that soybeans contain four classes of sterols: free, esterified, glucosylated, and acylated steryl glucosides. The major aglycone moieties consist of campesterol, stigmasterol, and β-sitosterol. Kiribuchi and co-workers also detected the presence of two triterpene alcohols— cycloartenol and cyclolaudenol. Soybean oil contains 7 triterpenes (Fedeli *et al.* 1966) 4 of which have been identified: a-amyrin, β-amyrin, cycloartenol, and cyclolaudenol. Palmitic, stearic, oleic, linoleic, and linolenic acids are the main acyl components of acylated steryl glucosides of soybeans. Trace amounts of myristic, arachidonic, and behenic acids were also present (Kiribuchi *et al.* 1966). The four classes of sterols can be extracted with acetone from ground whole soybeans and are found in crude phospholipid fractions (soybean foots) obtained as by-products in the refining of crude soybean oil. Except for the esterified form, the other three classes of soy sterols are present in small amounts in defatted soybean flakes (Honig *et al.* 1969).

GOITROGENS

VanEtten (1969) has made a thorough review of goitrogens and has evaluated the nature of the goitrogenic products of thioglucosides.

Soybean Goitrogenicity

Since 1933, many reports indicate that soybeans cause thyroid enlargement. Although the lack of iodine is the principle cause of soybean goiter, Block *et al.* (1961) believe that raw soybeans contain a goitrogen which is removed or destroyed during processing. The addition of 160 μg iodine (as KI) per 100 gm diet caused the hypertrophied thyroid in rats to return to normal. Nordsiek (1962) prevented the goitrogenic effect of raw soybean meal fed to rats by adding casein to the diet. The casein effect was not attributed to its iodine content. These results indicate that soybean goitrogenicity may be related to the poor nutritional state of the animal fed raw soybean meal. Changes that occur in the amount and distribution of iodoamino acids in the sera and in digests of the thyroid of rats produced by soybean goiter have been studied (Block *et al.* 1961). Socolow and Suzuki (1964) were unable to demonstrate positively the goitrogenicity of soybeans and other foods when added as supplements to a low iodine basal diet (Remington ration).

Certain fresh and oxidized vegetable oils contain goitrogens or substances that enhance thyroid enlargement in iodine deficient diets (Kaunitz and Johnson 1967).

Antithyrotoxic Factor (ATF)

Van Middlesworth (1957) suggested that thyroxine excretion may also be a cause of goiter. Whether such an excretion occurs with soybeans is unknown. Whether the antigoitrogenic effect of casein in raw soybean meal diets (Nordsiek 1962) may be related to an inhibition of thyroxine secretion is also unknown. However, soybean protein products, as well as many other food sources, contain an ATF which is effective in blocking the usual physiological responses of injected or orally administered thyroxine such as (a) decreased growth, (b) decreased survival time, (c) increased liver α-glycerophosphate dehydrogenase activity, and (d) increased metabolic rate (Westerfeld *et al.* 1962; Richert *et al.* 1964). The antithyrotoxic effect of soybeans and other substances is less effective in chicks (Westerfeld *et al.* 1968). Westerfeld *et al.* (1968) have shown that two factors may be responsible for the antithyrotoxic activity and chick growth-stimulating property of soybean protein products. Since ATF in soybeans is released by proteolytic hydrolysis, indications are that it is an amino acid or peptide bound to protein (O'Dell et al. 1955). Ruegamer et al. (1968) have devised a simulated human diet to determine the importance of ATF in human thyroid metabolism.

GROWTH-VITAMIN-MINERAL FACTORS

Much information exists on the effects of soybean meal and protein isolates on growth and the availability of vitamins and minerals. That some of these effects may be due, in part, to phytic acid has become increasingly more evident. In contrast to the relationship between growth inhibition and pancreatic hypertrophy discussed earlier, soybeans also increase the requirement for vitamin B_{12}, vitamins D_2 and D_3, calcium and phosphorus, zinc, and other essential trace minerals, as well as promote growth. Several studies with poultry, rats, swine, and monkeys have shown that poor growth, perosis, rickets, and other bone calcification problems occur in diets containing high levels of soy protein isolates. Occasionally, the same symptoms have been observed with raw and toasted soybean meal diets. Goitrogenicity has been discussed previously.

Growth-Promoting and Antiperotic Factors

Kratzer et al. (1959B) reported that methanol extracts of defatted meal contain an unidentified growth-promoting and an antiperotic factor and that commercially available soy phospholipids also promote growth. In a later study, Vohra et al. (1959) found that the methanol extract improved perosis only when the diet was supplemented with zinc and that genistin and soysterols present in the extract had no growth-promoting properties. In 1964, Kratzer et al. showed that a growth-promoting substance was present in the benzene-soluble portion of a methanolic extract of soybean meal as well as in the water-soluble phase. Vitamin D_3, added to diets containing the benzene-soluble fraction had little or no effect on growth. Wilcox et al. (1961A,B) reported that water extracts of soybean meal promoted growth of turkey poults fed isolated soy protein. Addition of ethylenediaminetetraacetic acid (EDTA), a chelating agent, to a zinc-deficient diet containing soy protein isolates, improved growth of turkeys, but perosis was improved in only some of the birds (Kratzer et al. 1959A). EDTA improved growth of rats fed zinc-deficient diets containing either casein-gelatin or isolated soy protein (Vohra and Heil 1969A). Soy phospholipids, however, significantly improved growth from a casein-gelatin diet but had no effect on the soy diet. It is believed that EDTA improved growth by increasing the availability of dietary zinc; the mechanism of action of the soy phospholipids is unknown.

Mineral Availability

Growth experiments have demonstrated that metal-binding constituents greatly increase the dietary requirements for zinc (O'Dell 1969), manganese and copper (Davis et al. 1961), iron (Davis et al. 1962; Fitch et al. 1964), and molybdenum (Norris et al. 1958) when soy protein isolates are the main source of protein in the ration. The essential features of the dietary components that affect zinc availability have been reviewed by O'Dell (1969). The source of protein in the diet, whether of plant or animal origin, has a marked effect on zinc utilization in the food. Synthetic and naturally occurring chelating agents decrease as well as increase availability of zinc and other essential minerals. An unknown organic substance in liver extract increases the availability of zinc (Scott and Zeigler 1960). No improvement in growth or bone formation in chicks fed isolated soy protein was observed when arginine and lysine, two amino acids that form zinc chelates, were added to the diet (Kratzer et al. 1959A). Kratzer et al. (1959A) were the first to report that EDTA decreases the zinc requirement of diets containing soy protein isolates. Autoclaving the soy protein also improves availability of zinc as well as other essential minerals. Lease (1966) has shown that phytate destruction during autoclaving may not be solely responsible for the increased availability of zinc in autoclaved plant protein. Absorption of zinc from the intestine is greatly reduced in diets composed of isolated soy protein compared with casein and egg white (O'Dell 1969). O'Dell and Savage (1960) provided the first experimental evidence that a zinc-phytate-protein complex was responsible for the increased requirement of zinc in animal diets. Similar zinc symptomatology has been observed by the addition of phytic acid to diets based on free amino acids (Likuski and Forbes 1964). In all diets containing phytic acid, supplementation with adequate zinc overcomes the adverse effect on growth rate and prevents the formation of deficiency symptoms.

Dehulled, defatted soybean meal contains almost 0.5-0.6% phosphorus, 70-80% of which is present as phytic acid. Soy protein isolates contain up to 1% phosphorus, most of which is present as a mixed salt (referred to as phytate). Several workers have indicated that phytic acid-phosphorus can be readily utilized by rats, chickens, and other animals. In a series of studies, Likuski and Forbes (1965) have shown that some factors, such as dietary calcium and inorganic phosphorus supplementation, can greatly affect utilization of phytate phosphrus in soy protein isolates. The strong interaction between calcium and soy protein greatly decreases the utilization of zinc and other trace minerals (Forbes 1964).

Additional information on the complex interrelationships of calcium on the utilization of minerals in the presence of phytate in various animals has been reviewed by O'Dell (1969). Chelating agents with stability constants for zinc in the range of 13-17 are effective in increasing zinc availability. The mechanism is not clearly understood, but O'Dell (1969) advances the concept that soluble zinc chelates, such as with EDTA, provide for an efficient exchange of zinc in the intestinal mucosa. Preliminary data indicated that only about 60% of the zinc in cereal and oilseed meals can be utilized by the animal. Vohra and Kratzer (1967) point out the importance of the source of isolated soy protein on zinc availability. The amount of zinc available from oilseed meals for growth and prevention of bone deformities in chicks also varies widely (Lease and Williams 1967). Lease (1967) reported that in sesame and safflower meals zinc was present as an insoluble complex, Ca-Mg-Zn-phytate complex at intestinal pH. In soybean meal, minerals and phytate phosphorus were soluble at intestinal pH in the form of a binding protein, referred to as "carrier." Three isolated soy proteins contained low amounts of carrier or none. As a consequence, the increasing use of soy products in food requires that greater knowledge is needed about factors that affect mineral requirements in humans.

The effect of soy protein-phytic acid interactions on the isolation and physical properties of proteins and on tofu-making is described in Chaps. 4 and 9.

In studying dietary interactions between essential minerals, Vohra and Heil (1969B) found that growth of turkey poults is significantly improved by the addition of Zn to a Mn-deficient diet but not by the addition of Mn to a Zn-deficient diet. An antagonism between Zn and Cu was also observed. Both Zn and Mn are needed at the same time for prevention of perosis.

Rachitogenicity

Raw and heated soybean meal or water extracts of these meals partially prevented rickets and markedly improved growth of turkey poults fed a glucose-isolated soy protein diet (Carlson et al. 1964A). The high levels of vitamin D did not have the antirachitic and growth-promoting properties exhibited by the meal fractions. These studies indicated that the antirachitogenic and growth-promoting factors were heat-stable but that the two factors may differ (Jensen and Mraz 1966A).

Isolated soy protein and raw soybean meal also possess heat-labile growth-depressing and rachitogenic factors when used in a glucose soy protein diet for turkey poults (Carlson et al. 1964B). The growth-depressing and rachitogenic effects of raw soybean meal were only partially overcome with a tenfold increase in vitamin D_3 supplementation, whereas autoclaving overcame both effects. β-Carotene and phytic acid were ruled out as factors that interfered with normal calcification in these studies. Isolated soy protein possesses the greatest rachitogenic activity. At least 60 min of autoclaving is required to destroy the rachitogenic factors in soy protein isolates (Thompson et al. 1968). Isolated soy protein diets are also rachitogenic to chicks (Jensen and Mraz 1966A) and swine (Hendricks et al. 1969, 1970).

There is conflicting evidence as to whether phytic acid in isolated soy protein is responsible for both mineral deficiency symptoms and calcification (rachitogenicity) problems in animals and humans (Jensen and Mraz 1966B). Both effects are overcome by autoclaving, however; EDTA improves zinc availability but apparently does not influence calcium and phosphorus availability.

PHENOLIC CONSTITUENTS

Genistein and daidzein (Walz 1931; Ahluwalia *et al.* 1953), which occur mainly as the glucosides, genistin, and daidzin, are the major phenolic compounds in soybeans (Ahluwalia *et al.* 1953). Walz (1931) reported 0.15% genistin and 0.007% daidzein, whereas Walter (1941) found 0.1% genistin in defatted meal. Both estimates were based on yields. A small amount of the genistein is present in soy protein isolates (Nash *et al.* 1967), and genistein, genistin, and daidzin occurred in crude lipid fractions extracted from defatted flakes (Honig *et al.* 1969). Genistin has been isolated from commercial, toasted, defatted soybeans (Walter 1941; Matrone *et al.* 1956; Magee 1963).

Substances exhibiting estrogenic activity are widely distributed in the plant kingdom (Bradbury and White 1954; Stob 1966) and in many vegetable oils, including soybean oil (Booth *et al.* 1960A). Since most of naturally occurring estrogenic substances show only weak activity (Bickoff *et al.* 1962), it is doubtful that normal consumption of foods that contain estrogens would provide sufficient amounts to elicit a physiological response in humans. Genistein is only 10^{-5} as active as the synthetic estrogen, diethylstilbestrol; and daidzein, the other isoflavone in soybeans, is only 1/4 as active as genistein (Wong and Flux 1962). Defatted soybean meal extracted with methanol has no estrogenic activity (Carter *et al.* 1953).

Other physiological effects of genistein and genistin have been reported by Matrone *et al.* (1956), Magee (1963), and Carter *et al.* (1953). Genistin and genistein, when fed to male rats at a level of 0.5% of the diet, inhibited growth, increased the iron content of the liver and spleen, elevated zinc content in the liver and bones, and increased deposition of calcium, phosphorus, and manganese in the bones (Magee 1963). No significant adverse effects were observed in rats fed 0.1% genistein in a diet containing 19% casein for 4 weeks. Whether these effects associated with high levels of genistein are also responsible for the growth-vitamin-mineral interrelationships of diets containing soy protein isolates discussed previously remains to be determined. Based on the amounts of genistin reportedly present in soybeans (Walz 1931; Walter 1941), apparently high levels of soybean meal would have to be incorporated into the rat diets to achieve the effects observed by Magee (1963).

Another factor to be considered regarding isoflavones is their possible modification during processing. Pieterse and Andrews (1956A,B) found that moldy corn and alfalfa silage had increased estrogenic activity as a result of the fermentation process. A new isoflavone, 6,7,4'-trihydroxyisoflavone (Factor 2) is present in tempeh, a fermented soy product (György *et al.* 1964). Although Factor 2 is a potent antioxidant *in vitro,* it did not prevent autoxidation of soybean oil and powder and hemolysis of red blood cells *in vivo* (Ikehata *et al.* 1968).

Methods are needed for quantitatively determining isoflavone compounds in soybean meal and other soy protein products, particularly in the textured vegetable protein and spun fiber products. Whether the isoflavones are degraded or converted into forms even more estrogenic than the parent compounds during processing is not known. Hertelendy and Common (1964) have isolated equol (7,4'-dihydroxyisoflavan) from the urine of fowl fed a commercial diet containing alfalfa. Cayen *et al.* (1964) have shown that a portion of tritiated genistein, injected intramuscularly into a hen, was recovered in urine extracts as (^{3}H) equol.

A chalcone (isoliquiritigenin) and an aurone (hispidol) are present in soybean seedlings (Wong 1966). Wang *et al.* (1969) have prepared fractions from defatted soybean meal that appear to contain leucoanthocyanins. Soybean meal contains small amounts of various phenolic acids (Arai *et al.* 1966).

OTHER FACTORS

Schultze (1953) has reported that certain isolated soy proteins cause lactation failure in rats. His studies, however, should be discounted since "alpha protein," a protein drastically modified chemically to produce an adhesive, was used. There may be a vitamin E antagonism related to the processing of soy protein isolates and its interaction with corn oil (Fisher *et al.* 1969). Chang and Yokota (1967) found that vitamin B_{12} and folic acid increased the availability of methionine in rats fed soybean protein. The addition of choline to soy isolate infant formulas fed to rats prevented the development of fatty livers, but did not improve weight gains and caloric efficiencies (Theuer and Sarett 1970).

Edelstein and Guggenheim (1969) demonstrated that a soy flour diet is deficient in vitamin B_{12} and contains a heat-labile substance that increases the requirement for vitamin B_{12}.

BIBLIOGRAPHY

AHLUWALIA, V. K., BHASIN, M. M., and SESHADRI, T. R. 1953. Isoflavones of soybeans. Current Sci. (India) *22,* 363–364.

ALBRECHT, W. J., MUSTAKAS, G. C., and MCGHEE, J. E. 1966. Rate studies on atmospheric steaming and immersion cooking of soybeans. Cereal Chem. *43,* 400–407.

ALLEN, J. C. 1968. Soybean lipoxygenase. I. Purification and the effect of organic solvent upon the kinetics of the reaction. European J. Biochem. *4,* 201–208.

ALTHOFF, J. D. 1970. Nitrogen balance in a nutritional test with TVP. Med. Klin. (Munich) *65,* No. 25, 1204–1207.

ALUMOT, E., and NITSAN, Z. 1961. The influence of soybean antitrypsin on intestinal proteolysis of the chick. J. Nutr. *73,* 71–77.

ANDRE, E. 1964. The lipoxidase of soybean seeds. The present state of our knowledge. Oleagineux *19,* 461–463.

ANDRE, E., and HOU, K. W. 1964. Lipoxidase. Its discovery in the residue from the preparation of soymilk. Oleagineux *19,* 187–193.

ARAI, S., SUZUKI, H., FUJIMAKI, M., and SAKURAI, Y. 1966. Studies on flavor components in soybean. II. Phenolic acids in defatted soybean flour. Agr. Biol Chem. (Tokyo) *30,* 364–369.

AXELROD, B., SALTMAN, P., BANDURSKI, R. S., and BAKER, R. S. 1952. Phosphokinases in higher plants. J. Biol. Chem. *197,* 89–96.

BADENHOP, A. F., and HACKLER, L. R. 1970. The effects of soaking soybeans in sodium hydroxide solution as a pretreatment for soy milk production. Cereal Sci. Today *15*, 84–88.

BAHADEN, K., and SAXENA, I. 1964. Study of a papain activation factor present in the dialyzable fraction of soybean extract. Biol. Plant Acad. Sci. Bohemoslov. *6*, 165–170.

BANKS, W., GREENWOOD, C. T., and JONES, I. G. 1960. Physicochemical studies on starches. Part XXI. Observations on Z-enzyme. J. Am. Chem. Soc. 150–155.

BARNES, R. H., FIALA, G., and KWONG, E. 1965A. Prevention of coprography in the rat and growth stimulating effects of methionine, cystine, and penicillin when added to diets containing unheated soybeans. J. Nutr. *85*, 127–131.

BARNES, R. H., and KWONG, E. 1965. Effect of soybean trypsin inhibitor and penicillin on cystine biosynthesis in the pancreas and its transport as exocrine protein secretion in the intestinal tract of the rat. J. Nutr. *86*, 245–252.

BARNES, R. H., KWONG, E., and FIALA, G. 1965B. Effect of penicillin added to an unheated soybean diet on cystine excretion in feces of the rat. J. Nutr. *85*, 123–126.

BEN-AZIZ, A., GROSSMAN, S., ASCARELLI, I., and BUDOWSKI, P. 1970. Linoleate oxidation induced by lipoxygenase and heme proteins: A direct spectrophotometric assay. Anal. Biochem. *34*, 88–100.

BERK, J. E. 1968. Gastrointestinal gas. Ann. N.Y. Acad. Sci. *150*, 1–190.

BICKOFF, E. M., LIVINGSTON, A. L., HENDRICKSON, A. P., and BOOTH, A. N. 1962. Relative potencies of several estrogen-like compounds found in forages. J. Agr. Food Chem. *10*, 410–412.

BIELORAI, R. 1969. Comparative digestibility of groundnut and soyabean meal *in vitro* and in chicks. J. Sci. Food Agr. *20*, 345–348.

BIELORAI, R., and BONDI, A. 1963. Relationship between antitryptic factors of some plant protein feeds and products of proteolysis precipitable by trichloroacetic acid. J. Sci. Food Agr. *14*, 124–132.

BIRD, R., and HOPKINS, R. H. 1954. Mechanism of β-amylase action. II. Multichain action on amylose fission products. Biochem. J. *56*, 140–146.

BIRK, Y. 1968. Chemistry and nutritional significance of proteinase inhibitors from plant sources. Ann. N.Y. Acad. Sci. *146*, 388–399.

BIRK, Y. 1969. Saponins. *In* Toxic Constituents of Plant Foodstuffs, I. E. Liener (Editor). Academic Press, New York.

BIRK, Y., and GERTLER, A. 1961. Effect of mild chemical and enzymatic treatment of soybean meal and soybean trypsin inhibitors on their nutritive and biochemical properties. J. Nutr. *75*, 379–387.

BIRK, Y., GERTLER, A., and KHALEF, S. 1963A. A pure trypsin inhibitor from soyabeans. Biochem. J. *87*, 281–284.

BIRK, Y., GERTLER, A., and KHALEF, S. 1963B. Separation of a *Tribolium*-protease inhibitor from soybeans on a calcium phosphate column. Biochim. Biophys. Acta *67*, 326–328.

BIRK, Y., GERTLER, A., and KHALEF, S. 1967. Further evidence for a dual, independent activity against trypsin and α-chymotrypsin of inhibitor AA from soybeans. Biochim. Biophys. Acta *147*, 402–404.

BIRK, Y., and WALDMAN, M. 1965. Amylolytic-, trypsin-inhibiting, and urease activity in three varieties of soybeans and in soybean plant. Qualitas Plant Mater. Vegetabiles *12*, 199–209.

BLAIN, J. A., and BARR, T. 1961. Destruction of linoleate hydroperoxide by soya extracts. Nature *190*, 538–539.

BLAIN, J. A., and SHEARER, G. 1965. Inhibition of soya lipoxidase. J. Sci. Food Agr. *16*, 373–378.

BLAIN, J. A., and STYLES, E. C. C. 1959. A lipoperoxidase factor in soya extracts. Nature *184*, 1141.

BLOCK, R. J. *et al.* 1961. The curative action of iodine on soybean goiter and the changes in the distribution of iodoamino acids in the serum and in thyroid gland digests. Arch. Biochem. Biophys. *93*, 15–24.

BOOTH, A. N., BICKOFF, E. M., and KOHLER, G. E. 1960A. Estrogen-like activity in vegetable oils and milk by-products. Science *131*. 1807–1808.

BOOTH, A. N., ROBBINS, D. J., RIBELIN, W. E., and DE EDS, F. 1960B. Effect of raw soybean meal and amino acids on pancreatic hypertrophy in rats. Proc. Soc. Exptl. Biol. Med. 104, 681–683.

BOOTH, A. N. et al. 1964. Prolonged pancreatic hypertrophy and reversibility in rats fed raw soybean meal. Proc. Soc. Exptl. Biol. Med. 116, 1067–1069.

BORCHERS, R. 1962. Digestibility of threonine and valine by rats fed soybean meal rations. J. Nutr. 78, 330–332.

BORCHERS, R. 1963. "Bound" growth inhibitor in raw soybean meal. Proc. Soc. Exptl. Biol. Med. 112, 84–85.

BORCHERS, R. 1964. Raw soybean feeding decreases transamidinase activity. Proc. Soc. Exptl. Biol. Med. 115, 893–894.

BORCHERS, R., ANDERSON, S. M., and SPELTS, J. 1965. Rate of respiratory carbon-14 dioxide excretion after injection of C^{14}-amino acids in rats fed raw soybean meal. J. Nutr. 86, 253–255.

BRADBURY, R. B., and WHITE, D. E. 1954. Estrogens and related substances in plants. Vitamins Hormones 12, 207–233.

BRAY, D. J. 1964. Pancreatic hypertrophy in laying pullets induced by unheated soybean meal. Poultry Sci. 43, 382–384.

BREDENKAMP, B. L. F., and LUCK, D. N. 1969. Effect of raw soybeans on levels of protein and nucleic acids in the rat pancreas. Proc. Soc. Exptl. Biol. Med. 132, 537–539.

BRONOVITSKAYA, Z. S., and KRETOVICH, V. L. 1970A. Molecular weight of lactate dehydrogenase from soybean cotyledons. Dokl. Akad. Nauk SSSR 190, 717–719.

BRONOVITSKAYA, Z. S., and KRETOVICH, V. L. 1970B. Molecular weight of malate dehydrogenase of soybean cotyledons. Dokl. Akad. Nauk SSSR 190, 461–463.

BUTTERY, B. R., and BUZZELL, R. I. 1968. Peroxidase activity in seeds of soybean varieties. Crop Sci. 8, 722–725.

CALLOWAY, D. H. 1966. Respiratory hydrogen and methane as affected by consumption of gas-forming foods. Gastroenterology 51, 383–389.

CALLOWAY, D. H., and BURROUGHS, S. E. 1969. Effect of dried beans and silicone on intestinal hydrogen and methane production in man. Gut 10, 180–184.

CALLOWAY, D. H., COLASITO, D. J., and MATHEWS, R. D. 1966. Gases produced by human intestinal microflora. Nature 212, 1238–1239.

CALLOWAY, D. H., and MURPHY, E. L. 1968. The use of expired air to measure intestinal gas formation. Ann. N.Y. Acad. Sci. 150, 82–95.

CAMACHO, Z., BROWN, J. R., and KITTO, G. B. 1970. Purification and properties of trypsin-like proteases from the starfish Dermasterias imbricata. J. Biol. Chem. 245, 3964–3972.

CARLSON, C. W., MCGINNIS, J., and JENSEN, L. S. 1964A. Antirachitic effects of soybean protein for turkey poults. J. Nutr. 82, 366–370.

CARLSON, C. W., SAXENA, H. C., JENSEN, L. S., and MCGINNIS, J. 1964B. Rachitogenic activity of soybean fractions. J. Nutr. 82, 507–511.

CARMEL, N., and BOHAK, Z. 1968. Reformation of soybean trypsin inhibitor on oxidation of the reduced protein. Israel J. Chem. 6, 1–110.

CARTER, M. W., SMART, W. W. G., Jr., and MATRONE, G. 1953. Estimation of estrogenic activity of genistein obtained from soybean meal. Proc. Soc. Exptl. Biol. Med. 84, 506–507.

CATSIMPOOLAS, N. 1969. Isolation of soybean lipoxidase by isoelectric focusing. Arch. Biochem. Biophys. 131, 185–190.

CATSIMPOOLAS, N., EKENSTAM, C., and MEYER, E. W. 1969A. Separation of soybean whey proteins by isoelectric focusing. Cereal Chem. 46, 357–372.

CATSIMPOOLAS, N., and LEUTHNER, E. 1969. Immunochemical methods for detection and quantification of Kunitz soybean trypsin inhibitor. Anal. Biochem. 31, 437–447.

CATSIMPOOLAS, N., and MEYER, E. W. 1969. Isolation of soybean hemagglutinin and demonstration of multiple forms by isoelectric focusing. Arch. Biochem. Biophys. 132, 279–285.

CATSIMPOOLAS, N., ROGERS, D. A., and MEYER, E. W. 1969B. Immunochemical and disc electrophoresis study of soybean trypsin inhibitor SBTIA-2. Cereal Chem. 46, 136–144.

CAYEN, M. N., CARTER, A. L., and COMMON, R. H. 1964. The conversion of genistein to equol in the fowl. Biochim. Biophys. Acta 86, 56–64.

CHANG, Y. O., and YOKOTA, F. 1967. Effect of vitamin B ^{12} and folic acid on utilization by rats of soybean protein and amino acids. J. Agr. Food Chem. 15, 444–447

CHRISTOPHER, J., PISTORIUS, E., and AXELROD, B. 1970. Isolation of an isoenzyme of soybean lipoxygenase. Biochim. Biophys. Acta 198, 12–19.

CIRCLE, S. J. 1950. Proteins and other nitrogenous constituents. In Soybeans and Soybean Products, Vol. I, K. S. Markley (Editor). John Wiley & Sons, New York.

COAN, M., and TRAVIS, J. 1969. Comparative biochemistry of proteases from a coelenterate. Comp. Biochem. Physiol. 32, 127–139.

COATES, M. E., HEWITT, D., and GOLOB, P. 1970. A comparison of the effects of raw and heated soybean meal in diets for germ-free and conventional chicks. Brit. J. Nutr. 24, 213–225.

COWAN, C. C., Jr. et al. 1969. A soy protein isolate formula in the management of allergy in infants and children. Southern Med. J. 62, 389–393.

CRAWFORD, L. V., ROANE, V. J., TRIPLETT, F., and HANISSIAN, A. S. 1965. Immunological studies on the legume family of foods. Ann. Allergy 23, 303–308.

CUNNINGHAM, W. L., MANNERS, D. J., and WRIGHT, A. 1962. Studies on carbohydrate-metabolizing enzymes. 8. The action of Z-enzyme on glycogen-type polysaccharides. Biochem. J. 85, 408–413.

DAL BORGO, G., SALMAN, A. J., PUBOLS, N. H., and MCGINNIS, J. 1968. Exocrine function of the chick pancreas as affected by dietary soybean meal and carbohydrate. Proc. Soc. Exptl. Biol. Med. 129, 877–881.

DAVIS, P. N., NORRIS, L. C., and KRATZER, F. H. 1961. Interference of soybean proteins with the utilization of trace minerals. J. Nutr. 77, 217–223.

DAVIS, P. N., NORRIS, L. C., and KRATZER, F. H. 1962. Iron deficiency studies in chicks using treated isolated soybean protein diets. J. Nutr. 78, 445–453.

DEMUELENAERE, H. J. H. 1964. Studies on the digestion of soybeans. J. Nutr. 82, 197–205.

DEMUELENAERE, H. J. H. 1965. Toxicity and haemagglutinating activity of legumes. Nature 206, 827–828.

DENBESTEN, L., CONNOR, W. E., KENT, T. H., and LIN, D. 1970. Effect of cellulose in the diet on the recovery of dietary plant sterols from the feces. J. Lipid Res. 11, 341–345.

D'HOLLANDER, F., and CHEVALLIER, F. 1969. Qualitative and quantititative estimation of free and esterified sterols in whole rat and in 23 of its tissues and organs. Biochim. Biophys. Acta 176, 146–162. (French)

DIBELLA, F. L., and LIENER, I. E. 1969. Soybean trypsin inhibitor cleavage and identification of a disulfide bridge not essential for activity. J. Biol. Chem. 244, 2824–2829.

DILLARD, M. G., HENICK, A. S., and KOCH, R. B. 1961. Differences in reactivity of legume lipoxidases. J. Biol. Chem. 236, 37–40.

DOLEV, A., ROHWEDDER, W. K., and DUTTON, H. J. 1967A. Mechanism of lipoxidase reaction. I. Specificity of hydroperoxidation of linoleic acid. Lipids 2, 28–32.

DOLEV, A., ROHWEDDER, W. K., MOUNTS, T. L., and DUTTON, H. J. 1967B. Mechanism of lipoxidase reaction. II. Origin of the oxygen incorporated into linoleate hydroperoxide. Lipids 2, 33–36.

EDELSTEIN, S., and GUGGENHEIM, K. 1969. Effect of raw soybean flour on vitamin B ^{12} requirements of rats. Israel J. Med. Sci. 5, 415–417.

ELDRIDGE, A. C., ANDERSON, R. L., and WOLF, W. J. 1966. Polyacrylamide-gel electrophoresis of soybean whey proteins and trypsin inhibitors. Arch. Biochem. Biophys. 115, 495–504.

ELDRIDGE, A. C., and WOLF, W. J. 1969. Polyacrylamide-gel electrophoresis of reduced and alkylated soybean trypsin inhibitors. Cereal Chem. 46, 470–478.

FEDELI, E., LANZANI, A., CAPELLA, P., and JACINI, G. 1966. Triterpene alcohols and sterols in vegetable oils. J. Am. Oil Chemists' Soc. 43, 254–256.

FEINSTEIN, G., and FEENEY, R. E. 1966. Interaction of inactive derivatives of chymotrypsin and trypsin with protein inhibitors. J. Biol. Chem. 241, 5183–5189.

FEINSTEIN, G., OSUGA, D. T., and FEENEY, R. E. 1966. The mechanism of inhibition of trypsin by ovomucoids. Biochem. Biophys. Res. Commun. 24, 495–499.

FINKENSTADT, W. R., and LASKOWSKI, M., Jr. 1967. Resynthesis by trypsin of the cleaved peptide bond in modified soybean trypsin inhibitors. J. Biol. Chem. 242, 771–772.

FISHER, H., GRIMINGER, P., and BUDOWSKI, P. 1969. Antivitamin E activity of isolated soybean protein for the chick. Z. Ernaehrungswiss. 9, 271–278.

FISHER, H., and SHAPIRO, R. 1963. Counteracting the growth retardation of raw soybean meal with extra protein and calories. J. Nutr. 80, 425–430.

FITCH, C. D., HARVILLE, W. E., DINNING, J. S., and PORTER, F. S. 1964. Iron deficiency in monkeys fed diets containing soybean protein. Proc. Soc. Exptl. Biol. Med. 116, 130–133.

FORBES, R. M. 1964. Mineral utilization in the rat. III. Effects of calcium, phosphorus, lactose and source of protein in zinc-deficient and in zinc-adequate diets. J. Nutr. 83, 225–233.

FRATTALI, V. 1969. Soybean inhibitors. III. Properties of a low molecular weight soybean proteinase inhibitor. J. Biol. Chem. 244, 274–280.

FRATTALI, V., and STEINER, R. F. 1968. Soybean inhibitors. I. Separation and properties of three inhibitors from commercial crude soybean trypsin inhibitor. Biochemistry 7, 521–530.

FRATTALI, V., and STEINER, R. F. 1969A. Interaction of trypsin and chymotrypsin with a soybean proteinase inhibitor. Biochem. Biophys. Res. Commun. 34, 480–487.

FRATTALI, V., and STEINER, R. F. 1969B. Soybean inhibitors. II. Preparative electrophoretic purification of soybean proteinase inhibitors on polyacrylamide gel. Anal. Biochem. 27, 285–291.

FREAR, D. S. 1968. Herbicide metabolism in plants. I. Purification and properties of UDP-glucose: arylamine N-glucosyltransferase from soybeans. Phytochemistry 7, 381–390.

FRIDMAN, C., LIS, H., SHARON, N., and KATCHALSKI, E. 1967. Isolation and characterization of soybean cytochrome C. Arch. Biochem. Biophys. 126, 299–304.

FRITZ, H., TRAUTSCHOLD, I., HAENDLE, H., and WERLE, E. 1968. Chemistry and biochemistry of proteinase inhibitors from mammalian tissue. Ann. N.Y. Acad. Sci. 146, 400–413.

FRITZ, H. et al. 1969. Identification of lysine and arginine residues as inhibitory centers of protease inhibitors with the aid of maleic anhydride and 2,3-butanedione. Hoppe-Seylers Z. Physiol. Chem. 350, 933–944.

FROST, A. B., and MANN, G. V. 1966. Effects of cystine deficiency and trypsin inhibitor on the metabolism of methionine. J. Nutr. 89, 49–54.

FRYDMAN, R. B., DE SOUZA, B. C., and CARDINI, C. E. 1966. Distribution of adenosine diphosphate D-glucose: α-1,4-glucan α-4-glucosyl transferase in higher plants. Biochim. Biophys. Acta 113, 620–623.

GALL, L. S. 1968. The role of intestinal flora in gas formation. Ann. N.Y. Acad. Sci. 150, 27–30.

GARLICH, J. D., and NESHEIM, M. C. 1966. Relationship of fractions of soybeans and a crystalline trypsin inhibitor to the effects of feeding unheated soybean meal to chicks. J. Nutr. 88, 100–110.

GATES, B. J., and TRAVIS, J. 1969. Isolation and comparative properties of shrimp trypsin. Biochemistry 8, 4483–4489.

GAVRICHENKOV, Y., YU, D., KAZAKOV, E. D., and STARODUBTSEVA, A. I. 1969. Activity of lipase in soybean seed. Prikl. Biokhim. Mikrobiol. 5, 324–328.

GEDDES, R. 1968. Rates of attack of some α-amylases upon various substrates. Carbohydrate Res. 7, 493–497.

GERTLER, A., and BIRK, Y. 1965. Purification and characterization of a β-amylase from soybeans. Biochem. J. 95, 621–627.

GERTLER, A., and BIRK, Y. 1966. The role of sulfhydryl groups in soybean β-amylase. Biochim. Biophys. Acta 118, 98–105.

GERTLER, A., BIRK, Y., and BONDI, A. 1967. A comparative study of the nutritional and physiological significance of pure soybean trypsin inhibitor and of ethanol-extracted soybean meals in chicks and rats. J. Nutr. *91*, 358–370.

GINI, B., and KOCH, R. B. 1961. Study of a lipohydroperoxide breakdown factor in soy extracts. J. Food Sci. *26*, 359–364.

GITZELMANN, R., and AURICCHIO, S. 1965. The handling of soya alpha-galactosides by a normal and a galactosemic child. Pediatrics *36*, 231–235.

GLASER, J., and JOHNSTONE, D. E. 1953A. Soybean milk as a substitute for mammalian milk in early infancy. Ann. Allergy *10*, 433–438.

GLASER, J., and JOHNSTONE, D. E. 1953B. Prophylaxis of allergic disease in the newborn. J. Am. Med. Assoc. *153*, 620–622.

GOLDBERG, A. 1970. Changes in carbohydrate metabolism in rats fed raw soybean flour. Nutr. Metab. *12*, 40–48.

GOLDBERG, A., and GUGGENHEIM, K. 1964. Effect of antibiotics on pancreatic enzymes of rats fed soybean flour. Arch. Biochem. Biophys. *108*, 250–254.

GORRILL, A. D. L., and THOMAS, J. W. 1967. Body weight changes, pancreas size and enzyme activity, and proteolytic activity and protein digestion in intestinal contents from calves fed soybean and milk protein diets. J. Nutr. *92*, 215–223.

GORRILL, A. D. L., THOMAS, J. W., STEWART, W. E., and MORRILL, J. L. 1967. Exocine pancreatic secretion by calves fed soybean and milk protein diets. J. Nutr. *92*, 86–92.

GREEN, L. J. *et al.* 1968. Two trypsin inhibitors from porcine pancreatic juice. J. Biol. Chem. *243*, 1804–1815.

GREENWOOD, C. T., MACGREGOR, A. W., and MILNE, E. A. 1965A. a-Amylolysis of starch. Staerke *17*, 219–225.

GREENWOOD, C. T., MACGREGOR, A. W., and MILNE, E. A. 1965B. Starch-degrading enzymes. II. The Z-enzyme from soybeans; purification and properties. Carbohydrate Res. *1*, 229–241.

GREENWOOD, C. T., MACGREGOR, A. W., and MILNE, E. A. 1965C. Starch-degrading enzymes. III. The action pattern of soybean Z-enzyme. Carbohydrate Res. *1*, 303–311.

GUSS, P. L., RICHARDSON, T., and STAHMANN, M. A. 1967. The oxidation-reduction enzymes of wheat. III. Isoenzymes of lipoxidase in wheat fractions and soybeans. Cereal Chem. *44*, 607–610.

GUSS, P. L., RICHARDSON, T., and STAHMANN, M. A. 1968. Oxidation of various lipid substrates with unfractionated soybean and wheat lipoxidase. J. Am. Oil Chemists' Soc. *45*, 272–276.

GYÖRGY, P., MURATA, K., and IKEHATA, H. 1964. Antioxidants isolated from fermented soybeans (tempeh). Nature *203*, 870–872.

HAINES, P. C., and LYMAN, R. L. 1961. Relationship of pancreatic enzyme secretion to growth inhibition in rats fed soybean trypsin inhibitor. J. Nutr. *74*, 445–452.

HALE, S. A., RICHARDSON, T., VON ELBE, J. H., and HAGEDORN, D. J. 1969. Isoenzymes of lipoxidase. Lipids *4*, 209–215.

HAMBERG, M., and SAMUELSSON, B. 1965. On the specificity of the oxygenation of unsaturated fatty acids catalyzed by soybean lipoxidase. J. Biol. Chem. *242*, 5329–5335.

HARRY, J. B., and STEINER, R. F. 1969. Characterization of the self-association of a soybean proteinase inhibitor by membrane osmometry. Biochemistry *8*, 5060–5064.

HAYNES, R., and FEENEY, R. E. 1968A. Properties of enzymatically cleaved inhibitors of trypsin. Biochim. Biophys. Acta *151*, 209–211.

HAYNES, R., and FEENEY, R. E. 1968B. Transformation of active-site lysine in naturally occurring trypsin inhibitors. A basis for a general mechanism for inhibition of proteolytic enzymes. Biochemistry *7*, 2879–2885.

HAYNES, R., OSUGA, D. T., and FEENEY, R. E. 1967. Modification of amino groups in inhibitors of proteolytic enzymes. Biochemistry *6*, 541–547.

HELLENDOORN, E. W. 1969. Intestinal effects following ingestion of beans. Food Technol. *23*, 795–800.

HENDRICKS, D. G. *et al.* 1969. Effect of level of soybean protein and ergocalciferol on mineral utilization by the baby pig. J. Animal Sci. *28*, 342–348.

HENDRICKS, D. G. *et al.* 1970. Effect of source and level of protein on mineral utilization by the baby pig. J. Nutr. *100*, 235–240.

HERTELENDY, F., and COMMON, R. H. 1964. Isolation and identification of equol from the urine of the domestic fowl. Poultry Sci. *43*, 954–957.

HIXSON, H. F., and LASKOWSKI, M., Jr. 1970. Evidence that cocoonase and trypsin interact with soybean trypsin inhibitor at the same reactive site. Biochemistry *9*, 166–170.

HOLMAN, R. T., EGWIM, P. O., CHRISTIE, W. W. 1969. Substrate specificity of soybean lipoxidase. J. Biol. Chem. *244*, 1149–1151.

HONIG, D. H., SESSA, D. J., HOFFMAN, R. L., and RACKIS, J. J. 1969. Lipids of defatted soybean flakes: Extraction and characterization. Food Technol. *23*, 803–808.

HOU, C. T., UMEMURA, Y., NAKAMURA, M., and FUNAHASHI, S. 1968. Enzymatic synthesis of steryl glucoside by a particulate preparation from immature soybean seeds. J. Biochem. (Tokyo) *63*, 351–360.

IKEHATA, H., WAKAIZUMI, M., and MURATA, K. 1968. Antioxidant and antihemolytic activity of a new isoflavone "Factor 2" isolated from tempeh. Agr. Biol. Chem. (Tokyo) *32*, 740–746.

ISHIDA, M., HAMAGUCHI, K., and IKENAKA, T. 1970. A study of the interaction of soybean trypsin inhibitor with trypsin by circular dichroism. J. Biochem. (Tokyo) *67*, 363–371.

ITO, T. 1960. Preparation of a hexokinase from soybean. Nippon Nogeikagaku Kaishi *34*, 275–278.

JAFFE, W. G. 1969. Hemagglutinins. *In* Toxic Constituents of Plant Foodstuffs, I. E. Liener (Editor). Academic Press, New York.

JENSEN, L. S., and MRAZ, F. R. 1966A. Rachitogenic activity of isolated soy protein for chicks. J. Nutr. *88*, 249–253.

JENSEN, L. S., and MRAZ, F. R. 1966B. Effect of chelating agents and high levels of calcium and phosphorus on bone calcification in chicks fed isolated soy protein. J. Nutr. *89*, 471–476.

JIRGENSONS, B. 1967. Effects of *n*-propyl alcohol and detergents in the optical rotatory dispersion of a-chymotrysinogen, β-casein, histone fraction F¹, and soybean trypsin inhibitor. J. Biol. Chem. *242*, 912–918.

JIRGENSONS, B., KAWABATA, M., and CAPETILLO, S. 1969. Circular dichroism of soybean trypsin inhibitor and its derivatives. Makromol. Chem. *125*, 126–135.

KAFATOS, F. C., LAW, J. H., and TARTAKOFF, A. M. 1967. Cocoonase, II. Substrate specificity, inhibitors, and classification of the enzyme. J. Biol. Chem. *242*, 1488–1494.

KAKADE, M. L., BARTON, T. L., SCHAIBLE, P. J., and EVANS, R. J. 1967. Biochemical changes in the pancreas of chicks fed raw soybeans and soybean meal. Poultry Sci. *46*, 1578–1580.

KAKADE, M. L., SIMONS, N., and LIENER, I. E. 1969. An evaluation of natural vs synthetic substrates for measuring the antitryptic activity of soybean samples. Cereal Chem. *46*, 518–526.

KAKADE, M. L., SIMONS, N., and LIENER, I. E. 1970A. Nutritional effects induced in rats by feeding natural and synthetic trypsin inhibitors. J. Nutr. *100*, 1003–1008.

KAKADE, M. L., SWENSON, D. H., and LIENER, I. E. 1970B. Note on the determination of chymotrypsin and chymotrypsin inhibitor activity using casein. Anal. Biochem. *33*, 255–258.

KASSELL, B. 1970. Naturally occurring inhibitors of proteolytic enzymes. *In* Methods in Enzymology, Vol. XIX, 839–890, G. E. Perlmann and L. Lorand (Editors). Academic Press, New York.

KATO, I., and TOMINAGA, N. 1970. Soybean trypsin inhibitor-C: an active derivative of soybean trypsin inhibitor composed of two noncovalently bonded peptide fragments. Federation European Biochem. Soc. Letters, *10*, 313–316.

KAUNITZ, H., and JOHNSON, R. E. 1967. Thyroid and pituitary pathology in iodine-deficient rats fed fresh and oxidized fats and oils. J. Nutr. *91*, 55–62.

KAWAMURA, S., and KASAI, T. 1968. Decomposition of soybean oligosaccharides by intestinal bacteria. III. Comparison of eighteen strains of *Escherichia coli* for consuming some mono- and oligosaccharides. Kagawa Univ. Fac. Agr. Tech. Bull. *20*, 41–48.

KHAYAMBASHI, H., and LYMAN, R. L. 1966. Growth depression and pancreatic and intestinal changes in rats force-fed amino acid diets containing soybean trypsin inhibitor. J. Nutr. *89*, 455–464.

KHAYAMBASHI, H., and LYMAN, R. L. 1969. Secretion of rat pancreas perfused with plasma from rats fed soybean trypsin inhibitor. Am. J. Physiol. *217*, 646–651.

KIES, M. W. *et al.* 1969. On the question of the identity of soybean "lipoxidase" and carotene oxidase. Biochem. Biophys. Res. Commun. *36*, 312–315.

KING, J. 1970. Isolation, properties and physiological role of lactic dehydrogenase from soybean cotyledons. Can. J. Botan. *48*, 533–540.

KIRIBUCHI, T., CHEN, C. S., and FUNAHASHI, S. 1965. Systematic analysis of sterols in soybeans and other oil seeds. Agr. Biol. Chem. (Tokyo) *29*, 265–267.

KIRIBUCHI, T., MIZUNAGA, T., and FUNAHASHI, S. 1966. Separation of soybean sterols by florisil chromatography and characterization of acylated steryl glucoside. Agr. Biol. Chem. (Tokyo) *30*, 770–778.

KIRIBUCHI, T., YASUMATSU, N., and FUNAHASHI, S. 1967. Synthesis of 6-O-palmitoyl-β-D-glucosyl β-sitosterol. Agr. Biol. Chem. (Tokyo) *31*, 1244–1247.

KOCH, R. B., STERN, B., and FERRARI, C. G. 1958. Linoleic acid and trilinolein as substrates for soybean lipoxidase. Arch. Biochem. Biophys. *78*, 165–179.

KONIJN, A. M., BIRK, Y., and GUGGENHEIM, K. 1970A. Pancreatic enzyme pattern in rats as affected by dietary soybean flour. J. Nutr. *100*, 361–368.

KONIJN, A. M., BIRK, Y., and GUGGENHEIM, K. 1970B. *In vitro* synthesis of pancreatic enzymes: Effects of soybean trypsin inhibitor. Am. J. Physiol. *218*, 1108–1112.

KONIJN, A. M., and GUGGENHEIM, K. 1967. Effect of raw soybean flour on the composition of rat pancreas. Proc. Soc. Exptl. Biol. Med. *126*, 65–67.

KONIJN, A. M., GUGGENHEIM, K., and BIRK, Y. 1969. Amylase synthesis in pancreas of rats fed soybean flour. J. Nutr. *97*, 265–270.

KONLANDE, J. E., and FISHER, H. 1969. Evidence for a nonabsorptive antihypercholesterolemic action of phytosterols in the chicken. J. Nutr. *98*, 435–442.

KOPEIKOVSKII, V. M., and KASHEVATSKAYA. 1966. Oxidizing processes occurring in soybeans during drying in dense layers. Maslob-Zhir. Prom. *32*, 15–16.

KRATZER, F. H. *et al.* 1959A. The effect of autoclaving soybean protein and the addition of ethylenediaminetetraacetic acid on the biological availability of dietary zinc for turkey poults. J. Nutr. *68*, 313–322.

KRATZER, F. H. *et al.* 1959B. Fractionation of soybean oil meal for growth and antiperotic factors. 1. Non-phospholipid nature of the factors. Poultry Sci. *38*, 1049–1055.

KRATZER, F. H., STARCHER, B., and MARTIN, E. W. 1964. Fractionation of soybean meal for growth and antiperotic factors. 3. Growth promoting activity in benzene soluble fraction. Poultry Sci. *43*, 663–667.

KRETOVICH, V. L., MORGUNOVA, E. A., and KARYAKINA, T. I. 1964. Enzymatic transamination of glyoxylate by various amino acids in plants. Dokl. Akad. Nauk. SSSR *159*, 928–930.

KUMAR, K. S. V. S., SUNDARARAJAN, V. S., and SARMA, P. S. 1957. Dephosphorylation of casein by plant phosphatases. Enzymologia *18*, 228–233.

KUNERT, G. 1962. Activation of soybean lipase. Flora (Jena) *152*, 168–170.

KUNITZ, M. 1946. Crystalline soybean trypsin inhibitor. J. Gen. Physiol. *29*, 149–154.

KUNITZ, M. 1947. Crystalline soybean trypsin inhibitor. II. General properties. J. Physiol. *30*, 291–310.

LANCHANTIN, G. F., FRIEDMANN, J. A., and HART, D. W. 1969. Interaction of soybean trypsin inhibitor with thrombin and its effect on prothrombin activation. J. Biol. Chem. *244*, 865–875.

LASKOWSKI, M., JR., and SEALOCK, R. W. 1971. Protein proteinase inhibitors—molecular aspects. *In* The Enzymes, Vol. III (Third Edition), 375–473, P. D. Boyer (Editor). Academic Press, New York.

LAUFER, S., TAUBER, H., and DAVIS, C. F. 1944. The amylolytic and proteolytic activity of soybean seed. Cereal Chem. *21*, 267–274.

LEARMONTH, E. M., and WOOD, J. C. 1960. The influence of soya flour in bread doughs. IV. Alpha-amylase of soya. Cereal Chem. *37*, 158–169.

LEASE, J. G. 1966. The effect of autoclaving sesame meal on its phytic acid content and on the availability of its zinc to the chick. Poultry Sci. *45*, 237–241.

LEASE, J. G. 1967. Availability to the chick of zinc-phytate complexes isolated from oilseed meals by an *in-vitro* digestion method. J. Nutr. *93*, 523–532.

LEASE, J. G., and WILLIAMS, W. P., Jr. 1967. Availability of zinc and comparison of *in-vitro* and *in-vivo* zinc uptake of certain oilseed meals. Poultry Sci. *46*, 233–241.

LEE, K. W., and ROUSH, A. H. 1964. Allantoinase assays and their application to yeast and soybean allantoinases. Arch. Biochem. Biophys. *108*, 460–467.

LEPKOVSKY, S., BINGHAM, E., and PENCHARZ, R. 1959. The fate of proteolytic enzymes from the pancreatic juice of chicks fed raw and heated soybeans. Poultry Sci. *38*, 1289–1295.

LEPKOVSKY, S. *et al.* 1965. The effect of raw soya beans upon the digestion of proteins and upon the function of the pancreas of intact chickens and of chickens with ileostomies. Brit. J. Nutr. *19*, 41–56.

LEPKOVSKY, S. *et al.* 1966. Pancreatic amylase in chickens fed on soya-bean diets. Brit. J. Nutr. *20*, 421–437.

LEPKOVSKY, S., and FURUTA, F. 1970. Lipase in pancreas and intestinal contents of chickens fed heated and raw soybean diets. Poultry Sci. *49*, 192–198.

LEPKOVSKY, S. *et al.* 1970A. Nutritional value of soybeans in the diet of depancreatized chickens. Poultry Sci. *49*, 942–948.

LEPKOVSKY, S., FURUTA, F., DIMICK, M. K., and YAMASHINA, I. 1970B. Enterokinase and the chicken pancreas. Poultry Sci. *49*, 421–426.

LEVITT, M. D., and INGELFINGER, F. J. 1968. Hydrogen and methane production in man. Ann. N.Y. Acad. Sci. *150*, 75–81.

LIENER, I. E. 1953. Soyin, a toxic protein from the soybean. I. Inhibition of rat growth. J. Nutr. *49*, 527–540.

LIENER, I. E. 1955. The photometric determination of the hemagglutinating activity of soyin and crude soybean extracts. Arch. Biochem. Biophys. *54*, 223–231.

LIENER, I. E. 1962. Toxic factors in edible legumes and their elimination. Am. J. Clin. Nutr. *11*, 281–298.

LIENER, I. E., and KAKADE, M. L. 1969. Protease inhibitors. *In* Toxic Constituents in Plant Foodstuffs, I. E. Liener (Editor). Academic Press, New York.

LIENER, I. E., and PALLANSCH, M. J. 1952. Purification of a toxic substance from defatted soybean meal. J. Biol. Chem. *197*, 29–36.

LIENER, I. E., and ROSE, J. E. 1953. Soyin, a toxic protein from the soybean. III. Immunochemical properties. Proc. Soc. Exptl. Biol. Med. *83*, 539–544.

LIKUSKI, H. J. A., and FORBES, R. M. 1964. Effect of phytic acid on the availability of zinc to amino acid and casein diets fed to chicks. J. Nutr. *84*, 145–148.

LIKUSKI, H. J. A., and FORBES, R. M. 1965. Mineral utilization in the rat. IV. Effects of calcium and phytic acid on the utilization of dietary zinc. J. Nutr. *85*, 230–234.

LIS, H., FRIDMAN, C., SHARON, N., and KATCHALSKI, E. 1966. Multiple hemagglutinins in soybean. Arch. Biochem. Biophys. *117*, 301–309.

LIS, H., SHARON, N., and KATCHALSKI, E. 1969. Identification of the carbohydrate-protein linking group in soybean hemagglutinin. Biochim. Biophys. Acta *192*, 364–366.

LIU, W. H. *et al.* 1968. Modification of arginines in trypsin inhibitors by 1,2-cyclohexanedione. Biochemistry *8*, 2886–2892.

LONGENECKER, J. B., MARTIN, W. H., and SARETT, H. P. 1964. Improvement in the protein efficiency of soybean concentrates and isolates by heat treatment. J. Agr. Food Chem. *12*, 411–412.

LYMAN, R. L. 1957. The effect of raw soybean meal and trypsin inhibitor diets on the intestinal and pancreatic nitrogen in the rat. J. Nutr. *62*, 285–294.

LYMAN, R. L., and LEPKOVSKY, S. 1957. The effect of raw soybean meal and trypsin inhibitor diets on pancreatic enzyme secretion in the rat. J. Nutr. *62*, 269–284.

LYMAN, R. L., WILCOX, S. S., and MONSEN, E. R. 1962. Pancreatic enzyme secretion produced in the rat by trypsin inhibitors. Am. J. Physiol. *202*, 1077–1082.

MA'AYANI, S., and KULKA, R. G. 1969. Amylase, procarboxypeptidase and chymotrypsinogen in pancreas of chicks fed raw or heated soybean diet. J. Nutr. 96, 363–367.
MAEKAWA, A., SUZUKI, T., and SAHASHI, Y. 1959. Glycine-ketoglutaric acid transaminase in plants. Tokyo Nogyo Daigaku Nogaku Shuho 5, 1–4.
MAGEE, A. C. 1963. Biological responses of young rats fed diets containing genistin and genistein. J. Nutr. 80, 151–156.
MAIER, V. P., and TAPPEL, A. L. 1959. Rate studies of unsaturated fatty acid oxidation catalyzed by hematin compounds. J. Am. Oil Chemists' Soc. 36, 8–12.
MANNERS, D. J. 1962. Enzymatic synthesis and degradation of starch and glycogen. In Advances in Carbohydrate Chemistry, Vol. 17, W. W. Pigman and M. L. Wolfrom (Editors). Academic Press, New York.
MARKERT, C. L., and MOELLER, F. 1959. Multiple forms of enzymes. Tissue, ontogenetic and species specific patterns. Proc. Natl. Acad. Sci. (U.S.) 45 753–763.
MARUO, B., and KOBAYASHI, T. 1951. VII. The relation between amylosynthease and amyloglucosidase. Proc. Natl. Acad. Sci. U.S. 45, 309–313.
MATRONE, G. et al. 1956. Effect of genistin on growth and development of the male mouse. J. Nutr. 59, 235–241.
MATTICK, L. R., and HAND, D. B. 1969. Identification of a volatile component in soybeans that contributes to the raw bean flavor. J. Agr. Food Chem. 17, 15–17.
MAYER, F. C., CAMPBELL, R. E., SMITH, A. K., and MCKINNEY, L. L. 1961. Soybean phosphatase. Purification and properties. Arch. Biochem. Biophys. 94, 301–307.
MELMED, R. N., and BOUCHIER, I. A. D. 1969. A further physiological role for naturally occurring trypsin inhibitors: The evidence for a trophic stimulant of the pancreatic acinar cell. Gut 10, 973–979.
MIETTINEN, T. A. 1967. Cholestanol and plant sterols in the adrenal gland of the rat. Acta Chem. Scand. 21, 286–290.
MILLAR, D. B., WILLICK, G. E., STEINER, R. F., and FRATTALI, V. 1969. Soybean inhibitors. IV. The reversible self-association of a soybean proteinase inhibitor. J. Biol. Chem. 244, 281–284.
MITSUDA, H., YASUMOTO, K., and YAMAMOTO, A. 1967A. Inactivation of lipoxygenase by hydrogen peroxide, cysteine and some other reagents. Agr. Biol. Chem. (Tokyo) 31, 853–860.
MITSUDA, H., YASUMOTO, K., and YAMAMOTO, A. 1967B. Inhibition of lipoxygenase by saturated monohydric alcohols through hydrophobic bonding. Arch. Biochem. Biophys. 118, 664–669.
MITSUDA, H., YASUMOTO, K., YAMAMOTO, A., and KUSANO, T. 1967C. Study on soybean lipoxygenase. I. Preparation of crystalline enzyme and assay by polarographic method. Agr. Biol. Chem. (Tokyo) 31, 115–118.
MOUSTAFA, E., and WONG, E. 1967. Purification and properties of chalcone-flavanone isomerase from soybean seed. Phytochemistry 6, 625–632.
MUSTAKAS, G. C. et al. 1969. Lipoxidase deactivation to improve stability, odor, and flavor of full-fat soy flours. J. Am. Oil Chemists' Soc. 46, 623–626.
NAKAMURA, M. 1951. VI. Distribution of phosphorylase, phosphatase and β-amylase in plants. 3. Mechanism of inhibition of phosphorylase activity by β-amylase. J. Agr. Chem. Soc. Japan 24, 302–309.
NAKAMURA, S., and WAKEYAMA, T. 1961. An attempt to demonstrate the distribution of trypsin inhibitors in the sera of various animals. J. Biochem. (Tokyo) 49, 733–741.
NASH, A. M., ELDRIDGE, A. C., and WOLF, W. J. 1967. Fractionation and characterization of alcohol extractables associated with soybean proteins. Nonprotein components. J. Agr. Food Chem. 15, 102–108.
NESHEIM, M. C., and GARLICH, J. D. 1966. Digestibility of unheated soybean meal for laying hens. J. Nutr. 88, 187–192.
NIEKAMP, C. W., HIXSON, H. F., Jr., and LASKOWSKI, M., Jr. 1969. Peptide-bond hydrolysis equilibria in native proteins. Conversion of virgin into modified soybean trypsin inhibitor. Biochemistry 8, 16–22.
NISHIZAWA, K. 1951. The enzymatic cleavage of galactosidases. III. Shinshu Univ., J. Fac. Text. Sericult. Ser. C 1, 1–73, 213–285.

NITSAN, Z., and ALUMOT, E. 1964. Overcoming the inhibition of intestinal proteolytic activity caused by raw soybeans in chicks of different ages. J. Nutr. *84*, 179–184.

NITSAN, Z., and BONDI, A. 1965. Comparison of nutritional effects induced in chicks, rats and mice by raw soyabean meal. Brit. J. Nutr. *19*, 177–187.

NITSAN, Z., and GERTLER, A. 1969. The effect of soybean trypsin inhibitors on the activity of digestive enzymes in chicks. Israel J. Chem. *7*, Proceedings.

NORDSIEK, F. W. 1962. Effects of added casein on the goitrogenic action of different dietary levels of soybeans. Proc. Soc. Exptl. Biol. Med. *110*, 417–420.

NORRIS, L. C., LEACH, R. M., Jr., and ZEIGLER, T. R. 1958. Recent research on the mineral requirements of poultry. Proc. Distillers Feed Conf. *13*, 63–73.

OBARA, T., KIMURA-KOBAYASHI, M., KOBAYASHI, T., and WATANABE, Y. 1970. Heterogeneity of soybean trypsin inhibitor. I. Chromatographic fractionation and poly-acrylamide-gel electrophoresis. Cereal Chem. *47*, 597–606.

O'DELL, B. L. 1969. Effect of dietary components upon zinc availability. A review with original data. Am. J. Clin. Nutr. *22*, 1315–1322.

O'DELL, B. L., and SAVAGE, J. E. 1960. Effect of phytic acid on zinc availability. Proc. Soc. Exptl. Biol. Med. *103*, 304–306.

O'DELL, B. L., STOLZENBURG, S. J., BRUEMMER, J. H., and HOGAN, A. G. 1955. The antithyrotoxic factor: Its solubilization and relation to intestinal xanthine oxidase. Arch. Biochem. Biophys. *54*, 232–239.

OFELT, C. W., SMITH, A. K., and MILLS, J. M. 1955A. Effect of soy flour on amylograms. Cereal Chem. *32*, 48–52.

OFELT, C. W., SMITH, A. K., and MILLS, J. M. 1955B. Proteases of the soybean. Cereal Chem. *32*, 53–63.

OZAWA, K., and LASKOWSKI, M., Jr. 1966. The reactive site of trypsin inhibitors. J. Biol. Chem. *241*, 3955–3961.

PAGE, A. C., Jr., GALE, P. H., KONIUSZY, F., and FOLKERS, K. 1959. Coenzyme Q. IX. Coenzyme Q_9 and Q_{10} content of dietary components. Arch. Biochem. Biophys. *85*, 474–477.

PEKAS, J. C. 1966. Zinc-65 metabolism: gastrointestinal secretion by the pig. Am. J. Physiol. *211*, 407–413.

PENG, C. M., and WANG, Y. L. 1954. Phytase from soybean sprouts purification and properties. Sheng Li Hsueh Pao *19*, 249–267.

PERL, M., and DIAMANT, Y. 1963. The lipase of the soya bean: Purification and properties of the enzyme. Israel J. Chem. *1* No. 3a, 192–193.

PERLMAN, F. 1965. Food allergy and vegetable proteins. Food Technol. *20*, 1438, 1442–1445.

PAPAIOANNOU, S. E., and LIENER, I. E. 1970. The involvement of tyrosine and amino groups in the interaction of tyrosine and soybean trypsin inhibitor. J. Biol Chem. *245*, 4931–4938.

PIETERSE, P. J. S., and ANDREWS, F. N. 1956A. The estrogenic activity of legume, grass and corn silage. J. Dairy Sci. *39*, 81–89.

PIETERSE, P. J. S., and ANDREWS, F. N. 1956B. The estrogenic activity of alfalfa and other feedstuffs. J. Animal Sci. *15*, 25–36.

PINSKY, A., and GROSSMAN, S. 1969. Proteases of the soybean. II. Specificity of the active fractions. J. Sci. Food Agr. *20*, 374–375.

POMERANZ, Y. 1963. The liquefying action of pancreatic, cereal, fungal and bacterial alpha-amylases. J. Food Sci. *28*, 149–155.

POMERANZ, Y., and LINDNER, C. 1960. A simple method for evaluation of heat treatment of soybean meal. J. Am. Oil Chemists' Soc. *37*, 124–126.

POMERANZ, Y., and MAMARIL, F. P. 1964. Effect of *N*-ethylmaleimide on the starch-liquefying enzyme from soy flour. Nature *203*, 863–864.

PUBOLS, M. H., SAXENA, H. C., and MCGINNIS, J. 1964. Pancreatic enzyme levels in chicks fed unheated soybean meal. Proc. Soc. Exptl. Biol. Med. *117*, 713–717.

PUDLES, J., and BACHELLERIE, D. 1968. Studies on proteolytic inhibitors. III. Stability of trypsin in trypsin-inhibitor complexes to specific chemical inactivating agents. Arch. Biochem. Biophys. *128*, 133–141.

RACKIS, J. J. 1965. Physiological properties of soybean trypsin inhibitors and their relationship to pancreatic hypertrophy and growth inhibition of rats. Federation Proc. 24, 1488–1493.

RACKIS, J. J. 1966. Soybean trypsin inhibitors: Their inactivation during meal processing. Food Technol. 20, 1482–1484.

RACKIS, J. J., and ANDERSON, R. L. 1964. Isolation of four soybean trypsin inhibitors by DEAE-cellulose chromatography. Biochem. Biophys. Res. Commun. 15, 230–235.

RACKIS, J. J., HONIG, D. H., SESSA, D. J., and STEGGERDA, F. R. 1970A. Flavor and flatulence factors in soybean protein products. J. Agr. Food Chem. 18, 977–982.

RACKIS, J. J., SASAME, H. A., ANDERSON, R. L., and SMITH, A. K. 1959. Chromatography of soybean proteins. I. Fractionation of whey proteins on diethylaminoethyl-cellulose. J. Am. Chem. Soc. 81, 6265–6270.

RACKIS, J. J. et al. 1962 Soybean trypsin inhibitors: isolation purification, and physical properties. Arch. Biochem. Biophysic 98, 471–478.

RACKIS, J. J. et al. 1970B. Soybean factors relating to gas production by intestinal bacteria. J. Food Sci. 35, 634–639.

RACKIS, J. J. et al. 1963. Feeding studies on soybeans. Growth and pancreatic hypertrophy in rats fed soybean meal fractions. Cereal Chem. 40, 531–538.

RANGNEKAR, Y. B., DE, S. S., and LUBRAHMANYAN, V. 1948. Soybean ascorbicase. Indian J. Med. Res. 36, 361–370.

RATNER, B., and CRAWFORD, L. V. 1955. Soybean: Anaphylactogenic properties. Ann. Allergy 13, 289–295.

RATNER, B., et al. 1955. Allergenicity of modified and processed foodstuffs. V. Soybeans—influence of heat on its allergenicity; use of soybean preparations as milk substitutes. J. Allergy 89, 187–193.

RHODES, M. B., BENNETT, W., and FEENEY, R. E. 1960. The trypsin and chymotrypsin inhibitors from avian egg whites. J. Biol. Chem. 235, 1686–1693.

RICHARDS, E. A., and STEGGERDA, F. R. 1966. Production and inhibition of gas in various regions in the intestine of the dog. Proc. Soc. Exptl. Biol. Med. 122, 573–576.

RICHARDS, E. A., STEGGERDA, F. R., and MURATA, A. 1968. Relationship of bean substrates and certain intestinal bacteria to gas production in the dog. Gastroenterology 55, 502–509.

RICHERT, D. A., SCHENKMAN, J., and WESTERFELD, W. W. 1964. An antithyrotoxic assay based upon the response of rat liver a-glycerophosphate dehydrogenase. J. Nutr. 83, 332–342.

ROCKLAND, L. B., GARDINER, B. L., and PIECZARKA, D. 1970. Stimulation of gas production and growth of Clostridium perfringens Type A (No. 3624) by legumes. J. Food Sci. 34, 411–414.

RUEGAMER, W. R., MIRISOLOFF, D., and PESCADOR, J. 1968. Rat assay of a simulated human diet for antithyrotoxic substances. J. Nutr. 96, 289–293.

SAKAKIBARA, S., SHIMAKAWA, J., IMASEKI, H., and URITANI, I. 1965. Relation between enzyme activities in vegetables and blanching. Nippon Shokuhin Kogyo Gakkaishi 12, 43–45.

SALMAN, A. J., DAL BORGO, G., PUBOLS, M. H., and MCGINNIS, J. 1967. Changes in pancreatic enzymes as a function of diet in the chick. Proc. Soc. Exptl. Biol. Med. 126, 694–698.

SALMAN, A. J., and MCGINNIS, J. 1969. Digestibility of autoclaved soybean meal after temporal adaptation of chicks to unheated soybean meal. Poultry Sci. 47, 1508–1513.

SALMAN, A. J., PUBOLS, M. H., and MCGINNIS, J. 1968. Chemical and microscopic nature of pancreata from chicks fed unheated soybean meal. Proc. Soc. Exptl. Biol. Med. 128, 258–261.

SAMBETH, W., NESHEIM, M. C., and SERAFIN, J. A. 1967. Separation of soybean whey into fractions with different biological activities for chicks and rats. J. Nutr. 92, 479–490.

SAXENA, H. C., JENSEN, L. S., and MCGINNIS, J. 1962. Influence of raw soybeans on oxygen consumption and liver and muscle glycogen content of chicks. Poultry Sci. 41, 1304–1305.

SAXENA, H. C., JENSEN, L. S., and MCGINNIS, J. 1963A. Pancreatic hypertrophy and chick growth inhibition by soybean fractions devoid of trypsin inhibitor. Proc. Soc. Exptl. Biol. Med. *112*, 101–105.

SAXENA, H. C., JENSEN, L. S., and MCGINNIS, J. 1963B. Influence of age of utilization of raw soybean meal by chickens. J. Nutr. *80*, 391–396.

SAXENA, H. C., JENSEN, L. S., MCGINNIS, J., and LAUBER, J. K. 1963C. Histo-physiological studies on chick pancreas influenced by feeding raw soybean meal. Proc. Soc. Exptl. Biol. Med. *112*, 390–393.

SCHINGOETHE, D. J., AUST, S. D., and THOMAS, J. W. 1970. Separation of a mouse growth inhibitor in soybeans from trypsin inhibitors. J. Nutr. *100*, 739–748.

SCHULTZE, M. O. 1953. Nutritional value of plant materials. III. Lactation failure of rats fed diets containing purified soybean proteins. J. Nutr. *49*, 231–243.

SCOTT, M. L., and ZEIGLER, T. R. 1960. Recent studies on organic and inorganic unidentified chick growth factors. Proc. Distillers Feed Conf. *15*, 20–25.

SEALOCK, R. W., and LASKOWSKI, M., Jr. 1969. Enzymatic replacement of the arginyl by a lysyl residue on the reactive site of soybean trypsin inhibitor. Biochemistry *8*, 3703–3710.

SEN, N. P., and SMITH, D. M. 1966. An improved enzymic-ultraviolet method for determination of uric acid in flours. J. Assoc. Offic. Anal. Chemists *49*, 899–902.

SERAFIN, J. A., and NESHEIM, M. C. 1970. Influence on dietary heat-labile factors in soybean meal upon bile acid pools and turnover in chicks. J. Nutr. *100*, 786–796.

SINGH, L., WILSON, C. M., and HADLEY, H. H. 1969. Genetic differences in soybean trypsin inhinitors separated by disc electrophoresis. Crop Sci. *9*, 489–491.

SMITH, A. K., BELTER, P. A., and ANDERSON, R. L. 1956. Urease activity in soybean meal products. J. Am. Oil Chemists' Soc. *33*, 360–363.

SMITH, W. L. and LANDS, W. E. M. 1970. The self-catalyzed destruction of lipoxygenase. Biochem. Biophys. Res. Commun. *41*, 846–851.

SMITH, A. K. et al. 1964. Tempeh: Nutritive value in relation to processing. Cereal Chem. *41*, 173–181.

SOCOLOW, E. L., and SUZUKI, M. 1964. Possible goitrogenic effects of selected Japanese foods. J. Nutr. *83*, 20–26.

SPIES, J. R. et al. 1951. The chemistry of allergens. XI. Properties and composition of natural proteoses isolated from oilseeds and nuts by the CS-1A procedure. J. Am. Chem. Soc. *73*, 3995–4001.

STEAD, R. H., DEMUELENAERE, H. J. H., and QUICKE, G. V. 1966. Trypsin inhibition, hemagglutination, and intraperitoneal toxicity in extracts of *Phaseolus vulgaris* and *Glycine* max. Arch. Biochem. Biophys. *113*, 703–708.

STEINER, R. F., and FRATTALI, V. 1969. Purification and properties of soybean trypsin inhibitors of proteolytic enzymes. J. Agr. Food Chem. *17*, 513–518.

STEGGERDA, F. R. 1968. Gastrointestinal gas following food consumption. Ann. N. Y. Acad. Sci. *150*, 57–66.

STEGGERDA, F. R., RICHARDS, E. A., and RACKIS, J. J. 1966. Effects of various soybean products on flatulence in the adult man. Proc. Soc. Exptl. Biol. Med. *121*, 1235–1239.

STOB, M. 1966. Estrogens in foods. *In* Toxicants Occurring Naturally in Foods. Natl. Acad. Sci.–Natl. Res. Council Publ. *1354*, Washington, D. C.

SURREY, K. 1964. Spectrophotometric method for determination of lipoxidase activity. Plant Physiol. *39*, 65–70.

SZILAGYI, E. 1968. Influence of soybean trypsin inhibitors on germination and growth of plants. Acta Biol. Debrecina *6*, 157–169.

TAGUCHI, H., and ECHIGO, T. 1968. Changes in soybean proteins during storage. II. Sulfhydryl groups and proteinase activity. Tamagawa Daigaku Nogakubu Kenkyu Hokoku, *7–8*, 111–118.

TAI, K. 1953. Distribution of urease in plant seeds. Shiga Agr. Coll. Sci. Rept. Ser. *1*, 55–59.

THEUER, R. C., and SARETT, H. P. 1970. Nutritional adequacy of soy isolate infant formulas in rats. J. Agr. Food Chem. *18*, 913–916.

THOMPSON, O. J., CARLSON, C. W., PALMER, I. S., and OLSON, O. E. 1968. Destruction of rachitogenic activity of isolated soybean protein by autoclaving as demonstrated by turkey poults. J. Nutr. *94*, 227–232.

TRAVIS, J., and ROBERTS, R. C. 1969. Human trypsin. Isolation and physical-chemical characterization. Biochemistry *8*, 2884–2889.

TROP, M., and BIRK, Y. 1970. The specificities of proteinases from *Streptomyces griseus* (Pronase). Biochem. J. *116*, 19–25.

VAN BUREN, J. B. *et al.* 1964. Indices of protein quality in dried soy milk. J. Agr. Food Chem. *12*, 524–528.

VANETTEN, C. H. 1969. Goitrogens. *In* Toxic Constituents of Plant Foodstuffs, I. E. Liener (Editor). Academic Press, New York.

VAN MIDDLESWORTH, L. 1957. Thyroxine excretion, a possible cause of goiter. Endocrinology *61*, 570–573.

VIOQUE, E., and HOLMAN, R. T. 1962. Characterization of ketodienes formed in the oxidation of linoleate by lipoxidase. Arch. Biochem. Biophys. *99*, 522–528.

VOGEL, R., TRAUTSCHOLD, I., and WERLE, E. 1968. Natural Proteinase Inhibitors. Academic Press, New York.

VOGELS, G. D., TRIJBELS, F., and UFFINK, A. 1966. Allantoinase from bacterial, plant, and animal sources. I. Purification and enzyme properties. Biochim. Biophys. Acta *122*, 482–496.

VOHRA, P., ALLRED, J. B., GUPTA, I. S., and KRATZER, F. H. 1959. Fractionation of soybean oil meal for growth and antiperotic factors. 2. Studies with genistin and soysterols. Poultry Sci. *38*, 1476–1477.

VOHRA, P., and HEIL, J. R. 1969A. The growth promoting properties of crude soy phospholipids. Poultry Sci. *48*, 1661–1667.

VOHRA, P., and HEIL, J. R. 1969B. Dietary interactions between ZN, MN and CU for turkey poults. Poultry Sci. *48*, 1686–1691.

VOHRA, P., and KRATZER, F. H. 1967. The importance of the source of isolated soybean protein in nutrition. Poultry Sci. *46*, 1016–1017.

WALKER, G. C. 1963. The formation of free radicals during the reaction of soybean lipoxidase. Biochem. Biophys. Res. Commun. *13*, 431–434.

WALTER, E. D. 1941. Genistin (an isoflavone glucoside) and its aglucone, genistein from soybeans. J. Am. Chem. Soc. *63*, 3273–3276.

WALZ, E. 1931. Isoflavone and saponin-glucosides in *Soja hispida.* Ann. Chem. *489*, 118–155.

WANG, L. C., and ANDERSON, R. L. 1969. Purification and properties of soybean allantoinase. Cereal Chem. *46*, 656–663.

WANG, L. C., SMITH, A. K., and COWAN, J. C. 1969. A note on the detection of leucoanthocyanins in defatted soybean flakes. Cereal Chem. *46*, 468–470.

WEIL, J., PINSKY, A., and GROSSMAN, S. 1966. The proteases of the soybean. Cereal Chem. *43*, 392–399.

WESTERFELD, W. W., HOFFMAN, W. W., and RICHERT, D. A. 1962. An antithyrotixic assay based upon the metabolic rate response. J. Nutr. *78*, 403–414.

WESTERFELD, W. W., RICHERT, D. A., and RUEGAMER, W. B. 1968. Thyroxine and antithyrotoxic effects in the chick. J. Nutr. *83*, 325–331.

WEYER, E. M. (Editor) 1968. Chemistry, pharmacology, and clinical applications of proteinase inhibitors. Ann. N.Y. Acad. Sci. *146*, 361–787.

WILCOX, R. A., CARLSON, C. W., and KOHLMEYER, W. 1961A. Separation by dialysis of a water soluble growth factor(s) found in soybean meal. Poultry Sci. *40*, 1766–1767.

WILCOX, R. A., CARLSON, C. W., KOHLMEYER, W., and GASTLER, G. F. 1961B. The growth response of turkey poults to a water extract of soybean oil meal as influenced by different sources of isolated soybean protein. Poultry Sci. *40*, 1353–1354.

WILD, G. M. 1954. Action patterns of starch enzymes. Iowa State J. Sci. *22*, 419–420.

WILKENS, W. F., MATTICK, L. R., and HAND, D. B. 1967. Effect of processing on oxidative off-flavors of soybean milk. Food Technol. *21*, 1630–1633.

WILKINSON, J. G. 1965. Isoenzymes. E & F. N. Spon London.

WOLF, W. J., and THOMAS, B. W. 1970. Thin layer and anion exchange chromatography of soybean saponins. J. Am. Oil Chemists' Soc. *47*, 86–90.

WONG, E. 1966. Occurrence and biosynthesis of 4',6-dihydroxyaurone in soybean. Phytochemistry *5*, 463–467.

WONG, E., and FLUX, D. S. 1962. Estrogenic activity of red clover isoflavones and some of their degradation products. J. Endocrinol. *24*, 341–348.

WU, Y. V., and SCHERAGA, H. A. 1962. Studies of soybean trypsin inhibitor. I. Physicochemical properties. Biochemistry *1*, 698–705.

YAMAMOTO, M., and IKENAKA, T. 1967. Studies on soybean trypsin inhibitors. I. Purification and characterization of two soybean trypsin inhibitors. J. Biochem. (Tokyo) *62*, 141–149.

YAMAMOTO, A., YASUMOTO, K. and MITSUDA, H. 1970. Isolation of lipoxygenase isozynes and comparison of their properties. Agr. Biol. Chem. (Tokyo) *34*, 1169–1177.

YASUMOTO, K., YAMAMOTO, A. and MITSUDA, H. 1970. Effect of phenolic antioxidants on lipoxygenase reaction. Agr. Biol. Chem. (Tokyo) *34*, 1162–1168.

YIN, H. C., and SUN, C. N. 1949. Localization of phosphorylase and of starch formation in seeds. Plant Physiol. *24*, 103–110.

YOSHIDA, A., UMAI, A., KURATA, Y., and KAWAMURA, S. 1969. Utilization of soybean oligosaccharides by the intact rat. Eiyo To Shokuryo *22*, 262–265.

CHAPTER 7

I. E. Liener | # Nutritional Value of Food Protein Products

INTRODUCTION

If the world is to have a realistic chance of meeting the problems created by an ever-expanding population, new and unconventional sources of protein must be sought. In a recent United Nations publication (1968) entitled "International Action to Avert the Impending Protein Crisis," the specific proposal is made for an increase in the use of oilseeds and oilseed protein concentrates as direct sources of protein in the human diet since "no other single source of unconventional protein could contribute so greatly and so promptly towards closing the protein gap". Among the oilseeds the soybean assumes a most prominent position; not only is its protein content high (30–46%), but this protein, when properly processed, is of good nutritional quality. The great potential which soybean protein has for meeting the protein requirements of man is illustrated in a very vivid fashion by the following figures. Bean (1966) has estimated that one acre of soybeans will provide enough protein to sustain a moderately active man for 2224 days. This figure may be compared with the number of days of protein requirement produced by 1 acre of wheat, 877 days; or of corn, 354 days. Even the latter values are considerably higher than the 77 days' supply of protein which would be derived from 1 acre of land used to support beef cattle.

The present chapter will be concerned primarily with the nutritional properties of the soybean and of various food products derived therefrom. Although primary consideration will be given to the protein, cognizance will also be made of other nutrients such as vitamins and minerals. Much of our knowledge concerning the nutritional properties of the soybean has been derived from experiments with animals, and such knowledge is frequently directly applicable to human nutrition. Nutritional experiments with human subjects pose special problems, the most difficult of which is acceptability. People will not eat certain foods simply "because it is good for them." Thus, the most serious hurdles to be overcome in the development of products containing soybean proteins is frequently not a nutritional one but one of consumer acceptance.

Since the older literature dealing with the nutritional aspects of the soybean has already been covered in a number of comprehensive reviews (Horvath 1938; Barnes and Maack 1943; Payne and Stuart 1944; Mitchell 1950), only brief reference will be made here to studies prior to 1950. Two excellent articles dealing with the role of vegetable proteins in human nutrition have been recently written by Swaminathan (1967) and Bressani and Elias (1967). The specific use of soybean products for protein in human foods has been the subject of two conferences sponsored by the USDA (1961, 1966).

PROTEIN AND AMINO ACID REQUIREMENTS OF MAN

Protein Requirements

Before being able to evaluate the potential role of the soybean for satisfying the protein requirements of man, some consideration should be given to the quantitative nature of these requirements. Although this, of course, is a very complex subject about which many books and monographs have been written [see, for example, Albanese (1959); Beaton and McHenry (1964); National Academy of Sciences (1961, 1968); FAO (1965)] it is nevertheless possible to make a number of generalizations. As might be expected, the requirements for protein are greatest during periods of most rapid growth. Protein needs, expressed on the basis of body weight, are greatest in babies, followed by young children, adolescents, and adults in that order. Protein requirements also depend on the physiological state of the individual. The pregnant or lactating mother requires an extra supply of protein to build up the tissues of the unborn child or for the production of milk.

The quantitative assessment of human dietary protein requirements is based partly on accurate scientific measurement and partly on the interpretation of the results of these measurements. For this reason, different expert bodies arrive at somewhat different estimates of man's requirements as exemplified by the figures shown in Table 7.1. The apparent discrepancy between these two sets of recommendations lies in the difference in the assumptions upon which they were based. The recommendations of the National Academy of Sciences are based on the assumption that the efficiency of utilization of protein in the average U.S. diet is 70% of that of an ideal protein (see below) and the inclusion of a 30% safety factor to take into account individual variability. The FAO recommendations are based on the amount of dietary nitrogen that is needed to replace obligatory losses of nitrogen in the urine and feces and from the skin, an allowance for nitrogen required for the synthesis of growing tissue, and a 10% increment to allow for the stress of ordinary life.

Amino Acid Requirements

The classical pioneer studies of Rose and his associates (Rose 1949, 1957) established the fact that the adult man requires a dietary source of eight amino acids for the maintenance of nitrogen equilibrium: isoleucine, leucine, lysine, methionine, phenylalanine, threonine, tryptophan, and valine.[1] Since methio-

---

[1] Histidine is required as a dietary essential for the infant. Arginine is believed to be required for maximum growth.

TABLE 7.1

DAILY PROTEIN REQUIREMENTS BY AGE AND STATE
OF DEVELOPMENT

Age and State of Development		Protein Requirements Gm/Kg Body Wt	
		NAS[1]	FAO[2]
Infants	0 – 3 months	2.2	2.3
	3 – 6 months	2.0	1.8
	6 – 9 months	1.8 – 2.0	1.5
	9 – 12 months	1.8 – 2.0	1.2
Children	1 – 3 yr	1.8	1.1
	4 – 6 yr	1.6	1.0
	7 – 9 yr	1.4	0.9
	10 – 12 yr	1.3	0.9
Adolescents	13 – 15 yr	1.0	0.8
	16 – 19 yr	0.9	0.8
Adults	> 20 yr	0.9	0.7
In pregnancy		1.1	0.8
In lactation		1.3	1.0

[1] Nat. Acad. of Sci. (1968).
[2] FAO (1965).

nine and phenylalanine are utilized by the body for the synthesis of cystine and tyrosine respectively, the requirement for these two amino acids can be partially met by any cystine and tyrosine present in the diet. In Table 7.2 are presented the amino acid requirements which have been proposed for infants and adults. More important than the quantitative requirements for each amino acid, however, is the pattern or ratio of each amino acid requirement to another. In other words, when a protein-containing food is fed at a level which meets the total protein requirement, the overall pattern of the essential amino acids is more important in determining the quality of the protein than the absolute amount of each essential amino acid. Thus, one can simply adopt as an ideal reference protein a hypothetical protein in which the essential amino acids are present in a pattern which reflects amino acid requirements. Such a reference protein or provisional pattern was in fact proposed by the FAO Committee on Protein Requirements (FAO 1957). The amino acid pattern of this hypothetical protein is shown in Table 7.2 where each amino acid is given in proportion to tryptophan taken as 1.0. At the time the 1957 FAO provisional pattern was adopted there was no good experimental evidence to prove that this pattern of amino acids was, in fact, superior to the pattern found in proteins of such foods as milk and eggs which are known to be of very high quality. Subsequent investigations have disclosed that protein blends and amino acid mixtures which mimic

TABLE 7.2

DAILY ESSENTIAL AMINO ACID REQUIREMENTS OF HUMANS
AND ESSENTIAL AMINO ACID PATTERNS OF REFERENCE PROTEINS

Amino Acid	Infants[1] Mg/Kg	Adults[1] Mg/Day Male	Adults[1] Mg/Day Female	1957 FAO Provisional Pattern[2] Tryptophan = 1.0	Whole Egg Protein Pattern[3] Tryptophan = 1.0	A/E Ratio of Whole Egg Protein[3] Mg/Gm Total Essential Amino Acids
Histidine	32	—	—	—	—	—
Isoleucine	90	450	700	3.0	4.1	129
Leucine	150	620	1100	3.4	5.5	172
Lysine	105	500	800	3.0	4.0	125
Phenylalanine	90	220	300	2.0	3.6	114
Tyrosine		900	1100[4]	2.0	2.6	81
Methionine		350	200	1.4	1.9	61
Cystine	85	200	810[5]	1.6	1.5	46
Threonine	60	305	500	2.0	3.2	99
Tryptophan	22	157	250	1.0	1.0	31
Valine	93	650	800	3.0	4.5	141

[1] Natl. Acad. Sci. (1959).
[2] FAO (1957).
[3] FAO (1965).
[4] If tyrosine is absent from the diet, 1100 mg/day of phenylalanine will satisfy the equivalent for total aromatic amino acids.
[5] If cystine is absent from the diet, 1100 mg/day of methionine will satisfy the requirement for total sulfur-containing amino acids.

the 1957 FAO provisional pattern are, in fact, nutritionally inferior to the proteins of eggs and milk (Nat. Acad. Sci. 1963; Clark 1964). This prompted a joint FAO/WHO expert group (FAO 1965) to recommend the adoption of the essential amino acid pattern of whole hen egg protein for reference purposes. Historically, it is interesting to note that the amino acid pattern of egg protein had been proposed almost 20 yr earlier by Mitchell and Block (1946) as a standard of reference for evaluating protein quality.

The essential amino acid pattern of whole egg protein may be expressed as the ratio of each amino acid to tryptophan taken as 1.0, or each essential amino acid expressed as a fraction of the total essential amino acids (A/E ratio). Both methods of expression are shown in Table 7.2. Expressing the essential amino acid pattern of proteins in this fashion, however, does not take into account the role of the nonessential amino acids. It should be recognized that in the absence of an adequate supply of the nonessential amino acids, the essential amino acids may be used for the synthesis of the nonessential amino acids, which, in spite of their designation, are needed for protein synthesis. For this reason the 1965 FAO/WHO group suggested that a useful index would also be provided by expressing the total essential amino acid content of a protein as a fraction of the total nitrogen (E/T ratio in Table 7.3). When this is done it becomes evident that whole egg protein assumes a rank which justifies its adoption as an ideal protein and that the 1957 FAO provisional pattern falls well below most food proteins. In general, proteins of animal origin rank highest and plant proteins lowest; soybean protein assumes an intermediate position.

TABLE 7.3

RATIO OF TOTAL ESSENTIAL AMINO ACIDS
TO TOTAL NITROGEN IN SELECTED FOODSTUFFS
AND IN THE 1957 FAO PROVISIONAL PATTERN

Protein Source	E/T Ratio Gm/Gm Total N
Whole egg protein	3.22
Cow's milk	3.20
Beef muscle	2.79
Fish	2.66
Soy flour	2.58
Sesame seed	2.47
Cottonseed	2.15
Peanut flour	2.08
White wheat flour	2.02
1957 FAO pattern	2.02
Wheat gluten	1.99
Cassava	1.31
Gelatin	1.05

Source: FAO (1965).

EVALUATION OF PROTEIN QUALITY

Amino Acid Composition

From what has already been said about the amino acid requirements of man, it is tempting to conclude that protein quality can be equated with amino acid composition, and, by and large, this turns out to be a reasonable approximation of the truth. Earlier compilations of the amino acid composition of soybeans (Circle 1950; Kuppuswamy *et al.* 1958; Aykroyd and Doughty 1964) were based largely on data obtained by microbiological assay procedures. With the introduction of ion-exchange chromatography and automated techniques for the determination of amino acids (Spackman *et al.* 1958), much more precise and reliable amino acid data on soybeans and soybean products have since appeared in the literature. The essential amino acid composition of a wide variety of soybean products is shown in Table 7.4, and frequent reference will be made to these data throughout this chapter. The amino acid composition of many different varieties and strains of soybeans has been compared, particularly with regard to methionine, the limiting amino acid of soybean protein; but thus far few significant differences have been found (Alderks 1949; Kuiken and Lyman 1949; Krober and Cartter 1966). Although methionine content appears to be likewise unaffected by season or location (Krober 1956), there is some indication that the level of methionine in the seed increases proportionately with protein content (Krober and Cartter 1966).

In Figure 7.1 a comparison is made of the amino acid pattern of soybean protein with the 1965 FAO amino acid pattern based on the composition of whole egg protein. Inspection of this figure reveals a close correspondence between these 2 patterns with the exception of the 2 sulfur-containing amino acids, cystine and methionine, both of which are lower in the soybean protein. One would predict, therefore, simply on the basis of its essential amino acid composition, that the sulfur-containing amino acids limit the nutritive value of soybean protein. The extent to which a protein supplies a limiting amino acid in comparison with a reference protein such as whole egg protein, is referred to as the "chemical score" of that protein. In the method originally proposed by Mitchell and Block (1946), the concentrations of amino acids are expressed as grams per 16 gm N, whereas in the 1965 FAO/WHO procedure each essential amino acid is calculated as milligrams per gram of total essential amino acids. Despite a number of suggested modifications (Kuhnau 1949; Oser 1951; Mitchell 1954; Rama Rao *et al.* 1964) this simple calculation still remains the method of choice for a rapid evaluation of protein quality based solely on amino acid data. On the basis of the data shown in Fig. 7.1 the chemical score of soybean protein would be calculated, according to the FAO/WHO procedure, as follows:

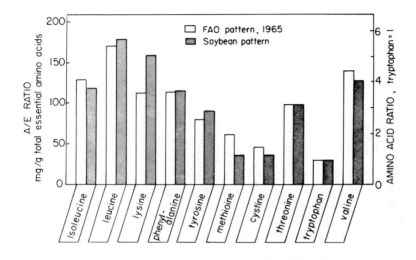

From Van Etten et al. (1967)

FIG. 7.1. THE ESSENTIAL AMINO ACID PATTERN OF SOYBEAN PRO-
TEIN COMPARED WITH THE WHOLE EGG PROTEIN

Histidine (not in Figure) is required as a dietary essential for the infant. Arginine
is believed to be required for maximum growth.

	Cystine	Methionine	Total
	(Mg/Gm Total Essential Amino Acids)		
Whole egg protein	46	61	107
Soybean protein	37	37	74

Chemical score $= \dfrac{74}{107} \times 100 = 69\%$

For comparison, the chemical scores of other proteins are shown in Table 7.5.

From the chemical score of soybean protein one might suppose that this
protein was about 70% as effective as egg protein in meeting the human

TABLE 7.4

ESSENTIAL AMINO ACID COMPOSITION OF VARIOUS SOYBEAN PRODUCTS[1]

Source of Soybean Protein	Protein[2] Content (%)	Essential Amino Acids[3] (Gm/16 Gm N)												Reference
		Ile	Leu	Lys	Met	Cys	Met + Cys	Phe	Tyr	Phe + Tyr	Thr	Trp	Val	
Whole soybeans (Chippewa)	34.3	4.2	7.4	6.4	1.1	...	2.3	4.5	...	...	3.6	1.7	4.3	Evans and Bandemer (1967)
Whole soybeans (Harsoy)	36.9	4.2	8.0	6.5	1.0	...	2.2	4.9	...	...	3.7	1.8	4.6	Evans and Bandemer (1967)
Whole meal, defatted, dehulled	52.6	4.4	7.2	5.4	1.2	0.8	2.0	4.6	3.4	8.0	3.6	1.3	4.0	Harmon et al. (1969)
Whole meal, defatted, hulled	45.7	4.6	7.7	5.6	1.3	0.9	2.2	5.0	3.6	8.6	4.1	1.4	5.0	Harmon et al. (1969)
Whole green soybeans (Edamame)	37.5	6.6	7.1	8.5	0.8	...	...	5.0	2.8	7.8	1.9	1.0	5.6	Standal (1967)
Dehulled soybeans	47.5	4.9	7.9	6.5	1.5	1.2	2.7	5.4	3.6	9.0	3.8	0.9	5.2	Smith et al. (1964B)
Whole soybeans, defatted	44.7	4.8	7.0	6.1	1.3	2.1	3.4	4.6	3.0	7.6	4.7	1.8	5.3	Tkachuk and Irvine (1969)
Full-fat chips	...	4.8	7.7	6.2	1.4	1.7	3.1	5.2	3.8	9.0	4.2	...	4.9	Iriarte and Barnes (1966)
Whole full-fat flour	46.6	4.8	7.8	6.5	1.4	1.6	3.0	5.1	3.9	9.0	4.2	...	5.0	Iriarte and Barnes (1966)
Dehulled, full-fat flour	42.0	4.6	7.6	6.3	1.3	1.6	2.9	5.0	3.8	8.8	4.2	...	5.2	Iriarte and Barnes (1966)
Defatted flour	59.0	4.6	7.7	6.2	1.3	1.2	2.5	5.3	...	...	4.2	1.4	4.9	Meyer (1968)
Defatted flour	52.0	4.6	7.7	6.7	...	...	3.2	...	...	9.0	4.0	1.4	5.1	Inglett et al. (1969)
Curd (tofu), Honolulu	57.4	4.1	6.7	5.6	1.3	...	...	5.5	3.2	8.7	2.2	2.1	4.1	Standal (1967)
Curd (tofu) Thailand	...	4.3	7.6	5.0	1.2	...	2.4	4.8	...	...	3.2	1.0	4.0	Evans and Bandemer (1967)
Curd, acid-precipitated	59.0	4.9	8.0	5.9	1.4	1.7	3.1	4.8	3.7	8.5	3.7	1.1	4.7	Hackler et al. (1967)
Curd, acid-precipitated	103.0	...	9.1	6.1	1.4	1.2	2.6	5.4	3.9	9.3	3.6	1.5	5.4	Van Etten et al. (1959)
Curd, acid-precipitated	102.0	5.0	7.9	5.7	1.3	1.0	2.3	5.9	4.6	10.5	3.8	1.0	5.2	Rackis et al. (1961)
Tempeh, laboratory product	53.1	4.9	7.9	6.3	1.5	1.2	2.7	5.0	3.7	8.7	3.9	1.0	5.2	Smith et al. (1964B)
Tempeh, laboratory product untreated	...	4.6	7.7	5.5	1.3	1.6	2.9	5.1	3.5	8.6	3.6	1.2	4.4	Stillings and Hackler (1965)
deep-fat fried, 7 min	...	4.6	7.6	4.0	1.2	1.2	2.4	4.7	3.1	7.8	3.5	1.3	4.5	
steamed, 2 hr	...	4.8	7.9	5.6	1.4	1.6	3.0	4.9	4.2	9.1	3.7	1.6	4.7	
Tempeh, Indonesian	55.4	5.7	9.1	6.6	1.3	1.9	3.2	4.8	2.6	7.4	4.5	1.4	5.5	Murata et al. (1967)
Natto	57.5	5.8	8.4	11.1	0.7	...	...	5.4	3.0	8.4	4.1	2.2	6.4	Standal (1967)
Miso	17.3	6.5	13.1	5.9	1.2	...	...	6.4	4.1	10.5	3.0	1.2	5.7	Standal (1967)

Concentrate	71.0	4.9	8.0	6.6	1.3	1.6	2.9	5.3	3.7	9.1	4.3	1.4	5.0	Huge (1961); Meyer (1966)
Isolate	96.0	4.6	7.6	5.4	1.2	0.8	2.0	5.5	3.6	9.1	3.5	...	4.0	Bressani et al. (1967)
Isolate (Promine)	90.0	5.0	7.9	6.3	1.1	1.0	2.1	5.5	3.8	9.3	3.7	1.3	5.2	Huge (1961); Meyer (1966)
Textured meat analogue	56.0	3.3	5.1	3.2	0.9	1.0	1.9	4.0	3.1	7.1	2.5	...	3.2	Bressani et al. (1967)
Sprouts	50.4	5.0	8.4	7.0	1.0	0.4	1.4	4.0	3.3	7.3	5.8	2.3	5.6	Standal (1967)

[1] The essential amino acid composition of soybean milk and various other fractions of soybeans are shown in Tables 7.11 and 7.13 respectively.
[2] On moisture-free basis.
[3] The following abbreviations have been used: Ile, isoleucine; Leu, leucine; Lys, lysine; Met, methionine; Cys, cystine; Phe, phenylalanine; Tyr, tyrosine; Thr, threonine; Trp, tryptophan; Val, valine.

TABLE 7.5

NUTRITIONAL VALUE OF SOME REPRESENTATIVE FOOD PROTEINS
BASED ON BIOLOGICAL EVALUATION WITH ANIMALS AS WELL
AS AMINO ACID COMPOSITION

Source of Protein	PER	BV	NPU	Limiting Amino acid	Chemical Score
Animal sources					
Whole egg	3.8	87–97	91–94	none	100
Cow's milk	2.5	85–90	86	S[1]	60
Beef muscle	3.2	76	71–76	S	80
Salmon	–	72	71	Tryptophan	75
Plant sources					
Soybeans[2]	0.7–1.8	58–69	48–61	S	69
Peanuts	1.7	56	43–54	S	70
Cottonseed	1.3–2.1	62	56–58	S	80
Rice	1.9	75	70	Lysine	57
Corn	1.2	60	49–55	Lysine	55
Wheat	1.0	52	52	Lysine	57

Source: Data compiled from Altschul (1965) and FAO (1965).

[1] The letter "S" denotes total sulfur-containing amino acids.
[2] Based on range of values shown in Table 7.6.

requirement for amino acids. Unfortunately, this simple method of evaluating protein quality has some rather severe limitations which are particularly evident in the case of soybeans. These limitations will be discussed in some detail below. In addition, although the chemical score discloses the fact that soybean protein is deficient in methionine and cystine, it does not reveal one of the most valuable attributes of soybean protein; namely, the fact that it has a much higher content of lysine than most plant proteins.

Biological Techniques Involving Animals

The chemical score, based as it generally is, upon the quantity of amino acids which can be recovered in an acid hydrolysate of a protein, assumes that the animal body can utilize all of each amino acid so measured. This is an assumption which does not take into account a number of factors which can alter the physiological availability of an amino acid such as: (1) the digestibility of the protein, that is, the extent to which amino acids are released from the protein by the digestive apparatus of the animal; (2) the rate at which amino acids may be absorbed from the gastrointestinal tract; and (3) complex interactions with other nutrients which may then affect (1) and (2). It follows therefore that, although a knowledge of the amino acid composition of a protein can provide a valuable index as to its potential nutritive value, it is the actual performance of that protein in an intact animal which must be ultimately assessed in some fashion or other.

Ideally, any answers that the nutritionist comes up with should be directly applicable to man. Since, however, experiments with humans are complicated, time-consuming, and expensive, the nutritionist must rely very heavily on extrapolated interpretations of data obtained by animal experimentation. It cannot be argued that human requirements for amino acids are identical with that of the rat, particularly if one compares the requirements of a rapidly growing animal with that of a fully mature adult human. Even a human infant does not grow on a percentage basis at rates comparable to that observed with experimental animals (Howe 1961). The requirements for the growth of an animal may bear little relevance to the requirements for repleting undernourished and convalescing bodies, for human lactation, or for generating immunological defenses against infection. Despite these reservations, experience has shown that protein evaluations based upon animal studies can be of value in assessing food proteins in human nutrition (Block and Mitchell 1946–1947; Frost 1959). Some of the various procedures used for the biological evaluation of the nutritive value of proteins may be found in publications by the National Academy of Sciences (1959, 1963) and FAO (1965).

Protein Efficiency Ratio (PER).–The PER, first proposed by Osborne and Mendel in 1917, remains the most widely used technique today for the biological evaluation of protein quality. It is defined as the weight gain of a growing animal (rat) divided by its protein intake, and, when conducted under standardized conditions, is capable of yielding fairly accurate and reproducible results (Derse 1962). By including casein as a control and relating the observed PER's of the experimental groups to casein (the PER value of casein is usually taken as 2.5), meaningful interlaboratory comparisons can be made. The PER values of soybean protein in relation to other sources of protein are shown in Table 7.5. It is evident from the data compiled in Table 7.6 that soybeans and soybean products may exhibit considerable variation in PER values depending on processing conditions, a point which is discussed in more detail later.

N-balance Studies.–The nutritive value of a protein may be evaluated in terms of its ability to provide amino acids for the synthesis or replacement of body tissue protein. This can be done most conveniently in the intact animal by comparing the quantity of N ingested (I) with that amount which is excreted in the urine (U) and feces (F). The difference between these two values, referred to as nitrogen balance (B), indicates whether the animal is gaining, losing, or maintaining its nitrogen resources:

$$B = I - (F + U)$$

If nitrogen intake is greater than the quantity of N excreted ($I > F + U$) the animal has retained N and is said to be in a state of positive N balance. If N intake is less than the N excreted ($I < F + U$), then the animal is losing N from the body and is said to be in negative N balance. If N intake should equal N excreted ($I = F + U$), the animal is maintaining N equilibrium. The ability of an

TABLE 7.6

BIOLOGICAL EVALUATION OF THE NUTRITIVE VALUE OF SOYBEAN PRODUCTS AS DETERMINED BY RAT FEEDING EXPERIMENTS

Source of Protein	PER[1] Range[2]	PER[1] Avg[3]	BV Range[2]	BV Avg[3]	Digestibility Range[2]	Digestibility Avg[3]	Reference
Whole soybean							
Raw, immature		1.1		49		88	Everson et al. (1944)
Autoclaved, immature		2.0					Everson et al. (1944)
Raw, vine-ripened		0.5		69		85	Everson and Heckert (1944)
							Mitchell and Beadles (1949)
Autoclaved, vine-ripened		1.5					Everson and Heckert (1944)
Raw, mature	0 –1.6 (25)	0.7	41–74 (14)	58	75–89 (13)	82	Kuppuswamy et al. (1958)
Dry heat, mature	0.4–1.1 (5)	0.7					Kuppuswamy et al. (1958)
Steamed, mature	1.2–1.4 (6)	1.3			90–94 (3)	92	Kuppuswamy et al. (1958)
Autoclaved, mature	0.4–2.0 (20)	1.3	64–67 (4)	64	83–94 (4)	90	Kuppuswamy et al. (1958)
Raw, germinated		1.4					Everson et al. (1944)
Autoclaved, germinated		1.9					Everson et al. (1944)
Grits, toasted		2.4					Huge (1961)
Meal, solvent extracted							
Uncooked	0.3–0.6 (6)	0.5	50–53 (2)	52	67–81 (2)	74	Kuppuswamy et al. (1958)
Autoclaved	1.1–2.9 (21)	1.9	61–68 (2)	65		84	Kuppuswamy et al. (1958)
Meal, expeller pressed	0.6–2.5 (22)	1.8					Kuppuswamy et al. (1958)
Flour, defatted	1.5–2.4 (13)	1.8	60–75 (5)	69	82–96 (6)	89	Kuppuswamy et al. (1958)
full-fat		1.6					Kon and Markuze (1931)
extrusion cooked		2.0					Smith (1969)
Milk[4]	1.6–2.3 (7)	2.0		79		91	Desikachar et al. (1946A); Hackler et al. (1965); Harkins and Saret (1967); Chang and Murray (1949); Shurpalekar et al. (1965)
Curd[5]	1.7–1.9 (2)	1.8	65–69 (3)	68		96	Pian (1930); Chang and Murray (1949);

Product					Reference
Tempeh	1.2–3.0 (8)	2.2			Kobatake et al. (1964); Standal (1967); Matsuno and Tamura (1964)
Steamed, 2 hr		2.1			György (1961); Smith et al. (1964B)
Deep fat-fried, 7 min		0.6			Hackler et al. (1964)
		2.6			Hackler et al. (1964)
Natto			55	72	Arimoto (1961); Standal (1963)
Miso				72	Matsuno and Tamura (1964)
Protein concentrate Unheated	0.3–2.5 (5)	1.4			Meyer (1966); Longenecker et al. (1964)
Heated		1.7			Longenecker et al. (1964)
Protein isolate	0.6–1.9 (6)	1.3			Huge (1961); Longenecker et al. (1964)
Textured meat analogue		2.3	65	92	Bressani et al.

[1] Wherever possible, PER values have been corrected on the basis of a PER of 2.5 for casein.
[2] Figures in parentheses refer to the number of observations which are included in the range of values shown.
[3] Where more than one observation is recorded, the figure shown is the average of the range of values shown in preceeding column.
[4] See also Table 7.11.
[5] PER values for other soybean fractions may be found in Table 7.13.

animal to maintain N equilibrium (as in a nongrowing animal) or to retain N (as in a growing animal) is dependent on the availability of a balanced assortment of essential amino acids in the diet. A diet deficient in one or more essential amino acids does not permit efficient N utilization, hence much of the dietary N would be lost in the urine and feces, a condition which would produce a negative nitrogen balance.

Because of its simplicity, the N balance technique can be effectively applied to studies with human subjects (see below). The use of animals, however, permits a further refinement in which an attempt is made to distinguish between that fraction of the dietary N which is not digested or absorbed and that fraction which is absorbed and retained. The *digestibility* of a protein (D) is defined as that percent of the N intake which is absorbed and is given by the equation:

$$D = \frac{I - (F - F_o)}{I} \times 100$$

Note the use of the additional term, F_o, which is the amount of N present in the feces, had not protein been added to the diet (sometimes called metabolic or endogenous fecal N). That percentage of the absorbed N which is actually retained by the animal body is termed biological value (BV) and is given by the expression:

$$BV = \frac{N \text{ retained}}{N \text{ absorbed}} \times 100 = \frac{I - (F - F_o) - (U - U_o)}{I - (F - F_o)} \times 100$$

The new term U_o refers to urinary N of metabolic origin.

Both digestibility and biological value may be combined to give a single value called net protein utilization (NPU) which is simply $D \times BV$. NPU may be calculated directly by combining the two expressions for D and BV:

$$NPU = \frac{I - (F - F_o) - (U - U_o)}{I} \times 100$$

Plasma Amino Acids.—Postabsorptive changes in the concentrations of the plasma amino acids have been found to be a useful index of the extent to which ingested protein can meet an animal's requirement for amino acids (Longenecker 1963). By means of this technique, soybean protein has been shown to be uniformly deficient in methionine for the rat (Goldberg and Guggenheim 1962), chick (Smith and Scott 1965), pig (Puchal *et al.* 1962), and humans (Longenecker 1963). Goldberg and Guggenheim (1962) also used this technique to demonstrate the fact that the availability of the amino acids methionine, lysine, and tryptophan from unheated soybean flour was significantly less than from the toasted flour.

Representative values for the nutritive value of food proteins of animal and plant origin as determined by the methods discussed here are given in Table 7.5 and, for various soybean products, in Table 7.6. In general, there is a high degree of correlation among these various parameters (Block and Mitchell 1946–1947) except in the case of soybeans where it is important to emphasize that the chemical score, which is based on amino acid composition rather than amino acid availability, may give misleading information. This arises from the fact that the nutritive value of soybean protein is significantly affected by processing under conditions which may have little effect on its composition. This point will be discussed in further detail later.

Experiments with Human Subjects

In theory at least, growth methods as well as N-balance techniques can be carried out with human subjects albeit with considerably greater technical difficulties. Experiments with human subjects, however, permit the measurement of biochemical parameters which provide valuable supplementary data, such as changes in the pattern of proteins and amino acids in the blood, blood urea N, urinary excretion of creatinine and sulfur, et. (Albanese 1959). The Protein Advisory group of WHO/FAO/UNICEF (1966) has published recommendations dealing with the human testing of protein-rich foods.

The measurement of the growth of infants in relation to protein intake is analogous to measurement of PER in animals. The level of protein in the diet must be carefully chosen since, with high intakes of protein, much of the protein may not be used for protein synthesis but is metabolized and excreted. Sufficient calories must also be provided in the diet to prevent the diversion of protein for energy production. This is particularly important in nutritional studies in the field where protein-rich foods are frequently fed under conditions in which the infants are suffering from a caloric deficiency as well as a protein deficiency.

The measurement of height or body length is of great value, particularly in older children, because height is usually less variable than weight. However, since height increases more slowly, the measurements have to be made over a fairly long period. Supplementary information relevant to skeletal growth can also be obtained from X-ray pictures and the urinary output of hydroxyproline (WHO/FAO/UNICEF 1966).

Nitrogen balance studies in humans has been employed for many years by nutritionists for evaluating the nutritive value of proteins. The classical studies of Rose and his coworkers (Rose 1949, 1957) which established man's requirements for essential amino acids were based on N-balance experiments with adults. The technique is relatively simple since it involves only the determination of nitrogen in the food, urine, and feces. From these data, N-balance, biological value, digestibility, and NPU can be readily calculated as already described for animals. The main problem is the fact that it is not practical to correct urinary and fecal N values for endogenous levels, although this can be done under certain conditions (Fomon 1961A; Snyderman *et al.* 1961). When endogenous levels of N excretion have not been determined, the "apparent" digestibility and "apparent" biological value of a protein are given by the expressions:

$$\text{"apparent" digestibility} = \frac{I - F}{I} \times 100$$

$$\text{"apparent" BV} = \frac{I - F - U}{I - F} \times 100$$

Mendel and Fine (1911 – 1912) made the observation rather early that soybeans produce positive nitrogen balance in a human subject. Since that time a considerable number of N-balance studies have confirmed the fact that the digestibility and biological value of soybean protein for humans are quite satisfactory and compare quite favorably with animal proteins. Some of these values are recorded in Table 7.7 along with comparable values for other food proteins. Further reference to these data will be made in discussions dealing with specific soybean food products.

Amino Acid Availability

Recognizing the fact that the amino acid composition of a protein does not always reveal the extent to which a particular amino acid may be available or utilized by the animal body, attempts have been made to determine the availability of the amino acids from soybean protein. One such technique involves the use of a basal diet deficient in a particular amino acid and then comparing the weight gains of a group of animals receiving the basal diet plus the test protein with other groups receiving the basal diet supplemented with graded levels of the missing amino acid (Guthneck *et al.* 1953; Schweigert and Guthneck 1953, 1954; Deshpande *et al.* 1957; Ousterhout *et al.* 1959; Smith 1968). Data relating to the availability of some of the amino acids of soybeans for the rat and chick as determined by this method are summarized in Table 7.8. Inspection of these data shows considerable variability in the degree to which the various amino acids are available to a given test animal. Generally speaking, it would appear that most of the amino acids of soybean protein are 65-100% available provided the soybean protein has been subjected to proper heat treatment. Particularly noteworthy is the fact that less than 1/2 of the methionine and lysine content of unheated soybean protein is available to the rat.

In Vitro Techniques

Numerous attempts have been made to develop relatively simple and rapid *in vitro* techniques which would serve as an indication of the nutritive value of proteins without resorting to time-consuming and expensive animal experiments. See reviews by Grau and Carroll (1958), Mauron (1961), Morrison and Narayana

TABLE 7.7

BIOLOGICAL EVALUATION OF SOYBEAN FOOD PRODUCTS BASED ON EXPERIMENTS WITH HUMAN SUBJECTS

Source of Protein	Soybean Protein				References
	Biological Value		Digestibility		
	Range[1]	Avg[2]	Range[1]	Avg[2]	
Immature bean		65			Smith and Van Duyne (1951)
Whole bean	95–97 (2)	96	90–91 (2)	91	Cahill et al. (1944); Smith (1944)
Soybean flour, defatted	61–92 (4)	81	88–94 (5)	92	Cahill et al. (1944); Smith (1944); Bricker et al. (1945); Murlin et al. (1946); DeMaeyer and Vanderbought (1961)
Soybean flour, full-fat + methionine		64 75		84 86	Parthasarathy et al. (1964A,B)
Soybean milk	83–95 (4)	91	80–97 (6)	89	Cahill et al. (1944); Smith (1944); Desikachar et al (1948); DeMaeyer and Vanderbought (1961)
Soybean curd		64	95–97 (2)	96	Oshima (1905); Cheng et al. (1941)
Protein isolate	60–81 (3)	71	81–89 (2)	85	Murlin et al. (1944); Supplee et al. (1946); DeMaeyer and Vanderbought (1961)
Textured meat analogue		81		92	Bressani et al (1967)
Other Animal and Plant Proteins					
Eggs		97		97	Block and Mitchell (1946–1947)
Milk (cow)		90		91	DeMaeyer and Vanderbought (1961)
Cottonseed flour		91		87	Murlin et al. (1944); DeMaeyer and Vanderbought (1961)
White flour		41		97	Bricker et al. (1945)
Corn		30			Block and Mitchell (1946–1947)

[1] Figures in parentheses refer to the number of observations which are included in the range of values shown.
[2] When more than one observation is recorded, the figure shown is the average of the range of values shown in preceeding column.

TABLE 7.8

AVAILABILITY OF ESSENTIAL AMINO ACIDS FROM SOYBEAN PROTEIN
BASED ON ANIMAL EXPERIMENTATION

Amino Acid	Source of Soybean Protein	Test Animal	Availability (%)	Reference
Isoleucine	Meal	Chick	65	Smith (1968)
	Isolate	Rat	80–88	Deshpande et al. (1957)
Leucine	Meal	Chick	88	Smith (1968)
Lysine	Unheated flakes	Rat	49	Schweigert and Guthneck (1953)
	Meal	Rat	76	Guthneck et al. (1953)
	Grits	Rat	80	Guthneck et al. (1953)
	Meal	Chick	98	Smith R. E. (1968)
	Isolate	Chick	71	Ousterhout et al. (1959)
	Isolate	Rat	92	Gupta et al. (1958)
Phenylalanine	Meal	Chick	100	Smith R. E. (1968)
Methionine	Unheated flakes	Rat	44	Schweigert and Guthneck (1954)
	Meal	Rat	71	Schweigert and Guthneck (1954)
	Grits	Rat	74	Schweigert and Guthneck (1954)
	Meal	Chick	100	Ousterhout et al. (1959)
Methionine + cystine	Unheated flakes	Chick	65	Ousterhout et al. (1959)
	Meal	Chick	100	Ousterhout et al. (1959)
	Isolate	Chick	69	Ousterhout et al. (1959)
Threonine	Meal	Chick	100	Smith R. E. (1968)
Tryptophan	Unheated flakes	Chick	93	Ousterhout et al. (1959)
	Unheated flakes	Rat	100	Lushbough et al. (1957)
	Meal	Chick	100	Ousterhout et al. (1959)
	Isolate	Chick	87	Ousterhout et al. (1959)
	Isolate	Rat	87	Gupta and Elvehjem (1957)
Valine	Meal	Chick	52	Smith (1968)

Rao (1966), and Sheffner (1967). Because of its traditionally important role in human foods as well as increasing use in animal feeds, the soybean has probably received more attention in this respect than any other protein. *In vitro* tests are particularly desirable in the case of soybeans since its nutritive value can be so markedly influenced by the processing conditions to which it is subjected.

Physical Tests.—A number of tests based on changes in the physical properties of soybean protein brought about by heat treatment have been employed as a means of controlling the processing of soybeans in order to produce products having optimal nutritional value. Heat treatment causes measurable changes in such physical properties of the protein as solubility in water and other solvent systems, refractive index, fluorescence, and dye-binding properties. Further details regarding the techniques used to measure these properties may be found in Chap. 9.

Available Lysine.—The most frequently employed chemical test for measuring the possible damaging effect of excessive heat treatment on the nutritional properties of soybean protein is "available lysine." Based on the reasonable assumption that only lysine molecules which have free ϵ-amino groups are biologically available, Carpenter and his associates (Carpenter and Ellinger 1955; Carpenter 1958, 1960) have developed a chemical technique for measuring the available lysine of plant foodstuffs. This method depends on the reaction of 1-fluoro-2, 4-dinitrobenzene (FDNB) with the free amino groups of an intact protein; the ϵ-dinitrophenyl (DNP) lysine that is produced by acid hydrolysis of the protein is measured colorimetrically and is presumed to represent the amount of lysine which is physiologically available. Although close correlations have been generally obtained between the results of Carpenter's procedure and the nutritive value of animal protein products, a number of difficulties are encountered with cereals and oilseed meals (Boyne *et al.* 1961). These are manifested as poor recoveries of ϵ-DNP-lysine largely due to its destruction by carbohydrate during hydrolysis and the formation of interfering substances which are not easily separated from ϵ-DNP-lysine. Several modifications of this technique have been proposed in order to circumvent these problems (Raghavendar Rao *et al.* 1963; Roach *et al.* 1967; Blom *et al.* 1967), including the use of 2,4,6-trinitrobenzenesulfonic acid in place of FDNB (Kakade and Liener 1969). It is questionable, however, whether one should really expect a correlation of available lysine with the nutritive value of a protein in which lysine is not a limiting amino acid. Such would be the case with soybean protein unless it were heated to the point where lysine was affected. This may explain why Hackler *et al.* (1967), for example, found no correlation between available lysine determined by Carpenter's method and the PER of various soybean fractions which were not deficient in lysine.

Tests for Biologically Active Components.—Soybeans contain a number of minor constituents which are characterized by having some specific kind of biological activity which is measurable *in vitro* (see Chapter 6). Changes in these activities as they might be affected by processing have been employed as possible indicators of concomitant effects on the nutritive value of the protein.

Urease.—Caskey and Knapp (1944) reported that the heat treatment required for producing maximum nutritional improvement of soybean oil meal was the same as that required to inactivate the enzyme urease. A more extensive collaborative study by Bird *et al.* (1947) confirmed the fact that meals of low nutritive value because of inadequate heating generally gave a positive urease test. Borchers *et al.* (1947) observed, however, that urease was more sensitive to heat inactivation than the trypsin inhibitor and should therefore not be regarded as being completely satisfactory for testing for the adequacy of heat treatment of soybean meals. The urease test is of little value for detecting overheated meals (Borchers *et al.* 1947; Balloun *et al.* 1953).

Trypsin Inhibitor.—Because of its implication as a factor contributing to the poor nutritive value of unheated soybean protein (Rackis 1965; Liener and Kakade 1969), the measurement of trypsin inhibitory activity is frequently used to assess the adequacy of heat treatment of soybean products (Borchers *et al.* 1947, 1948A). Westfall and Hauge (1948) found that the nutritive value of partially heated soybean flours was inversely proportional to their trypsin inhibitor content. Although trypsin inhibitor activity may reflect the amount of heat treatment a given sample of soybean protein may have received, this parameter should not be taken as an infallible index of the nutritive quality of the protein. For example, soybean fractions have been isolated which possess high levels of antitryptic activity but which have little or no growth-depressing activity (Rackis *et al.* 1963; Saxena *et al.* 1963; Garlich and Nesheim 1966; Sambeth *et al.* 1967). Kakade *et al.*, (1972) were unable to observe any correlation between the trypsin inhibitor activity of extracts from over 100 different varieties and strains of soybeans and their PER's.

Hemagglutinins.—The destruction of the hemagglutinin in soybeans (see Chap. 6) has been found to parallel the improvement in nutritive value effected by proper heat treatment (Liener and Hill 1953). A photometric procedure for measuring the hemagglutinating activity of soybean extracts has been described (Liener 1955).

Enzymatic and Microbiological Techniques.—Since the nutritive value of proteins depends to a large extent on the degree to which they are digested in the gastrointestinal tract, techniques have been developed in which an attempt is made to duplicate *in vitro* the digestive processes of the monogastric animal. Crude pancreatic extracts (pancreatin) or mixtures of purified digestive enzymes have been commonly used to digest the protein, and the release of free amino groups measured collectively by formol titration (Melnick *et al.* 1946) or individually by microbiological assay (Evans *et al.* 1947; Riesen *et al.* 1947; Hou *et al.* 1949; Ingram *et al.* 1949). Such studies have shown that the beneficial effect on the nutritive quality of soybean protein exerted by heat treatment is accompanied by an increase in the rate at which all of the amino acids are liberated *in vitro,* an effect which can be attributed to the destruction of the trypsin inhibitor (Liener and Fevold 1949). Excessive heat treatment of soybean protein, on the other hand, again reduces the rate of liberation of amino acids (Riesen *et al.* 1947). An index based on the pattern of essential amino acids liberated by pepsin digestion has been proposed for predicting the nutritive value of proteins (Sheffner *et al.* 1956). Mauron *et al.* (1965) have described a procedure which involves sequential digestion with pepsin and pancreatin with the digestion taking place in a dialysis bag in order to simulate the removal of amino acids from the intestinal tract. A critical comparison of a number of *in vitro* digestion techniques revealed many inconsistencies and led the authors (Szmelcman and Guggenheim 1967) to conclude that such techniques can at best serve only as guides to the amounts of amino acids which may be released by enzymatic digestion in the intestine.

Since a number of microorganisms require the same amino acids as needed by animal organisms for growth and metabolism, techniques have been developed by which the nutritive value of a protein may be evaluated by the growth response of a particular microorganism. In some cases, the intact protein is degraded by proteolytic enzymes elaborated by the microorganism itself; in other instances, preliminary digestion is effected by adding purified proteinases. As an example of the first type of technique, the protozoan, *Tetrahymena pyriformis,* has been employed (Anderson and Williams 1951; Pilcher and Williams 1954; Rosen and Fernell 1956; Fernell and Rosen 1956). In those techniques where preliminary digestion with added enzymes has been employed, the microorganisms which have been used subsequent to digestion include *Leuconostoc mesenteroides* (Horn *et al.* 1954), *Streptococcus fecalis* (Halevy and Grossowicz 1953; Terri *et al.* 1956), and *Streptococcus zymogenes* (Ford 1962). Using *S. zymogenes* in conjunction with protein samples which had been predigested with papain, Sheffner (1967) has reported that the growth response of this organism to a wide variety of proteins, including a number of soybean samples, was closely correlated with their corresponding NPU values.

NUTRITIONAL SIGNIFICANCE OF OTHER SOYBEAN CONSTITUENTS

Although the principal contribution which the soybean can make to the nutrition of man lies in the quantity and quality of protein which it contains, consideration should, nevertheless, be given to other soybean constituents which could have nutritional significance under special circumstances. Situations might arise, for example, in which the oil and carbohydrate components of soybeans might be called upon to serve to satisfy the caloric requirements of an under-nourished population. The vitamin and mineral constituents of soybeans might also prove to be decisive nutritional factors under conditions where the availability of these nutrients might be limiting or marginal.

Available Energy

The energy available for metabolism (metabolic energy) from the soybean may be calculated from its content of carbohydrate, fat, and protein taking into account the digestibility of each one of these components as well as the heat of combustion. Except under conditions of extreme caloric deprivation, the protein would not be expected to be utilized to any significant extent as a source of energy. The amount of energy theoretically available from soybean protein may be calculated by multiplying the protein content by the factor 3.47 Cal per gm (Watt and Merrill 1963). Little need be said here regarding the energy available from soybean oil except to indicate that soybean oil is highly digestible and has a caloric value of 8.37 Cal per gm of fat (Watt and Merrill 1963). Most soybean products intended for human consumption, however, have been defatted so that, unless indicated otherwise, the caloric contribution of soybean oil can generally be neglected. The carbohydrate content of soybean products will vary considerably depending on the extent to which carbohydrate material may have been removed during processing and will range from 22-29% for whole soybeans (Daubert 1950) to only 0.3-4.2% for the isolated protein (Wolf et al. 1966). Only a portion of this material is actually metabolizable for energy purposes. Such substances as galactans, pentosans, and hemicelluloses, which are utilized poorly, if at all, by the body, represent a substantial proportion of the total (Aspinwall et al. 1967). The actual percentage of the total carbohydrate which is available depends on the experimental animal and ranges from a figure of 14% determined with chicks (Lodhi et al. 1969) to 40% with rats (Adolph and Kao 1934). In the absence of conclusive data on human beings, the latter figure has been taken by the FAO (1949) to be the digestibility of carbohydrate from soybeans and soybean products, and, on this basis, the calorie factor becomes 1.68 Cal per gm of carbohydrate. The USDA Table of Composition (Watt and Merrill 1963), on the other hand, assumes the carbohydrate of soybeans to be almost completely digestible (97%) and bases its caloric calculations for soybeans and soybean products on a factor of 4.07 Cal per gm. In the case of soybean milk, soybean curd, and protein isolates, this factor may be applicable since the preparation of these products involves the removal of most of the insoluble and indigestible carbohydrates.

The fraction of total carbohydrate which may be classified as "crude fiber" will, of course, depend on the type of soybean product. Crude fiber values range from 0.1% or less for soybean curd and soybean milk to about 5% for the whole seed (FAO 1949; Watt and Merrill 1963). Since about 1/2 of the crude fiber in the latter case is attributable to the seed coat, the crude fiber content of soybean products is much reduced by removel of its hull. Aside from the fact that it is indigestible in itself, the crude fiber may interfere with the digestibility of other nutrients, particularly in monogastric animals such as man. Their presence in the cell wall may prevent the access of digestive juices to the protein within the cells. To what extent this may account for the poor digestibility of inadequately processed soybean products is not known. It is known, however, that the digestibility of plant protein concentrates, which are relatively low in crude fiber, is much higher than the crude plant materials which have a higher content of crude fiber (Swaminathan 1967).

Vitamins

When soybean products are consumed as part of a mixed diet, they can hardly be considered to be an important source of vitamins. However, when consumed in the form of a protein-rich supplement to a basal diet that may be deficient in vitamins as well as protein, the vitamin contribution of the soybean may assume a very decisive role in the maintenance of health. In the usual determination of PER the basal diet already contains an adequate supply of vitamins, so that any additional contribution of vitamins made by the soybean is obscured. In those few studies where the vitamins contributed by soybeans have been evaluated by employing a basal diet lacking vitamins of the B complex, it has been observed that if at least 10% of the protein is derived forom soybean flour then the latter provides enough of the B vitamins to satisfy the animal's (rat) requirement (Zucker and Zucker 1943; György and Prena 1964). Westerman et al. (1954) reported that soy flour at a level of 3% could replace wheat germ as a vitamin B supplement to nonenriched wheat flour. Zucker and Zucker (1943) have estimated that if approximately 1/2 of the daily protein requirement of an adult man was derived from soy flour, then 1/3-1/2 of his requirement for thiamine, riboflavin, and nicotinic acid would be met.

The vitamin content of various soybean products is summarized in Table 7.9. Rather wide variations in analytical values may be found in the literature which are most likely a reflection of the analytical techniques employed as well as the influence of processing conditions.

Fat-soluble Vitamins.—β-Carotene, the biological precursor of vitamin A, is present in green, immature soybeans to the extent of 2–7 μg per gm, while the mature bean contains a significantly lesser amount (Sherman 1940; Sherman and Salmon 1939). The β-carotene content of soybean milk is about 1/2 that of cow's milk (Shurpalekar et al. 1961). The feeding of raw soybeans to dairy calves at a level of 30% or more causes a marked lowering of the levels of vitamin A and carotene in the blood plasma (Shaw et al. 1951; Ellmore and Shaw 1954). Whether this effect is due to the enzyme lipoxidase, which is known to oxidize carotene (Sumner and Dounce 1939), is not certain.

TABLE 7.9

VITAMIN CONTENT OF SOYBEAN FOOD PRODUCTS

Soybean Product	β-Carotene (μg/Gm)	Thiamine[1] (μg/Gm)	Riboflavin[1] (μg/Gm)	Niacin[1] (μg/Gm)	Pantothenic Acid[1] (μg/Gm)	Pyridoxine[1] (μg/Gm)	Biotin[1] (μg/Gm)	Folic Acid[1] (μg/Gm)	Inositol[1] (Mg/Gm)	Choline[1] (Mg/Gm)	Ascorbic Acid (Mg/Gm)
Immature bean	2–7	6.4	3.5		12	3.5	0.5	1.3		3.0–3.3	0.2
Mature bean	0.2–2.4	11.0–17.5	2.3	20.0–25.9	12	6.4	0.6	2.3	1.9–2.6	3.4	0.2
Sprouts		11.9–21.9	4.8–7.0	29.9–48.0	18.8–34.4	14.1–17.7	1.1–1.7	3.7	2.5–3.9		0.4
Meal		12.0–44.1	2.7–3.3	19.0–40.0	13.3–16.0			4.0–4.9	1.8–2.1	3.5–3.8	
Flour		11.0–15.0	4.0–4.4	20.3–29.1	47.0–50.6	8.8	0.2	0.8–0.9			
Curd (tofu)		3.9	3.7	5.5							
Milk[2]	7.50	0.8	1.1	2.5							
Miso		1.3	1.4								21.6

Source: Data compiled from the following: Burkholder (1943); Burkholder and McVeigh (1945); Chang and Murray (1949); Engel (1943); Flynn (1949); Guggenheim and Szmelcman (1965); Hoff-Jorgensen et al. (1952); Kondo et al. (1954); Miller (1945); Mustakas et al. (1964); Sherman and Salmon (1939); Shurpalekar et al. (1961); Sugawara (1953); Van Duyne et al. (1945).

[1] Where a range of values is shown, average value is very closely given by taking the average of the two extreme values.
[2] Expressed as mg/liter with the exception of β-carotene which is expressed in terms of IU of vitamin A per liter.

Soybean and soybean products may be considered to be essentially devoid of vitamin D. Anderson (1961) has pointed out that if soy milk is to be used as a replacement for cow's milk it should be supplemented with vitamin D. The vitamin D requirements of turkey poults (Carlson et al. 1964A,B; Thompson et al. 1968), chicks (Jensen and Mraz 1966), and baby pigs (Miller et al. 1965), are intensified when soybean protein is included in the diet.

Any vitamin E present in soybean products is contained in the oil to the extent of 1.4 µg per gm of oil (Harris et al. 1950). During the course of the fermentation of soybeans, as in the preparation of tempeh, an antioxidant is produced which retards the development of rancidity and prevents a vitamin E deficiency in rats (György 1961). György et al. (1964) succeeded in isolating from tempeh three active antioxidants which were identified as daidzein (7,1′-dihydroxyisoflavone), genistein (5,7,4′-trihydroxyisoflavone), and "factor 2" (6,7,4′-trihydroxyisoflavone). Factor 2 was later reported to be ineffective in preventing the hemolysis of red blood cells from vitamin E deficient rats, nor did it prevent the autoxidation of soybean oil or full-fat soybean flour (Ikehata et al. 1968). Since factor 2 apparently does not occur in soybeans, it would appear that some of the pre-existing isoflavones of soybeans are converted to factor 2 as a consequence of the fermentative process.

Water-soluble Vitamins.—In comparison with cereal grains, soybean products are fairly good sources of the vitamins of the B complex. Because of its thermolability, however, thiamine is apt to be present in somewhat lower amounts in heat-processed soybean products. Thus, the toasting of solvent-extracted soybean flakes destroys about 1/2 of this vitamin (Weakley et al. 1961). Thiamine losses ranging from 10% to 75% have been reported for soybeans which had been cooked under varying conditions (Miller 1945; Ashikaga 1946; Bedford and McGregor 1948; McGregor and Bedford 1948; Smith and Van Duyne 1951; Wu and Fenton 1953). Clandinin et al. (1947) pointed out the necessity for supplementing the diet with thiamine and other B vitamins if overheated soybean meal is used as a source of protein in the basal diet of chicks. The B vitamins of extrusion-cooked full-fat soybean flour, however, seem to suffer little damage as a result of the cooking process (Mustakas et al. 1964). Appreciable losses in ascorbic acid also accompany the cooking of soybeans so that cooked soybean products should be regarded as a relatively poor source of vitamin C (Miller and Robbins 1934; Lee and Whitcombe 1945; Bedford and McGregor 1948; McGregor and Bedford 1948).

As will be noted in Table 7.9, a significant increase in the concentration of most of the B vitamins and ascorbic acid accompanies the germination process. Soybean sprouts may be considered to be a particularly good source of vitamin C (French et al. 1944). Miller et al. (1952) have estimated that soybean milk may retain 50–90% of the thiamine, 90% of the riboflavin, and 60–80% of the niacin found in the soybeans from which the milk was made. Desikachar et al. (1946B) found soy milk to be about 80% as potent as cow's milk with regard to its vitamin B-complex content. When the curd is separated from the milk only about 1/2 of the thiamine and 1/4 of the nicotinic acid is retained in the curd, while the riboflavin content is equivalent to the original bean (Chang and Murray 1949). The fermentation of soybeans, such as is involved in the preparation of natto and tempeh, has been observed to cause an increase in most of the B vitamins except thiamine which undergoes a significant decrease (Steinkraus *et al.* 1961; Arimoto *et al.* 1962; Roelofson and Talens 1964; Murata *et al.* 1967).

Since soybeans are devoid of vitamin B_{12}, it is not surprising that diets containing soybean protein require supplementation with this vitamin in order to produce maximum growth. Vitamin B_{12} supplementation, however, improves the growth of animals receiving raw soybeans to a greater extent than similar supplementation of diets containing heated soybeans (Fröhlich 1954A; Baliga and Rajagopalan 1954; Baliga *et al.* 1954). Using the urinary excretion of methyl malonic acid as an index of vitamin B_{12} deficiency, Edelstein and Guggenheim (1969) found that the excretion of this metabolite was reduced to a greater extent when vitamin B_{12} was given to rats on a raw soybean diet compared to rats administered this vitamin on a diet containing heated soybeans. These observations suggest that not only is raw soybean deficient in vitamin B_{12} but also contains a heat-labile substance which increases the requirement for this vitamin.

Minerals

Table 7.10 shows the mineral composition of soybeans and soybean products. Reference will be made to this table in the following paragraphs that deal with each mineral individually. It is questionable whether too much nutritional significance should be attached to these analytical values since there is ample evidence to indicate that the availability of most minerals from soybeans is quite low (Daniels and Nicols 1917; Harmon *et al.* 1969). As will also be noted below, there appear to be constituents in soybeans which interfere with the availability or utilization of certain minerals, particularly calcium and phosphorus.

Calcium.—Most of the nutritional interest in the calcium content of soybeans has involved a comparison of soybean milk with cow's milk. Analytically, the calcium content of soybean milk prepared in the traditional manner (0.08%) compares quite favorably with cow's milk (0.11%) (Shurpalekar *et al.* 1961). Schroeder *et al.* (1946) compared the availability of calcium from soybean milk with that from evaporated cow's milk for human subjects and concluded that 22.6% of the calcium of a proprietary preparation of soybean milk was available compared to 29.1% from cow's milk. These values agree fairly well with earlier experiments with human subjects which indicated that the utilization of the calcium of soybean milk prepared in the traditional manner is about 90% that of cow's milk (Desikachar *et al.* 1948). The calcium content of soybean curd is about four times higher than that of soybean milk because of the use of calcium salts for precipitating the protein. Adolph and Chen (1932) found little differ-

TABLE 7.10

MINERAL CONTENT OF SOYBEANS AND SOYBEAN FOOD PRODUCTS[1]

Soybean Product	Calcium (%)	Phosphorus (%)	Magnesium (%)	Zinc (Mg/Kg)	Iron (Mg/Kg)	Manganese (Mg/Kg)	Copper (Mg/Kg)
Immature bean	0.10	0.26			21.3		
Mature bean	0.16–0.47	0.42–0.82	0.22–0.24	37	90–150	32	12
Sprouts	0.40				100		
Meal	0.24–0.31	0.60	0.24–0.30	55–77	140	24–29	14–24
Flour	0.42–0.64	0.60			110–160		
Curd (tofu)	0.80	0.80–1.0			105		
Milk, traditional	0.76	0.15			68		
Milk, powdered proprietary	0.7–1.0	1.1			30–170		
Miso	0.11				35		
Natto	0.18	0.42			62		

Source: Date compiled from the following: Adolph and Chen (1932); Bailey et al. (1935); Cartter and Hopper (1942); Chang and Murray (1949); Dewar (1967); FAO (1954); Guggenheim and Szmelcman (1965); Harmon et al. (1969); Miller and Robbins (1934); Mitchell (1950); Morse (1950); Pant and Kapur (1963); Shurpalekar et al. (1961); Watt and Merrill (1963).

[1] All values corrected for moisture content. Where range of values is shown, average value is very closely given by taking the average of the two extreme values.

ence in the availability of the calcium of soybean curd and cow's milk for human subjects.

In contrast to the generally high availability of calcium from soybean milk and curd, the availability of calcium from the whole bean is quite low. Schroeder *et al.* (1946) reported that only 10% of the calcium of cooked soybeans could be effectively utilized by man. Experiments with animals strongly suggest that phytic acid contained in soybeans interferes with the availability of calcium (Nelson *et al.* 1968) which, in turn, may account in part for the rachitogenic properties of isolated soybean protein (Jensen and Mraz 1966; see also Chap. 6.)

Phosphorus.—Although soybeans contain almost twice as much phosphorus as most cereals, approximately 1/2–2/3 of it is present as phytic acid (Nelson 1967). The availability of phosphorus when present in the form of phytic acid depends on the species and age of the experimental animal. Thus, the phosphorus of soybeans appears to be well utilized by the rat (Spitzer and Phillips 1945A,B) but not by the chick (Nelson *et al.* 1968). Whether or not an animal can utilize phytate phosphorus seems to depend on the level of phytase activity in the intestinal tract (Nelson 1967). Even soybean protein which has been isolated by isoelectric precipitation seems to contain enough phytic acid, 0.5% (O'Dell and Savage 1960), to interfere with the availability of phosphorus for bone mineralization in the pig (Hendricks *et al.* 1969). Although no experiments with humans have been conducted to determine directly the availability of phosphorus from soybeans, it has been estimated that anywhere from 40 to 80% of the phytate phosphorus in cereals is available to man (McCance and Widdowson 1935).

Zinc.—Although soybeans contain an appreciable amount of zinc, from rat experiments it has been estimated that only 44% of it is actually available (Forbes and Yoke 1960). A number of investigators have reported that the dietary requirement of chicks and turkey poults for zinc is significantly increased when soybean protein is the main source of protein in the diet (Supplee *et al.* 1958; Kratzer *et al.* 1959; Kratzer 1965; Davis *et al.* 1962A; Morrison and Sarett 1958; Lease and Williams 1967). Autoclaving the protein or adding ethylenediaminetetraacetate (EDTA) to the diet eliminated this interference with the availability of zinc. *In vitro* experiments have provided direct evidence that isolated soybean protein can bind zinc, an effect which can be partially eliminated by autoclaving the protein or by adding EDTA (Allred *et al.* 1964). Phytic acid appears to be the specific component responsible for the zinc-bonding properties of soybean protein (O'Dell and Savage 1960; Allred *et al.* 1964; O'Dell *et al.* 1964), although other lines of evidence suggest that a glycoprotein might be involved (Lease 1967). The availability of zinc from soybean protein isolates varies from one source to another (Lease 1967; Vohra and Kratzer 1967).

Other minerals.—The ability of soybean protein to interfere with the availability of minerals can be extended to include manganese, copper, and molybdenum (Reid *et al.* 1956; Davis *et al.* 1962A). Here again this interference can usually be effectively counteracted by supplementing the diet with EDTA. Reports regarding the availability of iron from soybeans have been quite variable, and values for percentage of availability of iron have ranged from 28.5% to 80% (Sherman *et al.* 1934; Ranganathan 1938; Porter 1946). Experiments dealing with the effect of soybean protein on the availability of iron have likewise been inconclusive. In experiments with chicks, Davis *et al.* (1962A,B) found that isolated soybean protein did not interfere with the availability of iron for growth and hemoglobin synthesis. On the other hand, Fitch *et al.* (1964) observed reduced gastrointestinal absorption of iron and anemia due to an iron deficiency in monkeys fed a diet containing isolated soybean protein. Not only is the soybean lower than most cereals in chlorine and iodine content (Mitchell 1950; Morse 1950), but, in addition, one must take into consideration its goiterogenic properties which exacerbate an iodine deficiency (see Chap. 6). Anderson (1961) has emphasized the nutritional importance of supplementing infant soybean milk formulas with iodine.

Unknown Growth Factor(s)

There is evidence to suggest the possible presence of a still unidentified factor (or factors) in soybean meal which stimulates the growth of turkey poults (Wilcox *et al.* 1961A,B; Griffith and Young 1966). Isolated soybean protein itself has been reported to enhance the growth of ducklings (Richert and Westerfeld 1965), an effect which could not be accounted for on the basis of its amino acid content. Ershoff (1949) has postulated the existence of a factor in soybean flour which counteracts the growth depression evoked in young rats by dessicated thyroid powder or iodinated casein. This factor appears to be distinct from the growth factor in isolated soybean protein (Westerfeld *et al.* 1964).

FACTORS AFFECTING THE NUTRITIVE PROPERTIES OF SOYBEAN PROTEIN

Heat Treatment

Shortly after the soybean was introduced into this country as a commercial crop, Osborne and Mendel (1917), in a study of its potential value as a source of protein for animals, noted that soybeans would not support the growth of rats unless cooked in a steam bath. The very extensive body of literature which subsequently confirmed and extended this observation to many other species of animals including man has been reviewed elsewhere (Liener 1958). In general, the degree of improvement on nutritive value effected by heat treatment is dependent on the temperature, duration of heating, and moisture conditions. In the industrial production of various soybean products, these factors come into play during processing and exert an effect which is reflected in the nutritive value of the final product. Most of the studies reported in the literature, however, have been performed in the laboratory where maximum nutritive value can be achieved by careful adjustment of temperature, pressure, moisture content, and duration of heating. The curves shown in Fig. 7.2 are taken from one such study (Klose *et al.* 1948) which shows the effect which these variables can exert on the nutritive value of soybeans. These data also serve to illustrate the necessity for moist heat and the marked impairment in nutritive value which can accompany excessive heat treatment. It has been the observation of most investigators that, under laboratory conditions, maximum nutritive value of soybean protein is achieved by treatment with live steam for about 30 min or by autoclaving at 15 lb pressure for 15–20 min (Klose *et al.* 1948; Borchers *et al.* 1948A; Smith *et al.* 1964A; Rackis 1965, 1966). The improvement in protein efficiency effected by atmospheric steaming at a level of 19% moisture is somewhat higher than at a level of 5% (Rackis 1965, 1966).

An explanation for the beneficial effect of heat treatment on the nutritive value of soybean protein has been the object of innumerable studies, and the vast literature dealing with this subject has been reviewed elsewhere (Liener 1958, 1969). Suffice to state here that the improvement in the nutritive value of soybean protein appears to be related to the destruction of trypsin inhibitors and possibly other biologically active components (see Chap. 6). The precise mechanism, however, whereby the trypsin inhibitors of soybeans interfere with the availability of the sulfur-containing amino acids of the protein is not known, although numerous theories have been propounded (Liener 1969).

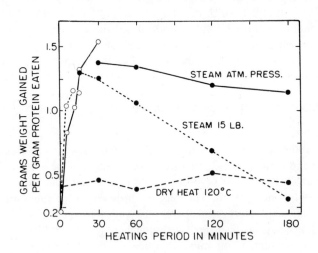

From Klose et al. (1948)

FIG. 7.2. EFFECT OF TYPE AND EXTENT OF HEAT
          TREATMENT   ON   NUTRITIONAL   VALUE
          OF SOYBEAN PROTEIN (GROUND WHOLE
          SOYBEANS)

Test period, 42 days; ● experiment 1; ○ experiment 2; 12 rats
per group.

As already indicated, an excessive amount of heat may adversely affect the nutritive value of the protein and the damage thus inflicted can usually be overcome by supplementation with lysine and the sulfur-containing amino acids (Fritz et al. 1947; McGinnis and Evans 1947; Klose et al. 1948; Clandinin et al. 1947; Iriarte and Barnes 1966; Taira et al. 1969). These deficiencies in overheated soybean meal can be explained by the vulnerability of cystine and lysine to destruction and/or inactivation by heat. Destruction, as used here, refers to a failure to recover an amino acid upon acid or alkaline hydrolysis. Inactivation refers to a situation where there may be complete recovery of the amino acid after acid hydrolysis, but nevertheless there has been a decrease in the biological availability of that amino acid. Cystine is particularly sensitive to heat, and as much as 1/2–2/3 of the cystine content of soybean protein may be destroyed by excessive heat (Evans et al. 1951; Iriarte and Barnes 1966; Taira 1966). Lysine not only undergoes destruction when soybean protein is overheated (Evans and Butts 1948; Iriarte and Barnes 1966; Taira et al. 1965A,B), but much of the lysine is also rendered unavailable (see previous discussion on "available lysine"). The reason that lysine becomes unavailable is due to the fact that the ε-amino groups of lysine interact with the reducing groups of sugars in what is referred to as a browning or Maillard reaction (Liener 1958). In the case of soybeans subjected to heat treatment, the hydrolysis of sucrose may give rise to appreciable levels of reducing sugars which interact with lysine. Lysine so modified is no longer physiologically available since the peptide bond containing the modified lysine is not susceptible to tryptic cleavage. Thus, the digestibility of soybean protein by pancreatic enzymes, whether measured in vitro or in the intact animal, is considerably reduced if the protein has been subjected to excessive heat (Evans and McGinnis 1946, 1948; Evans et al. 1947). A direct consequence of impaired digestion is a retardation in the rate at which all of the amino acids are released from the protein during digestion (Riesen et al. 1947; Evans and Butts 1949; Clandinin and Robblee 1952). According to Almquist (1951) since methionine is the limiting amino acid of soybean protein, a delay in digestion leading to an excretion of methionine would only serve to accentuate a deficiency of this amino acid. The destruction of cystine, noted before, further intensifies a deficiency of the sulfur amino acids in overheated soybeans.

In addition, cystine and lysine and a number of other amino acids including arginine, tryptophan, histidine and serine are either partially destroyed or inactivated by the excessive heating of soybean meal (Liener 1958). Since these amino acids are not limiting in soybean protein, their partial loss does not affect the nutritive properties of the protein. Although not all workers have been able to show that lysine becomes limiting in overheated soybean products (Iriarte and Barnes 1966), it is well to remember that one of the principal assets of soybean protein is its high lysine content which can be used to supplement lysine-deficient cereal products. Thus, the lysine deficiency of overheated soybean protein may not become apparent unless such products are used in combination with other proteins which are limiting in lysine.

Large amounts of water have been noted to prevent to a partial extent the damaging effects of undue heat on soybean protein (Renner et al. 1953). The partial protection provided by an excess of water may be related to the fact that a lesser destruction of lysine in soybean protein occurs under these conditions; cystine losses, however, remain unaffected (Taira et al. 1965B).

Supplementation with Amino Acids

Repeated reference has already been made to the fact that the principal nutritional defect of soybean protein is a deficiency of the sulfur-containing amino acids. The order in which other amino acids become limiting depends on the species of animal to which the protein is being fed. In rats and pigs, the next limiting amino acid is threonine followed by lysine (Berry et al. 1962); in chicks, threonine is also the next limiting amino acid but followed instead by valine and then lysine (Warnick and Anderson 1968). However, the supplementary effect of cystine or methionine on the nutritive value of raw soybeans is considerably greater than their effect on properly heated soybeans (Hayward et al. 1936A,B; Hayward and Hafner 1941; Almquist et al. 1942; Evans and McGinnis 1948; Barnes et al. 1962; Shurpalekar et al. 1963). This observation has led to the view that the effect of heat on raw soybean protein is to make the sulfur-containing amino acids more readily available to the animal (Hayward et al. 1936B), but the precise mechanism whereby heat exerts this effect is still not clear. Present evidence would suggest that hypertrophy of the pancreas caused by trypsin inhibitors leads to endogenous depletion of cystine in the animal tissue (Liener 1969).

Although cystine and methionine are the first limiting amino acids of raw soybean protein, the amino acids tyrosine, threonine, and valine have been reported to exert a further supplementary effect when added to diets containing unheated soybeans (Borchers 1959, 1961; Booth et al. 1960). However, even the supplementation of raw soybeans with an assortment of all of the essential amino acids will not increase its nutritive value to the same level as heated soybeans similarly supplemented (Saxena et al. 1962). This would indicate that an interference with the availability of essential amino acids cannot be the sole reason for the poor nutritive value of unheated soybeans.

Largely on the basis of animal experiments, it has been generally assumed that the nutritive value of soybean products for human consumption would be enhanced by supplementation with methionine. One of the few reports that actually demonstrates the beneficial effect of methionine on the utilization of soybean protein in humans is a study that was conducted on young girls in India (Parthasarathy et al. 1964A). The biological value of a processed full-fat soybean flour was increased from 64 to 75 when supplemented with DL-methionine at a level of 1.2 gm per 16 gm N. The hydroxy analogue of DL-methionine was equally effective at the same level of supplementation. The nutritive value of the methionine-supplemented soybean flour was almost as high as that of the spray-dried skim milk powder which had a BV of 83.

Storage

An adverse effect of storage on the nutritive value of soybean protein was first reported by Mitchell (1944) and Mitchell and Beadles (1949) who noted a definite deterioration in the digestibility and biological value of soybeans which had been stored for almost 3 yr at 25.5° C. This deterioration was particularly marked in the case of the whole raw bean but could be largely prevented by pretreatment with heat. For this reason the reactions responsible for this decrease in digestibility and biological value were believed to be largely enzymatic in nature. In a more recent study, Zimmerman et al. (1969) found that the NPU of the protein of defatted soybean meal decreased with prolonged storage time at temperatures of 20°, 30° and 40° C. The decrease in NPU was highly correlated with a decrease in available lysine, an increase in fluorescence, and an increase in dye absorption (Orange G). Isolated soybean protein, likewise, underwent a loss in available lysine (measured chemically) which was attributed to the formation of atypical peptide between lysine and glutamic or aspartic acid (Ben-Gera and Zimmerman 1964).

Germination

The nutritive value of soybeans for rats increases upon germination (Everson et al. 1944; Desikachar and De 1947), but, for some inexplicable reason, not for chicks (Mattingly and Bird 1945). This increase in nutritive value, at least for rats, occurs despite the fact that there is no change in trypsin inhibitor content (Desikachar and De 1947) or amino acid composition (Kasai et al. 1966; Standal 1967) (see Table 7.4). Standal (1963, 1967) concludes from her studies with rats that the nutritional value of soybean sprouts compares very favorably with other soybean products which are used as foods in the Orient, their NPU being intermediate between that of tofu and natto (see Table 7.6 for specific values).

Effect of Antibiotics

A wide spectrum of antibiotics has proved to be effective in overcoming the poor growth which is associated with the consumption of diets containing raw soybeans (Carroll et al. 1953; Hensley et al. 1953; Borchers et al. 1957; Braham et al. 1959). Among the various theories that have been advanced to explain the mode of action of antibiotics in this regard are the following: (1) an interference with the release of a growth inhibitor from raw soybeans in the intestinal tract (Borchers 1965, 1967); (2) an inhibition of the bacterial degradation of a growth-promoting substance (cystine ?) secreted by the pancreas (Barnes and Kwong 1965; Barnes et al. 1965A,B); (3) enhanced absorption of amino acids (Carroll et al. 1953); and (4) inhibition of pancreatic hypertrophy and the excessive pancreatic secretions which accompany the ingestion of raw soybeans (Goldberg and Guggenheim 1964). Although the influence of antibiotics on the utilization of soybean protein may have practical implications in the feeding of animals, there is no obvious relevance to human nutrition.

Dietary Source of Carbohydrate

There is some evidence to indicate that the type of carbohydrate used to supply some of the energy in diets containing soybean protein can influence the latter's nutritive value. Dal Borgo *et al.* (1967) noted that young chicks were able to utilize raw soybean meal more efficiently when it was fed in combination with glucose or sucrose than when starch was provided as the source of carbohydrate; no such effect was observed with heated soybean meal. It was suggested that the raw soybeans had an inhibitory effect on amylase production by the pancreas and thus interfered with the efficient utilization of starch.

SOYBEAN PRODUCTS USED FOR HUMAN CONSUMPTION

Soybeans as a Vegetable

Soybeans have not been readily accepted as a fresh, frozen, or canned vegetable because of their objectionable odor and the difficulty in shelling the green bean (Hale 1943). During World War II, when animal protein was in short supply, serious attempts were made to introduce certain varieties of soybean into the American diet as a fresh green vegetable. Salmon (1943) estimated that a 100-gm serving of fresh soybeans would supply 40% of the daily protein requirement of an adult. Smith and Van Duyne (1951) and Simpson (1943) have listed those varieties of soybeans which were found to be most suitable for use as a fresh vegetable or for freezing and canning.

The protein of green immature soybean has been reported to be superior in nutritive value to the mature bean, and, when properly cooked, the BV of the protein compared favorably with that of casein and beef liver (Parsons 1943; Everson and Heckert 1944). Edamame is a popular Oriental dish prepared from green, immature soybeans which have been washed, shelled, and steamed for about 40 min. The NPU and PER of this food as measured with rats was found to be superior to that of most Oriental soybean foods (Standal 1963) (see Table 7.6 for specific values). Smith and Van Duyne (1951) cite an experiment involving human subjects in which cooked green soybeans had a BV of 65, a value which seems rather low compared to most soybean products (see Table 7.7).

Mature soybeans have also been used to a limited extent in human foods (Hale 1943; Smith and Van Duyne 1951). The cooking methods used in the preparation of dishes made from mature soybeans have a marked influence on the nutritive value of the protein. Preliminary overnight soaking of the bean and cooking in water rather than in the dry heat of an oven serve to enhance the nutritive properties of the protein (Parsons 1943; Steele *et al.* 1947; Krishnamurthy *et al.* 1958). Dean (1958) attempted to treat kwashiorkor in children in Uganda by feeding them a mixture of cooked soybeans and bananas. The majority of the children were relieved of the signs of kwashiorkor and gained weight, but a few showed an aversion to this diet, experienced digestive difficulties and a tendency to vomit.

Soybean Flour

One of the most common forms in which soybean protein is used in Western type diets is as a flour. Details regarding the preparation of various soy flour, their composition, uses, and standards may be found in Chap. 9 and 10. Soybean flour may be used in the human diet in any number of ways: (1) As a separate item of the diet, although problems of acceptability frequently limit its consumption in this fashion. (2) As an ingredient of a wide variety of common dishes such as soups, stews, beverages, and desserts; a number of recipes for preparing palatable dishes containing full-fat soybean flour have been described by Schlosser and Dawson (1969). (3) In the formulation of bakery and cereal products or as a meat extender. (4) As a starting material for the preparation of infant formulas, protein concentrates, or isolates. (5) As a protein supplement to cereal grains and other foods.

Soybean flour has been used as the sole source of protein in human diets only under experimental conditions when more precise data regarding the nutritive value of the protein are desired. Representative data taken from such experiments have been included in Tables 7.6 and 7.7 based on animal and human experiments respectively. Of more practical interest is the use of soybean flour as a protein supplement to a well-accepted diet. In the early 1950's, a "Multi-Purpose Food" (MPF) in which toasted soybean flour provided the only source of protein was introduced as a dietary supplement for feeding in underdeveloped countries. The soybean flour was blended with essential vitamins and minerals in such a way that 1–2 oz portion was estimated to provide 1/3 of the daily requirement of these nutrients (Anon. 1959; Hafner 1961). Cooper and Bryan (1951) reported that MPF at a level of 1 oz per day was an effective supplement to the diets of school children as measured by gain in weight. A formula similar to that of MPF except that it included dextrin-maltose and hydrogenated peanut oil was almost equivalent to the nutritive value of skim milk as measured with rats and was recommended as a milk substitute for infants (Shurpalekar et al. 1965). Toasted full-fat soy flour (referred to in Japan as "kinako"), when fed as a main source of protein to weanling infants, was well accepted and supported good growth and N retention (Muto et al. 1963). Feeding trials on infants and children in a German orphanage conducted by Dean (1953) showed that about 1/2 of the milk in the diet of infants up to 1 yr of age could be replaced by a barley malt-soybean flour mixture and even more in the diet of older children without affecting their growth. In a study with African children, DeMaeyer and Vanderbought (1961) found that soybean flour when used as protein supplement to a basal rice diet had reasonably good nutritive value although its BV was somewhat less than that of milk (see Table 7.7). Panemangalore *et al.* (1964B), on the other hand, in similar studies with Indian children on a basal rice diet, found that defatted soybean flour when fortified with methionine was as good a protein supplement as that of skim milk. Biological evaluation with rats and chicks of full-fat soybean flour prepared by the extrusion-cooked process showed that its nutritive value was comparable to that of properly heated commercial full-fat and defatted soybean flours (Mustakas *et al.* 1964). This product, when used in conjunction with rice to feed infants, was not found to be appreciably different in nutritive value from whole milk powder (Huang *et al.* 1965).

The use of soybean flour in the formulation of bread and other baked goods and its value as a component of vegetable protein mixtures will be considered in the section dealing with soybean proteins as a protein supplement. The use of soybean flour as a starting material for the production of liquid formulas, protein isolates, and other products will be discussed in appropriate sections below.

Soybean Milk

Soybean milk, in the traditional sense, is simply an aqueous extract of whole soybeans. A detailed description of the techniques used for the preparation of soybean milk as well as its composition may be found in Chap. 10.

Soybean milk has been of considerable interest to nutritionists as a possible substitute for cow or human milk particularly in the feeding of infants who are allergic to animal milk or where cow's milk may be either too expensive or unavailable. Soybean milk and cow's milk have approximately the same protein content (3.5–4.0%), and a comparison of the amino acid composition of soybean and milk proteins (Table 7.11) shows a fairly close correspondence. The main deficiency of soybean protein as compared with the protein of cow or human milk is that of the S-containing amino acids. Animal experiments in general have shown that the nutritive value of soybean milk ranges anywhere from 60% to 90% of that of cow's milk (see Table 7.6); methionine supplementation raises its nutritive value to essentially the same level as that of cow's milk (Shurpalekar *et al.* 1965; Harkins and Sarett 1967; Dutra de Oliviera and Scatena 1967). Hackler *et al.* (1965) have pointed out the sensitivity of the nutritive value of soybean milk to the time and temperature of cooking and subsequent drying of the liquid product. The rather wide variations in the biological data relating to the nutritive value of soybean milk could very well be a reflection of differences in processing conditions. Work by Hackler and associates (Hackler *et al.* 1965, 1967; Van Buren *et al.* 1964) has shown that maximum nutritive value of the protein of soybean milk is attained within 5–10 min when the milk is heated at 121° C or in 60 min at 93° C, conditions which inactivate about 90% of the trypsin inhibitor. An impairment in nutritive value accompanies cooking

TABLE 7.11

A COMPARISON OF THE ESSENTIAL AMINO ACID COMPOSITION
OF SOYBEAN MILK WITH COW AND HUMAN MILK

Essential Amino Acid	Source of Milk			
	Soybean (Traditional)	Soybean (Commercial)	Cow	Human
	(Grams per 16 GM N)			
	Hackler and Stillings (1967)	Harkins and Sarett (1967)	Subrahmanyan et al. (1961)	Rice (1969)
Isoleucine	5.1	4.7	7.5	5.5
Leucine	8.3	8.1	11.0	9.1
Lysine	6.2	6.4	8.7	6.6
Methionine	1.4	1.2	3.2	—
Cystine	1.7	0.9	1.0	
Total sulfur AA	3.1	2.1	4.2	4.0
Phenylalanine		5.3		
Tyrosine		—		
Total aromatics	9.0		11.5	9.5
Threonine	3.8	3.9	4.7	4.5
Tryptophan	1.3	1.1	1.5	1.6
Valine	4.9	5.0	7.0	6.2
	Nutritive value[1]			
B V	80		87	100
Digestibility	95		91	90
N P U	76		79	90

[1] Based on studies with African children (DeMaeyer and Vanderbought 1961).

at higher temperatures or for longer periods of time. Care must also be exercised in the control of heating conditions employed in the spray or roller drying of the fluid milk product to prevent heat damage to the protein.

A considerable body of data obtained with human subjects may be found in the literature pertaining to the effectiveness of soybean milk as a replacement for cow or human in the nutrition of infants and young children. Much of the earlier work has been reviewed by Jones (1944) and Miller (1957), and a more recent review has been provided by Swaminathan and Parpia (1967). Theoretically, at least, the essential amino acids provided by the protein of soybean milk should satisfy the requirements of infants to the same extent as cow or human milk when adminstered at the same level of protein intake (Table 7.12). Criteria for measuring the ability of the protein of soybean milk to replace mammalian milk have included N-balance studies, weight gain, increase in body length, hemoglobin, and serum protein. Without going into the specific details, such studies have, in general, failed to reveal any significant margin of superi-

TABLE 7.12

ESSENTIAL AMINO ACID REQUIREMENTS OF INFANTS
COMPARED WITH INTAKES OF PROTEIN FROM SOYBEAN
MILK OR COW AND HUMAN MILK

Amino Acid	Minimum Requirement (Mg/Kg/Day)	Amino Acid Provided by Protein Fed at a Level of 2 Gm Protein/Kg/Day		
		Human Milk	Cow Milk	Soybean Milk
Histidine	34	32	45	33–57
Isoleucine	119	123	128	67–117
Leucine	150	230	216	91–159
Total S-amino acid	45	73	52	31–55
Total aromatic amino acid	90	92	104	65–115
Threonine	87	89	92	51–89
Tryptophan	22	31	30	11–20
Valine	105	128	138	67–117

Source: Holt and Snyderman (1961).

ority of cow or human milk over soybean milk suitably fortified with vitamins and minerals [see, for example, Kay et al. (1960); Fomon (1961A,B)]. These conclusions are somewhat at variance with the animal experiments cited earlier which showed soybean milk to be somewhat inferior in nutritive value to mammalian milk. This difference may perhaps be due to a less intense requirement for the sulfur amino acids by a growing child than by young rats (Block and Mitchell 1946-1947). Thus, although supplementation with sulfur-containing amino acids balances the protein value of soybean milk for rats, this deficiency appears to be of little consequence in practical infant feeding (Howard et al. 1956; Block et al. 1956).

Soybean Curd

A detailed description of the preparation of soybean curd or "tofu" may be found in Chap. 10. From the amino acid composition shown in Table 7.4, tofu does not appear to differ significantly from the protein found in most other soybean products. The biological evaluation of the nutritive value of soybean curd using animals or human subjects has given values (Tables 7.6 and 7.7), which are comparable to properly heated soybean flour. Since autoclaving does not enhance the nutritive value of the curd (Chang and Murray 1949), it may be concluded that tofu has received adequate heat treatment during processing for optimum nutritional value.

Tofu has been tested as a source of protein in the solid diet of weanling infants and its performance evaluated with respect to acceptability, weight gain, nitrogen balance, and serum protein level (Muto et al. 1963). On the basis of these criteria, tofu was judged to be nutritionally equivalent to the protein derived from a mixture of eggs, fish, and liver.

Other Fractions

The preparation of various protein fractions and their analyses may be found in Chap. 4 and 9. The nutritive value of the protein of various other fractions of soybean has been studied in detail by two groups of investigators (Rackis *et al.* 1961; Hackler *et al.* 1963. 1967). The principle differences between these two procedures may be summarized as follows:

	Rackis *et al.* (1961)	Hackler *et al.* (1963, 1967)
Starting material	Soybean meal	Whole soybeans
Acidification of the aqueous extract	With HCl	With acetic acid
Heat treatment of the aqueous extract	None	Cooked, 60 min, 100° C

The amino acid composition, protein scores, and PER's of these various fractions are presented in Table 7.13. A direct comparison between the PER values reported by these investigators for fractions derived form the aqueous extract is not possible because of the heat treatment used by Hackler's group in the preparation of these fractions. Both groups, however, concluded that the nutritive value of the water-insoluble residue was higher than the acid-precipitated curd. Rackis *et al.* found that the nutritive value of the residue could, by toasting, be further improved to that of toasted soybean meal, whereas the curd was not affected by subsequent heat treatment. There was no significant difference in PER's between the curd and the aqueous extract prepared by the Hackler procedure. The whey protein, in which most of the growth inhibitors reside, is low in nutritive value unless subjected to heat treatment which makes its protein quality higher than that of the original meal (Rackis *et al.* 1963). Although the soak water had a chemical score similar to that of the proteins of soybean meal, milk, curd, and whey, it did not support the growth of rats (Hand 1966). Amino acid composition data show the sulfur-containing amino acids to be limiting in all fractions, but the chemical scores based on these data seem to be of little value for predicting the nutritive value of the various fractions.

The relatively high nutritive value of the insoluble residue which remains after extraction of the bean or meal with water, particularly when toasted, is especially noteworthy. It should be realized, however, that this fraction would probably be of limited value for human feeding since it is very bulky and contains about 12% insoluble carbohydrate and fiber (Hackler *et al.* 1965). Rackis *et al.* (1963) have attributed the high nutritive value of toasted whey proteins to a good balance of essential amino acids, particularly tryptophan, lysine, and cystine plus methionine. This fraction is presently discarded as a waste product during the commercial processing of soybean meal into acid-precipitated protein.

Although the chemical score of soybean hulls is the lowest of the various soybean fractions, its PER is actually among the highest and is comparable to

TABLE 7.13

AMINO ACID COMPOSITION AND NUTRITIVE VALUE OF THE VARIOUS
SOYBEAN FRACTIONS SHOWN IN FIGURE 3.1

Measurement	Meal (Bean)	Hulls	Soybean Fraction				
			Milk	Residue	Curd	Whey Protein	Soak Water
				(Grams per 16 Gm N)			
Protein content, dry basis (%)	61 (45)	9.6 (17)	— (52)	52 (24)	102 (59)	101 (59)	— (19)
Percentage of original protein	100[1] (100)	— (3)	— (83)	26 (14)	61 (74)	6 (9)	— (0.5)
Amino acid composition							
Isoleucine	5.1 (4.8)	3.8 (3.9)	5.3 (4.8)	6.0 (4.5)	5.0 (4.9)	5.0 (2.9)	— (2.5)
Leucine	7.7 (7.8)	5.9 (6.7)	8.1 (7.9)	8.9 (8.0)	7.9 (8.0)	7.7 (4.0)	— (4.2)
Lysine	6.9 (6.5)	7.1 (6.0)	6.7 (6.1)	6.1 (5.1)	5.7 (5.9)	8.7 (8.8)	— (2.9)
Methionine	1.6 (1.4)	0.8 (1.0)	1.3 (1.4)	1.6 (1.0)	1.3 (1.4)	1.9 (2.2)	— (0.5)
Cystine	1.6 (1.6)	1.7 (1.4)	1.4 (1.6)	0.7 (2.3)	1.0 (1.7)	1.8 (2.7)	— (2.5)
Total S-AA	3.2 (3.0)	2.5 (2.4)	2.8 (3.0)	2.3 (3.3)	2.3 (3.1)	3.7 (4.9)	— (3.0)
Phenylalanine	5.0 (5.1)	3.2 (4.0)	5.6 (4.9)	5.2 (4.9)	5.9 (4.8)	4.5 (2.0)	— (3.2)
Tyrosine	3.9 (3.9)	4.7 (3.2)	4.4 (3.9)	3.3 (3.0)	4.6 (3.7)	4.7 (3.8)	— (2.4)
Total aromatic AA	8.9 (9.0)	7.9 (7.2)	10.0 (8.8)	8.5 (7.9)	10.5 (8.5)	9.2 (5.8)	— (5.6)
Threonine	4.3 (4.2)	3.7 (3.5)	4.0 (3.9)	4.7 (4.1)	3.8 (3.7)	6.2 (5.2)	— (3.3)
Tryptophan	1.3 (1.3)	— (1.3)	1.4 (1.4)	— (1.6)	1.0 (1.1)	1.3 (1.8)	— (0.6)
Valine	5.4 (5.0)	4.6 (4.6)	5.6 (4.8)	6.4 (5.0)	5.2 (4.7)	6.2 (3.2)	— (3.1)
Chemical score	73 (—)	57 (66)	— (75)	54 (75)	54 (73)	87 (71)	— (71)
PER[2]	0.85 (—)	2.15 (—)	— (1.93)	1.67 (2.37)	1.53 (1.92)	1.95 (1.68)	

Source: Data from Rackis et al. (1961, 1963); Van Etten et al. (1959); Smith et al. (1964A); Hackler et al. (1963, 1967). Values which have been taken from Hackler et al. Papers are shown in parentheses.

[1] Without the hulls.

[2] Corrected to a PER value of 2.50 for casein. Values taken from Rackis et al. (1963) are for fractions which have not been heated.

that of properly heated soybean meals (Smith *et al.* 1964A). This is rather surprising in view of the fact that the hulls have been reported to contain almost 40% crude fiber (Bailey *et al.* 1935), and large amounts of feces have, in fact, been observed in rats fed diets containing the hulls as a source of protein (Smith *et al.* 1964A). The removal of hulls from soybeans prior to processing does increase the protein content of the defatted meal from about 46% to 51% (Smith *et al.* 1964A; Harmon *et al.* 1969). Although the protein of the hulls contains somewhat lower amounts of phenylalanine, cystine, and methionine and somewhat greater quantities of lysine and tyrosine than in the meal (Rackis *et al.* 1961), these differences probably have little effect on the nutritive value of the meal since the hulls constitute only a small fraction (about 3%) of the total protein of the bean.

Protein Concentrates

A product containing about 70% protein can be produced by removing much of the non-nitrogenous constituents from soybean flakes or flour by extraction with ethyl alcohol, dilute acid, or with water following denaturation with heat (see Chap. 9). In addition to its high protein content, such a product is largely free of objectionable flavors and odors, and it has the functional properties which make it suitable for use in a variety of food products such as bread, cereals, and comminuted meat products.

On the basis of the amino acid composition of soybean protein concentrates (see Table 7.4) one would not expect its nutritive value to differ very much from that of soybean flour. Longenecker *et al.* (1964) found considerable variation in the PER values of a number of unheated commercial soy protein concentrates. When subjected to heat treatment (105° C for 30 min), however, values approaching that of soybean flour were obtained. The authors concluded from these observations that there must be marked differences in manufacturing processes for soybean concentrates so that optimal heat treatment is not always attained. Meyer (1966), on the other hand, was not able to demonstrate any improvement by heat in the nutritive value of the several commercial products examined. In general, one is forced to question the wisdom of trying to draw conclusions from feeding tests involving commercial soybean products whose processing history is not known or, if known, is not given. Furthermore, some commercial products may have deliberately received light or no heat treatment if they are intended for food applications where further heat treatment will be used.

Protein Isolates

The nutritive value of the protein isolated from soybeans received attention as long ago as 1912 when Osborne and Mendel found that the main protein of soybeans had about 2/3 of the growth promoting value of casein. Many years later De and Ganguly (1947) reported that the PER of this protein was about the same as that of autoclaved soybean meal. Most of the more recent nutritional studies have been conducted with soybean protein isolated by extracting soybean flakes or flour at an alkaline pH, followed by precipitation of the protein at an acid pH, and subsequent removal of the bulk of the water. Such preparations consist almost exclusively of protein (93–95%) and represent about 86% of the protein from unheated soybean meal or 60% from heated meals (Cogan *et al.* 1968). Data relating to the nutritional evaluation of soybean protein isolates with animals and human subjects may be found in Tables 7.6 and 7.7 respectively. In general, such data have shown that the nutritive value of isolated soybean protein as measured by PER or BV is approximately 65–85% that of isolated milk protein (casein), and ranks somewhat lower than the protein concentrates discussed earlier (Meyer 1966, 1968).

As the amino acid data indicate (Table 7.4), soybean protein isolate is more deficient in the sulfur amino acids than most other soybean preparations, a fact which has received biological confirmation with rats (Meyer 1966), pigs (Berry *et al.* 1966), and human subjects (Longenecker 1963). The supplementation of protein isolates with methionine raises its nutritive value to that of casein when assayed with rats (Huge 1961) and to about 85% of that of milk proteins when newborn pigs are used as the experimental animal (Schneider and Sarett 1969).

Several investigators (Longenecker *et al.* 1964, Bressani *et al.* 1967; Cogan *et al.* 1968) have reported that the nutritive value of soybean protein isolates can be improved by heat treatment, whereas others (Booth 1961; Huge 1961; Meyer 1966) have not noted such an effect. The most obvious explanation for this discrepancy would be that the technique employed for the isolation of soybean protein may not also always lead to the elimination of heat-labile growth inhibitors. This, in turn, may be a reflection of the thoroughness with which the isolated proteins have been washed during processing.

Alkaline extraction of soybean meal under more drastic conditions may yield an inferior product not because of the failure to eliminate growth inhibitors but because of damage to certain amino acids. DeGroot and Slump (1969) found that the protein isolated by acid precipitation from soybean meal which had been extracted at pH 12.2 for 4 hr at 40° C had a much lower NPU than the original meal. They attributed this decrease in nutritive value to the destruction of cystine which is accompanied by the formation of lysinoalanine. The latter is produced by the interaction of dehydroalanine, which is a decomposition product of cystine and serine, and the ε-amino groups of lysine. Supplementation of the alkaline-damaged protein with methionine partially restored its nutritive value but not fully to the level of the untreated protein source.

Use in Infant Foods.–The use of soybean protein isolates in the formulation of infant milk formulas has received considerable attention in recent years. Using normal infants as experimental subjects, Cherry *et al.* (1968) compared the nutritional properties of a milk formula containing 2.3% isolated soybean protein with a similar formula in which the protein was derived from skim milk at an equivalent protein concentration. Using the criteria of weight gain, levels of serum protein, amino acids, and cholesterol, and other hematologic data, these authors reported that growth was somewhat better on the skim milk formula although the biochemical and hematologic data were similar. Bates *et al.* (1968) subsequently pointed out that the inferior performance of the soybean formula may have been due to the fact that it provided only about 1/2 of the recommended level of methionine. When a milk formula containing soybean protein isolate as the sole source of protein was fortified with methionine, Bates *et al.* (1968) were then unable to find any significant difference from formulas prepared from either cow's milk or soybean flour. It is significant to note that these workers, as well as others (Cowan *et al.* 1969), have observed that the formulas containing soy protein produce a lower incidence of anal irritation than the ones made from soybean flour, an effect which can no doubt be attributed to the absence of fiber in the purified protein. Experiments with baby pigs have shown that the nutritional quality of an infant formula made from soybean protein isolate fortified with methionine was, in fact, far superior to formulas made with soybean flour (Schneider and Sarett 1969).

Use in Textured Foods.—One of the most dramatic and promising uses of soybean protein isolates is in the formulation of textured foods or meat analogues (Irmiter 1964; Thulin and Kuramoto 1967; Kiratsous 1969). In this process, the protein isolate is dissolved or suspended in alkali and then extruded through a die or forced through spinnerets into an acid or an acid-salt bath to form fibers. These can be manipulated to give products simulating the texture and flavor of a wide variety of meat foodstuffs. Such products contain, on the average, about 25% protein, or on a dry basis about 50% protein (Koury and Hodges 1968).

The amino acid composition of textured foods (Table 7.4) is not grossly different from the soybean protein from which it is derived. It will be noted in particular that the cystine has not been destroyed, presumably because the protein isolate has been treated at pH 12 for only 10 min at room temperature (Bressani *et al.* 1967), conditions which are considerably milder than those which had been employed by DeGroot and Slump (1969) in their studies (vide infra). Bressani *et al.* (1967) have made a comprehensive study of the protein quality of soybean protein textured foods using rats, dogs, and children as experimental subjects. From these studies, the authors concluded that the nutritive value of the textured foods was equivalent to that of beef protein and about 80% of that of casein. It was readily accepted by children and did not produce any adverse physiological effects.

Koury and Hodges (1968) found soybean textured foods to be quite acceptable to either hospitalized or out-patients, and clinical and biochemical measurements confirmed the maintenance of good health. The authors concluded that soybean protein, as the major source of protein in the human diet, is both nutritious and acceptable.

Fermented Products

Although fermented soybean products have been used for centuries in the Orient, up to quite recently information concerning the nutritional properties of such foods has been scanty. The nutritive properties of three fermented products will be considered here, namely tempeh, miso, and natto. The technological aspects dealing with the production of these products are considered in Chap. 11. Suffice to state here that miso and natto are products resulting from the fermentation of cooked soybeans by *Aspergillus oryzae* and *Bacillus subtilis (B. natto)*, respectively, and enjoy popular consumption in Japan. Tempeh is an Indonesian dish composed of soybeans which have been cooked and then fermented by the mold *Rhizopus oryzae.*

Tempeh.—The protein content of tempeh is about 20% on a wet basis and 50% when dried. It is very seldom eaten raw, but is usually roasted, cooked in soup, or fried in oil (Dean 1958). The amino acid composition of the protein of tempeh (Table 7.4) is not grossly altered by the fermentation process, although decreases in lysine and methionine have been reported to occur during fermentation (Steinkraus *et al.* 1961; Stillings and Hackler 1965). There is also a slight increase in free amino acids during fermentation (Murata *et al.* 1967), presumably due to the action of proteolytic enzymes elaborated by the mold. Steaming (100° C for 2 hr) of the fermented product had little effect on its amino acid composition, but as much as 20% of the cystine and lysine was destroyed during deep-fat frying (Stillings and Hackler 1965).

The earliest comment bearing directly on the nutritive value of tempeh is that of Van Veen and Schaeffer (1950) who claimed that tempeh is more easily digested than the unfermented bean and that the protein is excellent in quality. Specific data relating to the biological evaluation of the protein value of tempeh are to be found in Table 7.6. Most investigators have found tempeh to offer little, if any, nutritional advantage over unfermented soybean products (Steinkraus *et al.* 1961; Hackler *et al.* 1964; Smith *et al.* 1964B). On the other hand György (1961) noted considerable variation in the PER values of a number of tempeh preparations which had been processed under varying conditions, and some of those did have PER values which were higher than an unfermented control. Smith *et al.* (1964B) saw no evidence of pancreatic hypertrophy in animals fed tempeh, which would indicate that the trypsin inhibitor had been destroyed, probably during the preliminary cooking of the soybeans prior to inoculation with the mold. This perhaps explains why mild heating of tempeh (steaming for 2 hr) produces no further improvement in nutritive value (Hackler *et al.* 1964). Deep-fat frying (7 min), on the other hand, significantly reduced its PER and digestibility. These biological effects are in accord with the amino acid data discussed in the previous paragraph.

In view of the fact that tempeh is a product intended solely for human consumption, it is indeed surprising that so little work appears to have been done with human subjects. In this connection, the following quote from Grant (1952) is of interest: ". . . . many attempts were made (by prisoners in Japanese prison camps during World War II) to make soybeans palatable and digestible, the only satisfactory method proving to be one common in Indonesia, involving inoculation with a fungus. Otherwise these beans were likely to give much digestive disturbances when used in any quantity, even if first reduced to a fine meal." This statement would suggest that principal virtue of tempeh lies in the fact that some of the oligosaccharides believed to be responsible for flatulence (see Chap. 6) have been eliminated, this loss most likely occurring when the water used for cooking is discarded.

Natto.—Natto is used as a side dish or in combination with cooked rice. The protein content of natto is approximately the same as that of tempeh, and the essential amino acid composition of the protein (Table 7.4) does not differ appreciably from most other soybean products except for somewhat higher values for tryptophan and lysine (Standal 1967). Experiments with rats (Table 7.6) have shown that the nutritive value of natto is somewhat less than that of unfermented cooked beans (Hayashi and Ariyama 1960; Sano 1961; Arimoto 1961; Standal 1963). The product obtained from 8 hr of fermentation was superior in nutritive value to that which had been fermented for only 4 hr (Arimoto 1961). Supplementation with methionine and/or lysine had little effect on the protein quality of natto (Sano 1961).

Only limited studies have been conducted on natto with human subjects. Muto et al. (1963) tested natto as a source of protein in the diet of infants and concluded that it could substitute, at least in part, for animal protein with no adverse effects on growth, digestibility, and N retention. Natto in the form of a powder has also been used for making biscuits which were well accepted by children without adverse effects (Arimoto 1961).

Miso.—Miso differs from natto and tempeh in that it is prepared from a mixture of soybeans and rice in varying ratios with added salt. The rice is first inoculated with *A. oryzae* and, after mixing with soybeans, the fermentation is allowed to proceed for periods of up to a year. The final protein content ranges from 10% to 17%. The largest single use for miso is in soups. It is also spread on bread and on raw vegetables to add flavor, and is used, in combination with sugar and oil, for cooking fish, meat, and vegetables.

The amino acid composition of a preparation of miso made from a 1:1 mixture of soybeans and rice is shown in Table 7.4. The growth rate of rats fed this preparation of miso was comparable to that of natto and heated soybeans (Sano 1961). Matsuno and Tamura (1964) reported the biological value of miso to be 71–73 compared to a value of 69 and 80 for tofu and casein, respectively.

Diamant and Laxer (1968) have described the preparation of a miso-type product prepared from defatted soybean flakes and one of the following cereals:

corn, wheat, barley, sorghum, or potatoes. Such products, however, were very poor in nutritive value, and amino acid analysis showed methionine to be absent and inadequate levels of tryptophan and arginine. Supplementation with these three amino acids improved growth somewhat, but growth was still far less than that of the unfermented mixture of soybeans and cereal. The authors presented evidence to indicate that the high salt concentration (8%) of the miso-type product may have been partly responsible for its poor growth-promoting quality.

The paucity of reports dealing with the nutritive evaluation of miso with human subjects may, perhaps, be due to the fact that its high salt content precludes its testing at a protein level needed to produce meaningful results. It may also be that miso has been for so long such an important part of the Japanese diet with apparent impunity to health that its nutritive merits have gone unchallenged.

USE OF SOYBEAN PRODUCTS AS PROTEIN SUPPLEMENT

The deficiency of sulfur-containing amino acids which characterizes the protein of most soybean products can sometimes be minimized by combining such products with other proteins which are not deficient in these amino acids. But more important, perhaps, is the fact that any amino acid which may be limiting in the other protein may be present in excess amounts in the soybean protein. In other words, it is possible to combine soybean protein with other proteins in such a way as to provide a mixture which is nutritionally superior to each one alone. Since soybean protein contains a level of lysine which exceeds that found in the pattern of an ideal reference protein (see Fig. 7.1), the soybean provides an excellent means of correcting the lysine deficiency of most plant proteins. Examples of the supplementary effect of soybean protein on cereal proteins are shown in Table 7.14. It is evident that marked improvements in PER values are obtained when soybean protein is used to supplement protein derived from various cereals. This general conclusion, based largely on experiments with rats, has been confirmed in the case of wheat protein in N-balance studies with human subjects; Bricker et al. (1945) reported that the biological value of white flour, which was 41, could be increased to 55 if 19% of the wheat flour were replaced by soybean flour.

As Supplement to Wheat Protein

Bread.—Because of the marked supplementary effect which soybean protein exerts on lysine-deficient wheat protein, much attention has been directed to use of soybean flour or soybean protein concentrates in bread formulations [see reviews by Bailey et al. (1935); Horvath (1938); Burnett (1951); Diser (1961); Hayward and Diser (1961)]. As early as 1921, experiments conducted by the USDA showed that bread made with a mixture of 25 parts of soybean flour and 75 parts of wheat flour was adequate for the growth of rats (Johns and Finks

TABLE 7.14

EXAMPLES OF THE SUPPLEMENTAL RELATIONSHIP BETWEEN
SOYBEAN AND OTHER PLANT PROTEINS

Source of Test Protein	Level of Protein in Diet (%)	Test Protein	Soybean Protein	PER[1]	Reference
		(% of Total Protein in Diet)			
Corn meal	9	100	0	1.43	
	9	80	20	2.15	Bressani and Elias
	9	60	40	2.53	(1966)
	9	40	60	2.71	
	9	20	80	2.61	
Wheat flour	9	100	0	0.77	Sure (1948)
	9	91	9	1.03	
	9	65	35	2.16	Jones and Divine (1944)
Polished rice	6	100	0	1.76	
	6	92	8	1.84	Sure (1950)
	7	79	21	2.08	
	8	68	32	2.15	
Rye flour	6	100	0	1.29	Sure (1957)
	8	38	62	2.43	Kon and Markuze (1931)
Sesame	10	100	0	1.73	Tasker et al. (1962)
	10	35	65	2.17	Chang and Murray (1949)

[1] The rat was employed as the experimental animal in these studies.

1921). In 1931, Kon and Markuze reported that the PER of a bread formula could be markedly increased when 11-25% of the wheat flour was replaced by soybean flour. This improvement in the nutritive value of bread by adding soybean flour has been subsequently confirmed in many laboratories (Jones 1944; Jones and Divine 1944; Hove et al. 1945; Volz et al. 1945; Sure 1950; Guggenheim and Friedmann 1960; Bornstein and Lipstein 1962). The effect of substituting soybean flour for dried skim milk solids in bread formulas has also been studied. Carlson et al. (1946) reported that bread which contained 3% soybean flour was equal in nutritive value to bread containing 3% milk solids and significantly better than bread made from unsupplemented formulas. Harris et al. (1944), Volz et al. (1945), and Henry and Kon (1949) likewise showed that soy flour could not only completely replace the milk solids of bread without affecting its nutritive properties, but actually improved the nutritional value of bread already containing skim milk solids. Although the protein quality of bread continues to increase as the proportion of soybean flour in the formula is increased, there is a practical upper limit to the amount of soybean flour which can be used in this fashion. When more than 6% of the wheat flour is replaced by soybean flour there is some deterioration in bread character and alterations in dough handling techniques become necessary (Bayfield and Swanson 1946; Finney 1946; Burnett 1951; see also Chap. 10)

Only limited human studies on the nutritive value of breads containing soybean flour have been reported in the literature. Horvath in his review of 1938 refers to N-balance experiments which were conducted on human subjects in Italy and Germany in the 1920's. It was concluded from these studies that white or rye bread containing as much as 20% soybean flour was well tolerated, and utilization of the protein was better than that of breads containing wheat or rye flours alone.

Although soybean flour has been the most common vehicle for incorporating soybean protein into bread, the possibility of using more concentrated sources of soybean protein has not been overlooked. Wilding *et al.* (1968) have studied the nutritional consequences of supplementing the wheat protein of bread with a soybean concentrate containing 70% protein. Increasing the percentage of soy protein in relation to the wheat protein produced significant increases in PER with a maximum value being obtained with 75% soybean protein and 25% white bread protein. The supplementation of wheat flour with soybean protein isolates has likewise proved to be a very effective way of enhancing the nutritive value of bread (Howard *et al.* 1958; Ericson *et al.* 1961; Jansen and Ehle 1965; Mizrahi *et al.* 1967). The improvement in protein quality produced by supplementation with soybean protein isolate is proportional to the amount of lysine provided by the soybean protein (Ericson *et al.* 1961; Mizrahi *et al.* 1967).

Other Baked Goods.—Experimental data are also available to demonstrate that soybean flour can be used to enhance the nutritional quality of other baked goods containing wheat protein. Reynolds and Hall (1950) reported that various cakes and pastries to which soybean flour had been added were nutritionally superior to the nonsupplemented baked goods in terms of growth, PER, and N retention in experiments with rats. The addition of soybean protein in the form of grits to commercial formulas for cookies (Hayward and Diser 1961) or graham crackers (Carlson *et al.* 1947) markedly improved the nutritive quality of the protein.

As Supplement to Corn

It is evident from the data presented in Table 7.14 that soybean protein has a marked supplemental effect on the nutritive value of corn protein, particularly in a ratio of 40% corn protein to 60% soybean protein. This supplementary relationship is also evident from the data of Bressani and Marenco (1963) who found that the PER of lime-treated corn flour was increased from 1.0 to 2.5 by adding 8% soy protein isolate or 10% soybean flour. Along related lines, Cravioto *et al.* (1950) reported that the PER of tortillas made from lime-treated corn plus 10% soybean flour was 1.80 compared to only 1.0 when only lime-treated corn was used. Advantage has been taken of the supplementary relationship of corn and soybean proteins in the formulation of special mixtures of plant proteins for use in developing countries, such as Incaparina 14 which is described below in the section dealing with vegetable protein mixtures.

As Supplement to Rice

The supplementary effect of soybean flour on rice protein as demonstrated with rats (see Table 7.14) has been corroborated in studies with infants and children. Mixtures of soybean flour and rice have been fed to infants and found to support growth comparable to cow's milk although the digestibility of this mixture is somewhat less than that of milk (Huang *et al.* 1961, 1966; Snyderman *et al.* 1961). Panemangalore *et al.* (1964A) fed a group of children in India a basal diet consisting mainly of rice supplemented with suitably processed soybean flour or skim milk powder. Both supplements resulted in a significant increase in N retention, biological value, and NPU, with little difference between the soybean flour and milk. Experiments along similar lines were conducted by DeMaeyer and Vanderbought (1958) on African children. In this case, however, the soybean flour proved to be somewhat less satisfactory than the skim milk, which the authors attrituted to improperly processed soybean flour.

Use in Vegetable Protein Mixtures

Recognizing the many advantages to be derived from the use of properly processed soybean flour as a protein supplement to other plant proteins, considerable effort has gone into the formulation of blends of soybean protein with other plant proteins. These mixtures, when suitably fortified with vitamins and minerals, have great potential for the feeding of infants, children, and adolescents in developing areas of the world. Whether used as a sole source of food for the weaning of infants or as a protein supplement, the nutritional quality of these mixtures should be of the highest order because it must supply the nutrients missing in the diets being consumed by populations of low economic resources.

Of 17 vegetable protein mixtures now being manufactured for use in developing countries, 11 of these contain soybean flour as one of the ingredients (Orr and Adair 1967). Apart from its nutritive merits, the preference for soybean flour is probably due to the fact that there is a greater body of experience to draw upon in terms of processing for edible purposes than any other oilseed. The nutritive properties of some of these mixtures have been studied in great detail, although complete information on the formulation of some of these commercial products is not always available. In addition to vegetable protein mixtures which are being produced commercially, many formulations are still in the process of evaluation.

A vegetable protein mixture in which soybean flour provides the sole source of protein was developed by Dr. Henry Borsook in 1944 and has become known as the "Multiple-Purpose Food" (MPF). The nutritional properties of this product, and one similar to it which was developed in India, have already been briefly referred to. These products have been tested as protein supplements to poor vegetarian diets containing various cereals (Kuppuswamy et al. 1957). It is evident from the data presented in Table 7.15 that MPF produces a marked increase in the growth-promoting properties of such diets.

The much more practical approach to the formulation of vegetable protein mixtures has been the blending of soybean protein with the proteins derived from plant sources which may already be a traditional part of the diet in a particular country. Scientists at the Central Food Technological Institute (CFTI) in India, for example, have been particularly active in the development of mixtures in which soybean protein is used to supplement the protein of peanuts and indigenous legumes. At the Institute of Nutrition of Central America and Panama (INCAP) in Guatemala, corn and cottonseed meal have been used in conjunction with soybean protein. Such mixtures are fortified with essential vitamins and minerals, and, in some cases, essential amino acids, in order to insure an adequate intake of these nutrients. The biological evaluation of some of these plant protein mixtures, in terms of their PER as determined with rats, is presented in Table 7.16. It will be noted that with some of these mixtures, particularly those which have been fortified with methionine, PER values closely approaching that of milk protein have been obtained. Some of these plant protein mixtures provide excellent protein supplements to basal diets which are normally poor in protein quality (Table 7.15).

Peanut and Other Oilseed Proteins

As noted before most of the work in India on vegetable protein mixes has involved the blending of peanut protein with soybean protein. In addition to animal studies from which the data in Tables 7.15 and 7.16 were collected, a number of these protein blends have been evaluated in clinical trials with children. A blend composed of an equal mixture of full-fat soybean flour and peanut flour, and fortified with 1% L-lysine and 1% DL-methionine as well as vitamins and minerals, when fed as a supplement to children on basal diet of corn and tapioca, produced an increase in height, weight, and hemoglobin content of the blood comparable to skim milk powder (Panemangalore et al. 1946B; Narayana Rao et al. 1965). This same blend of soybean and peanut flours markedly increased the growth and N retention of school children on a basal diet of rice (Doraiswamy et al. 1964; Parthasarathy et al. 1964B). Blends containing sesame flour or coconut meal in addition to soybean and peanut flours have also been tested on children. A blend of soybean flour, peanut flour, and coconut meal (4:3:3), when used to supplement a basal wheat diet of young school children brought about a significant increase in the retention of nitrogen (Parthasarathy et al. 1964C). A microatomized blend of soybean flour, peanut flour, and sesame flour (2:2:1), administered as a liquid emulsion to children suffering from moderate to severe kwashiorkor, proved to be an effective cure (Prasanna *et al.* 1968).

Peanut flour has, in some instances, been replaced with peanut protein isolate and blended with an equal quantity of soybean flour to give a mixture which, when supplemented with methionine, had a PER almost equivalent to that of milk protein. (Korula *et al.* 1964A,B). When the blend was incorporated into a poor Indian rice diet or a low protein corn-tapioca diet so as to provide 2.5% and 5% extra protein, respectively, growth was comparable to that produced by skim milk powder.

A blend of soybean and peanut or cottonseed flours and corn meal fortified with minerals and vitamins is sold in South Africa under the name of ProNutro. The exact proportions of these ingredients have not been divulged. It is promoted mainly as a breakfast and baby food, and it is recommended for use in soups and stews. The PER value of ProNutro is almost 90% that of casein (DeMuelenaere 1969). When tested with human patients, no significant difference was found between ProNutro and skim milk powder in promoting recovery from kwashiorkor (Odondaal 1966). When fed as a 2-oz daily supplement to a group of 1200 children in a basal corn diet, no cases of kwashiorkor were reported and monthly weight gains were significantly increased.

Blends Containing Corn

Incaparina is the generic name given to a series of vegetable protein mixtures developed by INCAP, so formulated as to provide 25% or more of protein comparable in quality to those of animal origin. Originally the Incaparinas were mixtures of corn and cottonseed flour, but, in some of the more recent formulations, the cottonseed flour has been replaced wholly or partially by soybean flour. The formulation of Incaparina 14 was based on the observation previously noted (Table 7.14) that maximum complementation between corn protein and soybean protein is obtained when 20–40% of the protein in the diet is derived from corn and 60–80% from soybean. Incaparina 14 consists of 59% corn, 38% toasted soybean flour, 3% torula yeast, 1% $CaCO_3$, and 4500 IU vitamin A per 100 gm (Bressani 1966, 1969). In Incaparina 15, 1/2 of the soybean flour is replaced by cottonseed flour. These mixtures may be used directly to prepare a beverage by simply adding water or they may be incorporated into foods such as soups, puddings, cookies, precooked baby foods, etc.

PER values for Incaparina 14 and Incaparina 15, with and without amino acid supplementation, may be found in Table 7.16. The nutritive value of the unsupplemented formulas was somewhat inferior to casein. When supplemented with a combination of methionine, threonine, and lysine, however, their PER values exceeded that of casein. Incaparina 14 had a BV value of 73% and a digestibility coefficient of 80% when tested with dogs, values close to what was

TABLE 7.15

EFFECT OF ADDING PROTEIN-RICH SUPPLEMENTS CONTAINING SOYBEAN
PROTEIN TO LOW PROTEIN CEREAL DIETS

Basal Diet	Protein-rich Supplement	Final Protein Content of Diet (%)	Gain in Wt[1] (Gm/Wk)	Reference
Rice	No supplement	8.1	5.0	Kuppuswamy et al. (1957)
	Soybean flour (MPF)	12.4	16.1	Kuppuswamy et al. (1957)
	Peanut flour/soy flour (50/50)	10.1	10.7	Shurpalekar et al. (1964D)
	Peanut flour/soy flour (50/50) + 0.8% methionine	10.2	15.6	Shurpalekar et al. (1964D)
	Peanut flour/sesame flour/soy flour (40/20/40)	9.6	10.0	Tasker et al. (1967)
	Skim milk	9.6	11.0	Tasker et al. (1967)
Rice-tapioca	No supplement	5.5	3.0	Tasker et al. (1967)
	Peanut flour/sesame flour/soy flour (40/20/40)	10.5	12.5	Tasker et al. (1967)
	Peanut flour/coconut meal/soy flour (30/30/40)	15.5	18.1	Tasker et al. (1963B)
	Skim milk	15.5	18.1	Tasker et al. (1963B)
Corn-tapioca	No supplement	4.6	0.2	Narayana Rao et al. (1964)
	No supplement	6.1	2.9	Korula et al. (1964B)
	No supplement	6.5	4.2	Joseph et al. (1962)
	No supplement	10.5	13.0	Tasker et al. (1967)
	Peanut flour/soy flour (50/50)	14.3	17.5	Narayana Rao et al. (1964)
	Peanut flour/soy flour (50/50) + 1% methionine	14.8	18.3	Narayana Rao et al. (1964)

+ 1% methionine + 1% lysine	14.7	19.9	Narayana Rao et al. (1964)
Peanut flour/sesame flour/soy flour (40/20/40)	9.6	10.5	Tasker et al. (1967)
Peanut flour/chick pea/skim milk/soy flour (25/25/20/30)	17.2	14.2	Joseph et al. (1962)
Peanut flour/chick pea/skim milk/soy flour (25/10/25/30)	17.2	14.2	Joseph et al. (1962)
Peanut flour/sesame flour/chick pea/soy flour (30/10/30/30)	17.2	14.9	Joseph et al. (1962)
Peanut protein isolate/soy flour (50/50)	11.1	16.7	Korula et al. (1964B)
Peanut protein isolate/soy flour (50/50) + 0.4% methionine	11.0	18.6	Korula et al. (1964B)
Skim milk	11.2	19.2	Korula et al. (1964B)
Skim milk	14.1	18.6	Narayana Rao et al. (1964)
Skim milk	16.2	15.6	Joseph et al. (1962)

[1] Over an experiment period of 8 weeks.

TABLE 7.16

NUTRITIVE VALUE OF PROTEIN-RICH SUPPLEMENTS CONTAINING SOYBEAN PROTEIN

Sources of Proteins	Proportions[1]	Protein Content (%)	PER[2]	Reference
Corn/soy flour (Incaparina 14)	58/38	26	2.6	Bressani (1966)
Corn/soy flour + 0.2% methionine	58/38	26	2.9	Bressani (1966)
Corn/soy flour + 0.2% methionine + 0.2% threonine + 0.2% lysine	58/38	26	3.4	Bressani (1966)
Corn/extrusion cooked full-fat soy flour	43/40	21	2.5	Smith (1969)
Wheat/soy flour (WSB)	73/20	23	2.1	Senti (1969)
Wheat/soy flour	40/60		2.6	Parpia (1969)
Wheat/soy flour + methionine	40/60		3.0	Parpia (1969)
Peanut protein isolate/full-fat soy flour	50/50	26	2.3	Korula et al. (1964A); Shur-
Same as above + 0.4% methionine	50/50	26	2.9	palekar et al. (1964A,B,C)
Peanut flour/full-fat soy flour + 1% methionine	50/50	50	2.5	Shurpalekar et al. (1964B)
Peanut flour/full-fat soy flour + 1% methionine + 1% lysine	50/50	49	2.8	Narayana Rao et al. (1964, 1965)
Skim milk powder/soy flour	70/30	57	2.2	Prasanna et al. (1968)
Peanut flour/sesame flour/soy flour	48/20/30	16	2.3	Krishnamurthy et al. (1958)
Peanut flour/sesame flour/soy flour	40/30/30	44	2.4	Panemangalore et al. (1967)
Peanut flour/sesame flour/soy flour	40/20/40	54	2.4	Prasanna et al. (1968)
Corn meal/skim milk/soy flour (CSM)	64/5/24	19	2.5	Senti (1969)
Corn/cottonseed flour/soy flour (Incaparina 15)	58/19/19	26	2.2	Bressani (1966)
Corn/cottonseed flour/soy flour + 0.2%				

methionine + 0.1% lysine	58/19/19	26	2.7	Bressani (1966)
Peanut flour/wheat/soy flour	30/60/10		2.4	Parpia (1969)
Peanut flour/wheat/soy flour + 1% methionine	30/60/10		2.5	Parpia (1969)
Wheat/sesame/soy flour	60/15/25		2.6	Parpia (1969)
Peanut flour/coconut meal/soy flour	30/30/40	42	2.3	Tasker et al. (1963A)
Peanut flour/Bengal gram/soy flour	48/25/25	16	1.7	Krishnamurthy et al. (1959)
Sesame flour/chick pea flour/soy flour	35/47/18	38	2.9	Guggenheim and Szmelcman (1965)
Peanut flour/sesame flour/Bengal gram/soy flour	38/20/28/20	16	2.1	Krishnamurthy et al. (1959)
Peanut flour/sesame flour/chick pea flour/ soy flour	30/10/30/30	26	1.8	Joseph et al. (1962)

[1] Proportions shown in same sequence as presented in column 1.

[2] Wherever possible, PER values have been corrected on the basis that the PER of casein or skim milk powder = 2.5.

obtained with milk protein under the same conditions (Bressani and Elias 1966). The following data, taken from studies with children (Bressani and Elias 1966) likewise show that the nutritive value of Incaparina 14 was quite comparable to that of skim milk, whereas Incaparina 15, in which 1/2 of the soybean flour had been replaced by cottonseed flour, was of lower nutritive value:

	BV	Digestibility	Mg N Required for N Equilibrium
Skim milk	80.6	92.0	79
Incaparina 14	78.6	91.8	92
Incaparina 15	71.9	88.7	113

If high lysine corn *(opaque-2)* was used to replace common corn in formulas 14 and 15, there was a significant increase in the weight gain and PER values (with rats) in the case of formula 15 but not formula 14 (Bressani and Elias 1969). This difference was attributed to the fact that with cottonseed protein as part of the protein mixture there is a greater deficiency of lysine than when soybean protein alone is used to supplement the corn protein.

A food supplement containing 19% protein derived from a blend of 63.8% processed corn meal, 24.2% toasted defatted soybean flour, and 5% NFDM was developed by the American Corn Millers Federation in cooperation with USDA (Dimler 1967; Senti 1969). This mixture, referred to as CSM, also contains 5% soybean oil and is fortified with 2% of a vitamin-mineral premix. CSM was designed to provide a level of all the essential nutrients sufficient to render adequate the total diet of a preschool child if consumed at a level to supply approximately 25% of the energy needs. CSM may be served in the form of a gruel or porridge for school children and infants or used in baked goods, soups, and other recipes.

The amino acid pattern of CSM conforms rather closely to the 1965 FAO reference pattern based on whole egg protein except for a deficiency of the sulfur amino acids (Inglett *et al.* 1969). Little changes in levels of available lysine, vitamin A, and thiamine could be detected upon storage at elevated temperatures unless the moisture content was allowed to exceed 10% (Book-walter *et al.* 1968). Feeding experiments with rats have shown CMS to have a PER essentially equivalent to that of casein (Senti 1969). CMS has also been tested in a number of feeding trials with infants and children in various parts of the world. The results of these trials have uniformly showed CMS to be well accepted and capable of supporting satisfactory weight and N equilibrium.

Smith (1969) has described the preparation of an extrusion cooked mixture of full-fat soy flour and degerminated corn meal (40:43) as well as a variety of blends with other cereal flours and oilseed proteins. Only PER values with rats were reported, but these data indicated that the protein value of such mixtures is in many cases equivalent to that of casein.

Other Cereals and Legumes

Because of the world-wide popularity of wheat, a formulation (WSB) involving a blend of soybean flour with wheat protein in the form of straight-grade or bulgur flour or a concentrate has been recently developed by USDA (Senti 1969). The formula consists of 20% defatted soybean flour, 73.4% wheat fraction, 4% soybean oil, and 2.6% minerals and vitamins, and the final protein content is about 20%. The PER as determined in rat feeding tests is 2.1. Preliminary results of child feeding tests in Peru show that WSB maintains children in N balance when fed as a primary source of protein.

Guggenheim and Szmelcman (1965) were interested in formulating a protein-rich mixture of cereal proteins which are commonly available in the Middle East including sesame flour and chick peas *(Cicer arietinum)*. It was found that a mixture containing 37.3% protein derived from 47% autoclaved chick peas, 35% defatted sesame flour, and 18% defatted soybean flour had a higher nutritive value than any of its components alone when assayed with growth tests on rats. The authors refer to preliminary trials in which this protein mixture was administered to infants with good acceptance and tolerance. Sesame flour and chick pea (also known as Bengal gram) have also been used in combination with soybean flour by workers in India to produce a protein mixture of good quality (Krishnamurthy *et al.* 1959).

Inglett *et al.* (1969) have developed a computerized approach for predicting the combination of plant proteins which offers an optimal balance of essential amino acids. They predicted that the combinations (by weight) of two cereal proteins with soybean flour shown in Table 7.17 should give an amino acid pattern most closely resembling that of whole egg protein. A comparison of the

TABLE 7.17

OPTIMUM PROPORTIONS OF THREE-COMPONENT BLENDS OF CEREALS
WITH SOYBEAN FLOURS

Cereal-soybean Blends	Proportion of Blend (% Wt Basis)	Protein Contributions[1] (%)
Millet-sorghum-soybean flour	21–59–20	15–33–52
Millet-corn-soybean flour	45–36–19	32–17–51
Millet-wheat-soybean flour	33–52–15	25–35–40
Sorghum-corn-soybean flour	78–00–22	44–00–56
Sorghum-wheat-soybean flour	41–40–19	24–27–50
Corn-wheat-soybean flour	43–36–21	21–24–55

Source: Inglett *et al.* (1969).

[1] Final protein content of blend in all cases is 20%.

amino acid composition of these blends with the egg pattern showed the sulfur amino acids to be the limiting amino acids in all cases, but not limiting when compared with sulfur amino acid content of cow milk protein. It would be of interest to see if these mathematical predictions can be verified by animal experimentation. If this should be so, then it would appear possible to be able to predict the precise manner in which low protein cereals could be blended with soybean flour and other oilseed proteins in order to provide a mixture which is comparable in nutritive value to milk protein. In many parts of the world where milk protein is not available the simple expedient of supplementing the indigenous cereals with soybean protein could be an important first step in improving the nutritional status of the native population.

BIBLIOGRAPHY

ADOLPH, W. H., and CHEN, S.-C. 1932. The utilization of calcium in soybean diets. J. Nutr. *5*, 379–385.

ADOLPH, W. H., and KAO, H.-C. 1934. The biological availability of soybean carbohydrate. J. Nutr. *7*, 395–406.

ALBANESE, A. A. 1950. The protein and amino acid requirements of man. *In* Protein and Amino Acid Requirements of Mammals, A. A. Albanese (Editor). Academic Press, New York.

ALBANESE, A. A. 1959. Criteria of protein nutrition. *In* Protein and Amino Acid Nutrition, A. A. Albanese (Editor). Academic Press, New York.

ALDERKS, O. H. 1949. The study of 20 varieties of soybeans with respect to quantity and quality of the meal. J. Am. Oil Chemists' Soc. *26*, 126–132.

ALLRED, J. B., KRATZER, F. H., and PORTER, J. W. G. 1964. Some factors affecting the *in vitro* binding of zinc by isolated soya bean protein and by α-casein. Brit. J. Nutr. *18*, 575–582.

ALMQUIST, H. J. 1951. Nutritional applications of the amino acids. *In* Amino Acids and Proteins, D. M. Greenberg (Editor). Charles W. Thomas, Springfield, Ill.

ALMQUIST, H. J., MECCHI, E., KRATZER, F. H., and GRAU, C. R. 1942. Soybean protein as a source of amino acids for the chick. J. Nutr. *24*, 385–392.

ALTSCHUL, A. M. 1965. Proteins: Their Chemistry and Politics. Basic Books, New York.

ANDERSON, D. W. 1961. Problems in formulation of soy milk. Proc. Conf. Soybean Products for Protein in Human Foods, USDA, Peoria, Ill.

ANDERSON, M. A., and WILLIAMS, H. H. 1951. Microbiological evaluation of protein quality. I. A colorimetric method for the determination of the growth of *Tetrahymena geleii* W in protein suspension. J. Nutr. *44*, 335–343.

ANON. 1959. General Mills to market MPF. Soybean Dig. *20*, 20.

ARIMOTO, K. 1961. Nutritional research on fermented soybean products. *In* Meeting Protein Needs of Infants and Children. Natl. Acad. Sci.–Natl. Res. Council Publ. *843*.

ARIMOTO, K. *et al.* 1962. Nutritional value of soybean products. I. Nutritional value of "natto"-like product prepared by various processing conditions. Kokuritsu Eiyo Kenkyusho Kenkyu Hokoku, 40–45.

ASHIKAGA, C. 1946. Changes of vitamin B in foods during cooking. II. French soybean. J. Ferment. Technol. *24*, 85–88.

ASPINWALL, G. O., BEGBIE, R., and MCKAY, J. E. 1967. Polysaccharide components of soybeans. Cereal Sci. Today *12*, 223–228, 260–261.

AYKROYD, W. R., and DOUGHTY, J. 1964. Legumes in Human Nutrition. FAO Nutr. Studies *19*.

BAILEY, L. H., CAPEN, R. G., and LECLERC, J. A. 1935. The composition and characteristics of soybeans, soybean flour, and soybean bread. Cereal Chem. *12*, 41–472.

BALIGA, B. R. BALAKRISHNAN, S., and RAJAGOPALAN, R. 1954. Biological value of proteins as influenced by dietary vitamin B $_{12}$. Nature *174*, 35–36.

BALIGA, B. R., and RAJAGOPALAN, R. 1954. Influence on vitamin B $_{12}$ on the biological value of raw soya bean. Current Sci. (India) 23, 51–52.

BALLOUN, S. L., JOHNSON, E. L., and ARNOLD, L. K. 1953. Laboratory estimation of the nutritive value of soybean oil meals. Poultry Sci. 32, 517–527.

BARNES, R. H., FIALA, G., and KWONG, E. 1962. Methionine supplementation of processed soybeans in the cat. J. Nutr. 77, 278–284.

BARNES, R. H., FIALA, G., and KWONG, E. 1965B. Prevention of coprophagy in the rat and the growth-stimulating effects of methionine, cystine and penicillin when added to diets containing unheated soybeans. J. Nutr. 85, 127–131.

BARNES, R. H., and KWONG, E. 1965. Effect of soybean trypsin inhibitors and penicillin on cystine biosynthesis in the pancreas and its transport as exocrine protein secretion in the intestinal tract of the rat. J. Nutr. 86, 245–252.

BARNES, R. H., KWONG, E., and FIALA, G. 1965A. Effect of penicillin added to an unheated soybean diet on cystine excretion in feces of the rat. J. Nutr. 85, 123–126.

BARNES, R. H., and MAACK, J. E. 1943. Review of the literature on the nutritive value of soybeans. Univ. Minn.

BATES, R. D., BARRETT, W. W., ANDERSON, D. W. Jr., and SAPERSTEIN, S. 1968. Milk and soy formulas: a comparative growth study. Ann. Allergy 26, 577–583.

BAYFIELD, E. G., and SWANSON, E. C. 1946. Effect of yeast, bromate, and fermentation on bread containing soy flour. Cereal Chem. 23, 104–113.

BEAN, L. H. 1966. Closing the world's nutritional gap with animal or vegetable protein? FAO Bull. 6.

BEATON, G. H., and MCHENRY, E. W. 1964. Nutrition, A comprehensive Treatise. Academic Press, New York.

BEDFORD, C. L., and MCGREGOR, M. A. 1948. Effect of canning on the ascorbic acid and thiamine in vegetables. J. Am. Dietet. Assoc. 24, 866–869.

BEN-GERA, I., and ZIMMERMAN, G. 1964. Changes during storage in chemically determined lysine availability in soybean concentrate. Nature 202, 1007–1008.

BERRY, T. H. et al. 1962. The limiting amino acids in soybean protein. J. Animal Sci. 21, 558–561.

BERRY, T. H., COMBS, G. E., WALLACE, H. D., and ROBBINS, R. C. 1966. Responses of the growing pig to alterations in the amino acid pattern of isolated soybean protein. J. Animal Sci. 25, 722–728.

BIRD, H. R. et al. 1947. Urease activity and other chemical criteria as indicators of inadequate heating of soybean oil meal. J. Assoc. Offic. Agr. Chemists 30, 354–364.

BIRK, Y., and GERTLER, A. 1962. Effect of mild chemical and enzymatic treatments of soybean meal and soybean trypsin inhibitors on their nutritive and biochemical properties. J. Nutr. 75, 379–387.

BLOCK, R. J., ANDERSON, D. W., HOWARD, H. W., and BAUER, C. D. 1956. Effect of supplementing soybean proteins with lysine and other amino acids. Am. J. Diseases Children 92, 126–130.

BLOCK, R. J., and MITCHELL, H. H. 1946–1947. The correlation of the amino acid composition of proteins with their nutritive value. Nutr. Abstr. Rev. 16, 249–278.

BLOM, L., HENDRICKS, P., and CARIS, J. 1967. Determination of available lysine in foods. Anal. Biochem. 21, 382–400.

BOOKWALTER, G. N., MOSER, H. A., PFEIFER, V. F., and GRIFFIN, E. L. J. 1968. Storage stability of blended food products, Formula No. 2: a corn soymilk food supplement. Food Technol. 22, 1581–1584.

BOOTH, A. N. 1961. Physiological effects of feeding soybean meal and its fractions. Proc. Conf. Soybean Products for Protein in Human Foods, USDA, Peoria, Ill.

BOOTH, A. N., ROBBINS, D. J., RIBELIN, W. E., and DeEds, F. 1960. Effect of raw soybean meal and amino acids on pancreatic hypertrophy in rats. Proc. Soc. Exptl. Biol. Med. 104, 681–683.

BORCHERS, R. 1959. Tyrosine stimulates growth on raw soybean rations. Federation Proc. 18, 517.

BORCHERS, R. 1961. Counteraction of the growth depression of raw soybean oil meal by amino acid supplements in weanling rats. J. Nutr. 75, 330–334.

BORCHERS, R. 1965. Antibiotics and the anti-threonine effect of raw soybean meal. Life Sci. *4*, 1835–1837.

BORCHERS, R. 1967. Failure of antibiotics to stimulate growth of rats fed digested raw soybean meal. J. Agr. Food Chem. *15*, 362–363.

BORCHERS, R. A., ACKERSON, C. W., and MUSSEHL, F. E. 1948A. Trypsin inhibitor. VI. Effect of various heating periods on the growth promoting value of soybean oil meal for chicks. Poultry Sci. *27*, 601–604.

BORCHERS, R. A., ACKERSON, C. W., MUSSEHL, F. E., and MOEHL, A. 1948B. Trypsin inhibitor. VIII. Growth inhibiting properties of a soybean trypsin inhibitor. Arch. Biochem. *19*, 317–322.

BORCHERS, R., ANDERSON, C. W., and SANDSTEDT, R. M. 1947. Trypsin inhibitor. III. Determination and heat destruction of the trypsin inhibitor of soybeans. Arch. Biochem. *12*, 367–374.

BORCHERS, R., MOHAMMAD-ABADI, G., and WEAVER, J. M. 1957. Antibiotic growth stimulation of rats fed raw soybean oil meal. J. Agr. Food Chem. *5*, 371–373.

BORNSTEIN, S., and LIPSTEIN, B. 1962. Evaluation of the nutritive quality of different types of bread by feeding trials with chicks. Israel J. Agr. Res. *12*, 63–73.

BOYNE, A. W., CARPENTER, K. J., and WOODHAM, A. A. 1961. Progress report on an assessment of laboratory procedures suggested as indicators of protein quality in feeding stuffs. J. Sci. Food Agr. *12*, 832–848.

BRAHAM, J. E., BIRD, H. R., and BAUMANN, C. A. 1959. Effect of antibiotics on the weight of chicks and rats fed raw or autoclaved soybean meal. J. Nutr. *67*, 149–158.

BRESSANI, R. 1966. Soybeans as a source of protein for human feedings in Latin America. Proc. Intern. Conf. Soybean Protein Foods, USDA, Peoria, Ill. ARS-71–35.

BRESSANI, R. 1969. Formulation and testing of weaning and supplementary foods containing oilseed proteins. *In* Protein-Enriched Cereal Foods for World Needs, M. Milner (Editor). Am. Assoc. Cereal Chemists, St. Paul.

BRESSANI, R., and ELIAS, L. G. 1966. All-vegetable protein mixtures for human feeding. The development of INCAP vegetable mixture 14 based on soybean flour. J. Food Sci. *31*, 626–631.

BRESSANI, R., and ELIAS, L. G. 1967. Processed vegetable protein mixtures for human consumption in developing countries. Advan. Food Res. *16*, 1–78.

BRESSANI, R., and ELIAS, L. G. 1969. Studies on the use of *opaque-2* corn in vegetable protein-rich foods. J. Agr. Food Chem. *17*, 659–662.

BRESSANI, R., and MARENCO, E. 1963. The enrichment of lime-treated corn flour with proteins, lysine and tryptophan and vitamins. J. Agr. Food Chem. *11*, 517–522.

BRESSANI, R. *et al.* 1967. Protein quality of a soybean protein textured food in experimental animals and children. J. Nutr. *93*, 349–360.

BRICKER, M., MITCHELL, H. H., and KINSMAN, G. M. 1945. The protein requirements of adult human subjects in terms of the protein contained in individual foods and food combinations. J. Nutr. *30*, 269–283.

BURKHOLDER, P. R. 1943. Vitamins in edible soybeans. Science *98*, 188–190.

BURKHOLDER, P. R., and MCVEIGH, I. 1945. Vitamin content of some mature and germinated legume seeds. Plant Physiol. *20*, 301–306.

BURNETT, R. S. 1951. Soyabean protein food products. *In* soybeans and Soybean Products, Vol. II, K. S. Markley (Editor). John Wiley & Sons, New York.

CAHILL, W. M., SCHROEDER, L. J., and SMITH, A. H. 1944. The digestibility and biological value of soybean protein in whole soybeans and soy flour and soybean milk. J. Nutr. *28*, 209–218.

CARLSON, C. W., MCGINNIS, J., and JENSEN, L. S. 1964A. The antirachitic effects of soybean preparations for turkey poults. J. Nutr. *82*, 366–370.

CARLSON, C. W., SAXENA, H. C., JENSEN, L. C., and MCGINNIS, J. 1964B. Rachitogenic activity of soybean products. J. Nutr. *82*, 507–511.

CARLSON, S. C., HAFNER, F. H., and HAYWARD, J. W. 1946. Effect of soy flour and nonfat dry milk solids in white bread in the nutritional quality of the protein as measured by three biological methods. Cereal Chem. *23*, 305–317.

CARLSON, S. C., HERRMANN, E. C., BOHN, R. M., and HAYWARD, J. W. 1947. A nutritional study of the fortification of graham-type crackers with soy grits, calcium, and several vitamins. Cereal Chem. 24, 215–224.

CARPENTER, K. J. 1958. Chemical methods of evaluating protein quality. Proc. Nutr. Soc. (Engl. Scot.) 17, 91–100.

CARPENTER, K. J. 1960. The estimation of available lysine in animal-protein foods. Biochem. J. 77, 604–610.

CARPENTER, K. J., and ELLINGER, G. M. 1955. The estimation of 'available lysine' in protein concentrates. Biochem. J. 61, xi.

CARROLL, R. W. et al. 1953. Absorption of nitrogen and amino acids from soybean meal as affected by heat treatment or supplementation with aureomycin and methionine. Arch. Biochem. Biophys. 45, 260–269.

CARTTER, J. L., and HOPPER, T. H. 1942. Influence of variety, environment and fertility level on the chemical composition of soybean seed. USDA Tech. Bull. 787.

CASKEY, C. D., Jr., and KNAPP, F. C. 1944. Method for detecting inadequately heated soybean oil meal. Ind. Eng. Chem. Anal. Ed. 16, 640–641.

CHANG, I. C. L., and MURRAY, H. C. 1949. Biological value of the protein and the mineral, vitamin, and amino acid content of soymilk and curd. Cereal Chem. 26, 297–305.

CHENG, L. T., LI, H. C., and LAN, T. H. 1941. Biological values of soybean protein and mixed soybean-pork and soybean-egg proteins in human subjects. Chinese J. Physiol. 16, 83–90.

CHERRY, F. F., COOPER, M. D., STEWART, R. A., and PLATOU, R. V. 1968. Cow versus soy formulas. Am. J. Diseases Children 115, 677–692.

CIRCLE, S. J. 1950. Proteins and other nitrogenous constituents. In Soybeans and Soybean Products, Vol. I, K. S. Markley (Editor). John Wiley & Sons, New York.

CLANDININ, D. R., CRAVENS, W. W., ELVEHJEM, C. A., and HALPIN, J. G. 1947. Deficiencies in overheated soybean oil meal. Poultry Sci. 26, 150–156.

CLANDININ, D. R., and ROBBLEE, A. R. 1952. The effect of processing on the enzymatic liberation of lysine and arginine from soybean oil meal. J. Nutr. 46, 525–530.

CLARK, H. E. 1964. Utilization of essential amino acids. In New Methods of Nutritional Biochemistry, Vol. II, A. A. Albanese (Editor). Academic Press, New York.

COGAN, U., YARON, A., BERK, Z., and ZIMMERMAN, G. 1968. Effect of processing conditions on nutritive value of isolated soybean proteins. J. Agr. Food Chem. 16, 196–198.

COOPER, L. F., and BRYAN, M. D. G. 1951. Supplementing the school lunch. J. Home Econ. 43, 355–356.

COWAN, C. C. Jr. et al. 1969. A soy protein isolate formula in the management of allergy in infants and children. Southern Med. J. 61, 389–393.

CRAVIOTO, D. Y. et al. 1950. Comparison of the biological values of the protein of maize tortillas and tortillas made with maize and soybean flour. Cienca (Mex.) 10, 145–147.

DAL BORGO, G., PUBOLS, M. H., and MCGINNIS, J. 1967. Effect of using sugar or starch in the diet on biological responses in the chick to autoclaving hexane-extracted soybean meal. Poultry Sci. 46, 885–889.

DANIELS, A. L., and NICHOLS, N. B. 1917. The nutritive value of the soybean. J. Biol. Chem. 32, 91–102.

DAUBERT, B. F., 1950. Other constituents of the soybean. In Soybeans and Soybean Products, Vol. I, K. S. Markley (Editor). John Wiley & Sons, New York.

DAVIS, P. N., NORRIS, L. C., and KRATZER, F. H. 1962A. Interference of soybean proteins with the utilization of trace minerals. J. Nutr. 77, 217–223.

DAVIS, P. N., NORRIS, L. C., and KRATZER, F. H. 1962B. Iron deficiency studies in chicks using treated isolated soybean protein diets. J. Nutr. 78, 45–453.

DE, S. S., and GANGULY, J. 1947. Heat treatment and the biological value of soybean protein. Nature 159, 341–242.

DEAN, R. F. A. 1953. Plant proteins in child feeding. Med. Res. Council Spec. Rept. Ser. 279.

DEAN, R. F. A. 1958. Use of processed plant proteins as human food. *In* Processed Plant Protein Foodstuffs, A. M. Altschul (Editor). Academic Press, New York.

DEGROOT, A. P., and SLUMP, P. 1969. Effect of severe alkali treatment of proteins on amino acid composition and nutritive value. J. Nutr. *98*, 45–56.

DEMAEYER, E. M., and VANDERBOUGHT, H. L. 1958. A study of the nutritive value of proteins from different sources in the feeding of African children. J. Nutr. *65*, 335–352.

DEMAEYER, E. M., and VANDERBOUGHT, H. L. 1961. Determination of the nutritive value of different protein foods in the feeding of African children. *In* Meeting Protein Needs of Infants and Children. Natl. Acad. Sci.–Natl. Res. Council Publ. *843*.

DEMUELENAERE, H. J. H. 1969. Development, production, and marketing of high-protein foods. *In* Protein-enriched Cereal Foods for World Needs, M. Milner (Editor). Am. Assoc. Cereal Chemists, St. Paul.

DERSE, P. H. 1962. Evaluation of protein quality. II. Methods. J. Assoc. Offic. Agr. Chemists *41*, 192–194.

DESHPANDE, P. D., HARPER, A. E., COLLINS, M., and ELVEHJEM, C. A. 1957. Biological availability of isoleucine. Arch. Biochem. Biophys. *67*, 341–349.

DESIKACHAR, H. S. R., and DE, S. S. 1947. Role of inhibitors in soybean. Science *106*, 421–422.

DESIKACHAR, H. S. R., DE, S. S., and SUBRAHMANYAN, V. 1946A. Studies on the nutritive value of soya milk. I. Nutritive value of the protein. Ann. Biochem. Exptl. Med. (Calcutta) *6*, 49–56.

DESIKACHAR, H. S. R., DE, S. S., and SUBRAHMANYAN, V. 1946B. Studies on the nutritive value of soya milk. II. Comparison of the vitamin B complex content of soya milk and cow's milk. Ann. Biochem. Exptl. Med. (Calcutta) *6*, 57–60.

DESIKACHAR, H. S. R., DE, S. S., and SUBRAHMANYAN, V. 1948. Protein value of soya-bean milk: human feeding experiments. Indian J. Med. Res. *36*, 145–148.

DEWAR, W. A. 1967. The zinc and manganese content of some foods. J. Sci. Food Agr. *18*, 68–71.

DIAMANT, E. J., and LAXER, S. 1968. Nutritional evaluation of miso. Israel J. Chem. *6*, 147p.

DIMLER, R. J. 1967. Soybeans and corn join forces in food. Soybean Dig. *27*, 50–53.

DISER, G. M. 1961. Soy flour and soy grits as protein supplements for cereal products. Proc. Conf. Soybean Products for Protein in Human Foods, USDA, Peoria, Ill.

DORAISWAMY, T. R. *et al.* 1964. Studies on a processed protein food based on a blend of ground nut flour and full-fat soya flour fortified with essential amino acids, vitamins, and minerals. IV. Effect of supplementary protein food on growth and nutritional status of school children. J. Nutr. Dietet. *1*, 87–90.

DUTRA, DE OLIVIERA, J. E., and SCATENA, L. 1967. Nutritional value of protein from a soybean milk powder. J. Food Sci. *32*, 592–594.

EDELSTEIN, S., and GUGGENHEIM, K. 1969. Effect of raw soybean flour on vitamin B $_{12}$ requirement of rats. Israel J. Med. Sci. *5*, 415–417.

ELLMORE, M. F., and SHAW, J. C. 1954. The effect of feeding soybeans on blood plasma carotene and vitamin A of dairy calves. J. Dairy Sci. *37*, 1269–1272.

ENGEL, R. W. 1943. The choline content of animal and plant products. J. Nutr. *25*, 441–446.

ERICSON, L.-E., LARSSON, S., and LID, G. 1961. A comparison of the efficiencies of free lysine and of roller-dried milk, fish protein, and soya bean protein for the supplementation of wheat bread. Acta Physiol. Scand. *53*, 366–375.

ERSHOFF, B. H. 1949. Protective effects of soybean meal for the immature hyperthyroid rat. J. Nutr. *39*, 259–281.

EVANS, R. J., and BANDEMER, S. L. 1967. Nutritive value of legume seed proteins. J. Agr. Food Chem. *15*, 439–443.

EVANS, R. J., and BUTTS, H. A. 1948. Studies on the heat inactivation of lysine in soybean oil meal. J. Biol. Chem. *175*, 15–20.

EVANS, R. J., and BUTTS, H. A. 1949. Studies on the heat inactivation of methionine in soybean oil meal. J. Biol. Chem. *178*, 543–548.

EVANS, R. J., GROSCHKE, A. C., and BUTTS, H. A. 1951. Studies on the heat inactivation of cystine in soybean oil meal. Arch. Biochem. *30*, 414–422.

EVANS, R. J., and MCGINNIS, J. 1946. The influence of autoclaving soybean oil meal on the availability of cystine and methionine for the chick. J. Nutr. *31*, 449–461.

EVANS, R. J., and MCGINNIS, J. 1948. Cystine and methionine metabolism by chicks receiving raw or autoclaved soybean oil meal. J. Nutr. *35*, 477–488.

EVANS, R. J., MCGINNIS, J., and ST. JOHN, J. L. 1947. The influence of autoclaving soybean oil meal on the digestibility of the protein. J. Nutr. *33*, 661–672.

EVERSON, G., and HECKERT, A. 1944. The biological value of some leguminous sources of protein. J. Am. Dietet. Assoc. *20*, 81–82.

EVERSON, G., STEENBOCK, H., CEDERQUIST, dD. C., and PARSONS, H. T. 1944. The effect of germination, the stage of maturity, and variety upon the nutritive value of soybean protein. J. Nutr. *27*, 225–229.

FAO. 1949. Food Composition Tables for International Use. Food Agr. Organ. UN, Nutr. Studies *3*.

FAO. 1954. Food Composition Tables–Minerals and Vitamins–for International Use. Food Agr. Organ. UN Nutr. Studies. *11*.

FAO. 1957. Protein Requirements. Food Agr. Organ UN Studies. *16*.

FAO. 1965. Protein Requirements. Food Agr. Organ. UN Nutr. Meeting Rept. Ser. *37*.

FERNELL, W. R., and ROSEN, G. D. 1956. Microbiological evaluation of protein quality with *Tetrahymena pyriformis* W. 1. Characteristics of growth of the organism and determination of relative nutritive values of intact protein. Brit. J. Nutr. *10*, 143–156.

FINNEY, K. F. 1946. Loaf volume potentialities, buffering capacity, and other baking properties of soy flour in blends with spring wheat flour. Cereal Chem. *23*, 96–104.

FITCH, C. D., HARVILLE, W. E., DINNING, J. S., and PORTER, F. S. 1964. Iron deficiency in monkeys fed diets containing soybean protein. Proc. Soc. Exptl. Biol. Med. *116*, 130–133.

FLYNN, L. M. 1949. Report on determination of folic acid. J. Assoc. Offic. Agr. Chemists *32*, 464–478.

FOMON, S. J. 1961A. Factors influencing retention of nitrogen by normal full-term infants. *In* Meeting Protein Needs of Infants and Children. Natl. Acad. Sci.–Natl. Res. Council Publ. *843*.

FOMON, S. J. 1961B. Nitrogen balance studies with normal infants fed soya bean protein. Proc. Conf. Soybean Products for Protein in Human Foods, USDA, Peoria, Ill.

FORBES, R. M., and YOKE, M. 1960. Zinc requirement and balance studies with the rat. J. Nutr. *70*, 53–57.

FORD, J. E. 1962. A microbiological method for assessing the nutritional value of proteins. 2. The measurement of "available" methionine, leucine, isoleucine, arginine, histidine, tryptophan and valine. Brit. J. Nutr. *16*, 409–425.

FRENCH, C. E. *et al.* 1944. The production of vitamins in germinated peas, soybeans and beans. J. Nutr. *28*, 63–70.

FRITZ, J. C., KRAMKE, E. H., and REED, C. A. 1947. Effect of heat treatment on the biological value of soybeans. Poultry Sci. *26*, 657–661.

FRÖLICH, A. 1954. Relation between the quality of soybean oil meal and the requirements of vitamin B$_{12}$ for chicks. Nature *173*, 132–133.

FROST, D. V. 1959. Methods of measuring the nutritive value of proteins, protein hydrolyzates, and amino acid mixtures. The repletion method. *In* Protein and Amino Acid Nutrition, A. A. Albanese (Editor). Academic Press, New York.

GARLICH, J. D., and NESHEIM, M. C. 1966. Relationship of fractions of soybeans and a crystalline soybean trypsin inhibitor to the effects of feeding unheated soybean meal to chicks. J. Nutr. *88*, 100–110.

GOLDBERG, A., and GUGGENHEIM. K. 1962. The digestive release of amino acids and their concentrations in the portal plasma of rats after protein feeding. Biochem. J. *83*, 129–135.

GOLDBERG, A., and GUGGENHEIM, K. 1964. Effect of antibiotics on pancreatic enzymes of rats fed soybean flour. Arch. Biochem. Biophys. *108*, 250–254.

GRANT, M. W. 1952. Deficiency diseases in Japanese prison camps. Nature *169*, 91–92.

GRAU, C. R., and CARROLL, R. W. 1958. Evaluation of protein quality. *In* Processed Plant Protein Foodstuffs, A. M. Altschul (Editor). Academic Press, New York.

GRIFFITH, M., and YOUNG, R. J. 1966. Growth response of turkey poults to fractions of soybean meal. J. Nutr. *89,* 293–299.

GUGGENHEIM, K., and FRIEDMANN, N. 1960. Effect of extraction rate of flour and of supplementation with soya meal on the nutritive value of bread proteins. Food Technol. *14,* 298–300.

GUGGENHEIM, K., and SZMELCMAN, S. 1965. Protein-rich mixture based on vegetable foods available in Middle Eastern countries. J. Agr. Food Chem. *13,* 148–151.

GUPTA, J. D., DAKROURY, A. M., HARPER, A. E., and ELVEHJEM, C. A. 1958. Biological availability of lysine. J. Nutr. *64,* 259–270.

GUPTA, J. D., and ELVEHJEM, C. A. 1957. Biological availability of tryptophan. J. Nutr. *62,* 313–324.

GUTHNECK, B. T., BENNETT, B. A., and SCHWEIGERT, B. S. 1953. Utilization of amino acids from foods by the rat. II. Lysine. J. Nutr. *49,* 289–294.

GYÖRGY, P. 1961. The nutritive value of tempeh. *In* Meeting the Protein Needs of Infants and Children. Natl. Acad. Sci.–Natl. Res. Council Publ. *843.*

GYÖRGY, P., MURATA, K., and IKEHATA, H. 1964. Antioxidants isolated from fermented soybeans (tempeh). Nature *203,* 870–872.

GYÖRGY, P., and PRENA, C. 1964. Protein-rich foods in calorie-protein malnutrition. Problems of evaluation. Am. J. Clin. Nutr. *14,* 7–12.

HACKLER, L. R., HAND, D. B., STEINKRAUS, K. A., and VAN BUREN, J. B. 1963. A comparison of the nutritional value of protein from several soybean fractions. J. Nutr. *80,* 205–210.

HACKLER, L. R., STEINKRAUS, K. H., VAN BUREN, J. P., and HAND, D. B. 1964. Studies on the utilization of tempeh protein by weanling rats. J. Nutr. *82,* 452–456.

HACKLER, L. R. *et al.* 1965. Effect of heat treatment on nutritive value of soybean protein fed to weanling rats. J. Food Sci. *30,* 723–728.

HACKLER, L. R., and STILLINGS, B. R. 1967. Amino acid composition of heat-processed soymilk and its correlation with nutritive value. Cereal Chem. *44,* 70–77.

HACKLER, L. R., STILLINGS, B. R., and POLIMENI, R. J. 1967. Correlation of amino acid indexes with nutritional quality of several soybean fractions. Cereal Chem. *44,* 638–644.

HAFNER, F. H. 1961. Multi-purpose food. Soybean Dig. *21,* No. 8, 20–21.

HALE, J. K. 1943. Soybeans in the diet. J. Home Econ. *35,* 203–206.

HALEVY, S., and GROSSOWICZ, N. 1953. A microbiological approach to nutritional evaluation of proteins. Proc. Soc. Exptl. Biol. Med. *82,* 567–571.

HAND, D. B. 1966. Formulated soy beverages for infants and pre-school children. Proc. Intern. Conf. Soybean Protein Foods, USDA, Peoria, Ill. ARS-71-35.

HARKINS, R. W., and SARETT, H. P. 1967. Methods of comparing protein quality of soybean infant formulas in the rat. J. Nutr. *91,* 213–218.

HARMON, B. G., BECKER, D. E., JENSEN, A. H., and BAKER, D. H. 1969. Nutrient composition of corn and soybean meal. J. Animal Sci. *28,* 459–464.

HARRIS, P. L., QUAIFE, M. L., and SWANSON, W. J. 1950. Vitamin E content of foods. J. Nutr. *40,* 367–381.

HARRIS, R. S., CLARK, M., and LOCKHART, E. E. 1944. Nutritional value of bread containing soya flour and milk solids. Arch. Biochem. *4,* 243–247.

HAYASHI, Y., and ARIYAMA, H. 1960. Studies on pulses used as food in Japan. Tohoku J. Agr. Res. *11,* 171–185.

HAYWARD, J. W., and DISER, G. M. 1961. Soybean protein as flour and grits for improving dietary standards in many parts of the world. Soybean Dig. *21,* No. 10, 14–24.

HAYWARD, J. W., and HAFNER, F. H. 1941. The supplementary effect of cystine and methionine upon the protein of raw and cooked soybeans as determined with chicks and rats. Poultry Sci. *20,* 139–150.

HAYWARD, J. W., STEENBOCK, H., and BOHSTEDT, G. 1936A. The effect of heat as used in the extraction of soybean oil upon the nutritive value of the protein of soybean oil meal. J. Nutr. *11,* 219–234.

HAYWARD, J. W., STEENBOCK, H., and BOHSTEDT, G. 1936B. The effect of cystine and casein supplements upon the nutritive value of the protein of raw and heated soybeans. J. Nutr. *12*, 275–283.

HENDRICKS, D. G. *et al.* 1969. Effect of level of soybean protein and ergocalciferol on mineral utilization by the baby pig. J. Animal Sci. *28*, 342–348.

HENRY, K. M., and KON, S. K. 1949. The effect on the biological value of bread nitrogen of additions of dried skim milk and of soya flour. J. Dairy Res. *16*, 53–57.

HENSLEY, G. W., CARROLL, R. W., WILCOX, E. L., and GRAHAM, W. R., Jr. 1953. The effects of aureomycin and methionine supplements fed to rats receiving soybean meals. Arch. Biochem. Biophys. *45*, 270–274.

HOFF-JORGENSEN, E., MOUSTGAARD, J., and MÖLLER, P. 1952. The content of B vitamins in some ordinary Danish feedstuffs: the application of an improved medium for vitamin determinations with *Lactobacillus casei.* Acta Agr. Scand. *2*, 305–311.

HOLT, L. E., JR., and SNYDERMAN, S. E. 1961. The amino acid requirements of infants. J. Am. Med. Assoc. *175*, 100–103.

HORN, M. J., BLUM, A. E., and WOMACK, M. 1954. Availability of amino acids to microorganisms. 2. A rapid microbial method of determining protein value. J. Nutr. *52*, 375–381.

HORVATH, A. A. 1938. The nutritional value of soybeans. Am. J. Digest. Diseases *5*, 177–183.

HOU, H. C., RIESEN, W. H., and ELVEHJEM, C. A. 1949. Influence of heating on the liberation of certain amino acids from whole soybeans. Proc. Soc. Exptl. Biol. Med. *70*, 416–419.

HOVE, E. L., CARPENTER, L. E., and HARREL, C. G. 1945. The nutritive quality of some plant proteins and the supplemental effect of some protein concentrates on patent flour and whole wheat. Cereal Chem. *22*, 287–295.

HOWARD, H. W., BLOCK, R. J., ANDERSON, D. W., and BAUER, C. D. 1956. The effect of long time feeding of a soybean infant food diet to white rats. Ann. Allergy *14*, 166–171.

HOWARD, H. W., MONSON, W. J., BAUER, C. D., and BLOCK, R. J. 1958. Nutritive value of bread flour proteins as affected by practical supplementation with lactalbumin, non-fat dry milk solids, soyabean protein, wheat gluten, and lysine. J. Nutr. *64*, 151–165.

HOWE, E. 1961. Summary of progress on the use of purified amino acids in foods. *In* Meeting the Protein Needs of Infants and Children. Natl. Acad. Sci.–Natl. Res. Council Publ. *843.*

HSU, P. T., GRAHAM, W. D., CARVER, J. S., and MCGINNIS, J. 1949. Correlation of nutritive value for chicks of autoclaved soybean oil meal with fluorescence, browning and sticky droppings. Poultry Sci. *28*, 780.

HUANG, P.-C., CHEN, C.-H., and TUNG, T.-C., 1961. Studies of protein-rich foods for infants in Taiwan. II. Feeding experiments of soybean-cereal flakes in infants. J. Formosan Med. Assoc. *60*, 520–528.

HUANG, P.-C., TUNG, T.-C., LUE, H.-C., and WEI, H.-Y. 1965. Feeding of infants with toasted full fat soybean foods. J. Formosan Med. Assoc. *64*, 591–604.

HUANG, P.-C., TUNG, T.-C., and LUE, H.-C. 1966. Feeding of infants with full-fat soybean-rice foods. Proc. Intern Conf. Soybean Protein Foods, USDA, Peoria, Ill. ARS-71–35.

HUGE, W. E., 1961. Present and potential uses of soybean flour, grits, and protein concentrates in foods. Proc. Conf. Soybean Products for Protein in Human Foods, USDA, Peoria, Ill.

IKEHATA, H., WAKAIZUMI, M., and MURATA, K. 1968. Antioxidant and antihemolytic activity of a new isoflavone, "factor 2," isolated from tempeh. Agr. Biol. Chem. (Tokyo) *32*, 740–746.

INGLETT, G. E., CAVINS, J. F., KWOLEK, W. F., and WALL, J. S. 1969. Using a computer to optimize cereal based food composition. Cereal Sci. Today *14*, 69–74.

INGRAM, G. R., RIESEN, W. H., CRAVENS, W. W., and ELVEHJEM, C. A. 1949. Evaluating soybean oil meal protein for chick growth by enzymatic release of amino acids. Poultry Sci. *28*, 898–913.

IRIARTE, B. J. R., and BARNES, R. H. 1966. The effect of overheating on certain nutritional properties of the protein of soybeans. J. Food Technol. *20*, 131–134.

IRMITER, T. F. 1964. Foods from spun protein fibers. Nutr. Rev. *22*, 33–35.

JANSEN, G. R., and EHLE, S. R. 1965. Studies on breads supplemented with soy, non-fat dry milk, and lysine. Food Technol. *19*, 1439–1442.

JENSEN, L. S., and MRAZ, F. R. 1966. Rachitogenic activity of isolated soy protein for chicks. J. Nutr. *88*, 249–253.

JOHNS, C. O., and FINKS, A. J. 1921. Studies in nutrition. V. The nutritive value of soybean flour as a supplement to wheat flour. Am. J. Physiol. *55*, 455–461.

JONES, D. B. 1944. Nutritive value of soybean and peanut protein. Federation Proc. *3*, 116–120.

JONES, D. B., and DIVINE, J. P. 1944. The protein nutritional value of soybean, peanut, and cottonseed flour and their value as supplements to wheat flour. J. Nutr. *41*, 41–49.

JOSEPH, K. *et al.* 1962. The supplementary value of certain processed protein foods based on blends of groundnut, soya-bean, sesame, chickpea *(Cicer arietinum)* flours, and skim-milk powder to a maize-tapioca diet. Brit. J. Nutr. *16*, 49–57.

KAKADE, M. L., and LIENER, I. E. 1969. Determination of available lysine in proteins. Anal. Biochem. *27*, 273–280.

KAKADE, M. L., SIMONS, N. R., LIENER, I. E., and LAMBERT, J. W. 1972. Biochemical and nutritional assessment of different varieties of soybeans. J. Agr. Food Chem. *20*, 87–90.

KASAI, T., ISHIKAWA, Y., OBATA, Y., and TSUKAMOTO, T. 1966. Changes in amino acid composition during germination of the soybean. I. Changes in free amino acids, nitrogen compounds, and total amino acids. Agr. Biol. Chem. (Tokyo) *30*, 973–978.

KAY, J. L., DAESCHNER, C. W., Jr., and DESMOND, M. M. 1960. Evaluation of infants fed soybean and evaporated milk formulae from birth to three months, Am. J. Diseases Children *100*, 264–276.

KIRATSOUS, A. S. 1969. Meat analoges and flavors. Cereal Sci. Today *14*, 147–149.

KLOSE, A. A., HILL, B., and FEVOLD, H. L. 1948. Food value of soybeans as related to processing. Food Technol. *2*, 201–206.

KOBATAKE, Y., MATSUMO, N., and TAMURA, E. 1964. Nutritional value of Japanese soybean products. Ann. Rept. Natl. Inst. Nutr. (Tokyo), 8–9.

KON, S. K., and MARKUZE, Z. 1931. The biological values of the proteins of breads baked from rye and wheat flours alone or combined with yeast or soybean flour. Biochem. J. *25*, 1476–1484.

KONDO, K., IWAI, K., and YOSHIDA, T. 1954. Folic acid in plant tissues. III. Determination of folic acid and folinic acid in plant tissues. Food Sci., Res. Inst. Kyoto Univ. Bull. *13*, 67–69.

KORULA, S. *et al.* 1964A. Efficiency of a spray dried infant food formulation based on peanut protein isolate and soybean flour in meeting the protein reuirements of protein-depleted rats. J. Food Sci. Technol. *1*, 4–7.

KORULA, S. *et al.* 1964B. Studies on a spray dried infant food based on peanut protein isolate and full-fat soy flour and fortified with DL-methionine and certain vitamins and minerals. III. Supplementary value to maize-tapioca diet. Food Technol. *18*, 903–906.

KOURY, S. D., and HODGES, R. E. 1968. Soybean proteins for human diets? Wholesomeness and acceptibility. J. Am. Dietet. Assoc. *52*, 480–484.

KRATZER, F. H. 1965. Soybean protein-mineral interrelationships. Federation Proc. *24*, 1498–1500.

KRATZER, F. H. *et al.* 1959. The effect of autoclaving soybean protein and the addition of ethylenediaminetetracetic acid on the biological availability of dietary zinc for turkey poults. J. Nutr. *68*, 313–322.

KRISHNAMURTHY, K. *et al.* 1958. Effect of heat processing on the trypsin inhibitor and nutritive value of the protein of soybeans. Ann. Biochem. Exptl. Med. (Calcutta) *18*, 153–156.

KRISHNAMURTHY, K., GANAPATHY, S. N., SWAMINATHAN, M., and SUBRAH-

MANYAN, V. 1959. Studies on the nutritive value of protein foods based on blends of groundnut, soyabean, bengal gram *(Cicer arietinum)* and sesame flour. Food Sci. (Mysore) *8*, 388–389.

KRISHNAMURTHY, K. *et al.* 1960. Studies on the nutritive value of composite protein foods based on blends of groundnut, soybean and sesame flours. Food Sci. (Mysore) *9*, 86–88.

KROBER, O. A. 1956. Methionine content of soybeans as influenced by location and season. J. Agr. Food Chem. *4*, 254–257.

KROBER, O. A., and CARTTER, J. L. 1966. Relation of methionine content to protein levels in soybeans. Cereal Chem. *43*, 320–325.

KUHNAU, J. 1949. Biochemistry of food proteins. Angew Chem. *61*, 357–365. (German)

KUIKEN, K. A., and LYMAN, C. A. 1949. Essential amino acids composition of soybean meals prepared from twenty strains of soybeans. J. Biol. Chem. *177*, 29–36.

KUPPUSWAMY, S. *et al.* 1957. Supplementary value of Indian multi-purpose food to poor vegetarian diets based on different cereals and millets. Food Sci. (Mysore) *6*, 84–86.

KUPPUSWAMY, S., SRINIVASAN, M., and SUBRAHMANYAN, V. 1958. Proteins in foods. Indian Council Med. Res., New Delhi, India.

LEASE, L. G. 1967. Availability to the chick of zinc-phosphate complexes isolated from oilseed meals by an *in vitro* digestion method. J. Nutr. *93*, 523–532.

LEASE, L. G., and WILLIAMS, W. P., Jr. 1967. Availability of zinc and comparison of *in vitro* and *in vivo* zinc uptake of certain oilseed meals. Poultry Sci. *46*, 242–248.

LEE, F. A., and WHITCOMBE, J. 1945. Effect of freezing preservation and cooking on vitamin contents of green soybeans and soybean sprouts. J. Am. Dietet. Assoc. *21*, 696–697.

LIENER, I. E. 1955. The photometric determination of the hemagglutinating activity of soyin and crude soybean extracts. Arch. Biochem. Biophys. *54*, 223–231.

LIENER, I. E. 1958. Effect of heat on plant proteins. *In* Processed Plant Protein Foodstuffs, A. M. Altschul (Editor). Academic Press, New York.

LIENER, I. E. (Editor) 1969. Miscellaneous toxic factors. *In* Toxic Constituents of Plant Foodstuffs. Academic Press, New York.

LIENER, I. E., and FEVOLD, H. L. 1949. The effect of the soybean trypsin inhibitor on the enzymatic release of amino acids from autoclaved soybean meal. Arch. Biochem. Biophys. *21*, 395–407.

LIENER, I. E., and HILL, E. G. 1953. The effect of heat treatment on the nutritive value and hemagglutinating activity of soybean oil meal. J. Nutr. *49*, 609–620.

LIENER, I. E., and KAKADE, M. L. 1969. Protease inhibitors. *In* Toxic Constituents of Plant Foodstuffs, I. E. Liener (Editor). Academic Press, New York.

LIENER, I. E., KAKADE, M. L., and SIMONS, N. R. 1969. Unpublished observations.

LODHI, G. N., RENNER, R., and CLANDININ, D. R. 1969. Available carbohydrate in rapeseed meal and soybean meal as determined by a chemical method and a chick bioassay. J. Nutr. *99*, 413–418.

LONGENECKER, J. B. 1963. Utilization of dietary protein. *In* Newer Methods of Nutritional Biochemistry, A. A. Albanese (Editor). Academic Press, New York.

LONGENECKER, J. B., MARTIN, W. H., and SARETT, H. P. 1964. Protein quality improvement: improvement in the protein efficiency of soybean concentrates and isolates by heat treatment. J. Agr. Food Chem. *12*, 411–412.

LUSHBOUGH, C. H., PORTER, T., and SCHWEIGERT, B. S. 1957. Utilization of amino acids from foods by the rat. 4. Tryptophan. J. Nutr. *62*, 513–526.

MATSUNO, N., and TAMURA, E. 1964. Nutritional value of Japanese soybean products. Ann. Rept. Natl. Inst. Nutr. (Tokyo) 8–9.

MATTINGLY, J. P., and BIRD, H. R. 1945. Effect of heating, under various conditions, and of sprouting on the nutritive value of soybean oil meals and soybeans. Poultry Sci. *24*, 344–352.

MAURON, J. 1961. Concept of amino acid availability and its bearing on protein evaluation. *In* Meeting Protein Needs of Infant and Children. Natl. Acad. Sci.–Natl. Res. Council Publ. *843*.

MAURON, J., MOTTER, F., BUJARD, E., and EGLI, R. H., 1955. The availability of lysine, methionine, and tryptophan in condensed milk and milk powder. *In vitro* digestion studies. Arch. Biochem. Biophys. *59*, 433–451.

MCCANCE, R. A., and WIDDOWSON, E. M. 1935. Phytin in human nutrition. Biochem. J. *29*, 2694–2699.

MCGINNIS, J., and EVANS, R. J. 1947. Amino acid deficiencies of raw and overheated soybean oil meal for chicks. J. Nutr. *34*, 725–732.

MCGREGOR, M. A., and BEDFORD, C. L. 1948. Ascorbic acid and thiamine in fresh and frozen lima beans and soybeans. J. Am. Dietet. Assoc. *24*, 670–672.

MELNICK, D., OSER, B. L., and WEISS, S. 1946. Rate of enzymic digestion of proteins as a factor in nutrition. Science *103*, 326–329.

MENDEL, L., and FINE, M. S. 1911–1912. Studies in nutrition. IV. The utilization of the proteins of the legumes. J. Biol. Chem. *10*, 433–458.

MEYER, E. W. 1967. Soybean concentrates and isolates. Proc. Intern Conf. Soybean Protein Foods, USDA, Peoria, Ill. ARS-71-35.

MEYER, E. W. 1969. Soy-protein products for food. Conf. Protein-rich Food Products from Oilseeds, USDA, New Orleans. ARS-72-71.

MILLER, C. D. 1945. Thiamine content of Japanese soybean products. J. Am. Dietet. Assoc. *21*, 430–432.

MILLER, C. D., DINNING, H., and BAUER, A. 1952. Retention of nutrients in commercially prepared soybean curd. Food Res. *17*, 261–267.

MILLER, C. D., and ROBBINS, R. C. 1934. The nutritive value of green immature soybeans. J. Agr. Res. *49*, 161–167.

MILLER, E. R. *et al.* 1965. Comparisons of casein and soy proteins upon mineral balance and vitamin D_2 requirement of the baby pig. J. Nutr. *85*, 347–354.

MILLER, H. W. 1957. Review of literature on the nutritional value of soy milk. WHO/FAO/UNICEF Nutr. Panel *R. 1/Add. 4.*

MITCHELL, H. H. 1944. Determination of the nutritive value of the proteins of food products. Ind. Eng. Chem., Anal. Ed. *16*, 696–700.

MITCHELL, H. H. 1950. Nutritive Factors in Soybean Products. *In* Soybeans and Soybean Products, Vol. I, K. S. Markley (Editor). John Wiley & Sons, New York.

MITCHELL, H. H. 1954. *In* Symposium on Methods for the Evaluation of Nutritional Adequacy and Status, H. Spector, M. S. Peterson, and T. E. Friedemann (Editors). Natl. Res. Council, Washington, D.C.

MITCHELL, H. H., and BEADLES, J. R. 1949. The effect of storage on the nutritional qualities of the proteins of wheat, corn and soybeans. J. Nutr. *39*, 463–484.

MITCHELL, H. H., and BLOCK, R. J. 1946. Some relationships between the amino acid content of proteins and their nutritive values for the rat. J. Biol. Chem. *163*, 599–620.

MIZRAHI, S., ZIMMERMANN, G., BERK, Z., and COGAN, U. 1967. The use of isolated soybean protein in bread. Cereal Chem. *44*, 193–203.

MORRISON, A. B., and NARAYANA RAO, M. 1966. Measurement of the nutritional availability of amino acids in foods. Advan. Chem. Ser. *57*, 159–177.

MORRISON, A. B., and SARETT, H. P. 1958. Studies on zinc deficiency in the chick. J. Nutr. *65*, 267–280.

MORSE, W. J. 1950. Chemical composition of soybean seed. *In* Soybean and Soybean Products, Vol. I, K. S. Markley (Editor). John Wiley & Sons, New York.

MURATA, K., IKEHATA, H., and MIYAMOTO, T. 1967. Studies on the nutritional value of tempeh. J. Food Sci. *32*, 580–586.

MURLIN, J. R., EDWARDS, L. E., and HAWLEY, E. E. 1944. Biological values of some food proteins determined on human subjects. J. Biol. Chem. *156*, 785–786.

MURLIN, J. R., EDWARDS, L. E., HAWLEY, E. E., and CLARK, L. C. 1946. Biological value of proteins in relation to the essential amino acids which they contain. II. Interconvertibility of biological values illustrated by supplementing egg and soy protein with essential amino acids. J. Nutr. *31*, 555–564.

MUSTAKAS, G. C., GRIFFIN, E. L., Jr., ALLEN, L. E., and SMITH, O. B. 1964. Production and nutritional evaluation of extrusion-cooked full-fat soybean flour. J. Am. Oil Chemists' Soc. *41*, 607–614.

MUTO, S., TAKAHASHI, E., HARA, M., and KONUMA, Y. 1963. Soybean products as protein sources for weanling infants. J. Am. Dietet. Assoc. *43*, 451–456.

NARAYANA RAO, M. *et al.* 1964. A processed protein food based on a blend of peanut flour and full-fat soybean flour fortified with essential amino acids, vitamins, and minerals. I. Preparation, chemical composition, and shelf-life. J. Nutr. Dietet. *1*, 1–3; II. Amino acid composition and nutritive value of the proteins. *Ibid.* 4–7; III. Supplementary value to a maize-tapioca diet. *Ibid.* 8–13.

NARAYANA RAO, M., RAJAGOPALAN, R., SWAMINATHAN, M., and PARPIA, H.A.B., 1965. A vegetable protein mixture based on peanut and soybean flours. J. Am. Oil Chemists' Soc. *42*, 658–661.

NATL. ACAD. SCI. 1959. Evaluation of Protein Nutrition. Natl. Res. Council Publ. *711.*

NATL. ACAD. SCI. 1961. Progress in Meeting Protein Needs of Infants and Preschool Children. Natl. Res. Council Publ. *843.*

NATL. ACAD. SCI. 1963. Evaluation of Protein Quality. Natl. Res. Council Publ. *1100.*

NATL. ACAD. SCI. 1968. Recommended Dietary Allowances. Natl. Res. Council Publ. *1694.*

NELSON, T. S. 1967. The utilization of phytate phosphorus by poultry–a review. Poultry Sci. *46*, 862–871.

NELSON, T. S. *et al.* 1968. Effects of phytate on the calcium requirements of chicks. Poultry Sci. *47*, 1985–1989.

O'DELL, B. L., and SAVAGE, J. E. 1960. Effect of phytic acid on zinc availability. Proc. Soc. Exptl. Biol. Med. *103*, 304–305.

O'DELL, B. L., YOKE, J. M., and SAVAGE, J. E. 1964. Zinc availability in the chick as affected by phytate, calcium, and ethylenediamine-tetraacetate. Poultry Sci. *43*, 415–419.

ODONDAAL, W. A. 1966. Experiences in development of ProNutro in South Africa. *In* Preschool Child Malnutrition, Primary Deterrent to Human Progress. Natl. Acad. Sci.– Natl. Res. Council Publ. *1282*, Washington, D.C.

ORR, E., and ADAIR, D. 1967. The production of protein foods and concentrates from oilseeds. Trop. Prod. Inst. Rept. *G31*, London.

OSBORNE, T. B., and MENDEL, L. B. 1912. Observations on growth in feeding experiments with isolated food substances. Hoppe-Seylers Z. Physiol. Chem. *80*, 307–370. (German)

OSBORNE, T. B., and MENDEL, L. B. 1917. The use of soybean as food. J. Biol. Chem. *32*, 369–387.

OSER, B. L. 1951. Method for integrating essential amino acid content in the nutritional evaluation of protein. J. Am. Dietet. Assoc. *27*, 396–402.

OSHIMA, K. 1905. A digest of Japanese investigations of the nutrition of man. USDA Expt. Sta. Bull. *159.*

OUSTERHOUT, L. E., GRAU, C. R., and LUNDHOLM, B. D. 1959. Biological availability of amino acids in fish meals and other protein sources. J. Nutr. *69*, 65–73.

PANEMANGALORE, M. *et al.* 1964A. The metabolism of nitrogen and the digestibility coefficient and biological value of the proteins and net protein utilization in poor rice diet supplemented with methionine-fortified soya flour or skim milk powder. Can. J. Biochem. *42*, 641–650.

PANEMANGALORE, M. *et al.* 1964B. Studies on a processed protein food based on a blend of groundnut flour and full-fat soya flour fortified with essential amino acids, vitamins, and minerals. II. Amino acid composition and nutritive value of the proteins. J. Nutr. Dietet. (Coimbatore, India) *1*, 4–7.

PANEMANGALORE, M. *et al.* 1967. The relative efficacy of protein foods based on blends of groundnut, bengal gram, soybean, and sesame flours and fortified with limiting amino acids, vitamins, and minerals in meeting the protein needs of protein-depleted albino rats. J. Nutr. Dietet. (Coimbatore, India) *4*, 178–182.

PANT, R., and KAPUR, A. S. 1963. A comparative study of the chemical composition and nutritive value of some common Indian pulses and soya bean. Ann. Biochem. Exptl. Med. (Calcutta) *23*, 457–460.

PARPIA, H. A. B. 1969. Protein foods of India based on cereals, legumes, and oilseed meals. *In* Protein-enriched Cereal Foods for World Needs, M. Milner (Editor). Am. Assoc. Cereal Chemists St. Paul.

PARSONS, H. T. 1943. Effect of different cooking methods on soybean proteins. J. Home Econ. *35,* 211–213.

PARTHASARATHY, H. N. *et al.* 1964A. The effect of fortification of processed soya flour with dl-methionine hydroxy analogue or dl-methionine on the digestibility, biological value, and net protein utilization of the proteins as studied in children. Can. J. Biochem. *42,* 377–384.

PARTHASARATHY, H. N. *et al.* 1964B. Studies on a processed protein food based on a blend of groundnut flour and full-fat soya flour fortified with essential amino acids, vitamins, and minerals. V. Effect of supplementary protein food on the metabolism of nitrogen, calcium, and phosphorus in undernourished children subsisting on a rice diet. J. Nutr. Dietet. (Coimbatore, India) *1,* 91–94.

PARTHASARATHY, H. N. *et al.* 1964C. Effect of a supplementary protein food based on a blend of soyabean, groundnut and coconut flours on the retention of nitrogen, calcium and phosphorus in undernourished children subsisting on an inadequate diet. J. Nutr. Dietet. (Coimbatore, India) *1,* 285–287.

PAYNE, D. S., and STUART, L. S. 1944. Soybean protein in human nutrition. Advan. Protein Chem. *1,* 187–208.

PIAN, J. H. C. 1930. Biological value of the proteins of mung bean, peanut, and bean curd. Chinese J. Physiol. *4,* 431–436.

PILCHER, H. L., and WILLIAMS, H. H. 1954. Microbiological evaluation of protein quality. II. Studies of the responses of *Tetrahymena pyriformis* W to intact proteins. J. Nutr. *53,* 589–599.

PORTER, T. 1946. Michigan-grown soybeans, navy beans, and wheat as sources of iron. Mich. State Univ. Agr. Expt. Sta. Quart. Bull. *29,* 115–125.

PRASANNA, H. A., AMLA, I., INDIRA, K., and NARAYANA RAO, M. 1968. Studies on microatomized protein foods based on blends of low-fat groundnut, soya, and sesame flours, and skim milk powder and fortified with vitamins, calcium salts, and limiting amino acids. V. Relative efficacy in the treatment of kwashiorkor. Am. J. Clin. Nutr. *21,* 1355–1365.

PUCHAL, F. *et al.* 1962. The free blood plasma amino acids as related to the source of dietary proteins. J. Nutr. *76,* 11–16.

RACKIS, J. J. 1965. Physiological properties of soybean trypsin inhibitors and their relationship to pancreatic hypertrophy and growth inhibition of rats. Federation Proc. *24,* 1488–1493.

RACKIS, J. J. 1966. Soybean trypsin inhibitors: their inactivation during meal processing. J. Food Technol. *20,* 102–104.

RACKIS, J. J. *et al.* 1961. Amino acids in soybean hulls and oil meal fractions. J. Agr. Food Chem. *9,* 409–412.

RACKIS, J. J. *et al.* 1963. Feeding studies on soybeans. Growth and pancreatic hypertrophy in rats fed soybean meal fractions. Cereal Chem. *40,* 531–538.

RAGHAVENDAR RAO, S., CARTER, F. L., and FRAMPTON, V. L. 1963. Determination of available lysine in oilseed meal proteins. Anal. Chem. *35,* 1927–1930.

RAMA RAO, P. B., NORTON, H. W., and JOHNSON, B. C. 1964. The amino acid composition and nutritive value of proteins. 5. Amino acid requirements as a pattern for protein evaluation. J. Nutr. *82,* 88–92.

RANGANATHAN, S. 1938. The available iron in some common Indian foodstuffs, determined by the α,α'-dipyridine method. Indian J. Med. Res. *25,* 677–684.

REID, B. L., KURNICK, A. A., SVACHA, R. L., and COUCH, J. R. 1956. The effect of molybdenum on chick and poult growth. Proc. Soc. Exptl. Biol. Med. *93,* 245–248.

RENNER, R., CLANDININ, D. R., and ROBBLEE, A. R. 1953. Action of moisture on damage done during over-heating of soybean oil meal. Poultry Sci. *32,* 583–585.

REYNOLDS, M. S., and HALL, C. 1950. Effect of adding soy flour upon the protein value of baked products. J. Am. Dietet. Assoc. *26,* 584–589.

RICE, E. E. 1969. Nutritive values of oilseed proteins. Paper presented at Short Course on Oilseed Proteins: Chemistry, Technology, and Economics. Sponsored by Am. Assoc. Cereal Chemists and Assoc. Oil Chemists' Soc., French Lick, Indiana. July 13–16.

RICHERT, D. A., and WESTERFELD, W. W. 1965. Effect of casein and soybean protein diets on growth of ducklings. J. Nutr. *86,* 17–22.

RIESEN, W. H., CLANDININ, D. R., ELVEHJEM, C. A., and CRAVENS, W. W. 1947. Liberation of essential amino acids from raw, properly heated, and overheated soybean oil meal. J. Biol. Chem. *167,* 143–150.

ROACH, A. G., SANDERSON, P., and WILLIAMS, D. R. 1967. Comparison of methods for the determination of available lysine value in animal and vegetable protein sources. J. Sci. Food Agr. *18,* 274–278.

ROELOFSON, P. A., and TALENS, A. 1964. Changes in some B vitamins during molding of soybeans by *Rhizopus oryzae* in the production of tempeh kedelee. J. Food Sci. *29,* 224–226.

ROSE, W. C. 1949. The amino acid requirements of man. Federation Proc. *8,* 546–552.

ROSE, W. C. 1957. The amino acid requirements of adult man. Nutr. Abstr. Rev. *27,* 631–647.

ROSEN, G. D., and FERNELL, W. K. 1956. Microbiological evaluation of protein quality with *Tetrahymena pyriformis* W. 2. Relative nutritive values of proteins in foodstuffs. Brit. J. Nutr. *10,* 156–169.

SAKURAI, Y., and NAKANO, M. 1961. Production of high-protein food from fermented soybean products. *In* Meeting Protein Needs of Infants and Children. Natl. Acad. Sci.–Natl. Res. Council Publ. *843.*

SALMON, W. D. 1943. Soybeans for human food. J. Home Econ. *35,* 201–202.

SAMBETH, W., NESHEIM, M. C., and SERAFIN, J. A. 1967. Separation of soybean whey into fractions with different biological activities for chicks and rats. J. Nutr. *92,* 479–490.

SANO, T. 1961. Feeding studies with fermented soy products (natto and miso). *In* Meeting Protein Needs of Infants and Children. Natl. Acad. Sci.–Natl. Res. Council Publ. *843.*

SAXENA, H. C., JENSEN, L. S., and MCGINNIS, J. 1962. Failure of amino acid supplementation to completely overcome the growth depression effect of raw soybean meal in chicks, J. Nutr. *77,* 259–263.

SAXENA, H. C., JENSEN, L. S., and MCGINNIS, J. 1963. Pancreatic hypertrophy and chick growth inhibition by soybean fractions devoid of trypsin inhibitor. Proc. Soc. Exptl. Biol. Med. *112,* 101–105.

SCHLOSSER, G. C., and DAWSON, E. H. 1969. Cottonseed flour, peanut flour, and soy flour: formulas and procedures for family and institutional use in developing countries. USDA ARS Publ. *61–7.*

SCHNEIDER, D. L., and SARETT, H. P. 1969. Growth of baby pigs fed infant soybean formulas. J. Nutr. *98,* 279–287.

SCHROEDER, L. J., CAHILL, W. M., and SMITH, A. H. 1946. The utilization of calcium in soybean products and other calcium sources. J. Nutr. *32,* 413–422.

SCHWEIGERT, B. S., and GUTHNECK, B. T. 1953. Utilization of amino acids from foods by the rat. I. Methods of testing for lysine. J. Nutr. *49,* 277–287.

SCHWEIGERT, B. S., and GUTHNECK, B. T. 1954. Utilization of amino acids from foods by the rat. III. Methionine. J. Nutr, *54,* 333–343.

SENTI, F. R. 1969. Formulated cereal foods in the U.S. Food for Peace Program. *In* Protein-enriched Cereal Foods for World Needs, M. Milner (Editor). Am. Assoc. Cereal Chemists, St. Paul.

SHAW, J. C., MOORE, L. A., and SYKES, J. F. 1951. The effect of raw soybeans on blood plasma carotene and liver vitamin A of calves. J. Dairy Sci. *34,* 176–180.

SHEFFNER, A. L. 1967. *In vitro* protein evaluation. *In* Newer Methods of Nutritional Biochemistry, Vol. III, A. A. Albanese (Editor). Academic Press, New York.

SHEFFNER, A. L., ECKFELDT, G. A., and SPECTOR, H. 1956. The pepsin digest residue (PDR) amino acid index of net protein utilization. J. Nutr. *60,* 105–120.

SHERMAN, W. C. 1940. Chromatographic identification and biological evaluation of carotene from mature soybeans. Food Res. *5,* 13–22.

SHERMAN, W. C., ELVEHJEM, C. A., and HART, E. B. 1934. Further studies on the availability of iron in biological materials. J. Biol. Chem. *107*, 383–393.

SHERMAN, W. C., and SALMON, W. D. 1939. Carotene content of different varieties of green and mature soybeans and cowpeas. Food Res. *4*, 371–380.

SHURPALEKAR, S. R., CHANDRASEKHARA, M. R., SWAMINATHAN, M., and SUB-RAHMANYAN, V. 1961. Chemical composition and nutritive value of soyabean and soyabean products. Food Sci. (Mysore) *11*, 52–64.

SHURPALEKAR, S. R. *et al.* 1963. Effect of feeding diets containing raw soyabean fortified with methionine or heat-treated soyabean on growth, nitrogen retention, and liver enzymes of albino rats. Ann. Biochem. Exptl. Med. (Calcutta) *23*, 345–352.

SHURPALEKAR, S. R. *et al.* 1964A. Studies on a spray-dried infant food based on peanut protein isolate and full-fat soy flour and fortified with DL-methionine and certain vitamins and minerals. I. Preparation, chemical composition, and shelf-life. Food Technol. *18*, 898–900.

SHURPALEKAR, S. R. *et al.* 1964B. Studies on a spray-dried infant food based on peanut protein isolate and full-fat soy flour and fortified with DL-methionine and certain vitamins and minerals. II. Protein efficiency ratio and overall nutritive value. Food Technol. *18*, 900–902.

SHURPALEKAR, S. R. *et al.* 1964C. Preparation, chemical composition, and shelf-life of precooked roller-dried protein foods based on full-fat soya flour and low-fat groundnut flour, J. Sci. Food Agr. *15*, 370–371.

SHURPALEKAR, S. R. *et al.* 1964D. Supplementary value of precooked roller-dried protein foods based on full-fat soya flour and low-fat groundnut flour to a poor Indian rice diet. J. Sci. Food Agr. *15*, 373–377.

SHURPALEKAR, S. R., KORULA, S., CHANDRASEKHARA, M. R., and SWAMIN-ATHAN, M. 1965. Studies on the preparation, chemical composition, and nutritive value of a spray-dried soya food suitable for feeding weaned infants. J. Sci. Food Agr. *16*, 90–94.

SIMPSON, J. I. 1943. Soybean studies in Illinois. J. Home Econ. *35*, 207–210.

SMITH, A. H. 1944. Cited by Jones (1944).

SMITH, A. K. 1963. Foreign uses of soybean protein foods. Cereal Sci. Today *8*, 196, 200, 210.

SMITH, A. K. *et al.* 1964A. Growth and pancreatic hypertrophy of rats fed commercial and laboratory meals and hulls. Feedstuffs *36*, 46–47.

SMITH, A. K., *et al.* 1964B. Tempeh: nutritive value in relation to processing. Cereal Chem. *41*, 173–181.

SMITH, J. M., and VAN DUYNE, F. O. 1951. Other soybean products. *In* Soybeans and Soybean Products, Vol. II, K. S. Markley (Editor). John Wiley & Sons, New York.

SMITH, O. B. 1969. History and status of specific protein-rich foods extrusion-processed cereal foods. *In* Protein-enriched Cereal Foods for World Needs, M. Milner (Editor). Am. Assoc. Cereal Chemists, St. Paul.

SMITH, R. E. 1968. Assessment of availability of amino acids in fish meal, soybean meal, and feather meal by chick growth assay. Poultry Sci. *47*, 1624–1630.

SMITH, R. E., and SCOTT, H. M. 1965. Use of free amino acid concentrations in blood plasma in evaluating the amino acid adequacy in intact proteins for chick growth. II. Free amino acid patterns of blood plasma of chicks fed sesame and raw, heated, and over-heated soybean meals. J. Nutr. *86*, 45–50.

SNYDERMAN, S. E., BOYER, A., and HOLT, L. E., Jr. 1961. Evaluation of protein foods in premature infants. *In* Meeting Protein Needs of Infants and Children. Natl. Acad. Sci.–Natl. Res. Council Publ. *843*.

SPACKMAN, D. H., STEIN, W. H., and MOORE, S. 1958. Automatic recording apparatus for use in chromatography of amino acids. Anal. Chem. *30*, 1190–1205.

SPITZER, R. R., and PHILLIPS, P. H. 1945A. The availability of soybean oil meal phosphorus for the rat. J. Nutr. *30*, 117–126.

SPITZER, R. R., and PHILLIPS, P. H. 1945B. Enzymatic relationships in the utilization of soybean oil meal phosphorus by the rat. J. Nutr. *30*, 183–192.

STANDAL, B. R. 1963. Nutritional value of proteins of Oriental soybean foods. J. Nutr. *81*, 279–285.

STANDAL, B. R. 1967. Amino acids in Oriental soybean foods. J. Am. Dietet. Assoc. *50*, 397–400.

STEELE, B. F., SAUBERLICH, H. E., REYNOLDS, M. J., and BAUMANN, C. A. 1947. Amino acids in the urine of human subjects fed eggs or soybeans. J. Nutr. *33*, 209–220.

STEINKRAUS, K. H., HAND, D. B., VAN BUREN, J. P., and HACKLER, L. R. 1961. Pilot plant studies on tempeh. Proc. Conf. Soybean Products for Protein in Human Foods, USDA, Peoria, Ill.

STILLINGS, B. R., and HACKLER, L. R. 1965. Amino acid studies on the effect of fermentation time and heat-processing of tempeh. J. Food Sci. *30*, 1043–1048.

SUBRAHMANYAN, V. *et al.* 1961. Development and evaluation of processed foods based on edible peanut flour and protein. *In* Meeting Protein Needs of Infants and Children, Natl. Acad. Sci.–Natl. Res. Council Publ. *843*.

SUGAWARA, T. 1953. The formation of ascorbic acid in plants. V. The germination of seeds and production of ascorbic acid. Japan. J. Botany *14*, 125–146.

SUMNER, J. B., and DOUNCE, A. L. 1939. Carotene oxidase. Enzymologia *7*, 130–132.

SUPPLEE, G. C., CLARK, E. L., and DOOLITTLE, W. I. 1946. Differential nitrogen retention from casein, lactalbumin, and soy protein and hydrolysates therefrom. J. Dairy Sci. *29*, 717–726.

SUPPLEE, W. C. *et al.* 1958. Observations on zinc supplementations of poultry rations. Poultry Sci. *37*, 1245–1246.

SURE, B. 1948. Relative supplementary values of dried food yeast, soybean flour, peanut meal, dried nonfat milk solids, and dried buttermilk to the proteins in milled whole corn meal and milled enriched wheat flour. J. Nutr. *36*, 65–73.

SURE, B. 1950. Nutritional improvement of cereal grains with small amounts of foods of high protein content. Univ. Arkansas Coll. Agr., Agr. Expt. Sta. Bull. *493*.

SURE, B. 1957. The addition of small amounts of defatted fish flour to milled wheat flour, corn meal, and rice. Influence or growth and protein efficiency. J. Nutr. *61*, 547–554.

SWAMINATHAN, M. 1967. Availability of plant proteins. *In* Newer Methods of Nutritional Biochemistry, Vol. III, A. A. Albanese (Editor). Academic Press, New York.

SWAMINATHAN, M., and PARPIA, H. A. B. 1967. Milk substitutes based on oilseeds and nuts. World Rev. Nutr. Dietet. *8*, 184–206.

SZMELCMAN, S., and GUGGENHEIM, K. 1967. Availability of amino acids in processed plant protein food stuffs. J. Sci. Food Agr. *18*, 347–350.

TAIRA, H. 1966. Amino acid content of processed soybean. X. The influence of added sugars on the heat destruction of the basic and sulfur-containing amino acids in soybean products. Agr. Biol. Chem. (Tokyo), *30*, 847–855.

TAIRA, H., KOYANAGI, T., TAKANOHASHI, T., and OIKAWA, K. 1969. Studies on amino acid contents of processed soybean. XI. Evaluation of nutritional losses of overheated defatted soybean flour. Agr. Biol. Chem. (Tokyo) *33*, 1387–1398.

TAIRA, H., TAIRA, H., SUGIMURA, K., and SAKURAI, Y. 1965A. Studies on amino acid contents of processed soybean. VI. The heat destruction of amino acids in defatted soybean flour. Agr. Biol. Chem. (Tokyo) *29*, 1074–1079, 1965B. VII. The influence of added water on the heat destruction of amino acids in defatted soybean flour Agr. Biol. Chem. (Tokyo) *29*, 1080–1083.

TAIRA, H., TAIRA, H., SUGIMURA, K., and SAKURAI, Y. 1965B. Studies on amino acid contents of processed soybean. VIII. Effect of heating on total lysine and available lysine in defatted soybean flour. Agr. Biol. Chem. (Tokyo) *29*, 1080–1083.

TASKER, P. K. *et al.* 1962. Supplementary value of the proteins of coconut to bengal gram *(Cicer arietinum)* proteins and groundnut proteins. Ann. Biochem. Exptl. Med. (Calcutta) *22*, 181–184.

TASKER, P. K. *et al.* 1963A. The nutritive value of the proteins of a processed protein food based on a blend of a full-fat soya flour, groundnut flour, and coconut meal. Food Sci. (Mysore) *12*, 175–177.

TASKER, P. K. *et al.* 1963B. Supplementary value of a protein food based on a blend of soya, groundnut and coconut flours to a tapioca-rice diet. Food Sci. (Mysore) *12,* 178–181.
TASKER, P. K. *et al.* 1967. Studies on microatomized protein food based on blends of low-fat groundnut, soybean, and sesame flours and skim milk powder and fortified with vitamins, calcium salts, and limiting amino acids. IV. Supplementary value of the foods to diets based on rice and blends of tapioca, rice, and maize flours. J. Nutr. Dietet. (Coimbatore, India) *4,* 65–73.
TEERI, A. E., VIRCHOW, W. E., and LOUGHLIN, M. E. 1956. A microbiological method for the nutritional evaluation of proteins. J. Nutr. *59,* 587–593.
THOMPSON, O. J., CARLSON, C. W., PALMER, J. S., and OLSON, O. E. 1968. Destruction of rachitogenic activity of isolated soybean protein by autoclaving as demonstrated with turkey poults. J. Nutr. *94,* 227–232.
THULIN, W. W., and KURAMOTO, S. 1967. "Bontrae"–a meal-like ingredient for convenience foods. Food Technol. *21,* 168–171.
TKACHUK, R., and IRVINE, G. N. 1969. Amino acid compositions of cereals and oilseed meals. Cereal Chem. *46,* 206–208.
UNITED NATIONS. 1968. International action to avert an impending protein crisis. UN Publ. *E. 68 XIII 2,* New York.
USDA. 1962. Proceedings of Conference on Soybean Products for Protein in Human Foods, Sept. 13–15, Peoria, Ill.
USDA. 1967. Proceedings of International Conference on Soybean Protein Foods, Oct. 17–19, Peoria, Ill. ARS-71–35.
VAN BUREN, J. P. *et al.* 1964. Heat effects on soymilk: indices of protein quality in dried soymilks. J. Agr. Food Chem. *12,* 524–528.
VAN DUYNE, F. O., BRUCKART, S. M., CHASE, J. T., and SIMPSON, J. I. 1945. Ascorbic acid content of freshly harvested vegetables. J. Am. Dietet. Assoc. *21,* 153–154.
VAN ETTEN, C. H. *et al.* 1959. Amino acid composition of soybean protein fractions. J. Agr. Food Chem. *7,* 129–131.
VAN ETTEN, C. H., KWOLEK, W. F., PETERS, J. E., and BARCLAY, A. S. 1967. Plant seeds as protein sources for food or feed. Evaluation based on amino acid composition of 379 species. J. Agr. Food Chem. *15,* 1077–1089.
VAN VEEN, A. G., and SCHAEFFER, G. 1950. The influence of tempeh fungus on the soya bean. Doc. Nurl. Indones. Morbis Trop. *2,* 270–281.
VOHRA, P., and KRATZER, F. H. 1967. The importance of the source of isolated soybean protein in nutrition. Poultry Sci. *46,* 1016–1017.
VOLZ, F. E., FORBES, R. M., NELSON, W. L., and LOOSLI, J. K. 1945. The effect of soyflour on the nutritive value of the protein of white bread. J. Nutr. *29,* 269–275.
WARNICK, R. E., and ANDERSON, J. O. 1968. Limiting essential amino acids in soybean meal for growing chickens and the effect of heat upon the availability of the essential amino acids. Poultry Sci. *47,* 281–287.
WATT, B. K., and MERRILL, A. L. 1963. Composition of Foods, Raw, Processed and Prepared. USDA Handbook *8.*
WEAKLEY, F. B., ELDRIDGE, A. C., and MCKINNEY, L. L. 1961. The alleged thiamine-destroying factors in soybeans. J. Agr. Food Chem. *9,* 435–439.
WESTERFELD, W. W., RICHERT, D. A., and RUEGAMER, W. R., 1964. Thyroxine and antithyroxic effects in the chick. J. Nutr. *83,* 325–330.
WESTERMAN, B. D., OLIVER, B., and MAY, E. 1954. Improving the nutritive value of flour. VI. A comparison of the use of soya flour and wheat germ. J. Nutr. *54,* 225–236.
WESTFALL, R. J., and HAUGE, S. M. 1948. The nutritive quality and the trypsin inhibitor content of soybean flour heated at various temperatures. J. Nutr. *35,* 379–389.
WHO/FAO/UNICEF. 1966. Note on human testing of supplementary food mixtures. World Health Organ. Nutr. Doc. R 10/Add. 91/Rev. 1.
WILCOX, R. A., CARLSON, C. W., KOHLMEYER, W., and GASTLER, G. F. 1961A. Evidence for a water-soluble growth promoting factor(s) in soybean oil meal. Poultry Sci. *40,* 94–102.

WILCOX, R. A., CARLSON, C. W., KOHLMEYER, W., and GASTLER, G. F. 1961B. The growth response of turkey poults to a water extract of soybean oil meal as influenced by different sources of isolated soybean protein. Poultry Sci. *40,* 1353–1354.

WILDING, M. D., ALDEN, D. E., and RICE, E. E. 1968. Nutritive value and dietary properties of soy protein concentrates. Cereal Chem. *45,* 254–259.

WOLF, W. J., SLY, D. A., and KWOLEK, W. F., 1966. Carbohydrate content of soybean proteins. Cereal Chem. *43,* 80–94.

WU, C. H., and FENTON, F. 1953. Effect of sprouting and cooking of soybeans on palatability, lysine, and tryptophan, thiamine, and ascorbic acid. Food Res. *18,* 640–645.

ZIMMERMAN, G., BEN-GARA, I., WEISSMANN, S., and YANAI, S. 1969. Storage under controlled conditions of dry staple foods (defatted milk powder, wheat, soybeans, and defatted soybean meal) and its influence on their protein nutritive value. Deut. Ges. Chem. Apparatewesen. Monograph *63,* 347–380.

ZUCKER, T. F., and ZUCKER, L. 1943. Nutritive value of cotton, peanut and soy seeds. Ind. Eng. Chem., Ind. Ed. *35,* 868–872.

CHAPTER 8

C. M. Christensen

H. H. Kaufmann

Biological Processes in Stored Soybeans

INTRODUCTION

Maintenance of quality in stored grains and seeds of all kinds is essential if economic losses are to be avoided, and this means maintaining at as low a level as possible the biological processes which can result in damage. These processes result in various chemical and physical changes, the more conspicuous of which involve increased respiration and heating. In practice, those who store or process soybeans are interested in preventing heating and in preventing the discoloration and other damage that accompany heating; respiration as such is of little moment to them. However, in the study of biological processes which occur in moist stored seeds, it usually has been much easier in controlled tests in the laboratory to study respiration or, more specifically, production of carbon dioxide, than to study heat production. For some decades there have been conflicting views concerning the role of processes inherent in the seeds themselves versus processes engendered or carried on chiefly or solely by microflora on and within the seeds. This chapter presents a brief review of soybean storage problems and describes good storage practices. For those who wish a fuller account, the summary by Christensen and Kaufman (1969) should be consulted.

HEATING AND RESPIRATION

Seed Respiration

The first proponents of the theory that respiration of moist grains was due to the seeds themselves were Bailey and Gurjar (1918). Briefly, they stored samples of wheat at different moisture contents in sealed containers, kept these at different temperatures, usually for several days, then removed the air from the containers and determined the amount of carbon dioxide present. They assumed that this carbon dioxide was a product of seed respiration, although they did not test the germinability of the seeds before or after the tests. Among other things, they found that the highest respiratory rate was at 55° C.

One of the authors of this chapter (Dr. Christensen) stored certified seed grade hard red spring wheat seeds at moisture contents of 12, 14, 16, and 18% at 55° C and tested them for germinability each day for 4 days. The results are summarized in Table 8.1.

It seems highly probable that in many of their tests purporting to measure respiration of moist grain Bailey and Gurjar were working with dead seeds, and

278

TABLE 8.1

INFLUENCE OF MOISTURE CONTENT, TEMPERATURE, AND TIME ON
LOSS OF GERMINABILITY OF HARD RED SPRING WHEAT

Days Stored at 55° C	Moisture Content (Wet Wt) (%)	Germination (%)
1	12.0	95
	14.0	90
	16.0	72
	18.0	47 (all severely stunted)
2	12.0	89
	14.0	48
	16.0	0
	18.0	1 (severely stunted)
3	12.0	76
	14.0	0

Source: Unpublished work by C. M. Christensen.

so actually were measuring respiration of microflora. At the moisture contents they used, we now know that the predominant microflora would be species of *Aspergillus,* primarily *A. glaucus, A. candidus,* and *A. flavus.*

Malowan (1921) treated moist cottonseeds with formaldehyde, copper sulphate, and mercuric chloride, then measured respiration of the seeds. He concluded that since he had applied materials thought to be fungicidal or bactericidal to the seeds, respiration by microorganisms was eliminated, and that the rather rapid evolution of carbon dioxide by the moist seeds was due to the seeds themselves. He stated, "The results show that disinfection solutions do not prevent the generation of CO_2 and that bacteria, molds, or yeasts cannot be the cause of it." He was in error on two counts: 1) The compounds he used, in the concentrations in which they were applied, probably were not effective fungicides or bactericides under the conditions of his tests; this had been shown by Darsie *et al.* (1914) almost 10 yr before. Darsie *et al.* also had shown that seeds moist enough to germinate did not raise the temperature more than 3° C even at the most vigorous phase of germination. 2) Cottonseeds have a very dense, thick, and relatively impervious seed coat but, like most seeds, have a small opening through this coat at the point of attachment. Fungi readily grow through this opening into the interior of the seed and grow vigorously on the tissues of the meat; applying fungicides to the exterior of the seeds has no effect upon the fungi growing in the interior—it is approximately equivalent to applying fungicide to the outside of a house in an attempt to prevent fungi from growing on materials within the house. Malowan's totally invalid and erroneous conclusions would scarcely be worth refuting if it were not for the fact that as late as the

1950's they still were quoted by research workers in this field as positive proof that fungi could not be involved in respiration and heating of moist cottonseeds.

Fungus Respiration

Ramstad and Geddes (1943) were among the first to ascribe to fungi a role in the respiration of moist seeds, an opinion based partly on their finding that dead soybeans or broken soybeans respired more rapidly than sound soybeans, and that such dead soybeans were more rapidly invaded by fungi than sound, live ones. Milner and Geddes (1945, 1946) studied the respiration and heating of soybeans in the laboratory, using constant aeration and rather precisely controlled moisture content and temperature. They concluded that fungi, and especially *Aspergillus glaucus* and *A. flavus,* were responsible for much of the carbon dioxide production, heating, and associated chemical changes in the samples they studied. They also found that "sterilization" of soybeans by shaking them in a 1:1000 solution of mercuric chloride did not inhibit development of fungi when the beans subsequently were stored at a moisture content of 19.7%, in equilibrium with a relative humidity of 87%. *Aspergillus glaucus* and *A. flavus* developed just as luxuriantly on such "sterilized" beans as they did on soybeans that were inoculated with these same species of fungi, and resulted in just as much respiration and heating. They correctly ascribed heating of soybeans up to a temperature of 55° C to the action of fungi, principally to a species of *Aspergillus.*

Carter (1950) concluded that respiration and heating of moist wheat was due solely to the fungi growing on and within the grain, and that the physiological processes of the seeds themselves were not involved. This was supported by Hummel *et al.* (1954) who stored wheat known by actual test to be free of storage fungi, at moisture contents of 15-31%, at 35° C, with constant aeration, and measured the respiration over a period of 19 days. They found that the respiration of wheat free of storage fungi was low and constant with time—in wheat with 18% moisture and free of storage fungi the respiration was so low as to be almost unmeasurable. Since most seeds require a moisture content of at least 25-30%, wet weight basis, to germinate, it would be surprising if they would respire appreciably at moisture contents below this level, since it would mean that their respiratory systems would be active, and vital resources would be consumed in an environment where the seed could not develop—characteristics that would not make for evolutionary success or survival.

EXAMPLES OF SOYBEAN STORAGE LOSSES

Two cases of losses in storage will be described to illustrate the types of spoilage with which those in the soybean industries are chiefly concerned.

Case One

This was described by Ramstad and Geddes (1943). A bin had been filled in late November in Minneapolis with 12 carload lots of beans of grades 2, 3, and 4 with moisture contents ranging from 14.0 to 18.0% as determined by the Tag-Heppenstal moisture meter. Presumably only 1 representative sample from each carload lot was tested, and so the range in moisture content in the soybeans from any 1 carload lot was not known; it could have been and probably was considerable. The average moisture content of all the soybeans was 15.5%.

Within 37 days after the bin was loaded some rise in temperature occurred in various places, and after 62 days the temperature at a depth of 40 ft was 37° C (99° F) and at 45 ft 38° C (101° F). Because of a pronounced sour odor emanating from the beans it was thought that spoilage might be under way, and the bin was emptied. About 40,000 lb of beans were found to be damaged; some were severely damaged. One sample of the damaged beans had a moisture content of 20.3%, and another sample of severely damaged beans had a moisture content of 28.0%, approximately 13.0% above the average shown on the warehouse records. When 2 samples were tested by the Tag-Heppenstal meter and by oven drying, the meter gave moisture contents of 1.8% and 2.0% below those determined by oven drying. Obviously in this case, the elevator operator had no more than a general idea of the moisture content of the soybeans under his charge, and had been given little or no instruction or aid from management that would enable him to obtain more precise information on which preservation of the quality of the soybeans depended.

Case Two

This bin, in southeastern Kansas, was loaded in the fall of 1962 with truckload lots of soybeans brought in from the surrounding area. Supposedly none of the beans had a moisture content above 13.2%, which means that none of the representative samples from the truckload lots had a moisture content higher than this. Samples had not been withdrawn from this bin throughout the storage life of the soybeans in question to determine whether the moisture had transferred, or whether deterioration, or the early stages of deterioration, might be under way.

In the spring of 1963, the temperature began to rise slowly in portions of the bulk some distance below the surface, as indicated by the temperature detection system in the bin. The slow rise continued through the summer, without causing concern to the elevator manager or to his administrative superiors; and no attempt was made to find out the exact location and extent of the hot spot or spots, or to find out whether even higher temperatures than those recorded might prevail somewhere in the bulk. Almost certainly some of the soybeans at that time were in the final stages of spoilage, and the slow and continued rise in temperature should have been taken as a danger signal, as evidence of a possible emergency.

In late September of 1963, the temperature in portions of the bulk rose to over 93° C (200° F), and a state of emergency was recognized. A hole was broken in the wall near the bottom of the bin to admit air, and within hours the temperature rose to above 200° C (400° F). More than 90% of the soybeans in the bin were converted into a solid black, stinking mass that eventually, when it had cooled, had to be broken up by force to be removed from the bin. Samples taken from these spoiled soybeans are illustrated in Fig. 8.1.

Analysis of Losses

These two losses are representative of many similar ones encountered, not only in stored soybeans but in stored wheat, corn, rice, and sorghum. Nothing mysterious was involved: in portions of the mass the beans were of high enough moisture when loaded into the bins, or later acquired a high enough moisture through moisture transfer, so that storage fungi could grow. Perhaps growth at first was slow, at least in the bin in Kansas, where the temperature rose very gradually over a period of nearly six months; but finally growth was very rapid. Once these fungi begin to grow in a mass of moist grain, they are likely to create a self-perpetuating process, producing heat and moisture to stimulate their growth still more. These fungi can carry the temperature up to 55° C (131° F), which is the highest temperature that can be endured by *Aspergillus candidus* and *A. flavus,* the two species almost inevitably associated with and causing heating of stored grains and seeds. After these fungi have grown for some time they may increase the moisture enough so that thermophilic bacteria can grow. These bacteria can raise the temperature to about 70–75° C (158°–167° F), after which, if conditions are right, spontaneous heating may take over and raise the temperature to the point of combustion.

Soybeans are not difficult to store even though soybeans have the reputation

FIG. 8.1. SOYBEANS MATTED TOGETHER AND CAKED
(LEFT) AND BIN BURNED (RIGHT); THESE
ARE THE FINAL STAGES OF SPOILAGE CAUSED
BY STORAGE FUNGI

among grain men of being difficult to store; but if they are kept at a moisture content too low for fungi to grow, they are no more difficult to store than are other grains and seeds. This refers, of course, to the moisture content of the soybeans in the bin, not the moisture content as shown on the warehouse records. The combinations of moisture content, temperature, and time necessary for maintenance of quality in stored soybeans will be discussed below.

NATURE AND CAUSE OF STORAGE LOSSES

Nature of Losses

Stored soybeans are subject to the same kinds of loss in quality as other kinds of seeds: reduction in germinability, development of mustiness, discoloration, heating, and deleterious biochemical changes. In soybeans, however, deterioration during storage frequently is an "all or none" affair. This is due, in part, to the structure and composition of the seed (see Chap. 2).

Seed Structure Related to Losses

The seed consists of two hemispherical cotyledons that enclose an embryo, the whole surrounded by a seedcoat of several layers, the outer layer of which is made up of thick-walled palisade cells that are resistant to penetration or digestion by fungi. If the seedcoat is unbroken, storage fungi enter soybeans only through the hilum or through the micropyle. By the time soybeans have arrived at terminal storage most of them probably have sustained sufficient mechanical injury in the way of scratches and bruises so that they can be invaded readily by storage fungi. Even so, the seedcoat itself is not a very suitable substrate for the growth of storage fungi, and soybeans can be subjected to poor storage conditions for some time and still be bright and apparently sound. This is illustrated in Fig. 8.2.

The cotyledons, which constitute the major portion of soybean seeds, do not support a luxuriant growth of mycelium of storage fungi, or heavy sporulation. To the unaided eye the early or intermediate stages of invasion of soybeans by storage fungi are scarcely detectable.

When soybeans are stored in small quantities in the laboratory for the purpose of studying the relation of moisture content, temperature, and time to invasion of the beans by fungi, one can distinguish different stages of discoloration. In bulk storage, however, if the conditions are such that fungus invasion develops to the point of incipient discoloration, it is likely to progress within a very short time to the final stages of heating, obvious discoloration, and decay. This explains the "all or none" condition mentioned above.

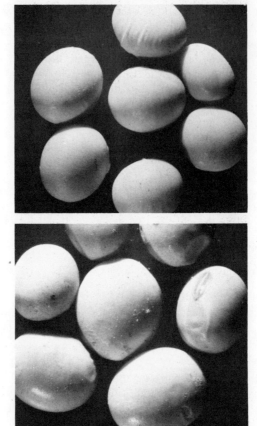

FIG. 8.2. SOYBEANS STORED FOR THREE MONTHS AT 22°–23° C WITH MOISTURE CONTENTS OF APPROXIMATELY 11% (TOP) AND 16% (BOTTOM)

Those stored with 11% moisture were free of storage fungi; whereas those stored at 16% moisture were moderately to heavily invaded by storage fungi and partly deteriorated, although this invasion was not visibiy evident and was not accompanied by any discoloration.

Damage and Heat Damage

The Official Grain Standards of the United States (USDA 1964) specify the following amounts of total and of heat-damaged soybeans permitted in each grade: Grade 1, 2.0% total and 0.2% heat-damaged; Grade 2, 3.0% total and 0.5% heat-damaged; Grade 3, 5% total and 1.0% heat-damaged; Grade 4, 8.0% total and 3.0% heat-damaged. The various kinds of damage are defined as follows:

Damaged Kernels.—Damaged kernels shall be soybeans and pieces of soybeans which are heat-damaged, sprouted, frosted, badly ground-damaged, badly weather-damaged, moldy, diseased, or otherwise materially damaged.

Heat-damaged Kernels.—Heat-damaged kernels shall be soybeans and pieces of soybeans which are materially discolored and damaged by heat.

Of the various types of damage defined in these Grain Standards, only the "moldy" and "heat-damaged" would occur in storage; "moldy" can also include discoloration or decay caused by field fungi before harvest. In practice, the chief type of damage encountered in stored soybeans is the dark brown to black discoloration known as "heat damage," illustrated in Fig. 8.1. It is a product of decay and heating caused by storage fungi.

Mycotoxins

Some fungi that invade seeds either before or during storage produce compounds toxic to animals. As a group these compounds are known as mycotoxins. Naturally, the presence of such compounds in food or feed is undesirable. The principal mycotoxin now known is aflatoxin (there are several aflatoxins, differing in chemical structure and in toxicity) produced by *Aspergillus flavus*. Howell (1968) tested 4121 samples of soybeans that were collected in the fields from 6 weeks before harvest until 6 weeks after the normal harvest time at maturity, and found no aflatoxin in any of them. He stated, "Aflatoxins of soybeans do not appear to be a serious problem." However, *A. flavus,* like other storage fungi, seldom invades seeds on plants in the field (Christensen and Kaufmann 1965) and so one would hardly expect to find aflatoxin in beans before harvest. Irving (1968) reported that of 866 samples taken from soybeans in commercial channels, only 7 contained detectable amounts of aflatoxin, and all of these had only very small amounts. Hesseltine *et al.* (1966) reported that soybeans were a poor substrate for the production of aflatoxin, even though they supported the growth of *A. flavus.* The evidence to date indicates that aflatoxins are not likely to be a common problem in soybeans or soybean products. If aflatoxins were present in the seeds, they would not be likely to be present in the oil extracted from the seeds, but would remain in the meal. Soybean meal ordinarily constitutes a relatively small proportion of the feed to which it is added; the possibility of soybean meal contributing significant amounts of aflatoxin to the feed would appear to be very remote.

Germinability

In soybeans kept for planting, invasion by storage fungi can reduce the percentage and vigor of germination, but usually soybeans of planting grade or certified seed grade are stored at too low a moisture content to permit damaging invasion by storage fungi. Germinability is not a grading factor in soybeans marketed for processing. Malformation of seedlings resulting from invasion of stored soybeans by storage fungi is indistinguishable from those caused by mechanical injury during threshing and subsequent handling.

Fungi and Storage

As with other kinds of seeds in storage, damage to soybeans, including heating, is caused primarily by storage fungi. These fungi do not invade the seeds before harvest; but inoculum of these fungi is ever present and, if storage conditions are favorable to the growth of the fungi, the soybeans will be invaded. The emphasis, therefore, must be, not on the avoidance of inoculum, but on the maintenance of conditions in storage that will not permit the storage fungi to develop. Primarily, this means maintenance of low moisture content and moderately low temperature. This is covered in more detail below.

STORAGE CONDITIONS AND QUALITY

The major conditions that lead to invasion of seed by storage fungi and subsequent loss of quality of soybeans in storage are moisture content, temperature, length of storage time, the degree to which the beans have been invaded before they arrive at a given storage site, and the amount of foreign material and debris present. Each of these will be discussed separately.

Moisture Limits

Storage fungi can slowly invade soybeans stored with moisture contents of 12.0–12.5% (Christensen and Dorworth 1966). The rate of invasion increases as the moisture content increases above this level. Invasion of soybeans with actual moisture contents of 12.5–13.0% is not likely to result in any loss of processing quality within a year even if the temperature is favorable for the growth of fungi, although it may cause some loss of germinability. Such invasion can slowly increase the moisture content of the beans being invaded; and may also slightly increase the temperature in pockets here or there. This may accentuate the normal differences in temperature that are likely to prevail if the bins are not forcibly aerated, and may lead to shifts in moisture that result in sudden increase in the growth of storage fungi and to damage from heating. Thus, the slow invasion of soybeans by storage fungi at moisture contents up to 13.0% is not in itself damaging, but it can be dangerous in that it may result in a sudden unexpected and perhaps uncontrollable increase in fungus growth and in heating.

Moisture Transfer

Soybeans will retain their original quality for years in bulk storage if the moisture content does not exceed 12.0% anywhere in the bulk, but those stored with a safe moisture may not retain that moisture uniformly throughout the bulk. With differences in temperature between different portions of a bulk, air currents are set up which transfer moisture from one portion of the bulk to another. This transfer may be slight and slow, or large and rapid, depending on the moisture content and on the temperature differential; the higher the moisture content, and the greater the temperature differential between different parts of a bulk, the more rapid the transfer. Johnson (1957) calculated that in corn stored with 14.5% moisture, in a bin 120 ft in diameter and 30 ft high, with a temperature differential of 22° C (40° F) between the interior of the bulk and the surface, the moisture content of a layer 6 in. deep on the surface could be increased to 20% in 21 days by transfer of moisture by air currents. Holman (1950) reported shifts in moisture content in soybeans stored in Illinois. Metal bins were filled in November, 1942, with soybeans of 12-13% moisture; in February, 1943, beans in the center of the surface of the bin had moisture contents of 16-17%, and in February, 1944, the moisture content of the beans in the center of the upper surface ranged from 20 to 24%. Christensen (1970) stored sorghum seeds with a moisture content of 14.3% in gallon containers in the laboratory, and maintained a temperature 12-14° C higher on one side than on the other. Within 3 days after the test was begun samples withdrawn from the cool side had a moisture content of 1.4% above that of the samples from the warm side, and after 6 days the difference in moisture content was 2.0%. After several weeks, the moisture content of samples from the cool side was over 19.0%, 7% higher than the moisture content of the samples from the warm side. Such shifts of moisture are likely to be larger and more rapid in large bulks of grain than in small bulks, partly because moisture may be removed from a relatively large portion of higher temperature and relocated in a relatively small portion of lower temperature. One of the major functions of aeration of bulk stored grains is to maintain a uniform temperature throughout the bulk and greatly reduce or entirely eliminate this transfer of moisture.

Temperature

The growth of storage fungi, and the damage they do, are greatly reduced by low temperature. To some extent, low temperature can be substituted for low moisture content in prolonging the storage life of seeds. Dorworth and Christensen (1968) stored soybeans at different temperatures and moisture contents and tested them at various intervals. Soybeans with 18.3% moisture retained a germinability of 100% for 12 weeks at 15° C, whereas at 20° C after 12 weeks the germinability was 90%; at 25° C, 10%, and at 30° C, zero. Their results are summarized in Fig. 8.3.

Time

As is evident from the foregoing, deterioration of stored soybeans is a function of moisture content, temperature, and time. These factors operate together and must be considered together. For long-time storage, the desirable combination is low moisture content and low temperature. We have samples of soybeans of the 1964 crop taken from bulks being loaded into ships for export in early 1965, that have been stored with a moisture content of 13-14% and at 4°-5° C for 6 yr with no deterioration whatever.

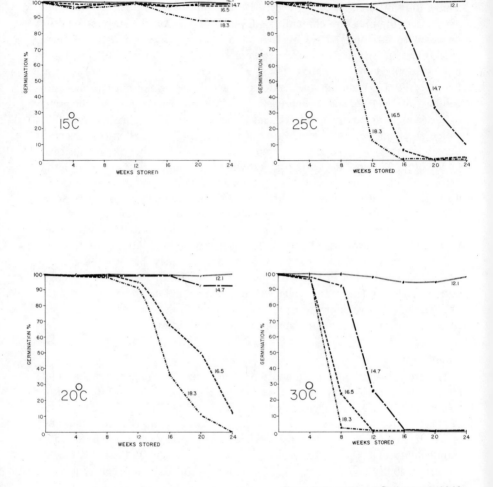

From Dorworth and Christensen (1968)

FIG. 8.3. INFLUENCE OF TEMPERATURE, MOISTURE CONTENT, AND TIME UPON LOSS OF GERMINABILITY OF SOYBEANS

In these tests, loss of germinability was preceded by and presumably due principally to invasion by storage fungi.

Condition of Seed on Storing

A lot of soybeans already lightly or moderately invaded by storage fungi is partly deteriorated, whether or not this is apparent to the naked eye, and is a poorer risk for continued storage than sound beans. Such lots, if placed under conditions that permit the fungi to develop further, will progress toward advanced spoilage more rapidly than perfectly sound beans. Also, once the seeds have been moderately invaded by storage fungi, the fungi may continue to grow and to cause damage at slightly lower moisture contents and temperatures than they would in perfectly sound beans.

Foreign Material

Pieces of pods, stems and other plant materials, including broken soybeans, are more readily attacked or invaded by storage fungi than are sound beans. Furthermore, the fine material collects in the "spout lines" when bins are loaded. In the transfer of soybeans with 2% foreign material (by definition, material that will pass through an 8/64 round-hole sieve) there may be more than 50% foreign material in the spout line. This portion may be so tightly packed that if the bin is aerated the air will flow around it; this is one reason why spoilage frequently starts in the spout line.

MAINTENANCE OF QUALITY

As emphasized by Christensen and Kaufmann (1969), the revolution that has come about in grain storage practices since about 1950 has been based, in part, upon a better understanding of the biological processes which can occur in stored grain which results in deterioration, and how these changes may be prevented, minimized, or circumvented. The major steps or ingredients of a program designed to avoid losses in storage are as follows.

Moisture Testing Problems

The moisture content limits for different grades of different kinds of grains as given in the Official Grain Standards of the United States (USDA 1964) were established to promote orderly marketing. Grain men have assumed that the moisture content of a representative sample taken from a given truckload or carload lot is the moisture content of all the grain in the lot, and that the moisture content can be measured rather accurately and precisely. A single sample taken from a given bulk tells us nothing about the *range* in moisture content within that bulk, and knowledge of the range in moisture content of grain going into a given bin is essential in evaluating storability. We commonly find a variation of 0.5-1.0% in moisture content of grain from a single truck-load, and the variation may be greater than that in a single carload or barge load. The range of moisture content between safe storage and high risk storage for soybeans is between 13 and 15% and in this range a difference of 0.5% can make a great difference in growth of storage fungi and in the eventual damage that may result. It is a good precaution to test as many samples as possible of the grain going into a given bin, each sample taken and tested separately.

Periodic Testing

Once the seeds have been binned it is desirable to remove samples periodically and to test them for moisture content. The samples also can be tested for number and kinds of fungi since an increase in fungi between successive test periods indicates impending deterioration (whether or not this is visible to naked eye inspection) and decrease in germinability. Decrease in germination percentage between successive test periods indicates incipient deterioration. Samples taken from different and known portions of the bin by means of a vacuum probe are preferable for storage evaluation to samples taken from the unloading spout or from the transfer belt, since these latter locations are mixtures of seeds from different and unknown portions of the bin. Samples taken at the same time from the belt adjacent to the unloading spout and by vacuum probe from known places in the bin have been found to have a difference of as much as 2–3% in moisture content. In one case, samples taken from the belt had a moisture content below that where damage could occur, whereas those taken later by probe had a moisture content high enough to permit damage to occur within a short time. The cause of heating in one portion of the bin puzzled the elevator superintendent when the evidence available to him (the average moisture content of the grain as it went into the bin, and the average moisture content of samples taken from the belt) indicated that spoilage from storage fungi was not to be expected. Proper sampling and testing are relatively inexpensive, and enable the elevator superintendent to know at all times the condition of the grains and seeds under his control. It makes the difference between assuming and knowing.

Accuracy of Moisture Meters,

Commonly, it is assumed that if readings on moisture meters can be taken to the second place to the right of the decimal point, the meters are accurate to that figure. Table 8.2 shows the range in moisture content as taken on the same samples of hard red winter wheat with the use of 4 meters, compared with the

TABLE 8.2

MOISTURE CONTENTS AS DETERMINED BY METERS AND
BY OVEN DRYING OF HARD RED WINTER
WHEAT, 1960 CROP, SAMPLES FROM ST. LOUIS,
MOISTURE CONTENT WET WEIGHT BASIS

Sample	Meter				Oven Drying
	Motomco	Radson	Steinlite	Weston	
1	14.20	14.20	13.31	13.51	14.2
2	13.70	13.40	13.85	12.70	14.3
3	14.30	14.00	13.44	12.76	14.5
4	13.30	14.60	16.35	13.97	14.5
5	14.77	14.50	15.00	13.52	15.1

Source: Christensen and Kaufmann (1969).

figure obtained by oven drying. Note that on a given sample, usually different for each meter, the moisture content as determined by oven drying was from 0.9 to more than 1% higher than the moisture content determined by the meter. All of these meters were new and in good working order, and the work was done in the laboratory. In practice, deviations greater than reported here are not uncommon. Grain men should recognize the fact that it is not possible to know the range in moisture content of a bulk of grain as precisely as once was thought. If the moisture content of stored grain is in the range where spoilage might be expected, again, it is only common sense to be suspicious and to take several samples for moisture determination. It is well, also, to be suspicious of the accuracy of the moisture meter and to check it, at least occasionally, by oven drying a number of samples.

Temperature Detection Systems

Temperature recording systems have been an integral part of good grain storage for so long they do not need emphasis here, but it is important to recognize their limitations. Some deterioration may occur before there is any detectable rise in temperature, and an appreciable rise in temperature is proof that spoilage is in progress somewhere in the bulk. Temperature detection cables usually are spaced 20-25 ft apart, and the thermocouples on each cable are spaced about 6 ft apart, so there are 2000-3500 cu ft of grain for each thermocouple. If heating develops in a volume of 20 bu equidistant from the nearest thermocouples this may be evident only by a slight rise in temperature at the thermocouples above and lateral to the hot spot. If this is disregarded as unimportant, a considerable volume of grain may be in the final stages of spoilage before trouble is recognized. In malting barley, in which high germinability is essential, any detectable rise in temperature is regarded as an indicator of trouble.

Fungi such as *Aspergillus restrictus* and *A. glaucus* growing in corn, wheat, or rice with a moisture content of 14.0-14.5%, or in soybeans with a moisture content of 12.5-13.0%, will not respire rapidly enough to cause any detectable rise in temperature. In the cereal grains, these fungi can cause severe germ damage, and in soybeans they may cause mustiness and some discoloration without any rise in temperature. They may also slowly increase the moisture content of the portion of grain in which they are growing until it reaches a level where *A. candidus* and *A. flavus* can grow rapidly and cause spoilage within a few days.

Aeration

Aeration serves two purposes: (1) To reduce the temperature of the bulk. A temperature of $4°-10°$ C $(40°-50°$ F) is usually preferred. It is low enough to inhibit the development of all storage insects and to greatly reduce the growth of fungi (Fig. 8.3), but not so low that when the grain is moved out in warm weather there is likely to be condensation. Naturally, reduction of the grain temperature to $4°-10°$ C ($40°-50°$ F) by natural air depends upon the outside air being at this temperature during the cooling period. (2) To maintain a uniform temperature throughout the grain mass, which greatly reduces or entirely eliminates transfer of moisture and so contributes to a greatly length-ened storage life of the grain. The aeration system chosen will depend on a number of factors, but some kind of effective aeration system is essential for long time safe storage of large bulks of grains and seeds. In regions of high storage risk, where the temperature and relative humidity both are high for long periods of time, aeration systems may be essential for maintenance of quality for any storage period longer than a few weeks.

Concerning cooling by aeration, Christensen and Kaufmann (1969) stated: "Aeration does not lower the temperature of all the grain throughout the bin gradually and uniformly. A rather sharply delimited cold front is developed where the air enters the grain. This front moves through the grain in the direction of the air flow. Behind this cold front, all the grain has the same temperature as the air moving through it; ahead of the cold front, the grain remains at the temperature it had when it was put into the bin. With wheat aerated at the rate of 0.1 cfm per bu of grain, about 120 hr are required for the front to move through, or about 5 days of continuous aeration (actually it may be 90-120 hr, depending on the difference in temperature between the outside air and the grain). The size of the grain mass does not matter, since the base of calculation here is cubic feet of air per bushel of grain per unit of time; a mass of 1 bu would require 0.1 cu ft of air per minute, and a mass of 100,000 bu would require 10,000 cu ft of air per minute. The last grain to cool is that at the exhaust side of the bin, and if damage is to be avoided this grain must be cooled before spoilage begins. This must be kept in mind when estimating the likely success of aeration in preventing spoilage."

Removal of Foreign Material

The accumulation of foreign material and fines in the spout line when bins are filled, as explained above, may result in trouble. One way to reduce this hazard is to remove as much of this material as possible. In a conventional tall and relatively narrow bin the spout line in which the foreign material has accumu-lated forms a relatively narrow cylinder. After the bin has been filled the center unloading port at the bottom of the bin can be opened; the first grain to be discharged is that in a column above the port, which will include most of the foreign material. The material so removed can be screened, the whole seeds returned to the bin and the foreign material stored separately or processed at once. With flat storage this procedure will not remove nearly as much of the foreign material.

BIBLIOGRAPHY

BAILEY, C. H., and GURJAR, A. J. 1918. Respiration of stored wheat. J. Agr. Res. *12*, 685–713.

CARTER, E. P. 1950. Role of fungi in the heating of moist wheat. USDA Circ. *838.*

CHRISTENSEN, C. M. 1970. Moisture content, moisture transfer, and invasion of stored sorghum seeds by fungi. Phytopathology *60*, 280–283.

CHRISTENSEN, C. M., and DORWORTH, C. E. 1966. Influence of moisture content, temperature, and time on invasion of soybeans by storage fungi. Phytopathology *56*, 412–418.

CHRISTENSEN, C. M., and KAUFMANN, H. H. 1965. Deterioration of stored grains by fungi. Ann. Rev. Phytopathology *3*, 69–84.

CHRISTENSEN, C. M., and KAUFMANN, H. H. 1969. Grain Storage: The Role of Fungi in Quality Loss. Univ. Minn. Press, Minneapolis.

DARSIE, M. L., ELLIOTT, C., and PEIRCE, G. J. 1914. A study of the germinating power of seeds. Botan. Gaz. *58*, 101–136.

DORWORTH, C. E., and CHRISTENSEN, C. M. 1968. Influence of moisture content, temperature, and storage time upon changes in fungus flora, germinability, and fat acidity values of soybeans. Phytopathology *58*, 1457–1459.

HESSELTINE, C. W., SHOTWELL, O. L., ELLIS, J. J., and STUBBLEFIELD, R. D. 1966. Aflatoxin formation by *Aspergillus flavus.* Bacteriol. Rev. *30*, 795–805.

HOLMAN, L. H. 1950. Handling and Storage of Soybeans. *In* Soybeans and Soybean Products, Vol. 1, K. S. Markley (Editor). John Wiley & Sons, New York.

HOWELL, R. W. 1968. The effects of crop production factors on the occurrence of *Aspergillus flavus* in soybeans. Proc. 1967 Mycotoxin Research Seminar. June 8–9. USDA, Washington, D. C.

HUMMEL, B. C. W., CUENDET, L. S., CHRISTENSEN, C. M., and GEDDES, W. F. 1954. Grain Storage Studies. XIII. Comparative changes in respiration, viability, and chemical composition of mold-free and mold-contaminated wheat upon storage. Cereal Chem. *31*, 143–150.

IRVING, G. W., JR. 1968. Perspectives on the mycotoxin problem in the United States. *In* Toxic Microorganisms: Mycotoxins–Botulism, M. Herzberg (Editor). U.S.-Japan Cooperative Program in Natural Resources Joint Panels on Toxic Microorganisms and the U.S. U.S. Dept. Interior. Govt. Printing Office, Washington, D. C.

JOHNSON, H. E. 1957. Cooling stored grain by aeration. Agr. Eng. *38*, 244–246.

MALOWAN, J. 1921. Some observations on the heating of cottonseed. Cotton Oil Press *4*, 47–49.

MILNER, M., and GEDDES, W. F. 1945. Grain Storage Studies. II. The effect of aeration, temperature, and time on the respiration of soybeans containing excessive moisture. Cereal Chem. *22*, 484–501.

MILNER, M. and GEDDES, W. F. 1946. Grain Storage Studies. III. The relation between moisture content, mold growth, and respiration of soybeans. Cereal Chem. *23*, 225–247.

RAMSTAD, P. E., and GEDDES, W. F. 1943. The respiration and storage behavior of soybeans. Minn. Agr. Expt. Sta. Tech. Bull. *156.*

USDA. 1964. Agr. Marketing Serv. Grain USDA. Official Grain Standards of the United States. Revised May 1964.

CHAPTER 9

S. J. Circle
A. K. Smith

Processing Soy Flours, Protein Concentrates, and Protein Isolates

INTRODUCTION

While the most popular and important sources of food protein in the United States and many other countries are meat, poultry and egg, dairy, and fishing industries, the steadily increasing costs of these animal proteins has compelled the food industry to focus attention on the low cost vegetable sources of protein, principally the oilseeds.

At present, the least expensive source in the United States of high quality food grade protein is defatted soy flour containing 50% protein which may be purchased at 6 1/2–7 ¢ per lb or at about 13–14 ¢ per lb of protein. This may be compared with beef containing about 20% protein at 80 c per lb or $4.00 per lb of protein. Dairy, fish, and poultry proteins can be obtained at a somewhat lower cost but still are quite expensive compared with vegetable sources of protein. Defatted soy flour (or flakes) is the starting material for both soy protein concentrates (containing 70% protein and selling at 18–28 ¢ per lb) and soy protein isolates (containing 90% protein and selling at 35–43 ¢ per lb). U.S. manufacturers of soy protein products are listed in the Appendix along with typical product specifications. The recent rise in the prices of nonfat dry milk and food grade sodium caseinate has given impetus to increasing interest in the various soy protein products available. The trend to extend or replace dairy food ingredients with analogous soy protein products will depend on how successful the soy processing industry will be in research to improve the flavor of their products.

Emphasis in this chapter will be given to work appearing in the literature of the past two decades, inasmuch as comprehensive reviews are available prior to this period (Circle 1950; Burnett 1951A,B). At the outset, let it be stated that almost all of the major processors of soybeans are geared in their oil extraction plants for the manufacture of soybean oil and meal, the latter primarily for animal feeds (Cravens and Sipos 1958; Burnett 1970). The need for efficient operation has impelled an increase in the capacity of these oil mills, so that modern plants may exceed 2000 tons of soybeans throughput daily through a single extractor (Becker 1968, 1971). The total annual U.S. processing capacity was estimated to be about 750 million bushels in 1967 and 1968, rising to 800 million in 1969, to 825 million in 1970, and projected to be 900 million bushels in 1971, and close to 100% was processed by solvent extraction (Kromer 1970); this capacity was utilized to the extent of 95% in 1969–1970. The actual crush in 1967-1968 was about 600 million bushels or close to 18 million tons. Eley (1968) estimated that food uses of soybean protein products in 1967 were 205-210 million pounds of soy flour and grits, 17-30 million pounds of soy protein concentrate, and 22-35 million pounds of soy protein isolate. Taking the larger figures in each case and applying conservative yield factors in processing, the sum total of soybeans representing the source of these products may be calculated to be about 0.27 million tons. This indicates that only approximately 1.5% of the processed soybeans in the United States were converted into edible protein products in 1967. The significance of this is that very few of the U.S. soybean processing plants can afford to design their operations specifically to produce edible protein products. The economics of the soybean crushing industry dictate large volume throughput for oil and feed meal production at a time when edible vegetable protein products bypassing the animal are just past the toddler age. The outcome is that the oil mill design usually is deficient in sophisticated sanitary features that would be desirably incorporated if the defatting operation were exclusively for food purposes. This is especially true of soybean seed storage and seed cleaning steps preceding bean preparation for the oil extraction step.

Soybeans in commerce selected for food operations normally conform to grades 1 or 2 which permit up to 3% damaged seed and 2% foreign material. Although these are more carefully cleaned than beans intended for feed meal, the same equipment may be used for both food and feed purposes. In U.S. practice, the seed cleaning is usually accomplished in relatively simple, dry mechanical screening devices, supplemented with air currents (Becker 1971), to give a stream of oversize beans contaminated with oversize foreign material, a stream of average size beans, and a stream of undersize beans contaminated with weed seeds and other foreign material. For making edible type products the average size beans are provided in a state as free of foreign matter as practicable. The seed coat may retain traces of soil not removed in the dry cleaning step. British practice in some plants is to wash the soybeans (Arkady) but so far as the authors are aware, this is not done in the United States. Quality of the edible soy protein products presumably would be improved if more stringent attention were given to seed cleaning and washing equipment. In addition to the seed, all raw materials entering the processing streams can affect quality of the finished products. Special attention must be given to provide that only potable water and culinary steam, as well as food grade chemicals, should contact the soy material in the processing system.

The rationale underlying processing of the soybean is fractionation, initially by mechanical separation of the seed parts (dry processing), followed by further fractionation by chemical means (wet processing), the latter permitting diffusion of molecular components. From a practical standpoint, soybean seeds may be considered to be composed of three seed parts, hull (including hilum), hypocotyl (germ), and paired cotyledons (Rackis *et al.* 1958). For food product purposes, the seeds should be thoroughly freed of all foreign matter such as soil residues, pods, stems, corn kernels, weed seeds, and other trash; they should then be split or cracked in a manner to permit complete removal of hull, hilum and hypo-cotyl, leaving pure cotyledons (approximately 90% of the seed weight; see Chap. 3) as the source of further fractions. The two important fractions for the food industry are the proteins and lipids residing in the cotyledons. Approximately 70% of the protein exists in the form of protein bodies, primarily ergastic as opposed to bioplasmic protein; the lipids are mainly present in smaller sphero-somes. Rothfus (1970) and others (see Chap. 4) have advocated the isolation of intact protein bodies as a means of fractionation, but practical means of achieving this have not been elucidated.

Traditionally, industrial practice in almost all soybean processing consists of a series of steps including seed cracking, separation of seed parts (frequently not entirely complete), flaking of the predominantly cotyledonous part (the preced-ing steps together are termed bean preparation), hexane extraction of the lipids in the defatting step, desolventizing of the lipid and defatted portions (see Fig. 9.1), followed by further fractionation of the lipid and nonlipid portions as desired. In common practice the hypocotyl portion is not deliberately separated from the cotyledons, although it would be advantageous to do so as there is evidence that the hypocotyl contains more "beany" components (Arai *et al.* 1970). Full-fat soybean products are made from cotyledons without extracting

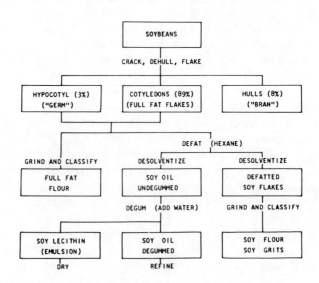

FIG. 9.1. PROCESS FLOW FOR SOY FLAKES,
FLOUR, GRITS, LECITHIN, AND OIL

the lipids, but represent a minor proportion of the commercial edible soybean protein products. Some other products, such as refatted soy flours, are made by recombining seed fractions in a desired manner. Protein isolate and protein concentrate are generally derived from defatted flakes. All the products of commerce are commonly made and sold according to specifications, including bacterial counts (see Appendix).

Practical and Commercial Fractions of Soybeans

Commercial comestible dry protein fractions of soybeans include full-fat soy flours, defatted soy flours, refatted soy flours, soy protein concentrates, soy protein isolates, and full-fat and defatted soy milks [the latter produced in Japan (Watanabe 1969)]. The lipid fractions will not be treated in this volume, but are separated initially into gums (lecithin) and degummed oils which are further refined. As a matter of convenience for discussion, we shall designate practical fractions derived from commercial processing (see Fig. 9.2) as H (hull and hilum), G (germ or hypocotyl), L (lipids), P (protein), I (insoluble residue), and S (solubles or whey solids other than P), with P being the alkali-soluble, acid-precipitable protein and I and S the predominantly carbohydrate portions obtained during isolation of P (see Fig. 9.3 illustrating the process). All commercial soy fractions may be considered to be composed of one or more of the practical fractions in varying amounts, as shown in Fig. 9.4.

It will be understood that the practical fractions will overlap to some extent. That is, in actual practice fractionation is not complete, and since hypocotyls are not usually separated as such, a part of these may be removed with the hulls, and the remainder with the cotyledons. The cotyledons may be contaminated to some small degree with hull and hilum, as well as with hypocotyl. The insoluble residue, due to incomplete removal of protein and solubles, may contain as much as 30–50% of protein. The protein itself may retain part of the solubles, depending on the degree of washing during its isolation. The solubles may vary in composition depending on whether the hexane-defatted flakes were also extracted with other organic solvents. If lipids removal during the defatting step is not thorough, then the protein and insolubles fractions will end up with small

SOYBEANS = PROTEIN + INSOLUBLES + SOLUBLES + LIPIDS + HULLS

H	=	HULLS
L	=	SOLVENT-EXTRACTABLE MATERIAL (HEXANE, ETHANOL)
P	=	ALKALI-EXTRACTABLE, ACID-PRECIPITABLE PROTEIN
I	=	INSOLUBLE RESIDUE FROM PROTEIN EXTRACTION
S	=	SOLUBLES OTHER THAN P (EXAMPLE: "WHEY")

FIG. 9.2. DESIGNATION OF PRACTICAL FRACTIONS OF SOYBEANS

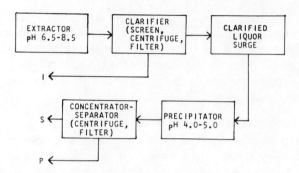

FIG. 9.3. PROCESS FLOW OF OPERATIONS IN
SOY PROTEIN ISOLATE PRODUCTION

AVOID USE OF TERM <u>SOY</u> BY ITSELF, WITHOUT QUALIFICATION

SOYBEANS	= P + I + S + L + H
SOY FULL-FAT FLOUR	= P + I + S + L
SOY DEFATTED FLOUR	= P + I + S
SOY PROTEIN-SOLUBLES	= P + S
SOY PROTEIN CONCENTRATE	= P + I
SOY PROTEIN(ATE) ISOLATE	= P (ISOELECTRIC OR SALT FORMS)
SOY SOLUBLES OR "WHEY"	= S
SOY "WHOLE MILK"	= P + S + L
SOY PROTEIN-LIPID COMPLEX	= P + L
TOFU	= P + L

FIG. 9.4. RELATIONSHIP OF COMMERCIAL SOY FRACTIONS

amounts of the lipids as part of their composition. The components of the lipids may differ, depending on the type of defatting solvent and moisture content of the flakes during the defatting step.

It must also be pointed out that all manufacturers possess a considerable amount of proprietary information not readily accessible to others. Thus, their processing methods and equipment for the same product may differ, giving rise to variations in the product from one manufacturer to another with respect to color, flavor, functional and nutritional characteristics, and other properties in greater or lesser degree.

DEFATTING OF SOYBEANS

Except for full-fat soy flour products, all the soy protein products of commerce are manufactured using defatted soy flakes or flour as the source material. This section covers steps involved in the defatting operation.

Bean Preparation

It has been pointed out in Chap. 8 that sound clean soybeans will retain their original quality for years in bulk storage if the moisture content does not exceed 12.0% anywhere in the bulk. Commonly, beans arriving at the oil mill may be higher than this in moisture. They are analyzed for moisture content (among other analyses), before being passed through a grain drier prior to storage. Design features of a modern soybean drier are discussed by Bunn (1970). According to him, it is important that the heated air is uniformly distributed through the grain and that each bean is exposed to an equal amount of heat and air for the same period of time to avoid overheating with possible adverse effect on solubility of the protein in the bean. Singer (1965) advocated drying to 12% moisture at a temperature below 180° F prior to cracking and dehulling, as higher temperature was detrimental to quality of the extracted oil. Becker (1971) states that in the past 20 yr drying of soybeans before storage has become standard practice in most plants, and that improperly dried soybeans are difficult to dehull and solvent extract. Soybeans are normally cracked into 6 or 8 parts prior to dehulling, then tempered or conditioned so that moisture is adjusted to 10–11% and temperature raised to about 160° F to insure proper plasticity so that acceptable flakes can be formed (Becker 1971). Singer (1965) recommended tempering to 13% moisture for optimal plasticity and minimal friability during the flaking step. Earlier reference to bean preparation for flaking is by Langhurst (1951). It is quite clear that in the bean preparation steps the processor must operate within narrow, carefully controlled temperature and moisture conditions to avoid overheating with resulting damage to protein and oil.

Solvent Extraction

Solvent extraction of oleaginous materials on a large scale has been practiced for more than a hundred years, initially in batch type extractors, but during the past 50 yr or more in extractors operating on the countercurrent principle. Since the early 1930's, continuous countercurrent extractors of many types have been designed (Langhurst 1951; Fincher 1958; Kuiken 1958; Becker 1968).

A great many of these are still in use, but in the last two decades the horizontal configuration has gained in favor over the vertical types, as capacity of single units has continued to increase in size. Discussion of oil extraction in detail is reserved for a companion volume on soybean lipids; but reference is made to two papers for recent developments (Vix and Decossas 1969; Becker 1971). Two types of modern extractors are illustrated in Fig. 9.5 and 9.6.

It is quite important to provide a high quality of solvent as extraction medium: n-hexane free of aromatics is commercially available, and this grade should be specified for food grade operations (Vix and Decossas 1969). Other solvents are discussed in Chap. 5.

Courtesy of Blaw-Knox Co.

FIG. 9.5. HORIZONTAL ROTATING BASKET EXTRACTOR

Desolventizing

It is also quite important to remove the solvent from the flake after defatting, and indeed this is the most critical stage of all bean processing, in which damage to the protein can either be kept minimal, or, conversely, can make the flakes unfit for further processing into protein isolate, as discussed in the following section on dispersibility characteristics of protein. To retain high solubility of protein, control of processing variables is essential with respect to temperature, pressure, presence of moisture and residence time (Becker 1971). For this, two systems among others have been described for modern installations, the flash desolventizer–deodorizer stripper (Becker 1971; Milligan 1969) as illustrated in Fig. 9.7, and the vapor desolventizer–vacuum deodorizer as shown in Fig. 9.8 (Becker 1971). For feed meals requiring thorough cooking, the desolventizer-toaster is used extensively (Becker 1971), but such treatment results in meal unsuitable for protein isolate processing.

Analytical methods for determination of residual solvents have been described (Sikes 1960; Dupuy and Fore 1970).

PROTEIN DISPERSIBILITY

It has been emphasized in the previous section that the conditions encountered in the desolventizing stage can profoundly affect the solubility of the

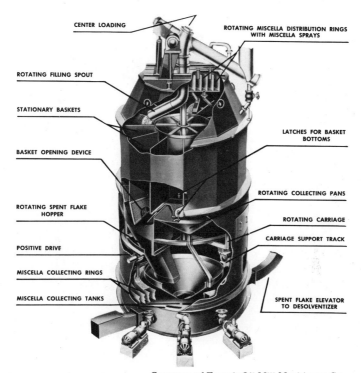

CENTER LOADING

ROTATING MISCELLA DISTRIBUTION RINGS
WITH MISCELLA SPRAYS

ROTATING FILLING SPOUT

STATIONARY BASKETS

LATCHES FOR BASKET
BOTTOMS

BASKET OPENING DEVICE

ROTATING COLLECTING PANS

ROTATING SPENT FLAKE
HOPPER

ROTATING CARRIAGE

CARRIAGE SUPPORT TRACK

POSITIVE DRIVE

MISCELLA COLLECTING RINGS

MISCELLA COLLECTING TANKS

SPENT FLAKE ELEVATOR
TO DESOLVENTIZER

Courtesy of French Oil Mill Machinery Co.

FIG. 9.6. HORIZONTAL   STATIONARY   BASKET   EXTRACTOR

protein in the defatted flakes. The further processing of defatted flakes into flour products, concentrates, isolates and "milks," and the use of these products as food ingredients require background knowledge of the dispersibility or solubility behavior of the protein in water as modified by changes in pH, salts, temperature, organic solvents, and other factors. This chapter is concerned with the application of such knowledge to the processing of these products on a practical scale, whereas Chap. 4 deals with the fundamental aspects of soy protein isolation, fractionation, and properties. Wolf points out in Chap. 4 that most investigations concerned with soy proteins have been based on preparations isolated from defatted flakes with aqueous solvents, and that many of these contain inclusions of low molecular weight nonprotein compounds which may have arisen as artifacts during initial water extraction of the defatted flakes, or were naturally present. Practical methods of removing some of these, such as phytic acid, remain to be worked out, as does the possibility of isolating proteins by segregating protein bodies through some sort of nonaqueous milling. Figure 4.4 of Chap. 4 illustrates the basis of the chemistry for some schemes of aqueous

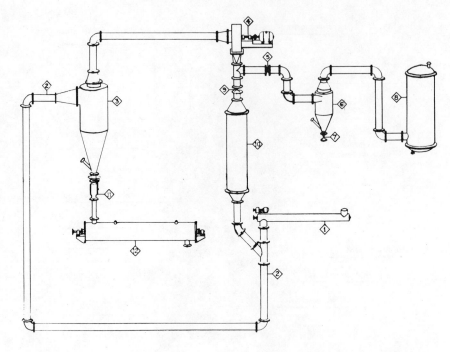

Courtesy of Engineering Management, Inc.

FIG. 9.7. FLASH DESOLVENTIZING SYSTEM

(1) Feed conveyor. (2) Desolventizing tube. (3) Flake separator. (4) Circulating blower. (5) System pressure control valve. (6) Vapor scrubber. (7) Scrubber discharge lock. (8) Solvent condenser. (9) Recirculating control valve. (10) Vapor superheater. (11) Separator discharge lock. (12) Flake stripper.

manipulation to separate protein from the flakes in the form of concentrate or isolate.

Water Dispersibility of Protein

Over the years a number of terms have been used to designate the solubility of the protein of soy flakes or flour in water at its natural pH of 6.5–6.7, such as water soluble protein (WSP), water dispersible protein (WDP), nitrogen solubility index (NSI), protein dispersibility index (PDI), nitrogen dispersed, nitrogen extracted, and others. Also, a number of empirical methods have been proposed as a measure related to the degree of heat treatment (or degree of protein denaturation) sustained by the soy flake products during their manufacture. Studies on water dispersion prior to 1950 have been reviewed (Circle 1950). For fundamental studies a number of investigators have preferred to use a phosphate buffer at pH 7.6 and 0.5 ionic strength (with and without mercaptoethanol) as cited by Wolf in Chap. 4. For commercial purposes, two analytical procedures have been developed and were adopted in 1965 (revised in 1969) as methods of

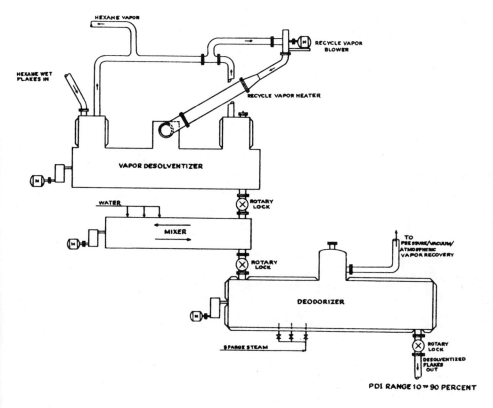

Courtesy of Blaw-Knox Co.

FIG. 9.8. VAPOR DESOLVENTIZER-DEODORIZER SYSTEM

the American Oil Chemists' Society (see Appendix): Ba 10–65, Protein Dispersibility Index (PDI); and Ba 11–65, Nitrogen Dispersibility Index (NSI). These are also known informally as the "fast stir" and "slow stir" methods, the "fast stir" giving higher results. Since the two give different results for the same material, although consistent in themselves, there remains some unresolved controversy as to which one to favor. Nevertheless, they are widely used in industry and in some of the product specifications. In view of the empirical nature of these methods, and of their not being applicable to every need, it may be advantageous for processors to devise their own analytical version for in-plant use, based on operating conditions in their processing.

The effects of solvent processing on protein dispersibility for five different soybean processing plants has been described by Belter and Smith (1952). Results showing changes in solubility of the protein at various stages in the defatting operation are reported in Table 9.1 in terms of NSI values. The results

TABLE 9.1

EFFECT OF SOLVENT EXTRACTION ON PROTEIN DENATURATION
AS MEASURED BY NITROGEN DISPERSIBILITY IN WATER AT PH 6.6

Sample From	Solvent Extraction Plant Number				
	1	2	3	4	5[1]
	Dispersible Nitrogen (%)				
Whole Beans	84.6	81.4	87.9	87.0	82.9
				84.5[2]	
Dehuller	84.9	79.3	87.1	84.1	79.3
Conditioner	84.9	70.7	79.5	82.2	80.7
Flaking roll	84.0	78.6	74.2	80.7	77.9
Extractor	87.9	78.4	....	79.9	81.1
Desolventizer	85.3	....	....	....	80.8
Deodorizer	43.7	46.6	48.0	51.6	65.4
Toaster	7.2	....	14.0	39.7	8.2

Source: Belter and Smith (1952).
[1] No dehulling was practiced at this plant.
[2] After preheating to reduce moisture in beans.

show that the preliminary steps of bean preparation (cracking, dehulling, conditioning, and flaking, followed by solvent extraction of the flakes with the currently-used hydrocarbon solvents) do not measurably decrease the dispersibility of the protein inasmuch as the flakes which were desolventized without heat had essentially the same NSI as the original beans. For comparative purposes Fig. 9.9 shows the decrease in NSI when raw flakes are treated with steam in an autoclave at atmospheric pressure, showing that the NSI decreases from 83 to 20 in a 20-min treatment. Thus, in processing soybean meal for producing protein isolates and many of the specialty food and industrial products, it is customary to desolventize the meal with a minimum or controlled amount of protein denaturation. In commercial operations this step should be effected in the vapor or flash desolventizers, described earlier in this chapter, which removes the solvent with only slight loss in NSI.

Smith et al. (1938) investigated the effects of neutral salts on the dispersibility of the nitrogenous constituents of finely ground soybean meal. They found that monovalent salts in the concentration range of 0.05–0.10 N had a retarding effect on the dispersibility of the protein as shown in Fig. 9.10, and the divalent salts of calcium and magnesium at 0.0175 N concentration had a much greater effect. In further exploring the effects of salts on the dispersibility of the nitrogenous compounds of the meal, they did not find any neutral salt concentration which gave as high dispersibility as water at the natural pH of the system.

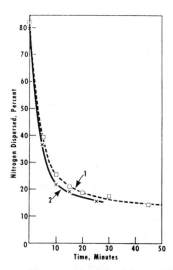

FIG. 9.9. CHANGE IN NITROGEN DISPERSI-
BILITY OF SOYBEAN MEAL IN
WATER WITH INCREASING TIME
OF ATMOSPHERIC STEAM TREAT-
MENT

Curve 1: flakes steamed after solvent extraction of
oil. Curve 2: flakes steamed before solvent extraction.

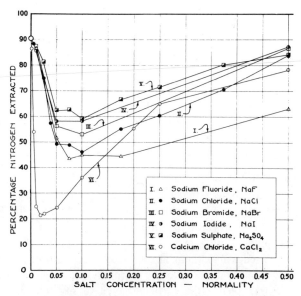

FIG. 9.10. PERCENTAGE NITROGEN DISPERSED
FROM DEFATTED SOY MEAL BY VARI-
OUS SALTS AND BY WATER

In another investigation, Smith *et al.* (1966) reported on the dispersibility of dehulled and defatted flakes derived from 23 varieties of soybeans, (ranging in protein content from 60.3 to 44.5%) and using a double extraction procedure at water to flake ratios of 20:1 and 10:1 at pH 7.2 and 30° C. They measured the NSI, yield of protein isolate, yield of residue and whey nitrogen as well as the nitrogen content of the isolated protein and residue. They found an average NSI of 93.7 with a range of 91–97, and that the NSI value was not influenced by the level of nitrogen in the flakes.

The yield of isolated protein in the above experiments ranged from 39.9–56.0gm per 100gm of flakes whereas the yield of insoluble residue, 21.1gm per 100gm of flakes, did not vary significantly. The nitrogen in the isolated protein ranged from 13.5 to 15.1% and the nitrogen in the residue from 2.07 to 3.01%; the nitrogen in the residue accounted for 5.6–7.5% of the total nitrogen of the flakes. The nitrogen in the whey solution usually accounted for 12–13% of the total nitrogen extracted although the extreme range was 9.0–15.3%. The nitrogen of the isolated protein and whey nitrogen accounted for 97% of the total nitrogen and the sum of the insoluble and soluble nitrogen accounted for 99–101% of the nitrogen of the flakes.

Effect of pH Variation

The effect of pH on the dispersibility of the nitrogenous constituents of ground defatted meal using hydrochloric, trichloroacetic, sulfuric, phosphoric, and oxalic acids and sodium and calcium hydroxides has been reported by Smith and Circle (1938) (see Fig. 4.4 of Chap. 4). The highest nitrogen dispersibility value in the acid range was 85% with HC1 at pH 1.4; this was the same value obtained with water at pH 6.7. In the alkaline range, the highest nitrogen dispersibility using NaOH at a pH of 12.2 was 96%. The minimum dispersibility in the acid range was obtained with oxalic acid at pH 4.2, giving a nitrogen dispersibility of 8%. Smith and Circle (1938) also investigated the effects of adding sodium and calcium chorides to the dispersion media and found that the depressing effect of the salts was more effectively overcome in the alkaline than in the acid range. Thus, in dispersing the protein from defatted meal the adverse effect of salts in hard water can be overcome by the use of caustic.

Effect of Temperature Variation

Beckel *et al.* (1946) showed that the nitrogen extracted by water from air- or vacuum-dried defatted soy flakes exhibits a maximum at about 70°–75° C for a given water to flake ratio; although this effect was relatively small, it was significant. This temperature effect was found to be greater for heat-treated flakes which exhibited a maximum at about 62° C (Circle *et al.* 1959). It is apparent that attention should be given by the processor to finding the optimum temperature of extraction for the particular type of soy source material provided, and conditions employed.

The above laboratory data do not necessarily predict the potential yield of isolated protein and other fractions in commercial processing, but they do indicate that higher yield of isolate may be anticipated from the use of high protein soybeans. Also, these data give an estimation of the amount of nitrogen lost in the whey solution which is a pollution problem unless it is recovered. From these data it is apparent that the extraction of thin flakes (0.008 in. in laboratory study) will yield as much protein isolate as ground meal.

The experiments of Nagel et al. (1938) and Smith et al. (1966) show that the dispersibility of the protein in a mild solvent, such as water, depends largely on the crushing of the cell structure of the seed and the fracture of any membrane-like structure which might prevent contact of the dispersion medium or water with the protein. The nature of the undispersed protein has not been investigated but it is possible that part of this protein has been insolubilized by reaction with the cations normally occurring in the seed.

Effect of Aging

Jones and Gersdorff (1938) investigated the decrease in dispersibility of nitrogenous compounds of high fat and defatted soy meal in 10% salt solutions at 30° and 76° F after storage periods of 1, 3, and 6 months. One set of samples was stored in jars and the other in bags. The results are recorded in Table 9.2.

TABLE 9.2

PERCENTAGE DECREASES IN VALUES FOR TRUE PROTEIN, SOLUBILITY, AND DIGESTIBILITY OF PROTEIN AFTER STORAGE OF SOYBEAN MEAL UNDER DIFFERENT CONDITIONS FOR ONE, THREE, AND SIX MONTHS

Storage Conditions		True Protein, %			Extractability, %			Digestibility, %		
		1 Mo	3 Mo	6 Mo	1 Mo	3 Mo	6 Mo	1 Mo	3 Mo	6 Mo
Stored in Jars										
High-fat meal	30°F.	0.31	2.88	4.35	1.18	3.08	4.53	0.92	5.61	7.26
High-fat meal	76°F.	2.15	4.83	7.16	2.92	5.91	8.71	4.53	9.49	12.05
Low-fat meal	30°F.	3.01	4.21	5.44	1.51	5.04	7.53	3.80	10.41	12.97
Low-fat meal	76°F.	3.95	6.36	8.71	3.82	8.47	10.72	5.57	15.30	17.25
Stored in Bags[1]										
High-fat meal	30°F.	..	..	5.85	..	..	8.92	..	...	10.43
High-fat meal	76°F.	..	..	8.75	..	..	13.11	..	...	14.48
Low-fat meal	30°F.	..	..	6.76	..	..	8.94	..	...	15.13
Low-fat meal	76°F.	..	..	9.97	..	..	12.39	..	...	18.94

Source: Jones and Gersdorff (1938).

[1] Analyses of meals stored in bags were made only after six months' storage.

Jones and Gersdorff used a single salt solution extraction procedure in obtaining the above results; using a double extraction, the loss in dispersibility was substantially reduced.

Smith and Circle (1938) investigated the loss in nitrogen dispersibility of defatted soy meal using water as the dispersing agent, and reported that for a single extraction procedure there was a decrease of about 1.1% per month for the first few months of storage at room temperature. This amount of loss was reduced nearly 50% if 3 successive extractions were made rather than 1. They found that the loss, when using salt solution extractions, was greater than with water, which corroborates the results obtained by Jones and Gersdorff.

Effect of Phytates

Smiley and Smith (1946), in an examination of the nitrogen content of proteins isolated by a variety of procedures, found that they contained substantial quantities of nonprotein components, especially phosphorus compounds. The phosphorus in the isolated proteins ranged from 0.8 to 1.0%, indicating a reaction of phytin, the principal phosphorus component of the soybean, with the protein when precipitated at pH 4.5.

Smith and Rackis (1957) removed the phytin from the water extract of the meal, and perhaps other anions as well, by passing the dispersion over an ion exchange resin (Dowex-1-x10), which reduced the phosphorus content of the precipitated protein to 0.2%. After the phytin was removed the pH of maximum precipitation of the protein was changed from 4.2 to 5.0 and the phytin-free protein was completely dispersible at pH 4.0 as shown in Fig. 9.11.

McKinney et al. (1949) also investigated methods which might be used for the removal of the phytin from the protein. Their work indicated that in the protein dispersions the phytin appeared to be associated with the native protein by a salt-labile linkage which was disassociated in weak alkaline solutions. They found it possible to remove the phytin by dialysis, although the time required was too long for a commercial operation. They found they could remove the phytin also by dialyzing the wet curd against 1 N sodium chloride solution or by extraction of the wet curd with a saturated solution of ammonium sulfate followed by dialysis to remove the salt. They found, also, that the phytase in the seed dissociates the phytin but that the rate is very slow and impractical as a procedure for the elimination of the phytin. They reported that divalent cations obtained by addition of calcium or barium salts reacted with the phytates in an alkaline solution but they were not able to remove practical quantities of the phytates by this procedure.

McKinney and Sollars (1949) found that protein extracted from the meal with sulfurous acid and precipitated with sodium hydroxide contained 1.7–3.0% phytic acid but they could not remove any appreciable amount by dialyzing the acid extract.

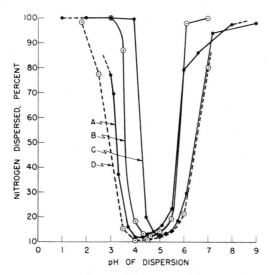

FIG. 9.11. THE EFFECT OF VARIOUS TREAT-
MENTS OF THE WATER EXTRACT
OF SOY MEAL ON THE PROTEIN
DISPERSIBILITY AT VARIOUS pH
VALUES

(A) Dispersibility of protein in water before special
treatment. (B) The water extract of the meal was dialyzed for 24 hr. (C) Water extract was dialyzed for 24 hr
and treated with Dowex-1-X10. (D) Same as C but
sodium phytate added.

Saio *et al.* (1969), in studies on the effect of adding phytic acid to soy milk in
the preparation of tofu, reported that the phytates reacted with both the
calcium and the protein to reduce the rate of protein coagulation to produce an
improved product.

PROCESSING SOY FLOUR AND GRITS

Quoting from Eley (1968), "Soy protein use in foods increased about 5-7%
annually during 1965-1967, according to industry sources." The production of
soy flour and grits for use in the bakery trade, which was marketed in the largest
tonnage, was growing the most rapidly at the annual rate of 7-10% per yr. The
use of the concentrate and isolate in the meat industry was reported as increasing at a rate of 5-6%.

The production of the various products was estimated by industry for Eley
for 1967 as follows: soy flour for baked goods (including 6 million pounds of
the enzyme active soy flour), 50 million pounds, for meat products, 30 million
pounds; for soy beverages, 10 million pounds; dry cereals and baby foods, 6
million pounds; brew flakes, 3 million pounds; pasta products, 1 million pounds;

and miscellaneous foods, 5-10 million pounds. The export market for soy flour was estimated at 10 million pounds with an additional purchase by the U.S. Government of 100 million pounds of soy flour for the preparation of corn-soy-milk (CSM) for use in developing countries, which brings yearly production to a total of about 210 million pounds. With the estimated increase for a 3-yr period, the production for soy flour and grits for 1970 would be about 250 million pounds.

Description of Soy Flour

Soy flour is a finely ground product processed from full-fat cotyledons or defatted flakes, which, according to standards established by the Soy Food Research Council, requires that 97% of the product shall pass through a No. 100 mesh U.S. Standard Screen. Finer grinds are also available.

The name soy flour may be somewhat misleading since a similarity to wheat flour is suggested; the soy flour does not contain gluten, a protein which is a major factor in characterizing wheat flour. Soy flour, in its defatted form and except for its fiber content, has more resemblance in physical and chemical properties to nonfat dry milk solids, and more properly might be called "defatted soy solids," or "soy powder," or "soy pulverate;" however, the defatted soy flour has nearly 20% more protein than the nonfat milk solids.

Soy grits are coarse-ground products made from full-fat cotyledons or defatted flakes and thus have the same composition as soy flour. The grits are graded in terms of U.S. Standard Screens; a screen size between No. 10 and 20 is graded as Coarse, between No. 20 and 40 is Medium, and between No. 40 and 80 is Fine. In order to obtain a desired texture in some food products the grits or even the flakes may be used rather than the soy flour. Other forms of unflavored soy grits in texturized extruded form are also available in various particle size ranges (see Appendix).

To meet the requirements of their many applications, the soy flours are produced at several different levels of fat and protein content and with different degrees of heat treatment.

Types and Composition

The principal types of soy flour are:

(1) Defatted soy flour, which is processed from dehulled and defatted flakes, usually containing less than 1% oil.

(2) Full-fat soy flour, which is processed from dehulled soybeans, and contains all the fat and protein normally in the cotyledons, with a minimum of 18% fat.

(3) Low-fat soy flour, which is processed by adding lipids to the defatted soy flour (refatting) to the desired lipid content, usually in the range of 4.5-9% by definition.

(4) Lecithinated soy flour, which is a special type processed by adding lecithin to the defatted flour (refatting) at some specified level up to 15%.

(5) Enzyme active soy flour, usually a full-fat type in England, and a defatted type in the United States (Johnson 1970B).

(6) Other types of soy flour are produced to meet special requirements. One company markets a soy flour containing 60% protein, prepared by combining the 70% protein concentrate with the 50% defatted soy flour (Turro 1970).

The composition of four types of soy flour is shown in Table 9.3. The carbohydrates, minerals and other components in soy flour are the same as reported for cotyledons in Chap.3; the percentage composition of these components may vary somewhat because of the natural variations in the composition of the soybean.

Moist Heat Treatment

In order to develop the functional requirements of soy flour for its many different uses, the controlled moist heat treatment used in processing covers a wide range: from a minimum treatment for a "white" product, through a moderate treatment for a "cooked" product, to a more intense treatment for a fully toasted product. The heat treatment affects the activity of the enzymes, flavor, color, nutritional value and dispersibility of the protein in water. The degree of heat treatment given the flakes, in the preparation of the soy flour, is indicated by the dispersibility of its nitrogenous components in water (NSI) and by its urease activity. The range of values in the following tabulation indicates the results of moist heat treatment given at 100° C:

Amount of Heat	Nitrogen Solubility Index (NSI)
Minimum heat	85-90
Light heat	40-60
Moderate heat	20-40
Fully toasted	10-20

When soy flakes have been processed with a minimum of heat treatment, that is, when they have been desolventized in a vapor or a flash desolventizer, they

TABLE 9.3

TYPICAL ANALYSES OF SOYBEANS AND SOY FLOUR

	Protein[1],[2] (%)	Moisture (%)	Fat[1] (%)	Fiber[1] (%)	Ash[1] (%)
Soybeans	42.6	11.0	20.0	5.3	5.0
Full-fat soy flour	46.6	5.0	22.1	2.1	5.2
Defatted soy flour	59.0	7.0	0.9	2.6	6.4
Lecithinated soy flour	48.6	5.5	16.4	2.2	5.3

[1] Moisture-free basis.
[2] % N x 6.25.

will retain essentially all of their enzyme activity as well as a high dispersibility of the protein in water; the trypsin inhibitors and hemagglutinin also will retain their activity. The light heat treatment inactivates nearly all of the biologically active components but retains a substantial part of the water dispersibility of the protein. The effect on protein dispersibility of autoclaving soy flakes or meal at atmospheric pressure is shown in Fig. 9.9. It will be observed that the dispersibility decreases very rapidly under these conditions; it is rather difficult to control the NSI precisely during the early stages of heat treatment.

Albrecht et al. (1966) have shown that steaming grits under 20 mesh in size and at 8% moisture will inactivate the urease and trypsin inhibitors and yet retain high protein solubility.

The usual moist treatment at about 100° C for 15 min serves to remove the beany flavor from the meal (Smith 1945) and at the same time increases nutritional value to its optimum.

The measurement in the loss of urease activity is a measure also of the inactivation of the other enzymes in the meal. Methods for measuring urease activity have been described by Croston et al. (1955) and by Smith et al. (1956).

Bean Selection

The important operations in processing soy flour include the selection of high quality, sound clean whole yellow beans as defined in Official Grain Standards of the United States published by USDA. Only yellow beans should be used since particles of the seed coat of black or dark colored beans or even beans with a dark hilum may inadvertently appear in the flour to cause undesirable dark specks.

Full-Fat Soy Flour

The selected beans are conditioned to a suitable moisture level, cracked between corrugated rolls and dehulled by aspiration and screening as in normal processing. For processing the full-fat flour the meats or grits are steam treated in conditioning equipment to meet the requirements of the desired type of soy flour, dried and ground in a hammer mill so that 97% will pass through a No. 100 U.S. Standard Screen. One of several types of pulverizing mills that may be used is shown in Fig. 9.12. Some pulverizing mills are designed to destroy any insects that may be present in the grain. For an untoasted or enzyme active soy flour the treatment with steam is omitted entirely. Full-fat products are very difficult to pulverize or to screen. It is customary to do the grinding in two steps and to separate the coarse from the fine particles in an air classifier between grindings. Full-fat soy flour usually is not screened except in small lots to measure particle dimensions.

Defatted Soy Flour

In processing the defatted soy flour commercially, the meats or grits are flaked to 0.009- to 0.010-in. thickness, solvent extracted, desolventized as in

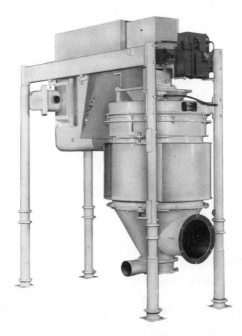

Courtesy of Entoleter, Inc.

FIG. 9.12. IMPACT   PULVERIZER-
CLASSIFIER MILL FOR PRO-
CESSING SOY FLOUR

normal processing, moist heat treated to the required NSI, dried, and ground. In
processing beans to make a flour with minimum protein denaturation (maximum
NSI) it is recommended that the solvent be removed from the flakes with a flash
desolventizer (Mustakas *et al.* 1962); however, vacuum desolventizing or other
methods may be used. When a flash desolventizer is used the residual solvent can
be stripped from the meal in a deodorizer using an inert gas. When debittering
and optimum nutritional value are required the treatment is accomplished by
using steam in the deodorizer. These last two steps can be accomplished also in
the normal processing in the desolventizer-toaster equipment (Sipos and Witte
1961).

After the moisture level in the flakes has been adjusted they are ground in a
hammer mill so that 97% will pass through a No. 100 U.S. Standard Screen.
When they are ground fine enough to pass a No. 100 screen, then 60% or more
will pass through a 300 mesh screen. Bolam and Earle (1951) have described a
method of sieve analysis using carbon tetrachloride to wash the ground sample
through the wire screen. Their method is reported to give more satisfactory
results than the earlier method which used shaking and brushing to force the
sample through the screen.

Pfeifer *et al.* (1960) have attempted to prepare a high protein soy flour fraction by grinding and air classification in a Pillsbury laboratory model classifier. They report that classification of the protein or any other fraction was not effective enough to have practical significance.

Paulsen (1968) has recommended a treatment of the meal with hydrogen peroxide and calcium chloride for improvement of flavor and color and for maintaining high protein dispersibility, especially for use of the flour in bread.

Horan (1967) has calculated the material balance in soy flour production using a batch of 60,000 lb of soybeans as follows: 1000 bu (or 60,000 lb) will yield 11,000 lb of oil, 43,300 lb of 50% protein meal, 4,200 lb of seed coat, and 1,500 lb of shrinkage.

Extruder-cooker Processing

A comparatively small scale extrusion process for the production of a fully toasted, full-fat soy flour has been developed by Mustakas *et al.* (1966) with the aid of UNICEF for use in developing countries. However, the process appears feasible for use wherever a limited scale production of full-fat soy flour is needed. The process uses a Wenger extruder (Fig. 9.13). A flow diagram of the continuous extruder process for making full-fat soy flour is shown in Fig. 9.14. Flakes or grits are prepared as in regular processing and then are preconditioned and cooked with sparged steam at 200°–212° F, the moisture adjusted to 18%, and then passed through a high speed mixer with additional steam added before entering the extruder section. The meal passes through the extruder in 1.0–1.5 min at a temperature in the range of 250°–290° F.

Courtesy of Wenger Mixer Mfg.

FIG. 9.13. EXTRUDER-COOKER-EXPANDER
PRESS FOR PROCESSING FULL-
FAT SOY FLOUR

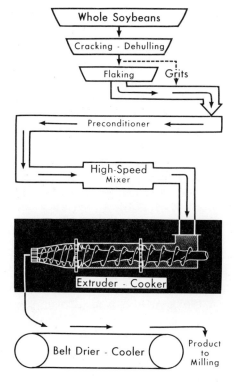

FIG. 9.14. FLOW DIAGRAM FOR EX-
TRUDER-COOKER-EXPANDER
PROCESSING   OF   SOY
FLOUR AND SNACK ITEMS

After the cooked and extruded product is cooled it is ground in an Alpine industrial size number 250 CWS Contraplex pin mill, 98% will pass through a 100 mesh screen and the remaining 2% appears to be fiber only. The product was tested for nutritive value, sanitation, flavor, stability, and other characteristics required of a food and was found to meet the necessary food grade standards.

Mustakas *et al.* (1970) and Bookwalter *et al.* 1971 extended their studies on the extrusion cooking of soybean with variations in time, temperature, and moisture in processing. After removal of the seed coat they found the lipoxidase was inactivated by treatment with dry heat for 6–8 min at 218°–220° F. The meats were remoistened to a predetermined level, cooked in the extruder for periods from 1/2–2 min, cooled, and ground.

They found that a high quality product was obtained when the extruder was operated in a temperature range of 250°–286° F in a calculated time-moisture range of 38–47. The soy flour had optimum nutritive value, good flavor, and a storage life of 12 months or longer.

Debittering Methods

A tremendous amount of effort has been expended on the part of a great number of investigators in the past 50 yr or more to improve the flavor of soybeans and soybean products. Most of this work is in the patent literature and reports of current work continue to be published. Much of this is difficult to evaluate without actually carrying out the described processes and subjecting the products to flavor-panel testing. Smith (1945) reviewed the patent literature through 1943, but concluded that the best treatment in his experience was extraction of the flaked soybeans with ethanol (Beckel and Smith 1944). Burnett (1951A) also reviewed various procedures for debittering full fat and defatted soy flours.

Soy flour products are highly competitive in the market, which precludes any expensive treatment to improve their flavor. As far as the authors are aware, all soy flour products of 50% protein content or less marketed in the U.S. are still made with moist heat treatment as the process of choice. One commercial product is treated additionally with alkaline earth salts, acids and/or peroxides (Paulsen 1968; Horan 1967). Another is mixed with soy protein concentrate (Turro 1970).

It is pertinent to point out that as soon as other treatments for debittering soy flour are weighed, such as wet processing to leach out low molecular weight components from defatted soy flakes, the complexity of handling, and the expense increase, it becomes advantageous to consider processing further into the concentrate or isolate.

PROCESSING SOY PROTEIN CONCENTRATE

Description and History

Soy protein concentrate is defined as "The product prepared from high quality, sound, clean, dehulled soybeans by removing most of the oil and water-soluble nonprotein components, that shall contain not less than 70% protein (% N x 6.25) on a moisture-free basis." In terms of the practical fractions of soybeans described earlier, it is composed of protein and insoluble residue (P + I).

Although they had not yet been so defined, products conforming to this definition were made in years past for industrial applications (Burnett 1951B), primarily for adhesive uses and as fillers in phenolic resin molded products. It is only since 1959 that edible products of this type became commercially available in response to the need for a blander derivative of soybeans than the soy flour products extant. Several companies have ventured into this field in the United

States, but only three at present are believed to be in commercial production. Each uses a different approach, based on wet processing, for obtaining the protein concentrate from defatted flakes or flour; other approaches have also been described. All of these depend on the concept of immobilizing the major protein fractions while permitting the soluble carbohydrates, salts, and other soluble low molecular weight components of the flakes or flour to diffuse or be leached from the cellular matrix comprising the defatted crushed cotyledons (see Fig. 9.15 for schematic of process flow). The protein is prevented from diffusing by one of several treatments: (a) leaching with 20–80% aqueous organic solvent, the concentration range in which the proteins are insoluble but which extracts the nonprotein solubles (Circle and Whitney 1968); (b) leaching with aqueous acids in the isolectric range of minimum protein solubility, pH 4–5 (see Fig. 4.4 of Chap. 4); (c) leaching with chilled water in the presence or absence of alkaline earth cations; or (d) leaching with hot water of cooked or toasted soy meal having a low NSI. In the first three methods the soy flakes may have a high NSI, which may or may not be retained in the end product, depending on the processing. General discussions of the processing are given by Circle and Whitney (1968), Meyer (1967), and Long (1962).

Composition.—A typical specification is listed in the Appendix. Commercially available products derived from the various processes have similar compositions, as shown in Table 9.4. Amino acid analyses are given in Chap. 3 and 7. Soluble carbohydrates are practically absent. The insoluble polysaccharides are mainly arabinogalactan and acidic pectin-type along with some galactomannan, xylan hemicellulose, and cellulose from the hull, if dehulling was inefficient (Meyer 1967).

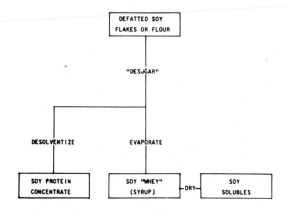

FIG. 9.15. PROCESS FLOW FOR SOY PROTEIN
CONCENTRATE PRODUCTION

TABLE 9.4

ANALYSES OF COMMERCIAL AND PILOT PLANT
SOY PROTEIN CONCENTRATES FROM SIX
DIFFERENT MANUFACTURERS

Sample Concentrate	Moisture (%)	Protein[1] As Is (%)	Protein[1,2] (%)	Fiber (%)	Ash (%)	pH[3]
A	7.5	66.6	72.0	3.5	5.5	6.8
B	7.1	68.1	73.3	2.8	5.5	6.8
C	3.8	69.5	72.3	5.0	2.8	6.8
D	5.2	68.3	72.1	3.8	4.5	7.3
E	9.0	64.1	70.5	5.3	2.3	4.7
F	5.5	73.8	78.1	—	1.7	5.7

[1] % N x 6.25.
[2] Moisture-free basis.
[3] pH of 10% aqueous dispersion.

Volume, Uses and Price.—According to Eley (1968), 1967 production of soy protein concentrate was estimated to be 17–30 million pounds per year, going mainly into the baked and meat-type products fields. It is extremely difficult to obtain more precise figures, as the manufacturers keep these closely guarded. Prices ranged from 18 to 28 ¢ per lb, according to Meyer (1967).

Processing

Yields of dried concentrate from the various processes are in the range 60–70%, based on the weight of defatted soy source materials (Meyer 1967). The first patent referring to an edible product of the acid-leached type appeared in 1959 (Sair 1959). The solubles other than alkaline-extractable, acid-precipitable protein were removed by aqueous leaching in the pH range approximately 4.0–4.8, separating the insoluble matter (P + I) from the soluble (S), in a batch operation repeated several times, and drying the (P + I). In a preferred embodiment, the pH was raised to the range 6–10.5, after leaching, and before drying. However, the commercial product has a pH usually close to neutrality (6.7–7.0), and a relatively high NSI (about 69), which is claimed to impart greater efficiency than a low NSI product as an emulsifier and water-binding agent in such applications as ground meat products.

Circle and Whitney (1968) described a process of concentrating the solubles (S) by countercurrent extraction in the isoelectric pH range of the soy protein, which they termed a "nonevaporative" means of economical recovery of the (S). They also employed aqueous organic solvent leaching and chilled water leaching (with or without alkaline earth cations) for the same purpose. In these media, meal with a high NSI may be used, and acidification is not required to immobilize the protein; however, the end products tend to have a low NSI.

Moshy (1964) used aqueous acid alone, or acid leaching followed by aqueous organic solvent leaching (Moshy 1965).

The process of leaching low NSI or toasted soy flakes with water or aqueous organic solvents in the temperature range 150°-200° F was described by Norris (1964): either batch or continuous countercurrent contact method and holding pH in the range 5.3-7.5 for purported best results to remove flavor and color bodies. McAnelly (1964) claimed to accomplish a similar purpose by first making a dough-like mass of soy flour and water, which was subjected to heat and pressure to denature the protein, and expanded to impart a porous structure, which was then leached with an aqueous solvent.

Mustakas *et al.* (1962) employed aqueous organic solvent leaching (50-70% lower alkanols) to produce a soy protein concentrate, and described a flash desolventizing system for recovering the alcohols. Other inventors have also resorted to aqueous organic solvents for treating soy flakes to leach out solubles (see Chap. 5).

Disposal or Use of Solubles

In the processes using water alone, (b), (c), and (d) described above, the solubles (S) differ very little from acid whey produced in the soy protein isolate processing described in the next section. Generally, current practice in the United States is to discard these in the absence of any directive to recover them. However, in the case of process (a) employing aqueous organic solvents, recovery and reuse of the organic solvent is mandatory from the point of its expense. During such recovery the solubles may be concentrated sufficiently to permit their economical recovery in the form of a syrup or in the dry state. In either form, the solubles may be used as an additive in feedstuffs. These aqueous organic solvent solubles differ from the acid whey. They have a lower ash content and lower nitrogen content; less inorganic matter is leached out; and they do not contain the so-called "albuminous" proteins of the acid whey.

Nutritive Value

The nutritive value of soy protein concentrates is discussed in Chap. 7.

Flour-concentrate Mixture

The practice of mixing 1 part soy protein concentrate with 1/2-2 parts soy flour having NSI over 66 is claimed to be advantageous for improved yield of bread and dough-handling aspects, when the mixture is used as an additive to dough in the process of making bread (Turro 1970).

PROCESSING SOY PROTEIN ISOLATE

Description and History

Soy protein isolate is defined as "the major proteinaceous fraction of soybeans prepared from high quality, sound, clean, dehulled soybeans by removing a preponderance of the nonprotein components, that shall contain not less than 90% protein (% N x 6.25) on a moisture-free basis." It is the practical fraction (P) referred to earlier.

Historically, as in the case of the concentrate, initial commercial interest in soy protein isolate in the early 1930's was for industrial purposes. So-called soy "casein" was used as a replacement for bovine casein in paper coatings, in which it served as a pigment binder. Significant quantities became available in 1937. The authors estimate that current consumption world wide for paper coatings is in excess of 50 million pounds annually. Reviews on industrial soy protein are available in the literature (Bain *et al.* 1961; Burnett 1951B). Unmodified edible soy protein isolate as a major article of commerce appeared in 1957 (Meyer 1969, 1967; Circle and Johnson 1958). But earlier reference (Burnett 1951A) had been made to relatively small quantities of neutral soy protein and enzyme-modified soy protein being sold as ingredients of confections, toppings, and other applications using minor amounts.

Volume and Price.—In 1967, it was estimated that production of the isolate was in the range of 22–35 million pounds annually, and of the modified, about 1 million pounds (Eley 1968). As in the case of the concentrates, several companies have been in and out of this field, and only two are believed to be currently active; production volume figures are closely guarded proprietary information. Current prices were in the range of 35–43 ¢ per lb. Despite predictions by Anson (1963, 1958) that production of the isolate would eventually be in the hundreds of millions of pounds annually, market uncertainties and the heavy investments needed for manufacturing plants on the scale required would appear to interdict any near term attainment of such amounts.

Composition.—Soy protein isolates made from defatted flakes have compositions as shown in Table 9.5. A typical specification is listed in the Appendix. Amino acid compositions will be found in Chap. 3 and 7.

Processing

Although soy protein isolates can be made from full-fat soybeans (Sugarman 1956; Chayen 1960), the products obtained contain appreciable and variable amounts of fat. Commercial quantities of a peanut protein-lipid complex or lipoprotein were made in England during the past decade; but the manufacturer has apparently ceased operations. Although the processes have an advantage in bypassing the hexane extraction step, these so-called lipoproteins have limited usefulness as a food ingredient due to their excessive fat content. At any rate, such lipoproteins can be readily made from nonfat soy protein isolate by homogenizing with controlled amounts of fat of any desired type (Circle and Meyer 1966), obviating the need for a full-fat soybean source material.

Source Material.—The usual starting material for soy protein isolate production is defatted soy meal or flour of high NSI. However, if a soy protein concentrate with a high NSI value were available at reasonable cost, it would

TABLE 9.5

ANALYSES OF COMMERCIAL AND PILOT PLANT
ISOLATED PROTEINS AND PROTEINATES FROM
SIX DIFFERENT MANUFACTURERS

Sample	Moisture (%)	Protein[1] As Is (%)	Protein[1,2] (%)	Fiber (%)	Ash (%)	pH[3]
Proteinate						
I	5.0	93.1	98.0	0.2	4.0	7.0
II	6.9	89.0	95.6	0.2	4.0	6.9
III	7.1	85.4	92.0	0.2	7.6	7.2
IV	8.5	91.4	99.8	−	3.1	7.0
Protein						
I	5.0	94.5	99.5	0.2	2.5	5.0
II	6.9	90.6	97.3	0.1	2.5	4.6
V	9.0	92.5	101.6	0.1	1.2	5.0
VI	10.3	87.7	97.8	0.1	2.1	4.6

[1] % N x 6.25.
[2] Moisture-free basis.
[3] pH of 10% aqueous dispersion.

present advantages in the processing, due to the absence of sugar and other nonprotein solubles. Circle (1950) has reviewed the equivalent processing of flakes into isolate by a preleaching step.

Schematic of Process Flow.—The steps involved in soy protein isolate production are delineated in Fig. 9.16. Reviews of processing protein isolate have been presented by Meyer (1967), Circle and Johnson (1958), Circle (1950), Burnett (1951A,B), and Smith (1958).

Aqueous Extraction.—The extraction is carried out in aqueous or mildly alkaline aqueous media under conditions with respect to temperature, liquid-solids ratio, pH, alkaline reagents, and other factors which vary from one manufacturer to another and which have been proposed or discussed by various workers (Alderks 1949; Anson and Pader 1958, 1957; Beaber and Obey 1962; Belter and Smith 1952; Circle *et al.* 1959; Circle 1950; Cogan *et al.* 1967; Eldridge and Nash 1965; Fontaine *et al.* 1946; McKinney *et al.* 1949; Obey 1951; Robbins *et al.* 1966; Sair 1961; Smiley and Smith 1946; Smith 1958, 1954; and Smith and Rackis 1957). Pilot plant operations were described by Smith (1958), Beckel *et al.* (1946), and Belter *et al.* (1944). For edible isolated protein production, aqueous alkaline extraction is preferably carried out at pH below 9 to avoid undue hydrolytic or rheological changes. Time and temperature of extraction, solids-liquid ratios, and other factors are selected to obtain optimally economic, and not necessarily maximum, yields.

Clarification.—The aqueous extract is separated from the insoluble residue by various screening, centrifuging, or filtering devices, or combinations of these.

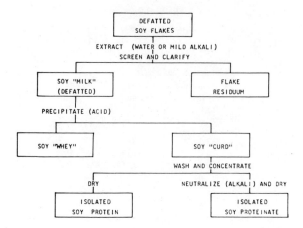

FIG. 9.16. PROCESS FLOW FOR SOY PROTEIN
ISOLATE PRODUCTION

Here the degree of clarification is based on considerations of end uses and economics, and varies from one manufacturer to another (see composition analyses in Table 9.5).

Precipitation.—Any food grade acid may be used. Among these the most common are sulfuric, hydrochloric, phosphoric, and acetic. The pH of the clarified extract is lowered to the range 4–5 at which solubility of the acid-precipitable protein is a minimum (Circle 1950). Temperature is not too critical a factor if kept below 60° C.

Curd-whey Separation.—This is accomplished by means of centrifuges (Fig. 9.17 and 9.18), or filters, or both, and is followed by a washing step the efficacy of which is dictated by economics and end uses, and is indicated by final composition (see Table 9.5). The curd is usually neutralized with food grade alkali to form the sodium proteinate salt before drying. However, isoelectric soy protein and the potassium and calcium proteinates have been and can be made available. The neutral proteinate is preferred for its water dispersibility.

Drying Step.—Various means of drying have been employed. The isoelectric form lends itself well to formation of a cake which may be granulated and dried in a forced draft oven or range type of drier, or it may be spray dried. The salt forms are preferably handled in spray driers, one type of which is illustrated in Fig. 9.19. Roller drying is rather difficult to control. Other possibilities include foam-mat and freeze drying, but these would not be economically feasible at present for commercial isolated soy protein production. The final products are usually packaged in multiple ply paper bags.

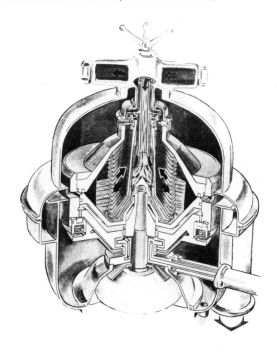

Courtesy of DeLaval Separator Co.

FIG. 9.17. DIAGRAM OF SELF-TRIGGERED
VARIABLE DISCHARGE DISC
CENTRIFUGE FOR SOY PROTEIN
CURD RECOVERY

Disposal and Utilization of By-products

Concomitant with isolated protein production (P) is the production of insoluble residue (I) and of whey solubles (S) as shown in Fig. 9.3. The three products are obtained commercially in varying ratios, depending on the processing, but may be roughly considered to be formed in equal amounts (Smith 1954). The I is obtained in the form of a wet cake or slurry, and is dewatered and dried in various types of equipment. Its protein content depends on the NSI or nitrogen solublity index of the original soy source material, and on the solids-liquid ratio, pH, time and temperature of aqueous extraction, and may vary considerably. Its main use is in prepared feeds for animals (defined as soybean extracted feed; see Appendix). The whey solubles are usually obtained in rather dilute form, 1–3% solids (Smith *et al.* 1962), depending on the solids-liquid ratio used in the extraction step. The published literature on problems of its complete economic recovery indicates that these problems are not yet solved (Belter and Smith 1952; Circle 1950; Kawamura 1955; Smith 1958, 1954; Smith *et al.* 1956, 1962; Smith and Wolf 1961; Wolf and Smith

Courtesy of Centrico, Inc.

FIG. 9.18. SELF-TRIGGERED   DISCHARGE DISC CENTRIFUGE FOR CURD RECOVERY

1961). However, part of the whey protein is recoverable by heat coagulation (Van Etten *et al.* 1959) or by treatment with edible gums or detergents (Eldridge *et al.* 1962; Smith *et al.* 1962). If these problems could be solved, properly processed soy solubles would readily find a place in foods or animal feeds, or could be used as a source material for further fractionation. Its high content of fermentable sugars make it a possible ingredient of various fermentation processes (Kawamura 1955).

Sanitation in Operating Practices

The analogy of soy processing with the milk industry to form "milk," "curd," and "whey" extends also to sanitation problems in large scale operations. Proper design calls for sanitary construction in stainless steel and using equipment, cleaning methods, and handling practices similar to those in a modern dairy. Use of heat and sanitizing agents coupled with Clean-in-Place (CIP) design of tanks, pipelines, and other equipment is recommended.

It is desirable to follow the microbiological picture with bacterial tests such as total plate, flat sour, and yeast and mold counts routinely carried out at various stages in the unit processes. Dairy type equipment is available in the market for those steps in Fig. 9.16 following clarification of the extracted liquor, but rather meager information on sanitary design and practice is available from the machinery and equipment designers and manufacturers concerning those stages prior to the clarified extract step.

Courtesy of DeLaval Separator Co.

FIG. 9.19. VERTICAL   SPRAY   DRYER   INSTALLA-
TION

Economics of Operation

Yield.—The literature references show yields of isolated protein to vary considerably. Pilot plan studies indicate yields of 33.1–42.7% (Alderks 1949), depending on variety of soybean used (calculated as bone-dry protein % N X 6.25, per 100 gm defatted flakes containing 9% moisture). The basis on which yield is reported is not always carefully spelled out, but is absolutely necessary to know if comparative figures are to have validity. Maximum yields were obtained when defatted flakes processed under ideal laboratory conditions were used, but varied with variety of soybean (Smith *et al.* 1966). State of comminution was not a factor for these optimally processed flakes, but was significant in other studies (Circle 1950). The yield of isolated protein based on defatted meal was stated to be 42% in the laboratory, but 30% was called good in commercial operation (Smith 1954).

Yield is directly dependent on the protein content of the soy source material, and also on the conditions of extraction as enumerated in previous sections. It is also affected by the amounts of associated and conjugated or complexed nonprotein constituents precipitating as impurities with the protein. These include lipids (see Chap. 4 and 5), phytic acid (see Chap. 4 and previously covered this Chapter), saponins (Chap. 4 and 5), and carbohydrates (Chap. 4 and 5). These nonprotein inclusions can affect the protein content of the isolate. Some, but not all, can be removed by alcohols (Chap. 5).

Other factors.—Along with yield, other factors which affect the economics of soy protein isolate processing include raw materials cost, labor, power, packaging, sales and advertising expenses, overhead, depreciation, and amortization. Johnson (1969) has presented estimates of some of these costs.

MODIFIED SOY PROTEIN ISOLATES

Spun Fiber Textured Protein "Tows"

Soy protein isolate is commonly marketed as a creamy-white, spray-dried powder, usually the sodium salt with a pH in the range of 6.7–7.2 when dispersed in water (see Appendix for typical specification). This neutral proteinate form, or alternatively the dry or wet curd isoelectric form, can be spun by appropriate manipulation into fibers of 0.01-in. diameter or less, which are gathered into "tows" and converted into meat analogs, as described in the next chapter. Several years ago in the United States, such tows were available for sale in the unflavored wet wrung acid condition and could be shipped without spoilage, being preserved by the acid and high salt content retained from the coagulating bath. Although inedible as such, they were convertible into foods resembling meat after removing the salts by washing and adding suitable neutralizing, flavoring, coloring, and binding agents. However, to the authors' knowledge, the unflavored wet tows no longer are an article of commerce in the United States, although in Japan they are distributed frozen in 5-kg packages as an ingredient for meat processors (Watanabe 1969). It would seem to be more feasible from the standpoint of the economics involved to carry out the conversion of the wet tow into meat analogs on the same premises where the fiber is spun; this is apparently the case in the United States. Details of the history and processing of the spun fiber protein food products are given in Chap. 10, in the section on meat analogs (these are also called spun fiber textured protein products).

Enzyme Modified Proteins

Unmodified soy protein isolate as usually marketed is composed of several different components with molecular weights in the range of 200,000–350,000, and is insoluble at pH 4–5 (See Chap. 4). Partial hydrolysis by pepsin treatment (Burnett 1951A) reduces the molecular weight to the range 3000–5000 and makes the hydrolysate soluble in water over the entire pH scale including pH 4-5. Such products have been given the misnomer "soy albumen," although they more properly should be classified as "peptones." They are sold in the United States to the extent of about 1 million pounds annually, at prices ranging from 70 ¢ to $1.20 per lb, according to Eley (1968). Their chief application is in confections (nougat, fudge, caramel, etc.), in marshmallow toppings, icings, and dessert mixes as a whipping agent, and in beverages as a foaming agent. Details of processing and uses are reviewed in Chap. 10 in the section on aerating agents.

FUNCTIONAL PROPERTIES OF SOY PROTEIN PRODUCTS

Introduction

The processing of soybeans into the various edible soy protein products described above has developed into viable commercial enterprises only because of the acceptance by the food industry of these products. It is emphasized that these commercial edible soy protein products are not consumed as such, but are used as ingredients in food formulations. Except for some consumer-ready textured foods (meat analogs), the soy protein products have little gustatory appeal or palatability in themselves, being sold for the most part in the form of unflavored dry powders, grits, granules or chunks. The current interest in these products on the part of food manufacturers lies not in their appearance, but in those properties which enable their use as ingredients to impart favorable, desirable, and acceptable changes in structure, texture, and composition to the finished foods, at an attractive price.

In many cases, they are employed to supplement, extend, or replace more costly ingredients derived from animal protein sources including meat, eggs, and milk; in other cases, they compete with grain or other plant derivatives. In any event, the choice of soy protein products as food ingredients depends on their providing economical performance on a par with what they replace without detracting from finished product quality. These soy products (regular and extruded flour products, concentrates, isolates, and modified isolates) are normally marketed in the United States conforming to various specifications of color, particle size, bulk density, solubility, and other characteristics. These are usually classified as physical properties, although solubility, involving solvation, may also be considered as a chemical property. Physical and chemical properties of soy protein fractions have been discussed by Wolf in Chap. 4 (see also Wolf and Smith 1961; Wolf 1969, 1970). Rather arbitrarily, a tacit understanding has evolved among workers in this field to categorize separately the physical, chemical, sensory (flavor), nutritional, and functional properties of these soy protein products without attempting to define these precisely. Although the nutritional value of these products as proteins does approach closely that of animal-derived proteins, and generally is better than that of most other vegetable proteins, the chief interest at present is focused on their functional properties;

that is, those which contribute some performance aspect, especially on manipulation in aqueous dispersion, to affect structure and texture of the formulation favorably. (In order to incorporate soy protein products into finished foods, at some point in the processing, hydration is required; this is true even for dry blends such as prepared mixes, which usually call for addition of water in order to put them to use.)

Delineation of Functional Properties

These were first listed in a separate category by Circle and Johnson (1958) for soy protein isolate aqueous dispersions, to encompass physicochemical properties such as dough-forming, moisture-holding, fat-binding, emulsifying, foaming, film-forming, thickening, stabilizing, gelling, cohesiveness, and adhesiveness (see also Johnson and Circle 1959). But the appellation "functional" was coined later (Johnson and Circle 1962). Smith (1958) referred to several similar properties of protein isolates intended for industrial use. Anson (1958) mentioned combinations of oilseed protein and oilseed fat for simulating dairy products, and discussed meatlike textures achieved by suitable manipulation of oilseed protein, namely by heat gelation and fiber-spinning of gel precursors.

Functional Properties of Soy Flour and Protein Concentrate Products

Soy flour and grits (P + I + S in terms of practical fractions) and soy protein concentrate (P + I) contain insoluble carbohydrate (I) in addition to protein (P). Although the protein is considered the component most responsible for the functional value of soy protein products, the insoluble carbohydrate may also have influence, both by diluting the protein, and on its own account. This has been pointed out by Johnson (1970B) and Wolf (1970). The insoluble carbohydrate may be detrimental to rheological characteristics (viscosity and heat gelation). It may affect water absorption or moisture-holding capacity. In any event, the soy flour and protein concentrate products are less versatile than the protein isolate in their functional properties, but nevertheless find many applications in food formulations.

Soy Flour and Grit Products (Lemancik and Ziemba 1962; Ziemba 1966). —Soy flour is said to make bread and cookie doughs more pliable, easier to handle and machine, and less sticky. After baking, bread has enriched crust color, better eating quality, and increased shelf life through enhanced moisture retention. Piecrust is claimed to be more tender, cookie dough to release more readily from forming dies, and breading mix and doughnuts to have reduced fat absorption on frying and improved browning, all with fewer cripples, on addition of soy flour. Baby cereals are less pasty with better pickup and release on drying rolls, and with improved texture, color, and appearance after drying. Soy flour in fudge retards crystallization and drying out. Caramels handle better on the wrapping machine and do not stick to the wrapper; oil bleeding is checked, and moisture retention improved. Soy flour can be added to soup mixes, gravies, and chili products as a thickening agent. In molasses, spice oils, and garlic powders it has value as an anticaking agent. The enzyme-active flour is useful for its bleaching effect on carotene. Coarse soy grits can be used as nut-like textured particles in toppings and cookies.

In macaroni, soy flour imparts a firmer texture (Paulsen 1961). Soy flour and grits are used in comminuted meat products to improve fat emulsification, fat-binding, and water-binding, but soy protein isolate is considered superior for these properties (Rock et al. 1966). Many of these and other applications are treated in greater depth in appropriate sections of Chap. 10.

Soy Protein Concentrates (Ziemba 1966).—The advantage of concentrates over soy flours lies in their blander flavor, lighter color, and higher protein content. In general, they can be used in food formulations in almost the same manner as the soy flours, but their lower flavor profile permits a higher use level without adverse effect on flavor of the final products. However, they do entail a higher cost.

Functional Properties of Soy Protein Isolate

In comminuted meat products, soy protein isolates are used as emulsifiers, binders, moisture retainers, and stabilizers (Rakosky 1970, Rock et al. 1966); they are said to impart better slicing and chewing qualities to these products. In dairy type products they provide emulsifying, suspending, and stabilizing action, and can be used to formulate almost the complete line, including margarine, imitation milk and creams, imitation cream cheese, frozen dessert, and others. In the formulation of puddings and spreads, they are useful as thickening, gelling, and suspending agents (Circle and Johnson 1958).

They are the chief structural material of meat analogs prepared by fiber-spinning and heat-gelation techniques. Although competitively attractive for use in dairy type and meat type products, their relatively high cost is a drawback to their usage in cereal-based products. The functional use of soy protein isolate in food products is discussed in more detail in Chap. 10.

Wolf (1970) has summarized in tabular form the functional properties of soybean proteins in food systems; this is reproduced in Table 9.6.

Tests for Functional Properties

The desirability of having available relatively simple physical or chemical tests which could be used to predict how soy protein products will affect food systems in which they are incorporated has occurred to several workers (Briskey 1970; Circle et al. 1964; Eldridge et al. 1963A,B; Kelley and Pressey 1966; Catsimpoolas and Meyer 1970; Johnson 1970 A,B; Meyer 1970; Wolf 1970; Mattil 1971). Quoting Mattil (1971) concerning "the problems of those who would evaluate the functional properties of various proteins. The question must always be asked, 'In what system; and what is the effect of the other components of the system?' How do the data obtained from an emulsifiability test run

TABLE 9.6

FUNCTIONAL PROPERTIES OF SOY PROTEIN PRODUCTS AS FOOD INGREDIENTS

Functional Property	Protein Form Used[1]	Food System	References
Emulsification			
Formation	F, C, I	Frankfurters, bologna, sausages	Rock et al. (1966); Pearson et al. (1965); Inklaar and Fortuin (1969)
		Breads, cakes, soups	Wood (1967); Tremple and Meador (1958)
Stabilization	F, C, I	Whipped toppings, frozen desserts	Circle and Johnson (1958)
		Frankfurters, bologna, sausages	Rock et al. (1966); Pearson et al. (1965); Inklaar and Fortuin (1969)
		Soups	Wood (1967)
Fat absorption			
Promotion	F, C, I	Frankfurters, bologna, sausages, meat patties	Rock et al. (1966); Wood (1967)
Prevention	F, I	Doughnuts, pancakes	Ziemba (1966); Eley (1968); Johnson (1970A,B)
Water absorption			
Uptake	F, C	Breads, cakes	Wood (1967); Tremple and Meador (1958); Turro and Sipos (1968)
		Macaroni	Paulsen (1961)
		Confections	Ziemba (1966)
Retention	F, C	Breads, cakes	Wood (1967); Tremple and Meador (1958)
Texture			
Viscosity	F, C, I	Soups, gravies, chili	Wood (1967); Ziemba (1966)
Gelation	I	Simulated ground meats	Anson and Pader (1958); Circle et al. (1964); Frank and Circle (1959)

	[1]	Application	References
Chip and chunk formation	F	Simulated meats	Ziemba (1966, 1969)
Shred formation	F, I	Simulated meats	Rusoff et al. (1962)
Fiber formation	I	Simulated meats	Ziemba (1966, 1969); Thulin and Kuramoto (1967)
Dough formation	F, C, I	Baked goods	Circle and Johnson (1958)
Film formation	I	Frankfurters, bologna	Circle and Johnson (1958); Ziemba (1966)
Adhesion	C, I	Sausages, lunch meats, meat patties, meat loaves and rolls, boned hams	Rock et al. (1966); Ziemba (1966)
		Dehydrated meats	Coleman and Creswick (1966)
Cohesion	F, I	Baked goods	Circle and Johnson (1958)
		Macaroni	Paulsen (1961)
		Simulated meats	Rusoff et al. (1962)
Elasticity	I	Baked goods	Circle and Johnson (1958)
		Simulated meats	Rusoff et al. (1962)
Color control			
Bleaching	F	Breads	Wood (1967)
Browning	F	Breads, pancakes, waffles	Wood (1967); Eley (1968)
Aeration	I	Whipped toppings, chiffon mixes, confections	Ziemba (1966); Eldridge et al. (1963A,B); Circle and Johnson (1958)

Source: Wolf (1970).

[1] F, C, and I represent flours, concentrates, and isolates, respectively.

in a beaker relate to the complex ionic environment in a frankfurter? Do thickening properties measured by traditional viscosity tests really predict how a protein will act in a meat system? It has been demonstrated that solubility profiles can be used to detect protein denaturation, and to predict modes of preparation of protein concentrates and isolates, but to what extent can they be used to predict functionality in real life systems?" Johnson (1970A,B) states that physical and chemical tests for functionality of a protein product will give useful information, but that in the final analysis it is necessary to incorporate the product as an ingredient into the food formulation and produce the finished product, in order to determine its performance beyond doubt. Nevertheless, it is considered worthwhile to enumerate several tests for functional performance which have been described, and which have found use value in the food industry. It should be pointed out that there must exist a sizable number of unpublished tests developed over the years, and reserved as proprietary information for in-house use by the larger food firms, especially in the baked goods, dairy, and meat fields.

Water Absorption and NSI.—The degree of water retention by soy flour and grits and soy protein concentrate is considered to be useful as an indication of performance in several food formulations, especially some involving dough handling. One such test is given in the Appendix. The relationship of water absorption to NSI was indicated by Lemancik and Ziemba (1962) and Johnson (1970A). The NSI itself is used as a predictor for selection and performance of heat-treated soy flour products for use in certain food formulations (Johnson, 1970A). Extruder-expanded textured soy flour chunks are said to reach 3-1/2 times their original dry weight in the fully-hydrated, drained condition (Martin and LeClair 1967). Paulsen (1961) described a water absorption test for macaroni, and applied it to compare macaroni products with and without incorporated soy flour.

Aerating Capability.—Whipability or foaming action is a property desirable for whip toppings, whipped desserts, and frozen desserts, among others. Eldridge *et al.* (1963A,B) described foam preparation from soy protein isolate dispersions, along with methods of measuring foam volume and stability.

Emulsifying Capacity of Soy Protein Isolate.—The use of soy protein isolate as a minor additive in sausage emulsions for its moisture-binding, fat-emulsifying, and emulsion-stabilizing properties has been discussed by several workers (Briskey 1970; Meyer 1970; Rakosky 1970; Rock *et al.* 1966). Although Pearson *et al.* (1965) claimed that soy protein isolate was not as good an emulsifier as milk proteins in their model test procedure, this was refuted in actual sausage tests reported by Inklaar and Fortsin (1969). Further work on the model test method would appear to be warranted.

Rheological Properties of Soy Protein Isolate.—These include viscosity and gelation. Circle *et al.* (1964) studied the rheological behavior of relatively concentrated aqueous dispersions of a commercially available soy protein isolate. The viscosity was found to rise exponentially with increase of concentration in the absence of heat. Heat caused thickening and then gelation in concentrations above 7% by weight with a temperature threshold of 65° C. Rate of gelling and gel firmness were found to be dependent on temperature, time of heating, and protein concentration. At 8–14% protein concentration, gels formed within 10–30 min at 70°–100° C, but were disrupted if overheated at 125° C. The gels were firm, resilient, and self-supporting at protein concentrations above 16–17%, and less susceptible to disruption by overheating. Fig. 9.20 shows the effect of temperature on apparent viscosity of dispersions containing 8 to 18% soy protein isolate.

Circle *et al.* (1964) also investigated the influence of additives on the heat-gelation of soy protein isolate dispersions. They found that the viscosities of both unheated and heated 10% dispersions were raised by added soy oil, soy lecithin, wheat starch, carboxymethylcellulose or carrageenan. The addition of

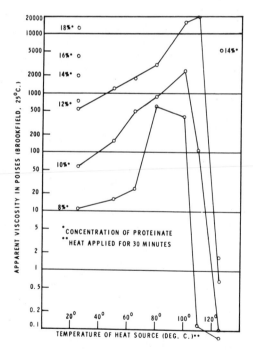

FIG. 9.20. EFFECT OF TEMPERATURE ON APPARENT VISCOSITY OF 8, 10, 12, 14, 16, and 18% SODIUM SOY PROTEINATE AQUEOUS DISPERSIONS HELD FOR 30 MINUTES AT TEMPERATURE SHOWN

salts to 10% dispersions lowered the viscosity of the unheated dispersion, but raised that of the heated. The specific disulfide reducing agents, sodium sulfite and cysteine, profoundly lowered the viscosity of both unheated and heated dispersions, and prevented gelation.

Anson (1958) advocated the use of soy protein isolate gels to simulate meat in the form of tough chewy gel particles held in a matrix of a more tender gel binder to create an inhomogeneous structure by proper manipulation through heating their gel precursors.

The mechanisms of gel formation are discussed by Wolf in Chap. 4. Catsimpoolas and Meyer (1970) proposed the concept of conversion of a protein sol to a "progel" state on heating, with formation of a gel on cooling. The gel-progel transition was considered reversible, with the sol-progel change being irreversible. Both progel and gel are subject to disruption on excessive heating to what they term a "metasol" which no longer gels on cooling. Further discussion on gelation is given in Chap. 10.

BIBLIOGRAPHY

ALBRECHT, W. J., MUSTAKAS, G. C., and MCGHEE, J. E. 1966. Rate studies on atmospheric steaming and immersion cooking of soybeans. Cereal Chem. *43*, 400–407.

ARKADY. HI-SOY brochure. British Arkady Company, Arkady Soya Mills, Manchester, England.

ARAI, S. *et al.* 1970. Studies on flavor components in soybean. V. n-Hexanal and some volatile alcohols; their distribution in raw soybean tissues and formation in crude soy protein concentrate by lipoxidase. Agr. Biol. Chem. (Tokyo) *34*, 1420–1423.

ALDERKS, O. H. 1949. The study of 20 varieties of soybeans with respect to quantity and quality of oil, isolated protein, and nutritional value of the meal. J. Am. Oil Chemists' Soc. *26*, 126–132.

ANSON, M. L. 1958. Potential uses of isolated oilseed protein in foodstuffs. *In* Processed Plant Protein Foodstuffs, A. M. Altschul (Editor). Academic Press, New York.

ANSON, M. L. 1963. The future of isolated oilseed protein in foods. Proc. Seed Protein Conf., USDA–ARS. USDA Southern Regional Research Laboratory, New Orleans, Jan. 21–23.

ANSON, M. L., and PADER, M. 1957. Extraction of soy protein. U.S. Pat. 2,785,155. Mar. 12.

ANSON, M. L., and PADER, M. 1958. Protein food products and method of making same. U.S. Pat. 2,830,902, Apr. 15; 2,833,651, May 6.

BAIN, W. M., CIRCLE, S. J., and OLSON, R. A. 1961. Isolated soy proteins for paper coating. *In* Synthetic and Protein Adhesives for Paper Coating. Tappi Monograph Ser. *22* 206–241.

BEABER, N. J., and OBEY, J. H. 1962. Method for producing organoleptically bland protein. U.S. Pat. 3,043,826. July 10.

BECKEL, A. C., BELTER, P. A., and SMITH, A. K. 1946. Soybean protein production. Effects of temperature and water-flake ratio. Ind. Eng. Chem. *38*, 731–734.

BECKEL, A. C., and SMITH, A. K. 1944. Alcohol extraction improves soya flour flavor and color. Food. Ind. *16*, No. 8, 71, 119.

BECKER, K. W. 1968. Advances in solvent extraction; its meaning to the food industry. Chem. Eng. Progr. Symp. Ser. *64*, No. 86, 60–65.

BECKER, K. W. 1971. Processing of oilseeds to meal and protein flakes. J. Am. Oil Chemists' Soc. *48*, 299–304.

BELTER, P. A., BECKEL, A. C., and SMITH, A. K. 1944. Soybean protein production. Comparison of the use of alcohol extracted with petroleum ether extracted flakes in a pilot plant. Ind. Eng. Chem. *36*, 799–803.

BELTER, P. A., and SMITH, A. K. 1952. Protein denaturation in soybean meal during processing. J. Am. Oil Chemists' Soc. *29*, 170–174.

BOLAM, I. J., and EARLE, F. R. 1951. Sieve analysis of ground soybeans and soy flour. J. Am. Oil Chemists' Soc. *28*, 194–195.

BOOKWALTER, G. N., et al. 1971. Full-fat soy flour extrusion cooked: properties and food uses. J. Food Sci. *36*, 5–9.

BRISKEY, E. J. 1970. Functional evaluation of protein in food systems. *In* Evaluation of Novel Protein Products. 303–312. (Wenner-Gren Center Symposium, September, 1968). Pergamon Press, New York.

BUNN, J. W., JR. 1970. Moisture extraction concepts, design, and practices from soybeans. J. Am. Oil Chemists' Soc. *47*, 366A, 368A.

BURNETT, R. S. 1951A. Soybean protein food products. *In* Soybeans and Soybean Products, Vol. 2, K. S. Markley (Editor). John Wiley & Sons, New York.

BURNETT, R. S. 1951B. Soybean protein industrial products. *In* Soybeans and Soybean Products, Vol. 2, K. S. Markley (Editor). John Wiley & Sons, New York.

BURNETT, R. S. 1970. Soybean Meal. Production, Technology, Evaluation, Use in Feeds. American Soybean Inst. Hudson, Iowa.

CATSIMPOOLAS, N., and MEYER, E. W. 1970. Gelation phenomena of soybean globulins. I. Protein-protein interactions. Cereal Chem. *47*, 559–570.

CHAYEN, I. H. 1960. Synthetic protein-lipid complex and a method of making the same from raw vegetable materials. U.S. Pat. 2,928,821. Mar. 15.

CIRCLE, S. J. 1950. Proteins and other nitrogenous constituents. *In* Soybeans and Soybean Products, Vol. 1, K. S. Markley (Editor). John Wiley & Sons, New York.

CIRCLE, S. J., and JOHNSON, D. W. 1958. Edible isolated soybean protein. *In* Processed Plant Protein Foodstuffs, A. M. Altschul (Editor). Academic Press, New York.

CIRCLE, S. J., JULIAN, P. L., and WHITNEY, R. W. 1959. Process for isolating soya protein. U.S. Pat. 2,881,159. Apr. 7.

CIRCLE, S. J., and MEYER, E. W. 1966. Protein and soy lecithin composition. U.S. Pat. 3,268,335. Aug. 23.

CIRCLE, S. J., MEYER, E. W., and WHITNEY, R. W. 1964. Rheology of soy protein dispersions. Effect of heat and other factors on gelation. Cereal Chem. *41*, 157–172.

CIRCLE, S. J., and WHITNEY, R. W. 1968. Process of non-evaporative countercurrent concentration of solids in the processing of protein and carbohydrate-containing materials from soybean. U.S. Pat. 3,365,440. Jan. 23.

COGAN, U., YARON, A., BERK, Z., and MIZRAHI, S. 1967. Isolation of soybean protein: Effect of processing conditions on yields and purity. J. Am. Oil Chemists' Soc. *44*, 321–324.

COLEMAN, R. J., and CRESWICK, N. S. 1966. Method of dehydrating meat and product. U.S. Pat. 3,253,931. May 31.

CRAVENS, W. W., and SIPOS, E. F. 1958. Soybean oil meal. *In* Processed Plant Protein Foodstuffs, A. M. Altschul (Editor). Academic Press, New York.

CROSTON, C. B., SMITH, A. K., and COWAN, J. C. 1955. Measurement of urease activity in soybean oil meal. J. Am. Oil Chemists' Soc. *32*, 279–282.

DUPUY, H. P., and FORE, S. P. 1970. Determination of residual solvent in oilseed meals. II. Volatilization procedure. J. Am. Oil Chemists' Soc. *47*, 231–233. I. Dimethylformamide extraction procedure. *Ibid.* 17–18.

ELEY, C. P. 1968. Food uses of soy protein. *In* Marketing and Transportation Situation. USDA–ERS *388, MTS-170*, 27–30.

ELDRIDGE, A. C., HALL, P. K., and WOLF, W. J. 1963B. Stable foams from unhydrolyzed soybean protein. Food Technol. *17*, 120–123.

ELDRIDGE, A. C., and NASH, A. M. 1965. Process of producing soybean proteinate. U.S. Pat. 3,218,307. Nov. 16.

ELDRIDGE, A. C., NASH, A. M., and SMITH, A. K. 1962. Soybean whey protein-poly-saccharide complex. U.S. Pat. 3,069,327. Dec. 18.
ELDRIDGE, A. C., WOLF, W. J., NASH, A. M., and SMITH, A. K. 1963A. Alcohol washing of soybean protein. J. Agr. Food Chem. *11*, 323–328.
FINCHER, H. D. 1958. Processing of oilseeds. *In* Processed Plant Protein Foodstuffs, A. M. Altschul (Editor). Academic Press, New York.
FONTAINE, T. D., PONS, W. A., JR., and IRVING, G. W., JR. 1946. Protein-phytic acid relationship in peanuts and cottonseed. J. Biol. Chem. *164*, 487–507.
FRANK, S. S., and CIRCLE, S. J. 1959. The use of isolated soybean protein for non-meat simulated sausage products. Frankfurter and bologna types. Food Technol. *13*, 307–313.
HORAN, F. E. 1967. Defatted and full-fat soy flours by conventional processes. Proc. Intern. Conf. Soybean Protein Foods, Oct. 1966, USDA, Peoria, Ill. USDA–ARS *71–35*.
INKLAAR, P. A., and FORTUIN, J. 1969. Determining the emulsifying and emulsion-stabilizing capacity of protein meat additives. Food Technol. *23*, 103–107.
JOHNSON, D. W. 1969. Isolated soy proteins and soy protein concentrates. UN Ind. Develop. Organ. Meeting on Soya Bean Processing and Use, Peoria, Ill. Nov. 17–21. United Nations, N. Y.
JOHNSON, D. W. 1970A. Oilseed proteins–properties and applications. Food Prod. Develop. *3*, No. 8, 78, 80, 84, 87.
JOHNSON, D. W. 1970B. Functional properties of oilseed proteins. J. Am. Oil Chemists' Soc. *47*, 402–407.
JOHNSON, D. W., and CIRCLE, S. J. 1959. Multipurpose quality protein offers 'plus' factors. Food Process. *20*, No. 3, Mar.
JOHNSON, D. W., and CIRCLE, S. J. 1962. Food uses and properties of protein fractions from soybeans. First Intern. Congr. Food Sci. Technol. London, England, Sept.
JONES, D. B., and GERSDORFF, E. F. 1938. Changes that occur in the protein of soybeans as a result of storage. J. Am. Chem. Soc. *60*, 723–724.
KAWAMURA, S. 1955. Studies on the utilization of the by-products of soybean protein manufacture. VII. Summarizing discussions. Kagawa Agr. Coll. Tech. Bull. *6*, No. 3, 227–239.
KELLEY, J. J., and PRESSEY, R. 1966. Studies with soybean protein and fiber formation. Cereal Chem. *43*, 195–206.
KROMER, G. W. 1970. U.S. soybean processing capacity expanding. Fats and Oil Situation (Nov.) USDA *FOS–255*, 27, 29, 31, 33.
KUIKEN, K. A. 1958. Effect of other processing factors. *In* Processed Plant Protein Foodstuffs, A. M. Altschul (Editor). Academic Press, New York.
LANGHURST, L. F. 1951. Solvent extraction processes. *In* Soybeans and Soybean Products, Vol. 2, K. S. Markley (Editor). John Wiley & Sons, New York.
LEMANCIK, J. F., and ZIEMBA, J. V. 1962. Versatile soy flours. Food Eng. *34*, No. 7, 90–91.
LONG, J. E. 1962. A bland protein from soybeans. Presented at 47th Natl. Meeting Am. Assoc. Cereal Chemists, St. Louis, May 20–24.
MCANELLY, J. K. 1964. Method for producing a soybean protein product and the resulting product. U.S. Pat. 3,142,571. July 28.
MCKINNEY, L. L., SOLLARS, W. F., and SETZKORN, E. A. 1949. Studies on the preparation of soybean protein free from phosphorus. J. Biol. Chem. *178*, 117–132.
MCKINNEY, L. L., and SOLLARS, W. F. 1949. Extraction of soybean protein with sulfurous acid. Ind. Eng. Chem. *41*, 1058–1060.
MARTIN, R. E., and LECLAIR, D. V. 1967. For creating products–textured proteins. Food Eng. *39*, No. 4, 66–69.
MATTIL, K. F. 1971. Functional requirements of proteins for foods J. Am. Oil Chemists' Soc. *48*, 477–480.
MEYER, E. W. 1967. Soy protein concentrates and isolates. Proc. Intern. Conf. Soybean Protein Foods, Oct. 1966. USDA–ARS *71–35*.
MEYER, E. W. 1969. Soy protein products for food. Conf. Protein-Rich Food Products from Oilseeds, New Orleans, La. May 15–16, 1968. USDA–ARS *72–71*.

MEYER, E. W. 1970. Soya protein isolates for food. *In* Proteins as Human Food, R. A. Lawrie (Editor). Butterworth's, Whitefriars Press, London, England.

MILLIGAN, E. D. 1969. Flash desolventizers. Oil Mill Gazetteer, Nov., 10–15.

MOSHY, R. J. 1964. Process for treating soybean flour. U.S. Pat. 3,126,286. Mar. 24.

MOSHY, R. J. 1965. Process for treating soybean flour to improve its flavor. U.S. Pat. 3,168,406. Febr. 2.

MUSTAKAS, G. C., GRIFFIN, E. L., JR., and SOHNS, V. E. 1966. Full fat soybean flours by continuous extrusion cooking. World Protein Resources, Am. Chem. Soc. Advan. Chem. Ser. *57,* 101–108.

MUSTAKAS, G. C., KIRK, L. D., and GRIFFIN, E. L., JR. 1962. Flash desolventizing defatted soybean meals washed with aqueous alcohols to yield a high protein product. J. Am. Oil Chemists' Soc. *39,* 222–226.

MUSTAKAS, G. C. *et al.* 1970. Extruder processing to improve nutritional quality, flavor and keeping quality of full fat soyflour. Food Technol. *24,* 102–108.

NAGEL, R. A., BECKER, H. C., and MILNER, R. T. 1938. Some physical factors affecting the dispersion of soybean protein in water. Cereal Chem. *40,* 463–471.

NATIONAL SOYBEAN PROCESSORS ASSOCIATION. 1946. Handbook of analytical methods for soybeans and soybean products.

NORRIS, F. A. 1964. Protein recovery from seed meal. Can. Pat. 697,264. Nov. 3.

OBEY, J. H. 1951. Method and apparatus for extracting proteins. U.S. Pat. 2,559,257. July 3.

PAULSEN, T. M. 1961. A study of macaroni products containing soy flour. Food Technol. *15,* 118–121.

PAULSEN, T. M. 1968. Process for treating soybean particulates. U.S. Pat. 3,361,574; 3,361,575. Jan. 2. (also U.S. Pat. 3,100,709, Aug. 13, 1963).

PAULSEN, T. M., and HORAN, F. E. 1965. Functional characteristics of edible soya flours. Cereal Sci. Today *10,* 14–17.

PEARSON, A. M., SPOONER, M. E., HEGARTY, G. R., and BRATZLER, L. J. 1965. The emulsifying capacity and stability of soy sodium proteinate, potassium caseinate, and nonfat dry milk. Food Technol. *19,* 1841–1845.

PFEIFER, V. E., STRINGFELLOW, A. C., and GRIFFIN, E. L., JR. 1960. Fractionating corn, sorghum and soy flours by fine grinding and air classification. Am. Miller Processor *88,* No. 8, 11–13, 24.

RACKIS, J. J., SMITH, A. K., and SASAME, H. A. 1958. Protein in soybean hypocotyl. Arch. Biochem. Biophys. *78,* 180–187.

RAKOSKY, J. 1970. Soy products for the meat industry. J. Agr. Food Chem. *18,* 1005–1009.

ROBBINS, F. M., BONAGURA, A. G., and YARE, R. S. 1966. Process of preparing heat gelable soybean protein. U.S. Pat. 3,261,822. July 19.

ROCK, H., SIPOS, E. F., and MEYER, E. W. 1966. Soy protein—its role in processed meat production. Meat *32,* 52–56.

ROTHFUS, J. A. 1970. Newer techniques in protein isolation and characterization. J. Am. Oil Chemists' Soc. *47,* 316–325.

RUSOFF, I. I., OHAN, W. J., and LONG, C. L. 1962. Protein food product and process. U.S. Pat. 3,047,395. July 31.

SAIO, K., KOYAMA, E., YAMAZAKI, S., and WATANABE, T. 1969. Protein-calcium-phytic acid relatioship in soybean. III. Effect of phytic acid on coagulative reaction in tofu-making. Agr. Biol. Chem. (Tokyo) *33,* 36–42.

SAIR, L. 1959. Proteinaceous soy composition and method of preparing. U.S. Pat. 2,881,076. Apr. 7.

SAIR, L. 1961. Method of extracting protein from defatted soybean material. U.S. Pat. 3,001,875. Sept. 26.

SIKES, J. K. 1960. Determination of residual hexane in solvent-extracted meal. J. Am. Oil Chemists' Soc. *37,* 84–87.

SINGER, P. A. 1965. Soybean drying. Soybean Dig. *25,* No. 8, 8–9; No. 9, 10.

SIPOS, E., and WITTE, N. H. 1961. The desolventizer-toaster process for soybean oil meal. J. Am. Oil Chemists' Soc. *38,* 11, 12, 17–19.

SMILEY, W. G., and SMITH, A. K. 1946. Preparation and nitrogen content of soybean protein. Cereal Chem. *23*, 288–296.

SMITH, A. K. 1945. Debittering soybeans. A list of patents for removing the bitter taste from soybeans. Soybean Dig. *5*, No. 7, 25–26, 28.

SMITH, A. K. 1954. Isolation and utilization of vegetable proteins. Econ. Botany *8*, 291–315.

SMITH, A. K. 1958. Vegetable protein isolates. *In* Processed Plant Protein Foodstuffs, A. M. Altschul (Editor). Academic Press, New York.

SMITH, A. K., BELTER, P. A., and ANDERSON, R. L. 1956. Urease activity in soybean meal products. J. Am. Oil Chemists' Soc. *33*, 360–363.

SMITH, A. K., and CIRCLE, S. J. 1938. Peptization of soybean protein: extraction of nitrogenous constituents from oil free meal by acids and bases with and without added salts. Ind. Eng. Chem. *30*, 1414–1418.

SMITH, A. K., CIRCLE, S. J., and BROTHER, G. H. 1938. Peptization of soybean protein: effect of neutral salts on the quantity of nitrogenous constituents extracted from oil free meal. J. Am. Chem. Soc. *60*, 1316–1320.

SMITH, A. K., NASH, A. M., ELDRIDGE, A. C., and WOLF, W. J. 1962. Recovery of soybean whey protein with edible gums and detergents. J. Agr. Food Chem. *10*, 302–304.

SMITH, A. K., and RACKIS, J. J. 1957. Phytin elimination in soybean protein isolation. J. Am. Chem. Soc. *79*, 633–637.

SMITH, A. K., and WOLF, W. J. 1961. Food uses and properties of soybean protein. I. Food uses. Food Technol. *15*, 4–6, 8, 10.

SMITH, A. K. *et al.* 1966. Nitrogen solubility index, isolated protein yield and whey nitrogen content of several soybean strains. Cereal Chem. *43*, 261–270.

SUGARMAN, N. 1956. Process for simultaneously extracting oil and protein from oleaginous materials. U.S. Pat. 2,762,820. Sept. 11.

THULIN, W. W., and KURAMOTO, S. 1967. Bontrae—a new meat-like ingredient for convenience foods. Food Technol. *21*, 168–171.

TREMPLE, L. G., and MEADOR, R. J. 1958. Soy flour in cake baking. Baker's Dig. *32*, No. 4, 32.

TURRO, E. J., 1970. Bakery formulation. U.S. Pat. 3,529,970. Sept. 22.

TURRO, E. J., and SIPOS, E. 1968. Effect of various soy protein products on bread characteristics. Baker's Dig. *42*, 44–50, 61.

VAN ETTEN, C. H. *et al.* 1959. Amino Acid composition of soybean protein fractions. J. Agr. Food Chem. *7*, 129–131.

VIX, H. L. E., and DECOSSAS, K. M. 1969. Processing oilseeds into edible products: solvent extraction. *In* Engineering of Unconventional Protein Production, H. Bieber (Editor). Chem. Eng. Progr. Symp. Ser. *65*, No. 93, 49–56.

WATANABE, T. 1969. Industrial production of soybean foods in Japan. United Nations Industrial Development Organization. Meeting on Soya Bean Processing and Use, Peoria, Ill. Nov. 17–21. United Nations, N. Y.

WOLF, W. J. 1969. Chemical and physical properties of soybean protein. Baker's Dig. Oct. 30–37.

WOLF, W. J. 1970. Soybean proteins. Their functional, chemical and physical properties. J. Agr. Food Chem. *18*, 969–976.

WOLF, W. J., and SMITH, A. K. 1961. Food uses and properties of soybean protein. II. Physical and chemical properties of soybean protein. Food Technol. *15*, No. 5, 12, 13, 16, 18, 21, 23, 26, 28, 31, 33.

WOOD, J. C. 1967. Soy flour in food products. Ingredient Survey. Food Manuf. *42*, 11–15.

ZIEMBA, J. V. 1966. Let soy proteins work wonders for you. Food Eng. *38*, No. 5, 82–89.

ZIEMBA, J. V. 1969. Simulated meats—how they're made. Food Eng. *41*, No. 11, 72–75.

CHAPTER 10

A. K. Smith
S. J. Circle | # Protein Products as Food Ingredients

FLAVOR

In the United States and many other countries great interest in food uses of soybeans has been stimulated by the high nutritive value of their proteins, their high content of polyunsaturated fatty acids, the example of their successful use in the Orient, and the low cost of their production. Despite the great interest in soybean foods, progress in their development has been slower than anticipated. This is attributed to the characteristic flavor of most soy products which is very difficult to remove. Some of the experimental work directed to this goal of improving flavor and color has been reported in Chap. 5.

The flavor problem in the oriental countries was resolved to their satisfaction early in the history of soybean utilization (Chap. 1) through the development of fermentation products which contributed acceptable flavors to the soybean as well as other foods. However, the foods used in the Orient have not been generally accepted in other countries, and it is now apparent that it will be necessary to develop soybean flours, concentrates, and isolates with a bland flavor to obtain the wider acceptance which they deserve. Although present uses in the United States are numerous and varied, they are usually incorporated at a rather low concentration and more for functional uses than for nutritional value.

The flavor problem has been difficult to solve (1) because of the low concentration of the flavor components naturally occurring in the raw bean and the difficulty involved in their isolation in sufficient quantities for characterization and other investigational purposes; (2) when the bean cell structure is crushed in processing, the natural enzymes of the bean, such as the lipoxidases and lipases, become active and introduce new flavor components in the system from lipid precursors; (3) the solvents used for fat extraction may leave residues; and (4) the heat used in processing may stimulate interactions between many of the minor components and the protein and form other artifacts, thereby greatly increasing the difficulty of their identification and elimination.

It has already been pointed out in Chap. 5 that organic solvent extraction of flavor and other minor components reveals the principal flavors to be more readily soluble in polar than in nonpolar organic solvents, and that high moisture in the beans, or the presence of some water in the solvent, assists in removing flavor and color components. Thus, present information suggests that one approach to the flavor problem may be to use a combination of polar and nonpolar solvents containing some water, which will extract the flavor and color

339

components along with the oil; another may be to follow hexane extraction with an aqueous organic (polar) solvent.

Taste Panel Results

Moser *et al.* (1967) developed a taste panel procedure for comparing the flavor level in soybeans and soy flour products, and used it in estimating the effectiveness of various solvents and heat treatment for removing flavor. They used a scoring system based on a 10-point scale with 1 representing the strongest and 10 the blandest flavor. Tests were made on unheated full-fat and defatted soy flour products, prepared from several soybean varieties, after treatment with steam in an autoclave at atmospheric pressure for various periods of time. It was found that treatment in the autoclave for 10 min was as effective in removing flavor as 40 min. The highest flavor score obtained by steam treatment was 6.3.

Flavor evaluations were made also on hexane-extracted flakes after they were washed with (1) water (pH 4.6), (2) 80% methanol, (3) 80% ethanol, (4) 80% ethanol, followed by 20 min of steam treatment, and (5) 80% isopropanol. The results were as follows:

Treatment Given Hexane-defatted Flakes	Flavor Score
Water wash, pH 4.6	3.6
80% methanol	6.3
80% ethanol	7.3
80% ethanol-20 min steam treatment	8.0
80% isopropanol	7.2

It is apparent that washing with alcohols was more effective in removing flavor than water washing or the use of steam alone. In comparing several soybean varieties and several locations of the same variety no differences in flavor were observed.

Flavor Components

Fujimaki *et al.* (1969) and Arai *et al.* (1970) studied the flavor problems of soybeans in detail, using paper, thin layer, column, and gas chromatography, as well as ultraviolet and infrared light and mass spectrometry. They reported identification of a rather large number of minor organic components which they found contribute to the flavor of soybeans and the processed products. Table 10.1 gives a list of the components they identified and the methods used in their identification. Although 35 or more compounds were isolated, because of the difficulties encountered in their isolation only a limited number could be positively identified as contributing a particular flavor to the raw or processed bean products.

In the raw soybeans the alcohols were in highest concentration, especially *n*-hexanol, which was found to make a significant contribution to the raw beany

TABLE 10.1

IDENTIFIED OR ESTIMATED FLAVOR COMPONENTS
IN SOYBEAN

	Origin				
Component	Raw Soy-bean	Defat-ted Soybean	Soy Pro-tein	Method for Identification or Estimation[1]	Flavor Profile
n-Alkanals					
Methanal		+		PPC,UV	
Ethanal	+	+	+	IR,UV,GLC,EA,mp	
Propanal			+	GLC	
n-Hexanal	+	+++	++	IR,UV,MS,EA,GLC,mp	Grassy
2-Alkenals					
2-Heptenal		+		PPC,UV	Grassy
2,4-Alkadienals					
2,4-Decadienal		+		PPC,UV	Wet napkin-like
2-Alkanones					
2-Propanone	+	+	+	IR,UV,GLC,EA,mp	
2-Pentanone		+		IR,UV,PPC,EA,mp	
2-Hexanone		+		PPC	Grassy
a-Diketones					
Glyoxal	+			IR,TLC	Stimulus
Methyl glyoxal	+			IR,TLC	Stimulus
a-Keto acids					
Glyoxylic	+			TLC	Stimulus
Pyruvic	+			TLC	Stimulus
a-Ketoisocaproic	+			TLC	Sour
a-Ketoglutaric	++			TLC	Sour
Aromatic aldehydes					
Benzaldehyde	+			IR,TLC	Fruity
Protocatechuic	+			IR,TLC	Astringent
Fatty acids					
Acetic	+			GLC	
Propionic	+			GLC	
Isovaleric	+			GLC	Butter-like
n-Valeric	+			GLC	Butter-like
Isocaproic	+			GLC	Green bean-like
n-Caproic	++			GLC	Green bean-like
n-Caprylic	+			GLC	Green bean-like
n-Nonanoic	+			GLC	Waxy
n-Capric	+			GLC	Waxy
Phenolic acids					
Syringic		++		IR,PPC,UV	Astringent
Vanillic		+		PPC,UV	Astringent
Ferulic		+		PPC,UV	Astringent
Gentisic		+		PPC,UV	Astringent
Salicylic		+		PPC,UV	
p-Coumaric		+		PPC,UV	
p-Hydroxybenzoic		+		PPC,UV	
Isochlorogenic		+		PPC,UV,CAC	Bitter, sour
Chlorogenic		+		PPC,UV,CAC	Bitter, sour

TABLE 10.1 (continued)

IDENTIFIED OR ESTIMATED FLAVOR COMPONENTS IN SOYBEAN

Component	Origin			Method for Identification or Estimation[1]	Flavor Profile
	Raw Soybean	Defatted Soybean	Soy Protein		
Amines					
Monomethylamine	+			GLC,PPC	Fishy
Dimethylamine	++			IR,GLC,PPC	Fishy
Piperidine	+			GLC,PPC	Cereal flour-like
Cadaverine	?			PPC	Cereal flour-like
Aliphatic esters					
n-pentanol acetate	+			GLC	Fruity
Alkanols					
Methanol	+++	+	+	GLC,TLC,CR	
Ethanol	+++	+	+	GLC,TLC	
2-Pentanol	+			GLC,TLC	
Isopentanol	+	+	+	IR,GLC,TLC	Green bean-like
n-pentanol	+	+	+	MS,GLC,TLC	Green bean-like
n-Hexanol	+++	++	++	IR,MS,GLC,TLC	Green bean-like
n-Heptanol	++	++	++	GLC,TLC	Green bean-like
Inorganic compounds					
Ammonium hydroxide	++			PPC,CR	
Hydrogen sulfide	+			CR	

Source: Fujimaki *et al.* (1969).

[1] CAC—column adsorption chromatography; CR—color reaction; EA—elemental analysis; GLC—gas (liquid) chromatography; IR—infrared; mp—melting point; MS—mass spectrometry; MW—molecular weight; PPC—paper chromatography; TLC—thin layer chromatography; UV—ultraviolet.

flavor. Although the alcohols were found to be located in all seed parts they were in the highest concentration in the hypocotyl. Both alcohols and carbonyl compounds appeared to contribute importantly to the flavor of the defatted products and to the isolated protein. In another experiment they found that lipoxigenase activity accelerated the formation of n-hexanal in the isolated protein. The n-hexanal and n-hexanol were capable of binding with the soybean protein, and thereby became resistant to solvent extraction and to vacuum and sweeping distillation. The binding of the flavor components to the protein was attributed to hydrophobic bonding.

Fujimaki *et al.* (1968) investigated the removal of the flavor components from soybean protein isolates (Promine-D) by treatment with proteolytic enzymes. The effectiveness of 12 proteolytic enzymes of plant, animal, and microbial origin were tested; the list included Molsin, a crude preparation of aspergillo-peptidase A (APase A) produced by *Aspergillus saitoi,* pepsin, papain, bromelin,

Rapidase, Coronase, Prozyme and others; of the group tested, Molsin was found to be the most effective of the enzymes for liberating flavor components, which were then extracted from the system with 90% ethanol. Not only flavor components but also residual lipids or fatty acids were liberated from the soybeans and defatted flour through enzymatic digestion. According to the authors, the resulting protein has less flavor and a greater flavor stability in storage.

The enzymatic digestion was carried out on a 1% solution at optimum pH of the enzymes, and samples were taken at appropriate intervals. The proteolysis was stopped by adding trichloroacetic acid to precipitate the protein, and proteolysis was determined. For testing, samples were neutralized to pH 6.0, held in a boiling water bath for 15 min to inactivate the enzyme, centrifuged, and the supernatant compared with the control which was incubated without the enzyme. The intensity of the beany, bitter, astringent, and other flavors was evaluated by a panel of six members. The bitter flavor was shown to be related mostly to peptides which were found to contain leucine as a terminal amino acid.

Of the 12 enzymes investigated Molsin was the most effective in liberating flavor components from the protein, and at the same time it liberated bitter peptides. It was found that purified APase A displayed almost the same deodorizing effect as Molsin but the proteolysate from the APase A was much more bitter than that from Molsin; thus, it was concluded that Molsin contained another enzyme which lessened the bitterness produced by APase A. On comparing the free amino acids and the terminal amino acids of the peptides in the APase A hydrolyzate with those in the Molsin hydrolyzate, it was concluded that the enzyme related to the debittering was aspergillus acid carboxypeptidase (AACPase). Accordingly, almost simultaneous deodorizing and debittering were possible by using a combination of APase and AACPase in the system. The procedure was to adjust a 1% protein solution (1 kg) to pH 2.8 with HC1 and incubate with APase A (5 mg) at 40° C for 2 hr. This solution was then readjusted to pH 3.5 with NaOH, AACPase (5 mg) was added, and incubation at 40° C continued for another 2 hr. The liberated flavor components were removed with 90% ethanol. Investigations with pepsin showed it to be one of the most effective enzymes for producing bitter peptides.

Mattick and Hand (1969), using modern analytical methods, have investigated further the effects of lipoxidase activity on the flavor of soy milk. They isolated 80 volatile compounds which were said to be the results of lipoxidase activity, and 40 of these compounds were identified. The majority of these compounds were aldehydes, ketones, and alcohols and correspond to many of the compounds reported by Fujimaki *et al.* (1969) in Table 10.1. They reported that 31 of these compounds have some effect on flavor and one, ethyl vinyl ketone, was reported as having a typical beany note. These studies indicate that the flavor can be improved by inactivation of the lipoxidase prior to its reaction with the lipids in the soybeans. A similar conclusion was reached by Mustakas *et al.* (1969).

It was shown that heating the water extract of the bean immediately after grinding was not effective, thus the new procedure was to grind the beans in water above 80° C. Bourne (1970) has reported that the boiling-water, grinding method of processing soy milk is being tested for use in the Philippines at the University of the Philippines, Laguna.

Rackis *et al.* (1970) have reported that removal of 99.8% of the oil from the soybean flakes, with pentane-hexane solvent, removed little if any, of the beany, bitter, grassy flavor. Further extraction of the defatted flakes with an aqueous ethanol solution of hexane-alcohol azeotrope removed the complex mixtures of lipids and most of the flavor.

Honig *et al.* (1971) in studies on the flavor components of the soybean have reported the isolation and identification of ethyl-*a*-D-galactopyranoside and pinitol from dehulled, full-fat soybean flakes with an azeotropic mixture of hexane:ethanol:water 80:19:1. They found that full-fat and defatted flakes contain 0.6 and 0.3% of pinitol, respectively.

Nearly 0.2% ethyl galactoside was formed in full-fat flakes during extraction with the hexane-ethanol azeotrope but only 0.03% during extraction of defatted flakes. They report that the ethyl-*a*-D-galactopyranoside is bitter and the pinitol somewhat sweet. Taste panel tests indicated that the bitter threshold of the galactoside is 0.5% in water solution which indicates an insignificant contribution to the bitter flavor of soy products.

Plastein Formation and Flavor

In their investigation on the soybean flavor problem, Fujimaki *et al.* (1970) studied the plastein reaction of soy protein as well as other food proteins. When a protein is hydrolyzed enzymatically and the hydrolyzate is then incubated with certain proteolytic enzymes under appropriate conditions, the hydrolysis is reversed and a high molecular weight protein-like substance is formed whose properties are somewhat different from the original protein; the new protein-like substance is called "plastein."

When soy protein is partially hydrolyzed with pepsin the hydrolysate is bitter because of the presence of bitter peptides; on treatment of the hydrolysate with *a*-chymotrypsin under suitable conditions a plastein is formed and the bitterness gradually disappears.

As an example of the plastein reaction, Fujimaki *et al.* (1968), working with acid precipitated soy protein, dissolved 10 gm in dilute HCl (pH 1.6) and incubated it with pepsin (50 mg) at 37° C until the 5% trichloracetic acid (TCA) insoluble fraction reached 20% of the total nitrogenous substances. This part of the operation liberated the flavor components of the soy protein and the system became very bitter. The proteolyzate was then adjusted to pH 7.0 with NaOH, concentrated under vacuum at 40° C to 20% concentration, mixed with *a*-chymotrypsin (100 mg), and incubated at 37° C without agitation. The hydrolysis reaction was reversed and the TCA insoluble fraction increased from 5% to 75% and the bitter taste disappeared.

Of the many enzymes investigated for plastein production the a-chymotrypsin was the most effective, followed by Bioprase and Prozyme which were almost as good. The TCA insoluble fraction of plastein was washed with 50% ethanol and freeze dried to give a powdered protein-like product almost free from bitterness or other undesirable flavors.

In other experiments, Yamashita et al. (1970A) prepared soy plastein which contained 13.22% nitrogen on a dry basis; its amino acid composition was similar to that of the original protein. Based on its essential amino acid pattern, its chemical protein score was 76 compared with a milk chemical score of 67. The in vitro digestibility of the soy plastein with pepsin and trypsin was 84.8 and 76.5, respectively, which was almost the same as that of the original denatured soy protein.

In further studies on the enzymatic modification of proteins in foods, Yamashita et al. (1970A) found that the peptides glycyl-L-leucine (Gly-Leu) and L-leucyl-L phenylalanine (Leu-Phe) were very bitter. In the formation of the plastein the Gly-Leu participated mostly at its leucine terminal, with the results that the glycyl-L-leucyl residues were located at or near the N-termini of the plastein molecule, and the Leu-Phe were at or near the N- and C- termini. In reforming into the plastein molecule, the bitter flavor of the Gly-Leu and Leu-Phe was lost.

Plastein Formation and Nutrition

In another communication, Yamashita et al. (1970B) reported experiments on using the plastein reaction as a means of incorporating methionine and cystine into the structure of soy protein plastein, thus producing a product of potentially greater nutritive value. Since free amino acids will not react with the peptides in a protein hydrolyzate, it was necessary to prepare methionine and cystine containing peptides for the reaction. They were prepared by a pepsin hydrolysis of an ovalbumin preparation (Difco) and from a powdered, acid-treated, wool-keratin. These peptides were used in the plastein reaction on pepsin-hydrolyzed and acid-precipitated soy protein, catalyzed with Nagarse. This experiment was quite successful in the incorporation of cystine in the molecule but not for incorporating methionine. In another experiment, a methionine ethyl ester was used with papain as the catalyzing agent. This combination was successful in introducing methionine in the new molecule. These results indicate the possibility of preparing plastein products having an amino acid composition similar to an ideal protein.

Some Food Uses Tolerant of Soy Flavor

Although much research and testing remain to be done before the flavor problem is adequately solved there are substantial areas of food utilization which do not have a serious problem. Where soy products are used in combination with meat products, such as in sausage type meats, in ground meats, or when they are cooked in combination with fat or oil, the problem is minor. In fact, the cooking of full-fat grits in oil produces a tasty product. Soy products can be incorporated in sweet goods, pastries, and baked wheat products in limited amounts without a serious flavor problem. However, development of new processing procedures to effectively produce a truly bland product without significant loss of solubility of the protein will greatly extend the food uses of the protein.

BREAD AND PASTRIES

Soy Flour History

A crude form of soy flour was developed in the United States as early as 1926 (Brock 1951) but was sold only as a "health flour." However, the food shortage brought on by the second World War stimulated a strong interest in soy flour as a new source of food protein, especially for improving the nutritive value of bread and other bakery products. A number of soybean processors, using both the screw press and solvent methods, instituted production of soy flour for the bakery trade.

At that early period in the food uses of soybeans in the United States there was a serious lack of experience in the processing of soy flour and on how it should be used in bakery products. The crudely produced and improperly used soy flour failed in almost every effort to make a satisfactory loaf of bread; and this first attempt to introduce soy flour to the bakery trade was an unfortunate experience for both the soy processing industry and the baking industry.

Following the war period and with initiation of research on soy problems, the early mistakes of the processing industry were corrected and new and improved procedures for processing soy flour were developed (Chap. 9). A better understanding of the functional uses of soy flour in bakery products was developed, and now there is an ever-increasing number of successful uses for soy flour in the baking and confection industries.

Lemancik and Ziemba (1962), Johnson (1970), and Wolf (1970) have reviewed the functional properties of the various types of soy flour and their many applications in baked products.

The high protein content of soy flour and its high lysine content have suggested its use as a supplement for improving the nutritional value of wheat products (Diser 1962). However, marketing experience in the United States has demonstrated that except for baby foods and school lunch programs, "good nutrition" does not usually attract the buying public. It was found that marketing of soy flour is based more realistically on its functional properties such as water absorption, dough mixing tolerance, color, texture, flavor, and other physical properties. However, it is interesting to note that the governments of Israel and Colombia require by law the addition of 5% soy flour to their bread.

Effect of Soy Flour on Baking Characteristics

An important characteristic of wheat flour is its gluten content, since the gluten determines the extensibility of the dough and loaf volume of the bread. Soy flour is devoid of gluten; when added to bread dough it places an added stress on the gluten, and if too much is added, it will modify the structure of the bread and also reduce loaf volume; thus, there is a limit to the amount of soy flour which can be incorporated in bread. The high protein wheat flours will tolerate a larger addition of soy flour than the weak flours.

Ofelt *et al.* (1954A,B), using a sponge and dough baking procedure, investigated the baking behavior, oxidation requirements, and bread quality of a series of commercial defatted and full-fat soy flours at a level of 5% based on the wheat flour. Samples of full-fat soy flour were collected each month for 6 months from the production of 4 processors, and defatted samples were collected for 5 months. The NSI of the protein (which is a measure of the heat treatment given the flour) for the best grades of the full-fat flour was in the range of 35-45, and for the defatted soy flour was 50-60. The wheat used for all tests was an unbleached commercially milled baker's patent flour containing 0.44% ash, and 11.4% protein on a 14% moisture basis. Preliminary tests showed the need of an oxidizing agent to maintain normal loaf volume. Thus, a series of baking tests with different levels of potassium bromate was made which showed that bromate in the range of 1.0-3.0 mg% on the wheat flour basis resulted in satisfactory dough handling properties and normal loaf volume at the 5% level of soy flour. Fig. 10.1 shows cross sections of two loaves (1) containing 4% NFDM and (2) 4% NFDM fortified with addition of 3% soy flour.

The defatted soy flour had bromate requirements in the range of 3.0-4.0 mg% for normal loaf volume and good dough handling properties. It was reported also for both types of soy flour that an increase of 1% water absorption was desirable for each 1% of soy flour added.

Because some of the countries of Eastern Europe as well as other areas of the world use high extraction wheat flour, without the addition of sugar and shortening, studies were made by Ofelt and Smith (1955) on the importance of oxidation when defatted soy flour, at the 5% level, was added to a high extraction flour. The straight dough method was used to yield 1 lb loaves. The formula for the basic control loaf included high extraction wheat flour, water, 2% yeast, and 2% salt. A reference loaf containing 5% sugar and 5% shortening in addition to 5% soy flour with optimum bromate was included. The results for 1 series of baking tests using a 90% extraction flour, containing 14.2% protein, are shown in Fig. 10.2.

FIG. 10.1. CROSS SECTION OF TWO BREAD LOAVES: (LEFT) CONTAINS 4% NFDM and (RIGHT) 3% SOY FLOUR WITH 4% (NFDM)

FIG. 10.2. LOAVES BAKED FROM 90% EXTRACTION FLOUR (14.2% PROTEIN) MILLED FROM HIGH PROTEIN HARD RED WINTER WHEAT

Loaf 2 is the basic loaf without sugar and shortening. Loaves 18, 19, 20, 21, and 22 contain an added 5% defatted soy flour and 0, 4, 6, 8, and 10 mg percentage potassium bromate, respectively. Loaf 23 shows the effect of adding 5% sugar and shortening to the soy loaf 21.

Loaf No. 2 is the basic loaf without sugar or shortening, and loaves 18, 19, 20, 21, and 22 contain 5% defatted soy flour and 0, 4, 6, 8, and 10 mg% of potassium bromate respectively. Loaf 23 shows the effect of adding 5% each of sugar and shortening to soy loaf 21. These data show that the deleterious effects of soy flour on dough handling properties and bread characteristics are eliminated for all practical purposes by the addition of optimum amounts of oxidizing agents, thus giving a bread equal or superior to the control loaf.

Turro and Sipos (1968) and Turro (1970A,B) have reported on the effects of various soy protein products on bread characteristics including a special process soy flour containing 60% protein prepared from a combination of 70% soy protein concentrate and 50% protein soy flour. They found that the higher the protein in the soy flour the higher the water absorption capacity, and that the higher the water-soluble protein the better the grain and crumb color of the bread. Their work also demonstrated that in using the sponge and dough baking procedure the soy flour should be added to the dough. The soy flour has a weakening effect when added to the sponge.

Using rats, the nutritional value of bread containing 3% NFDM was compared with bread containing 3% of the special process soy flour. A semisynthetic ration was used in which the crumb comprised 90% of the diet. For diets at 12.6% protein level the weight gains were equal for both types of protein supplement; the PER for the soy protein was 1.02 and for the NFDM was 0.97. Thus, at the 3% level the special soy flour improves nutritional value as much as the NFDM.

Adler and Pomeranz (1959) investigated the effect of lecithin on loaf volume of bread supplemented with soy flour, using different levels of flour extraction and wheat strength, without sugar and shortening. They found that loaf volume can be maintained by increasing the level of oxidizing agent to the optimum and adding lecithin to the dough. The amount of lecithin for 3, 6, and 9% defatted soy flour based on the wheat flour was determined.

Paulsen (1968) patented the treatment of soy flour with hydrogen peroxide and a calcium salt or acid. Paulsen and Horan (1965) have reported on the comparative baking quality of the chemically-treated and heat-treated soy flours in bread. They found that the functional properties of the chemically-treated soy flour, such as texture, color, flavor, and other baking characteristics, were superior to those of the heat-treated flours.

Pollock and Geddes (1960A,B) investigated laboratory preparations of soy flour and flour fractions on white bread characteristics. They found that 1 hr of heat treatment of soy flour containing 7.9% moisture at 75° C did not appreciably affect the water dispersibility of its protein. They reported, also, that inclusion of raw soy flour in farinograph doughs, at levels of 1–5%, imparted to normal and rest period curves the characteristics of a stronger flour, the effect increasing with the soy flour level. They found that the dialysate of the whey fraction had a very injurious effect on bread characteristics whereas the protein from the whey had excellent baking quality.

Two soy fractions, Gelsoy (DeVoss *et al.* 1950) and soy whey solids, were investigated by Ofelt *et al.* (1953) for their effect on crumb softness. Gelsoy is a water-soluble protein and solubles concentrate (P + S) separated from ethanol washed soybean flakes; whey solids (S) is the product occurring in the water-soluble fraction recovered from the isolation of soy protein when precipitated at pH 4.5. They reported that 1% Gelsoy or 1.0–1.25% of whey solids gave as much crumb softness as 0.3% polyoxyethylene monostearate (POEMS). They suggest that the bread softening effect of whey solids may be due to a nonprotein component of the soybean.

A recent report by Hoover and Tsen (1970) indicates further advancement in methods of supplementing bread with soy protein products without loss of loaf volume and other physical properties. They found that by using sodium stearoyl-2-lactate (SSL), calcium stearoyl-2-lactate (CSL) or ethoxylated monoglyceride (EM) they can add up to 12% of soy flour to the dough without loss of loaf volume. The use of these compounds also permits a substantial reduction in the use of shortening. The SSL, CSL, and EM are effective, also, in the use of soy flour and oilseed protein concentrates in pastries as well as bread.

When the continuous baking process was introduced into the U.S. baking industry it was found that the use of nonfat dry milk NFDM had a deleterious effect on dough handling characteristics, and use of this additive had to be discontinued. However, it was soon discovered that combinations of certain types of soy flour with cheese whey and other ingredients made a satisfactory replacement for the NFDM, as reported by Guy *et al.* (1969).

Soy Protein Isolate in Bread

Mizrahi *et al.* (1967B) investigated the functional and nutritional properties of soybean protein isolate and of calcium-precipitated protein as a bread supplement. They found that isolated protein can be used in bread up to a level of 6% without impairment of loaf volume, taste, color, and odor, providing 1% lecithin is used in the premix. They estimated that addition of 6% isolate is equivalent, in protein supplementation, to 10–12% of defatted soy flour or 14–16% of full-fat soy flour. They found that potassium bromate, which was effective in maintaining loaf volume for soy flour, was not effective for the isolate. The calcium-precipitated protein was superior to the acid-precipitated protein in mixing tolerance. The protein isolate increased water absorption 1%, and calcium-precipitated protein, 1.5% for each percentage added; and there was no change in crumb texture.

Although isolated protein is inferior in nutritive value to soybean meal when used as a sole source of protein, the results of feeding tests by Mizrahi *et al.* (1967B) showed no difference in the PER of bread containing isolated protein and that of bread supplemented with soybean flour at isonitrogenic levels. Increasing the supplement of isolated protein from 6% to 10% did not increase nutritional value.

Soy Flour and Flavor

It is well recognized that present types of soy flour have a characteristic flavor and when used in bread at too high a level will introduce a foreign or unusual flavor, which has a serious influence on its acceptance.

Ofelt *et al.* (1952) compared the flavor of bread containing 5% defatted soy flour with bread containing 4% NFDM using a taste panel of 225 people of the USDA Northern Regional Research Laboratory.

For making the comparison the triangle taste test was used in which the taster receives three samples, 2 containing soy flour and 1 containing NFDM, or 2 with NFDM and 1 with soy flour. The taster is asked to identify the two samples which are alike.

Two tests were made on each bake; the first was on 1-day-old bread and the second, on 4-day-old bread. The probability of choosing the correct pair is 33.3%; the overall average correct identification was 33.67%. Thus, no significant difference between the flavor of the 2 breads was detectable at the 5% level of soy flour content. Firmness tests on the crumb of the two breads indicated that the rate of firming was about the same. It appeared probable from these results that some of the modern types of soy flour could be used at a level of 10–12% of the wheat flour and yet impart an acceptable flavor.

Enzyme Active Soy Flour

The enzyme active or raw soy flour when added to a dough made from unbleached wheat flour at a level of 0.5–1% will bleach the color of the bread whereas at levels above 3% it will modify adversely the dough handling properties. This color improvement is effected by the active lipoxidase in the soy flour. When the lipoxidase is coupled with an unsaturated soap such as sodium linoleate it is an effective bleach for the carotene in the wheat flour (Haas and Bohn 1934A,B; and Haas 1934A,B). The British bakeries add 0.7% of the enzyme active soy flour to about 90% of their bread, according to Wood (1967), and report an improvement in both color and flavor.

The α-amylase is another enzyme in unheated soy flour, which has been reported by Ofelt *et al.* (1955) to affect dough characteristics adversely. They compared the amylograms of starch paste containing α- or β-amylase with amylograms of starch paste to which they added raw or heat-treated soy flour. Their tests demonstrated that the α-amylase in the soy flour was responsible for the reduction in the viscosity of the starch. From these tests it appears that the α-amylase in unheated soy flour, when used in bread dough above 2–3%, is responsible for slack or sticky doughs. The change in viscosity produced by the addition of 5% of raw soy flour was estimated to be equivalent to the effect obtained by the addition of 12 Sandstedt-Kneen-Blish units of α-amylase. It was shown that 5 min of steam treatment at atmospheric pressure inactivated the α-amylase. In a survey of 20 commercial soy flours only 1 demonstrated bleaching action, which indicated that it had not had sufficient heat treatment to inactivate the enzymes.

Learmonth and Wood (1960) repeated the experiments of Ofelt and Smith (1955) on the effect of raw soy flour on the amylograms of starch paste and obtained the same reduction in viscosity. However, Learmonth and Wood compared the activity of the a-amylase of malt with that of raw soy flour and concluded that the activity in raw soy flour was not sufficient to affect the viscosity of wheat flour dough. Thus, it appears that if the a-amylase of malt acts the same as the a-amylase in soy flour, then some other explanation must be found for the softening of wheat flour dough on the addition of 5% of raw soy flour.

Learmonth (1952) has shown that aqueous extracts of soybeans inhibit the proteolytic enzymes in malt on a gelatin substrate and, also, that the extract of wheat flour and papain inhibit the proteolytic enzymes on a gluten substrate. The inhibiting effect is compared to that of potassium bromate. Learmonth (1958) has reported also that the papain inhibiting factor in soybeans is concentrated in the germ. The addition of 0.2–0.4% of raw soy germ was shown to alter the Chopin alveogram of wheat flour in the same way as the addition of oxidizing improvers. He reports, also, that the antiproteolytic factor is insoluble in water below pH 5.66 and that the effect in bread doughs is to reduce or prevent proteolytic breakdown during fermentation. A similar proteolytic inhibitor was reported to be present in other beans and peas.

Soy Flour in Britain

Wood (1967) has reviewed the use of soy flour in bread in Britain and states that 90% of the bread produced in his country contains soy flour. He describes the two types of soy flour used in Britain as enzyme-active and heat-treated, or enzyme-inactive. He describes their bread as (1) fermented and (2) mechanically developed, and states that soy flour has important functions in each type. Brown (1966) reported on uses of soy flour in cake and other formulations, as well as bread, in the United Kingdom.

In the fermented bread, the enzyme active soy flour when used at the recommended level of 2–3 lb per sack of flour with 1 1/2–2 times its weight of additional water improves the color, flavor, and keeping quality of the bread through the action of the lipoxidase. The bread also has a brighter crust and lower production costs.

In the mechanically developed bread, part of the fermentation is replaced by mechanical action in a mixer designed for the purpose. A high level of oxidizing agent (75 ppm of ascorbic acid) and 0.7% of raw soy flour by weight is required. They found that use of soy flour with the added water gave much the same improvement to their bread as in the fermentation process.

Detection of Soy Flour in Wheat Flour

Pomeranz and Miller (1960) have described a method of detection and estimation of soy flour in wheat flour by examination of the sample under low magnification using ultra violet light. The presence of a canary yellow fluorescence from the soy flour particles can be used to detect or to estimate various quantities in wheat flour by counting and comparing standards of known composition. The fluorescence is unaffected by heat treatment of the soy flour and detection of 0.01% is possible. On wetting or on wetting and subsequent redrying in air, the fluorescence of the soy flour particles is quenched, probably by leaching of the fluorescing substance.

OTHER BAKED GOODS

General

The various types of soy flour are used in a wide range of baked goods and confections at levels up to 10% of the weight of the wheat flour, and in some instances a 20% level has been recommended. The addition of soy flour in baked goods requires the addition of 1 to 2 times its weight of water.

The solvent extraction of soybeans with hydrocarbons leaves a portion of the phospholipids of the bean in the flakes. These phospholipids, or lecithins as they are called in industry, improve oil emulsifying capacity of the soy flour and help to produce smoother doughs and mixes. However, a higher emulsifying capacity is obtained by use of lecithinated soy flour (a refatted flour). Other advantages in using soy flour in baked goods include increased mixing tolerance, easier machining, improved moisture retention, improved crust color and longer shelf-life of most products, while at the same time reducing the requirements of shortening and NFDM.

Soy flour can be used in most formulations or recipes by replacing a predetermined percentage of wheat flour with soy flour. When exactly balanced formulations are used and full-fat soy flour is to be replaced with the defatted flour, then only 0.8 parts of the defatted is used in place of the full-fat product.

A partial list of bakery products in which soy flour is incorporated are bran muffins, coffee cake, devil's food cake, pound cake, piecrust, pancake mixes, cookies, fruit bars, doughnuts, and others. Turro and Sipos (1970) have described the use of a special high protein 60% soy flour in sponge cake, pound cake, devil's food cake, and yellow layer cake. They observed that NFDM can be replaced at levels between 50 and 100% by the high protein special soy flour. At the 50% level they found that, except for an increased water absorption, no formula changes were necessary. Formulas or recipes for using soy flour in cakes and other baked goods are available from soy flour producers (see Appendix for list of manufacturers).

Doughnuts

Full-fat or lecithinated soy flour (5–15% lecithin) with a high NSI when added to a doughnut mix reduces fat absorption and enhances the eating quality of the doughnut.

Rusoff *et al.* (1961) prepared doughnuts supplemented with toasted soy flour, cottonseed and peanut flours, whole fish, and fish fillets as a potential high protein food for countries that are short of inexpensive protein. The formula contained approximately 20% protein on a dry solids basis and 15% protein in the finished product or "as is" basis. Mixes were prepared which required only the addition of water at the dough mixing stage. The doughnuts were deep fat fried at 375° F for 90 sec. The short frying period was used to diminish nutritional loss through destruction of lysine, methionine, and cystine. The nutritional value of these products was assessed by feeding rats at a 10% protein level. Table 10.2 gives a tabulation of PER (Protein Efficiency Ratio) values of these doughnuts with casein as a control. There was very little loss in essential amino acids with the short period of frying, as determined by amino acid analyses of the doughnut mix and the fried doughnuts.

Snack Products

The snack products industry in the United States has grown very rapidly in recent years and was reported in a statistical survey by Snack Food (Anon. 1970C) to have reached $3.5 billion in annual sales in 1969, with an average annual increase of 5.9% over the decade 1960–1969. The popularity of snacks was attributed, in part, to more leisure time in modern living.

The basic raw materials for most snack items are corn meal, wheat flour, potato flour, oat flour, tapioca, milo, and some modified starch products. These high starch products can be made easily into a wide variety of textures from light, fragile, highly-puffed items to dense crisp products (Feldberg 1969; and

TABLE 10.2

RESULTS OF RAT FEEDING TESTS[1]
ON HIGH PROTEIN DOUGHNUTS

	Wt Gain (Gm)	Protein Quality Value (Relative to Casein=100)	PER at 4 Weeks
Casein control	57	100.0	2.12
Soy doughnut mix	95	119.3	2.53
Soy doughnut	75	101.9	2.16
Cottonseed doughnut	44	69.3	1.47
Peanut doughnut	42	65.1	1.38
Whole fish doughnut	73	105.2	2.23
Fish fillet doughnut	71	104.4	2.13

Source: Rusoff *et al.* (1961)

[1] Diet of 10% protein.

Schaeder *et al.* 1969). Soy flour or other soy protein products can be added in snack formulas up to about 15% to make the dough pliable and easy to handle, to modify crust color, and to increase shelf-life. The use of soy protein products at the 10–15% level can substantially upgrade the nutritional value of the cereal-based products, which is a desirable improvement since, customarily, snacks are not eaten at the regular dining periods.

Sanderude (1969) has pointed out the many advantages of the use of the extruder in processing the cereals into snacks. The extruder is a means of cooking cereal products under controlled conditions of moisture, time, and temperature into expanded products having various shapes, sizes, and colors. Fig. 10.3 shows an experimental extruder for cooking soybeans or cereal products or a combination of the two for research on the production of new food products including snacks.

Zick (1969) has reported that cheese is one of the most important snack flavors but because of the steady increase in the cost of dehydrated cheddar cheese the trend is toward dilution with other sources of flavor including artificial cheese flavor. Zick has reported that autolyzed yeast and hydrolyzed vegetable proteins with low levels of lactic acid, to which may be added salt and monosodium glutamate, are used to obtain a cheese-like flavor.

Since imitation dairy products are increasing in popularity it is to be expected

Courtesy of USDA Northern Regional Res. Lab.

FIG. 10.3. AN EXPERIMENTAL EX-
TRUDER FOR COOKING
SOYBEAN OR CEREAL PRO-
DUCTS OR A COMBINATION
OF THE TWO IN RESEARCH
ON SNACKS AND OTHER
PRODUCTS

that imitation cheese will increase in importance for use as a snack flavor. It is anticipated that a good cheddar type cheese can be made by a combination of milk curd and soy flour or other soy protein products by normal cheese processing procedures or suitable modifications if legally permitted. For flavor purposes selective fermentation of soy flour with added casein or whey protein might be one pathway to produce a good cheese flavor at a cost below that of the dehydrated cheddar cheese.

BREAKFAST CEREALS

Soy flour and other protein concentrates have found a limited use in breakfast cereals. Penty (1947) patented the use of solvent-extracted soy flour in breakfast cereals, advocating the addition of 20–25% soy flour with 70–80% yellow corn. Luke (1948) claimed the manufacture of a flaked cereal-soy flour combination for food use.

In more recent years new processes have been developed in Europe as well as the United States using mixtures of cereal grains with added flavor, color, vitamins, minerals, and proteins to improve flavor and eye appeal as well as nutritive value. However, the use of soy flour products in cereals has been quite limited because of their characteristic flavor. New soy protein processing methods and projected flavor improvement can be expected to eliminate the flavor problem in breakfast cereals and increase the use of soy protein products in this application.

MACARONI PRODUCTS

The wheat flour used in making macaroni and noodles can be supplemented with soy flour or other soy protein products in the range of 12.5–25% to obtain products which will retain additional moisture although the color will be darkened somewhat. When soy flour is added to macaroni, U.S. regulations require a minimum of 12.5%.

Paulsen (1961) added soy flour to pasta products at levels of 12.5, 17, and 25%, and described methods for evaluating water absorption and losses of solids and nitrogen on cooking. The combination of soy flour with semolina increased the firmness of spaghetti subjected to long cooking periods and improved its nutritional quality.

DAIRY TYPE PRODUCTS

Imitation Milk

Agricultural products in marketing are always subject to competition from synthetic products or replacement with lower cost farm products, exemplified by the appearance in the marketplace over the years of synthetic fibers, urea in animal feeds, and oleomargarine, to name a few. The use of soy milk and other soy protein products to extend or replace dairy products is also a striking example of this phenomenon.

The dairy industry is now concerned with increasing competition from imitation milk, imitation cheese, nondairy frozen desserts, nondairy coffee whiteners, nondairy yogurt, and others (USDA 1969B; Hetrick 1969; Knightly 1969). The dairy industry itself has been forced to develop an interest in processing filled milk, low fat milk, and imitation products to meet competition outside their industry. However, present U.S. Federal and State dairy laws greatly restrict competition with modified or imitation dairy products and thus retard new developments in this area (USDA 1969B). For example, filled milk, in which the butterfat of whole milk has been replaced with a nondairy fat, cannot be marketed either interstate or within more than 30 States, and the marketing of any product containing milk or nonfat milk solids is governed by various State and Federal dairy regulations. However, the casein and milk whey fractions have been classified as chemical products and can be used to detour these regulations. In 1968–1971 (Federal Register 1968; Anon. 1970A, 1971) proposals appeared to establish standards for governing the production quality and composition of imitation dairy products. If these come into the market in competition with dairy products, it is likely that government standards of nutritional quality also will be established for the benefit of the producer as well as the consumer.

Imitation or synthetic milks should not be confused with—but do include—the category of "filled milk" (USDA 1969B), a product made by replacing the butterfat in cow's milk with a vegetable oil, usually with coconut oil. U.S. Government regulations, promulgated in 1923, forbid the use of nondairy products, except for some vitamins, in filled milk. Presently interstate trade is not permitted, nor do present regulations specify the solids concentration or the nutritional quality of the product. However, it was reported (Anon. 1971) that at long last the U.S. Food and Drug Administration was considering asking the U.S. Congress to repeal the Filled Milk Act, Filled Cheese Act, Dry Milk Solids Act, and other related laws, steps recommended at the White House Conference on Food, Nutrition and Health of December, 1969, and discussed at the follow-up conference of February, 1971. Such a repeal, if enacted, would generate accelerated activity in the field of extended or simulated dairy products.

Soy Milk

In 1925, Dr. Harry Miller, while a medical missionary in China, produced and used in nursing homes and hospitals a milk-like product made from soybeans for feeding babies, children and nurses (Miller 1962). The first commercial development was in Shanghai in 1935. On Dr. Miller's return to the United States, he introduced into this country the use of soy milk, fortified with vitamins and minerals (Smith and Beckel 1946). Although his product was intended for general use, its most successful area has been for feeding babies. In the United

States Heiner *et al.* (1964) have reported that up to 7% of U.S. babies are allergic to cow's milk. The human infant finds soy milk palatable; and modern formulations have found a specialty market for feeding babies.

Soy milk can be processed from whole soybeans or full-fat soy flour. However, the traditional product is made from good quality whole soybeans. The beans are thoroughly washed and soaked in water 3 hr or more depending on the temperature of the water. The beans are then ground wet with the addition of enough water to give the desired solids content in the final product. The ratio of water to beans on a weight basis should be approximately 10:1. The slurry is heated at near its boiling point for 15–20 min to improve its nutritional value and flavor and to sterilize the product. Upon removing the insoluble residue the resulting milk is a highly stable oil emulsion.

Hand *et al.* (1964) and Van Buren *et al.* (1964) have made extensive pilot plant investigations for selecting equipment and procedures for producing soy milk, with special studies on heat processing and flavor. In a taste evaluation of their products, they found that the classical method of water extraction of soaked soybeans gave the best product with respect to flavor and consistency. In comparing nutritional value of soy milk prepared by several different procedures, they found essentially the same PER values for milks when they were spray dried, vacuum roller dried, atmosphere roller dried, or freeze dried.

Hackler *et al.* (1965) stated that the time and temperature of heat treatment of soy milk from 1 to 6 hr at 93° C had no adverse effects on growth of rats, PER, or available lysine. However, cooking at 121° C for a short period decreased available lysine, and spray drying with an inlet temperature of 277° C or above caused a drop in available lysine and nutritional value.

Wilkens *et al.* (1967) found information which indicates that the lipoxidase system in the soybeans promotes off-flavors in soy milk. They found that by using high temperatures and a rapid hydration grinding process on dehulled soybeans they can produce a nearly bland soy milk. They state that an acceptable flavor was produced by grinding dehulled dry beans with water at a temperature ranging between 80° and 100° C and maintaining that temperature for 10 min. Adding antioxidants is said, also, to permit the use of lower temperature in controlling lipoxidase activity.

Fukushima and Van Buren (1970) investigated the effect of several factors, such as heat treatments, pH, and reducing agents on the redispersibility of the protein in dried soy milk. When the soy milk was dried in an oven at 50° C they found that the initial heating decreased redispersibility but that further heating reversed the effect. Increasing the concentration of solids decreased redispersibility and reducing agents increased dispersibility. They attributed the decrease in dispersibility to the exposure of -SH groups which had been masked in the interior of the molecule.

Badenhop and Wilkens (1969) have reported that when high temperature grinding of soybeans is preceded by an initial soaking in water, 1-octen-3-ol is formed in the milk; however, the amount varied with the pH and other factors, and an alkaline condition decreased the formation of the 1-octen-3-ol. In another investigation, Badenhop and Hackler (1970) made a more detailed study of the effects of soaking soybeans in several concentrations of sodium hydroxide between 0.048 and 0.097 N on the flavor, nutritive value, niacin content, and other properties prior to the extraction of the milk. They found that pretreatment in 0.05 N sodium hydroxide produced a milk with a pH of 7.37, which a taste panel judged to have a significantly better flavor. Assays showed it to have a higher niacin content than when the pretreatment was in water. There was a slight loss in cystine, which was not thought to be serious considering the beneficial effect on flavor.

A recent development in processing soy milk is the procedure recommended by Mattick and Hand (1969). After soaking the beans in water, they are ground at a temperature above $80°$ C to inactivate the lipoxidase before it can have a significant effect on flavor. This new process is presently being market tested under the supervision of Bourne (1970) in the Philippines at the University of the Philippines, Laguna.

Lo *et al.* (1968A,B) have investigated the changes in the viscosity of soy milk as the solids concentration is increased by evaporation. Using a Brookfield viscometer they found that at 5% solids concentration the viscosity is dependent upon the force of shear as it is directly related to the speed of rotation of the spindle; this indicates that soy milk is a non-Newtonian fluid. At 20% solids concentration the apparent viscosity decreased with increasing time of shear, thus demonstrating the thixotropic nature of soy milk. When the soy milk was evaporated to 27% solids it became a thick, resilient gel. The addition of sucrose before concentration had the effect of increasing the viscosity of the milk. From this work it appears unlikely that an evaporated milk of more than 15% solids concentration would be possible.

Filled Milk

Modler *et al.* (1970) have reported on the possible use of a lightly hydrogenated soy oil rather than the very stable coconut oil in filled milk. Their objective was to prepare a filled milk with a fat or oil that would be reasonably stable to oxidation and yet contain significant quantities of unsaturated and essential fatty acids.

Filled milks were formulated using fresh skim milk, monoglyceride emulsifiers, and vegetable oils. All samples were warmed to $130°$ F, pasteurized at $170°$ F in a high-temperature short-time (HTST) heat exchanger at 500/5000 psi and cooled to $36°$ F. The milk was examined organoleptically and chemically.

Filled milks were prepared and tested according to the above procedure from corn, cottonseed, peanut, olive oil, safflower oil and lightly hydrogenated soybean oil. All of the milks except that containing the lightly hydrogenated soy oil were rejected because of the objectionable oxidized off-flavors.

The studies indicate the possibility of using lightly hyrogenated soy oil in filled milk and yet retaining sufficient unsaturation of the fatty acids of the oil to be effective in a diet for maintaining low cholesterol. They point out that the major advantage of using soybean oil in filled milks lies in its high content of linoleic acid. It is stated also that soybean oil is one of the cheapest vegetable oils and is nutritionally superior to coconut oil or milkfat.

Soybean Cheese

Hang and Jackson (1967A,B) prepared a fermented cheese-like product from soy milk using *Streptococcus thermophilus* as the fermenting organism. The cheese starter was developed by adding 200 mg of freeze-dried culture to a flask containing 200 gm of autoclaved soy milk which was incubated at 32° C for 15 hr. One milliliter of this mother starter was then added to 200 gm of autoclaved soy milk.

The autoclaved and inoculated soy milk was converted into soy curd or cheese by three different methods of coagulation: (1) by the addition of calcium sulfate, (2) by addition of acetic acid to pH 4.5, and (3) by lactic acid fermentation at 41° C. In the lactic fermentation, after cutting the curd, it was cooked at 48° C, placed in the hoop and pressed for 24 hr.

The major differences in the three curds was in the yield of precipitated protein, and in moisture content and hardness. The acetic acid coagulation gave a yield of 67.8% of the total protein, whereas the yields for the calcium sulfate precipitation and the lactic fermentation were 54.1% and 55%, respectively. The curd from the lactic acid fermentation produced cheese with the best body and texture.

In the second investigation, Hang and Jackson (1967B) modified their product with the addition of a rennet extract and with skim milk. The main role of the starter organism appeared to be acid production, which is essential for this operation. The rennet did not reduce the time of coagulation but it did improve the body and flavor of the cheese. The skim milk, along with the rennet and starter organisms, combined to reduce the time from addition of the starter to cutting the curd. The cheese was aged 63 days at 20° C and the flavor was reported to change from clear fresh to clear mild. During the aging period there was a loss of 8-10% moisture. The cheese made with soy milk and starter only, remained fresh and elastic, while that made with the rennet extract generally became softer and mild. Including the skim milk along with the rennet extract seemed to improve the flavor of the finished product. The authors indicate that "this product can be preserved without refrigeration as long as it has been coated with melted paraffin."

Obara (1968) investigated the making of a cheese-like product by treating soy curd directly with proteinases rather than by using traditional fermentation processes. To prepare the curd he made a water extract of the beans at the temperature of boiling water for 2 min at a water to bean ratio of 10:1. The curd was precipitated at 70° C with calcium sulfate in the concentration range of 0.03–0.04 N, based on the soy milk. The enzymes investigated were papain, Pronase, Bioprase, Molsin and trypsin, and the ripening process was at a temperature of about 17° C. Also, he investigated enzyme combinations consisting of papain and Bioprase, papain and Pronase, and papain, Bioprase, and Pronase.

In comparing the results of single enzyme treatments the papain gave the best results with respect to flavor and texture whereas the trypsin had almost no activity. The activity of Molsin, which has maximum activity at pH 2.3, was unsatisfactory. A combination of enzymes had a higher rate of activity than was obtained with a single enzyme, and the combination of papain, Bioprase, and Pronase gave the highest rate of ripening as well as the best flavor and texture. His soy cheese on a 54% moisture basis contained 24% crude protein and 15% fat.

Using rat feeding tests, the nutritional value of the cheese produced under optimum conditions with the use of papain was compared with raw and boiled soybeans and with natto.

The protein efficiency ratio (PER) and biological value (BV) for the Obara cheese was reported as 2.7 and 63, respectively; whereas the corresponding values for casein were 3.1 and 83.1

Imitation Cream Cheese

Recently, imitation cream cheese formulations based on sodium caseinate have appeared in the institutional market (Ziemba 1970), and also in the retail market. However, Circle and Johnson (1958) described a similar product using soy protein isolate as the protein base.

Coffee Whiteners, Whip Toppings, and Frozen Desserts

In recent years a number of companies have produced and marketed coffee whiteners, whip toppings, and frozen desserts in competition with their dairy counterparts (cereal and coffee creams, whipping cream, and ice cream). Formulations for these and other imitation dairy products are listed by Arenson (1969). A dry mix suitable for soft serve frozen dessert based on soy protein isolate was described recently (Anon. 1967A). Several whip topping patents have issued, claiming to contain soy protein derivatives, in dry mix form (Katz 1969; General Foods 1969) and in frozen form (Lorant 1969). Liquid whip toppings containing soy protein derivatives have been available in retail stores, packaged in aerosol dispenser cans.

Most formulations of coffee whiteners, especially those in spray-dried powder form, are based on sodium caseinate, but other liquid products have appeared in the market containing soy protein derivatives. Some formulations may have a tendency to "feather" (precipitate), or "fat-off" (form fat globules) on addition to coffee. Melnychyn and Stapley (1969A,B) claimed to alleviate this problem by a process of acylating the protein.

Yogurt Type Products

Yamanaka *et al.* (1970) have developed a yogurt type of beverage using a combination of *Lactobacillus bulgaricus* and *Streptococcus thermophilus* for their fermentation. However, they state in their report that specific organisms are not critical since substantially the same results were obtained with other organisms now in use for processing yogurt and other sour milk beverages. The two organisms were cultured separately in a 10% dispersion of skim milk solids at 37° C for 18 hr and then combined for the yogurt fermentation.

They have found, also, that the addition of certain amino acids to their beverage will mask the characteristic flavor of the soy protein, which makes possible the use of a substantial level of soy protein in their product. One of the amino acid combinations used for masking the soy flavor, made up in grams, is as follows: L-alanine 3.5, L-arginine 3.7, L-aspartic acid 2.5, sodium L-glutamate 34.0, L-lysine 6.0, L-methionine 5.0, and glycine 7.5.

One of several formulations shown by the authors for yogurt fermentation is made up, in grams, of NFDM 30, isolated soy proteinate 42, sucrose 26, artificial sweetener 0.57, amino acid mixture 0.2. These ingredients are dissolved in about 300 ml deionized water and the pH adjusted to 5.8 with citric acid and additional water to a total volume of 388 ml. The liquid medium is sterilized at 60–65° C for 30 min, cooled to 37° C, and artificial flavoring is added. The composition is then inoculated with 12 ml of the mixed culture, fermented at 31° C for 6 hr, and then refrigerated.

In a second example, the 0.2 gm mixture of amino acids used in the first preparation was replaced with 0.2 gm of L-proline. In a third example, they used a mixture of 0.12 gm of proline and 0.2 gm of L-alanine with results that were equally as good in masking the soy flavor as in the first preparation. After these several tests the authors were still uncertain as to which of the amino acids or amino acid combinations were responsible for masking the soy flavor.

COMMINUTED MEAT PRODUCTS AND MEAT ANALOGS

Comminuted Meat Products

Soy flour, grits, protein concentrates, and isolates are used extensively in comminuted meat products. They are used in processing frankfurters, bologna, nonspecific meat loaves, meat balls, meat patties, salisbury steak, chili con carne, luncheon meats, and similar items.

The market for frankfurters and bologna is one of the largest outlets for soy protein and this market is far from saturated. It has been reported (Rakosky 1970) that approximately 1.5 billion pounds of frankfurters were processed in the United States in 1968. If 2% soy protein isolate were to be used in this quantity of frankfurters, 30 million pounds would be required. Although the amount of soy protein used in this area is not known, it is quite apparent that present production would satisfy only part of the potential market.

Present use of soy protein products in comminuted meats is considered to be primarily functional. However, in many of these products the soy protein level is high enough to have a substantial effect on nutritional value. By increasing the protein level the proportion of animal fat can be reduced, a factor of interest to people having a high cholesterol problem. Table 10.3 lists the comminuted meat products in which soy products are used, and gives the permitted level according to the present Federal regulations as summarized by Rakosky (1967).

TABLE 10.3

MEAT TYPE FOODS IN WHICH SOY PROTEIN
PRODUCTS ARE USED

Manufactured Product	Soy Product and Permitted Level	Comments
Cooked sausage[1]	Soy flour 3 1/2% Soy protein concentrate 3 1/2% Isolated soy protein 2%[2]	Individually or collectively with other approved extenders. Where isolated soy protein is used, 2% is equivalent to 3 1/2% of others.
Fresh sausage[1]	Same as cooked.	Same as above.
Chili con carne	Soy flour 8% Soy grits 8% Soy protein concentrate 8% Isolated soy protein 8%[2]	Individually or collectively with other approved extenders.
Spaghetti with meat balls[1] Salisbury steak[1]	Soy flour 12% Soy grits 12% Soy protein concentrate 12% Isolated soy protein 12%[2]	Same as above.
Imitation sausage, soups, stews, non specific loaves, scrapple, tamales, meat pies, pork with barbecue sauce, beef with barbecue sauce, patties	Sufficient for the purpose.	Provided meat and moisture requirements are met where such requirements may exist.

Source: Rakosky (1967)

[1] Note: The use of soy protein products in both cooked and fresh sausage, meat balls, and salisbury steak must be shown on the label in a prominent manner, contiguous to the name of the product.

[2] Isolated soy protein must contain 0.1% TiO_2 (as Ti) as a tracer for analytical purposes.

Rock *et al.* (1966) reported that many of the problems of sausage making are related to the wide variation in the raw material or meat trimmings used in the process. These variations are the result of age, diet, weight and sex of the animal, anatomical origin of the trimmings, and whether the latter are fresh or frozen. They state that the proper use of soy protein products in the formulations as binders compensates for these variations.

Rakosky (1967) has described the processing of frankfurters and bologna as a problem of finely comminuting the meat materials and additives by chopping, flaking, or grinding into a continuous matrix and encapsulating the finely divided fat particles. In this emulsion the salt-soluble muscle proteins, myosin and actomyosin, function as a heat coagulable continuous phase. The function of the soy protein is to supplement the myosin and actomyosin as an emulsifying and encapsulating agent, to prevent fat separation, and to hold the meat juices, especially during cooking. In federally inspected plants, soy flour and protein concentrate are permitted in frankfurters to the extent of 3.5%; soy protein isolate to the extent of 2%. The latter must be tagged with titanium dioxide at the level of 1000 ppm Ti for analytical detection purposes.

Circle *et al.* (1964) studied the effects of heat, protein concentration, and other factors on the rheology and gelation characteristics of protein isolate in water dispersions, properties which contribute to its functional use in comminuted meats. They found that in water dispersion the viscosity increases exponentially with increasing concentration. At a concentration of 7% or higher, heat causes thickening and gelation, the temperature threshold being 65° C. The rate of gelling and gel firmness depends primarily on temperature, time of heating, and protein concentration.

The functional effectiveness of soy flour, concentrate, and isolate in comminuted meats is dependent to a degree on the content of water-dispersible protein, but not entirely. Soy protein isolate, which is almost completely soluble in water at its pH of 6.9-7.2, and which contains more than 90% protein, is, accordingly, the most effective of the several soy products used in comminuted meats, but also the most expensive. There exists the possibility in sausage making that the emulsion in the finished cooked frankfurter will break, and the separated white fat particles will create an undesirable appearance, requiring that the product be reprocessed. Thus, it is important to use additional emulsifier-binder as insurance against a processing failure.

A good example of both the functional and nutritional use of soy protein is in making meat patties where soy flour grits, or the concentrate may be used as described by Rakosky (1970). The amount of these products which can be used is designated by the Meat Inspection Division of USDA as "sufficient for the purpose." In making patties it is necessary to add water at 2 to 3 times the weight of the soy protein. The primary function of the protein is to improve dimensional stability of the pattie while cooking and to decrease cooking losses. When properly used the patties will be tastier, will have higher protein and lower fat, and thus be better balanced nutritionally. Huffman and Powell (1970) have reported that patties made with as little as 2% soy grits had higher tenderness scores than without soy.

Rakosky (1970) has discussed, also, the use of soy products in preparation of chili con carne. In the usual method of making chili, much of the fat is rendered out of the meat, and this may be retained by the addition of corn meal or soy protein products. When a soy product is used it increases the protein content more than by the addition of corn meal, and if soy grits are used, they give a pleasing grainy texture. Soy flour, grits, concentrate, and isolate are permitted at a level of 8% in chili products, in federally inspected plants. Other foods related to the meat industry which can use soy protein products at a level sufficient for the purpose are stews, scrapple, tamales, meat pies, and beef and pork barbecue sauce.

Ziemba (1969, 1966A,B) reviewed the uses of the various types of soy protein products in a variety of food products, with emphasis on meat applications.

The problems involved in achieving acceptable meat-like flavors for meat type products and meat analogs have been discussed by Downey and Eiserle (1970A,B) and Kiratsous (1969).

Meat Analogs

Meat analogs, also called textured protein products (TPP), are products processed from vegetable proteins to resemble meat in chewiness and flavor. Presently the two most important methods are (1) by spinning soy protein isolate into a "tow" of fibers, and shaping and flavoring into the desired meat-like products; and (2) by processing a formulated soy flour mass through an extruder into different sizes and shapes to give a chewy product of the desired flavor. A third method is to form a chewy gel by heat gelation of a soy protein isolate dispersion under the proper conditions. A fourth method of achieving textures uses graded coarsely ground granular particles of isoelectric soy protein isolate which are soaked in water containing calcium hydroxide or trisodium phosphate to raise the pH from 4.5 to the range 5.5–7.5, depending on the type of "bite" desired (Circle and Johnson 1958).

Noyes (1969) has reviewed in detail more than 20 patents concerning textured protein products.

Recently, the U.S. Food and Drug Administration (Federal Register 1970) published a proposed revised standard of identity for this new class of foods, "textured protein products," including nutritional requirements. This aroused a good deal of controversy in the food industry, which was not yet resolved at the time this volume went to press. Martin (1969) discussed possible legal problems arising from the use of soy protein products in meat and simulated meat foods.

Spun Fiber Type Meat Analog

Peanut protein, casein, and zein (Croston *et al.* 1945) were processed into textile fibers and marketed in the period of 1935–1945. Textile fibers were developed also from soy protein by Boyer *et al.* (1945) but did not reach the commercial market. At that time, because of insufficient knowledge of the basic protein chemistry involved, the tensile strength of the vegetable protein fibers could not be improved well enough to meet the competition of the rapidly developing synthetic fibers and the vegetable protein textile fibers disappeared from the market. Somewhat later, Boyer (1954, 1956) modified the spinning process to produce a fibrous mass simulating meat in texture and flavor. This fibrous mass can be molded, colored, and shaped to resemble many of our common meats such as beef, poultry, pork, ham, bacon, and fish, and now are called meat analogs.

The fiber spinning process for making meat analogs is essentially the same as that used for spinning textile fibers, except that fat, flavor, color, and fiber-binding agents are incorporated into the fibers during processing. Kelley and Pressey (1966) reported their studies of several factors affecting viscosity of the "spinning dope" and fiber formation. Briefly, the process consists of dispersing the soy protein isolate in sodium hydroxide at a concentration of 14-18% at a pH of 10-11, and aging at $40°-50°$ C until the dispersion reaches a "spinnable" stage. The alkaline dispersion, usually referred to as "spinning dope," is forced by means of a metering pump through the holes in a spinnerette into a coagulating bath containing acid and salts. Spinnerettes and coagulating bath are shown in Fig. 10.4. The spinnerette has 15,000 or more holes 0.008-0.01 in. in diameter, and there may be several spinnerettes operating in the same coagulating bath. The tow, as it is now called, goes through a second heated bath where the fibers are given additional stretching while adding fat, egg albumen, or other adhesives for binding the fibers together, as well as adding flavor and other ingredients. According to Boyer and Saewert (1956) the toughness of the fibers is controlled by the pH, salt concentration, and temperature of the bath. A web of spun fibers is shown in Fig. 10.5 and meat analogs in Fig. 10.6.

Considerable literature is now extant on these spun fiber protein products. The following articles of a review nature are of interest: Ziemba (1971), Hartman (1966), Giddey (1965), Buller and Klis (1965), National Provisioner (Anon. 1965), Irmiter (1964), and Kyd (1963). Still others are discussed below.

Kuramoto *et al.* (1965) reported that the flavor of protein fibers and other simulated meat products can be improved by treatment with sulfur dioxide. Giddey (1960) found that polysaccharides, such as carrageenin, can be used for binding the fibers together; he recommended a ratio of protein to carrageenin of about 10:1.

According to Odell (1967), the spun fiber foods, on a dry basis, contain 1/3-2/3 of their weight in monofilament fiber. Although the composition of the final product can be made to vary widely, a typical example reported by Thulin and Kuramoto (1967) contains approximately 40% protein fiber, 10% binder, 20% fat, and 30% flavor, color, and supplemental nutrients. Another analysis on

Courtesy of General Mills, Inc.

FIG. 10.4. FIBER SPINNING: SPINNERETTE IN ACID COAGULATING BATH FOR MAKING PROTEIN FIBERS

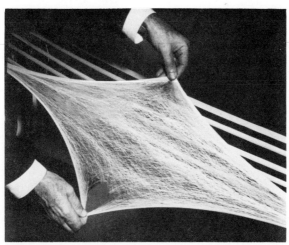

Courtesy of General Mills, Inc.

FIG. 10.5. SPUN FIBERS FROM ISOLATED PROTEIN FOR USE IN PREPARATION OF MEAT ANALOGS

Courtesy of Worthington Foods, Inc.

FIG. 10.6. A VARIETY OF SIMULATED MEAT-TYPE PRODUCTS – ALL VEG-
ETARIAN FOODS MADE FROM FIBROTEIN, WORTHINGTON'S
BRAND OF SPUN SOY FIBER

(Upper left) a variety of luncheon meat-type products. (Upper right) chicken-style, smoked
beef-style, and beef-style entrées. (Center) bowl of all-vegetable "meat loaf mix." (Lower
left) a plate of country sausage-style Fibrotein product. (Lower right) a platter of simulated
fried chicken and sausagettes.

a dry basis was reported as 60% protein, 20% fat, 17% carbohydrates, and 3%
ash. As initially manufactured they have a moisture range of 50–70%. The final
products have flavors resembling ham, bacon, chicken, turkey, beef, and other
meat products and usually are more tender than the natural product. They can
be refrigerated, canned, frozen, or dried. Several products made from spun fibers
are shown in Fig. 10.6.

In comparing the cost of analogs with meat, Odeli (1967) has estimated the
cost of unflavored spun fibers as 50 ¢ per lb. He points out that the analogs are
boneless, cooked, and completely in an edible form, and that price comparisons
with meat should be made only on that basis. With this type of comparison the
cost of the analog will be much below the average price of meat. Also, this price

advantage is expected to improve since the price of meat will probably increase whereas the price of soy protein may be expected to decrease from its present level.

The nutritional tests reported for this type of analog by Odell (1967) indicate that their PER is about 90% of that of casein or 88% of that of milk protein.

Extrusion–Cooked Type Meat Analog

Several companies have developed meat analogs by the method of continuous extrusion under heat and pressure of a prepared soy protein product formulation (Atkinson 1969, 1970; Anon. 1967A,B, 1970B). The process makes small chunks which, when hydrated, have a chewy texture and taste like meat. The principal advantage of this method is the use of low cost soy flour in the formulation rather than the more expensive protein isolate.

A premix of soy flour with added fat, flavoring, carbohydrates, coloring, and other desirable components is made, preconditioned, and passed through an extruder under predetermined conditions of moisture, temperature, time, and pressure. When the product comes from the extruder to atmospheric pressure it usually will expand. The shape of the chunk will be determined by the shape of the die at the extruder outlet and the speed of the revolving knife which cuts the extruded product to the desired length.

According to Martin and LeClair (1967) an unflavored, dried product made from defatted soy flour at a moisture content of 6–8%, has a composition of protein 50–53%, fat 1.0%, fiber 3.0%, and ash 5–6% which is the same as the original soy flour. The composition of the product can be modified by the addition of fat up to 30% and of carbohydrates and other food components.

Mustakas *et al.* (1970) described the operation of an extruder for making full-fat soy flour and the techniques described can be useful in making extruded meat analogs. It should be possible to use the full-fat soy flour as well as the defatted for making meat analogs. Fig. 9.13 in Chap. 9 illustrates one type of extruder which can be used for such processing.

Jenkins (1970) has patented the addition of sulfur and sulfur compounds to soy meal when processed in an extruder to produce meat-like products. Ralston Purina (1970) claimed the use of an extruded mixture of defatted soybean, peanut, and corn meal treated with a sulfur reducing agent to produce a meat analog.

Heat-gelled Type Meat Analog

A food product developed by Anson and Pader (1958, 1959) comprised a protein system in the form of a chewy protein gel, and a process entailing three steps: (1) adjusting concentration and pH of a protein-water system, (2) shaping of the system, and (3) applying heat to the system to set it to a chewy gel.

Frank and Circle (1959) prepared meat analogs simulating bologna and frankfurters in structure, texture, and flavor from isolated soy protein. In their process the isolated protein (isoelectrically precipitated) was added to a hot solution of trisodium phosphate in a mixer. While mixing, the other ingredients consisting of fat, emulsifier, soy sauce, smoke flavor, spices, and color were added and pH was adjusted to about 6.3. The properly stirred mixture was stuffed into cellulose frankfurter casings, linked, tied, and steamed at 10–15 psig for about 10 min. With this heat treatment, the protein dispersion is changed from a viscous sol to a gel which binds the fat, water, and other components together. Formulations for soy "frankfurters" and "bologna" are as follows:

Ingredients	Soy Protein "Frankfurters"	Soy Protein "Bologna"
	(Gm)	(Gm)
Water at 60° C	596	640
Na_3PO_4 $12H_2O$	24	16
Protein isolate, isolectric	220	200
Fat (with emulsifier)	120	100
Hydrolyzed vegetable protein	24	24
Smoke flavor	8	12
Spices	8	8
Certified color	optional	optional

On a wet basis the cooked frankfurters had a proximate analysis of protein 18.6%, fat 16.4%, and moisture 57%.

Countries that are short in meat supplies have indicated an interest in this type of product.

Meat Fibers in Heat-gelled Protein Matrix

A novel approach to forming a dehydrated meat product with improved rehydration characteristics has been patented by Coleman and Creswick (1966). Natural meat tissue (striated skeletal muscle of mammalian, avian, piscine, crustacean, or molluscan origin) is reduced to short, discontinuous rod-like fibers (1/8–1/2 in. in length) by shredding or tearing, with or without prior cooking. These are mixed with an aqueous paste of edible heat-coagulable protein of vegetable or animal origin (neutral soy protein isolate or egg albumin, for example; or mixtures of the two). A dough formed by thorough mixing is forced through an extruder as a ribbon (aligning the fibers in parallel form), and the ribbon cut into shaped pieces, which are cooked and dried.

Assay of Soy Protein Products in Meat-Type Foods

The analytical method used by the USDA Meat Inspection Laboratory (MIL) for policing the use of soy flour and cereal products in meats has been published by Bennett (1948). The method determines the amount of hemicellulose in the product using an alcoholic-KOH digestion procedure and final solubilizing in hydrocholoric acid. It applies also to soy protein concentrate.

The isolated protein does not contain any specific component on which an analytical test can be based. Thus the present procedure used by MIL is based on a so-called "tag," placed in the protein when it is processed, which consists of 0.1% titanium in the form of titanium dioxide. Although the method has not been published it is reported by Rakosky (1970) to be based on the formation of yellow titanium peroxide which is determined spectrophotometrically.

A quantitative immunochemical method for detecting soy protein in raw and cooked meat products was advocated by Kamm (1970). Immunoelectrophoresis and gel electrophoresis methods have also been proposed.

GELLING AND AERATING AGENTS

Gelsoy as Gelling Agent

Eldridge (Chap. 5) described a product called Gelsoy, made by washing hexane-defatted, high NSI soy flakes with 90–95% ethanol to remove the beany flavor and about 1% of lipids, which were flash desolventized to minimize denaturation of the protein, then extracted with water. The extract is clarified and spray dried to recover the dry product (based on practical fractions, it is composed of P+S; see Chap. 9). On an 8% moisture basis Gelsoy is 50–60% protein, 1% reducing sugars, 15–20% hydrolyzable sugars, and 9% ash.

On heating a 10% solution of Gelsoy at 90° C it forms an irreversible self-supporting opaque gel (DeVoss et al. 1950). Gelsoy also has foaming and whipping properties. When 20–40 gm are added to 300 ml of water and heated to near boiling temperature, about 5 liters of foam is developed. The addition of locust bean gum increased the stability of the foam; and on adding syrup, the foam developed into a meringue. Glabe et al. (1956) reported on the use of Gelsoy in sausage-type meats, meat loaves, canned meats, potted meats, and for the control of overrun in prepared premix formulations of low fat frozen confections.

Soy Protein Isolate as Gelling Agent

Conditions were described for converting aqueous dispersions of unhydrolyzed soy protein isolate from the viscous sol state to the gel state (see Chap. 4 and 9) (this heat gelation has been discussed above in the section on meat analogs). This gel structure can be used as a matrix for other structures also, such as high protein puddings, for example (Circle and Johnson 1958).

Soy Protein Isolate as Aerating Agent

Eldridge (Chap. 5) pointed out that isolates washed with 80–90% aqueous alcohols will form stable foams.

Soy Whey Protein as Aerating Agent

Watts (1937) showed that a pH 4.5 water extract of unheated soybean meal, also known as soybean whey, can be whipped into a stable foam. The water extract accounts for about 30% of the soy meal and includes the soluble sugars as well as about 11% of the total nitrogenous components of the dehulled meal. The nitrogenous components are divided approximately equally between non-protein nitrogen and low molecular weight proteins. Watts reported that this soybean fraction made a whip similar to egg white. Although these early studies did not develop into a commercial product they encouraged further research on aerating agents.

Enzyme Modified Isolates as Aerating Agent

It has been pointed out (Chap. 4, 5, and 9) that the major proteins of the soybean have molecular weights in the range of 200,000 to 350,000 and are insoluble in water in the acid range. They denature readily with moist heat treatment and can be whipped into a foam, especially if they have been treated with alcohol.

Burnett (1951) has reviewed methods of reducing the molecular dimensions of soybean protein by a partial pepsin hydrolysis; this treatment reduces the protein molecule to peptone-like products. After enzymatic treatment the solution is clarified in a centrifuge, concentrated by evaporation, and recovered by spray drying. These products are made in several slightly different modifications to satisfy the varied requirements of the confection and pastry industries. In trade channels, these partially hydrolyzed proteins are called "soy albumens" probably because they replace or extend egg albumen in some food applications. However, it should be noted that these products do not have the physical and chemical characteristics of an albumen as defined in the generally accepted classification of proteins. They more nearly resemble smaller molecular weight molecules called peptones.

In contrast to the unhydrolyzed proteins the soy albumens are soluble in water in the acid as well as the alkaline pH range, soluble in hot syrup, and can be pasteurized without coagulation. They can be used alone in many products as an aerating agent, but in some applications are used in combinations with egg albumen or with whole eggs to improve the whipping rate, volume, and stability of whips. Their rate of whipping is somewhat slower than that of the egg white. But volume is usually greater and they can be beaten for an extended period, in the absence of fat, without losing volume. These modified proteins have found a limited but important place in the food industry for the preparation of confections and desserts. These soy albumens are used for making nougats, creams, divinity, kisses, fudge, and similar types of candy as well as meringue powders, icings, and other confections. They are used also in sponge and angel food cakes.

Foam-mat Drying Adjunct

Berry *et al.* (1965), in work on evaluating citrus foams for foam-mat drying, found that a combination of soy albumen (enzyme-modified protein) and methylated cellulose at 0.9% concentration, in a ratio of 2.3:1.0, at a Brix 50 concentration, and a whipping time of 30 min at 40° F, were the best conditions for preparing foams to produce suitable orange powders.

Foaming Agent for Soda Water

Enzyme-modified soy protein is listed as one of the optional foaming agents in the identity standard for soda water (Federal Register 1969).

MISCELLANEOUS FOOD APPLICATIONS

Brew Flakes

Soy flakes, grits, and peptones have been used since about 1937 or earlier (Burnett 1951) as adjuncts in brewing beer. Grits and ground meal from screw press processing were the first products used in brewing but later they were replaced by solvent-extracted flakes. The best results are obtained with flakes or flour having a high NSI with a minimum of heat treatment in processing. Up to 0.75 lb of flakes per barrel of beer has been recommended by Hayward (1941).

The flakes may be used in the normal mashing operation to provide amino acids, peptides, minerals, and vitamins as nutrients for the yeast. It was reported by Wahl (1944) and Wahl and Wahl (1937) that addition of hydrolyzed soybean protein directly to the beer improves foam stability, flavor, and body of the beer.

Soups, Gravies and Sauces

Other minor uses of soy protein products include soups, gravies, and sauces in which they increase viscosity or improve body. Soy flour may be added to pan grease to a level of 20% along with an emulsifying agent to prevent baked products from sticking to the pan or griddle surface.

Confections, Imitation Nut Meats, and Nut Butters

Rakosky (1969) discussed the use of soy protein products in candy and confectionery products. A simulated nutmeat from spun isolate fiber was patented (Andregg 1957). Other products resembling nuts have been made from whole soybeans; and several workers have described spreads resembling peanut butter.

Spray Drying Adjunct

Mizrahi et al. (1967A) investigated the use of isolated soy protein as an aid in spray drying of bananas using both the soluble sodium proteinate (NaP) and the insoluble isoelectric precipitated protein (ISP). In their tests they used 4, 10, 20, and 40% of each protein on a dry basis with an inlet temperature of 200° C and an outlet temperature of 90°-100° C.

At the end of 6 weeks of storage in low density polyethylene bags, they report that shakes prepared from all powders except at the 40% level were judged as acceptable. The ISP was somewhat superior to the NaP as an anti-caking agent and had a lighter color. It was suggested that in some instances the banana-protein product might be desirable as a nutritional supplement for infant feeding where milk is not available.

NONFERMENTED ORIENTAL SOYBEAN FOODS

Introduction

The more important traditional nonfermented soybean foods of the Orient include soy milk, tofu or bean curd, kori-tofu or dry tofu, aburage or fried tofu, yuba, kinako, and soybean sprouts (Watanabe 1969; Smith 1958, 1949). These products, however, are not of equal popularity in the oriental countries; for example, soy milk is presently used only in Mainland China and Hong Kong although recent reports indicate that it is spreading to other countries. Tofu and its modified forms of aburage and kori-tofu are consumed in larger tonnage than any of the other soybean food products in Japan; and this is probably true throughout the Orient. Table 10.4 gives the proximate composition in protein, fat, carbohydrates, and ash of the oriental nonfermented foods. The composi-tion of each food will vary with the composition of the original bean as well as with the method of preparation. It will be noted that all of them are high in protein and fat.

Because the Orient is very short in meat and dairy products, the food uses of the soybean make an important contribution to their protein and fat require-ments. For example, Japan's yearly per capita consumption of meat (USDA 1969A) is 18 lb, which is one of the lowest of the countries on record. For comparison, the per capita consumption of meat in the Philippines is 28 lb and the highest is in Argentina at 220 lb. However, despite the low level of meat consumption in Japan, the people have adequate protein in their diet because of

TABLE 10.4

COMPOSITION OF SOME NONFERMENTED ORIENTAL
SOY FOODS

	Moisture (%)	Protein N × 6.25 (%)	Fat (%)	CHO (%)	Ash (%)
Tofu	88.0	6.7	3.5	1.9	0.6
Aburage	44.0	20.4	31.4	...	1.4
Kori-tofu (dry)	10.4	58.8	26.4	7.0	2.6
Yuba	8.7	57.6	24.1	11.9	3.0
Kinako	5.0	42.1	19.2	29.5	5.0

Source: Watanabe (1969)

their substantial use of fish and soybeans. To further increase their protein supplies, Japan is steadily increasing its poultry production through the use of soybean meal in poultry feeds. Japan has attained one of the best diets of any country.

Although soybean production in Japan has been decreasing for the past 15 yr, the shortage of domestic beans is compensated by imports from the United States and other countries. In the period of 1950-1954, the average yearly production of soybeans in Japan was 16.5 million bushels, whereas in 1968 their production was down to 6 million bushels. During this period their imports from the United States increased from less than 20 million bushels annually to nearly 100 million bushels in 1970. Since Japan must import about 25% of her food requirements, and U.S. soybeans can be purchased in Tokyo for less than domestic beans, the importation of U.S. beans is a favorable economic factor in the Japanese food supply.

A trend similar to that in Japan appears to be developing in Taiwan, which has many of the same food uses for soybeans as in Japan. While the domestic soybean production in Taiwan has increased in recent years, "acreage of land suitable for the cultivation of soybeans has reached its limit and production may decrease in the future in favor of imported soybeans" (USDA 1969C). The soybean processing industry in Taiwan is expanding rapidly and soybeans are recognized "as one of the few imported agricultural commodities essential to the welfare of Taiwan." In 1968, the Taiwan soybean supply was made up of approximately 2.7 million bushels of domestic production plus 14 million bushels of imports from the United States to give a total of 16.7 million bushels. In comparing the per capita annual consumption of soybeans in the United States, Taiwan, Japan, and Mainland China the values are 3.5, 1.7, 0.8, and 0.5 bushels, respectively. The United States is highest on the list because of our extensive use of soybean meal in animal feeds. It appears now that Taiwan will follow the pattern set by Japan in the importation and use of U.S. soybeans.

Although the original seed stock of U.S. soybeans came from the Orient, because of our plant breeding program as well as differences in climate and soil, some changes have occurred in the composition and physical properties of U.S. beans. For example, U.S. beans average about 3% higher in oil than Japanese beans and, probably because of climatic differences during the ripening and harvesting period, U.S. beans frequently contain a substantial quantity of "hard beans" (Smith *et al.* 1961). Hard beans are defined as "beans which are much smaller in size than the average and have a very low rate of water absorption." This hardness results from the character of the seed coat, and hard beans do not absorb water and cook the same as normal beans. Because many of the oriental food processing operations, such as making tofu, shoyu, miso, and others, start with soaking the beans in water and cooking, the differences between U.S. and Japanese beans raised some objections to the use of U.S. soybeans in Japan

(Smith 1958). However, because of subsequent research on soybeans in the United States and Japan there is a better understanding of the differences and most of the problems have been eliminated.

Chinese Soy Milk

The traditional soy milk of China is a simple water extract of whole soybeans. In present day processing it is customary to improve the traditional product by supplementing with oil, sugar, vitamins, and minerals to obtain a composition similar to cow's milk or mother's milk. The oil-soluble vitamins are incorporated in the oil. Many feeding experiments with animals and babies (Chap. 7) testify to the high nutritional value of the modified product. The soy milk can be prepared as fresh fluid milk, as milk concentrate, or spray-dried products at a very low cost in comparison with the milk of various animals.

Because of its characteristic flavor its use in China has been largely limited to the low income population. More recently, a modified form has found a good market as a soft drink in Hong Kong.

Dried Soybean Whole and Defatted Milks

According to Watanabe (1969), spray-dried forms of whole soy milk (from water extract of whole beans), and of defatted soy milk (from water extract of defatted meal) are being manufactured in Japan.

Tofu

Tofu is the most important of the soybean foods in the Orient in supplying protein nutrition for the people. The traditional tofu is a highly hydrated, gelatinous product containing about 88% water, sometimes said to resemble cottage cheese. It is made daily, and sold and consumed as a fresh product. In 1958, it was reported (Smith 1958) that Japan had more than 40,000 fresh tofu plants; many of them processed less than 1 bu of beans per day. Since that time the trend has been to modernize the process and build larger plants to reduce labor costs. One of the recent developments is a process for producing dry tofu in plants having a daily capacity of 10 tons or more of soybeans; the yield in these plants is approximately 50% on a dry bean basis.

When the Japanese started making tofu from imported soybeans they found that the U.S. beans processed differently than Japanese beans and showed differences in the final product (Smith 1958). The differences were demonstrated to be a varietal characteristic. After testing a number of U.S. varieties, the U.S. Hawkeye was found to be the best of those tested. To investigate these differences in greater detail, Smith et al. (1960) set up a tofu pilot plant and process to study varietal differences; this process also illustrates the traditional method of making tofu. Fig. 10.7 is a flow diagram for processing both fresh and dry tofu.

Fresh Tofu.—In the pilot process, 1.8 kg of soybeans were washed, soaked overnight in water, and ground while adding a small stream of water. The mash was placed in a pressure cooker and water added to make the total water to bean ratio 10:1. The mash was cooked in the pressure cooker for about 30 min starting at room temperature and heating until the pressure gauge reached 4 lb. The pressure in the kettle was used to force the milk through an outlet into a coarse filter bag laid inside a press. On filtering and pressing the mash, 16.5 kg of milk were obtained. When the milk had cooled to 65° C, the curd was precipitated by adding 50 gm of calcium sulfate suspended in 25 ml of water. In precipitating the curd, the milk is first stirred vigorously, the precipitant is then added and the stirring stopped immediately to permit the curd to form slowly. The slow formation of the curd favors the formation of a gelatinous product. After the curd has settled, part of the whey is siphoned away and the curd poured into a special wooden filter box lined with a coarse filter cloth. After the remaining whey has been drained off, the curd is pressed lightly while still in the box by using a weight on a board which fits inside the box. The cake is then removed from the box by turning it upside down in a tank of water where it is allowed to wash for an hour or more. On removing the curd from the tank it is ready for the market.

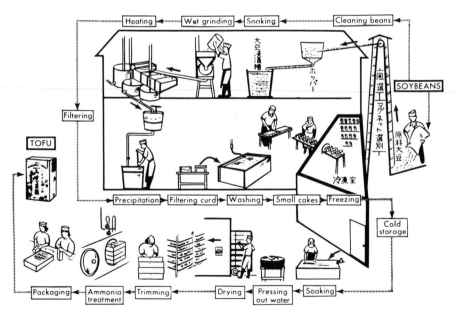

FIG. 10.7. FLOW DIAGRAM FOR PROCESSING FROZEN (KORI) TOFU

The yield of tofu from the original 1.8 kg of beans is about 5.55 kg, containing approximately 88% water. On a dry basis the yield is 46% and its composition is 55% protein and 28% oil.

Saio *et al.* 1969 investigated the effect on tofu making of fractionating soy protein into 7S and 11S components.

The processors of Japan noted that some varieties of soybeans are superior to others for making tofu with respect to yield and food characteristics such as texture, color, and flavor. Smith *et al.* (1960) investigated a number of Japanese and U.S. soybean varieties to find the most suitable for making tofu. They found that, in general, the high protein soybeans gave higher yields. Yields, based on the weight of the beans, for the Japanese varieties ranged from 39.7% to 52.0% and for the U.S. varieties from 46.1% to 50.1%. Tofu from U.S. beans was light yellow in color whereas from Japanese beans it was gray. The average proximate analysis of all samples was 88% water, 6.0% protein, 3.5% fat, and 1.9% carbohydrates. While the search for the most suitable varieties of U.S. soybean for making tofu continues, the ones presently preferred are the Hawkeye and Harosoy.

Bagged Tofu.—One of the recent innovations in tofu processing is called "bagged tofu." In this process, the milk described above and calcium salt are introduced at the same time into a plastic bag, the bag is closed and dropped into a tank of hot water for a period of about 1 hr to coagulate and sterilize the packaged tofu. However, in this process there is no filtration step. Thus, the whey remains in the bag and the solids content of the product is lower than in the standard procedure. The whey, which on standing separates from the curd, is eliminated by the consumer by making a pinhole in the bottom of the bag.

Dried Tofu.—The dried tofu, also called frozen or kori-tofu has the advantage of a shelf-life of 6 to 12 months; the yield on the basis of the weight of beans is about 50%.

The processing of dry tofu starts out by making fresh tofu except that the curd is coagulated with calcium chloride rather than the sulfate and is pressed to contain about 78% water. The curd is cut into pieces 3 1/2 × 3 3/4 in. and frozen in the temperature range of $-10°$-$20°$ C and then maintained at a temperature of about $-3°$ C for 20 days. The frozen cakes are thawed with a spray of water and the water content reduced to about 60% by passing the cakes between smooth press rolls or in a centrifuge. The remaining water is removed in a tunnel drier. The cakes are trimmed to a uniform size, treated with ammonia gas to increase their water absorption properties, and packaged. The final product contains about 56% protein, 27% fat, 6% carbohydrates, and 8% moisture and may be fortified to contain certain vitamins. Preliminary experiments have indicated this product can be flavored to resemble meat.

Fried Tofu.—About 1/3 of the soybeans consumed for tofu making in Japan is for deep fat fried tofu of various types. Aburage, the chief type, is made from thin-sliced tofu by two stages of frying. It is yellow in color, hard-surfaced, and contains more oil than fresh tofu on a dry basis (Watanabe 1969).

Yuba

On heating soybean milk in a shallow vat to near boiling for 20–30 min, a film composed of protein and oil forms on the surface (Watanabe 1969). After the film has become toughened it can be lifted with two sticks from the surface of the milk and hung on a line or spread on a galvanized wire mesh for drying; 10–20 such sheets are removed before the pan is refilled with soy milk. The product is known as yuba (see Figure 10.8). Reports indicate that it was one of the ancient foods of China.

When dry, yuba is quite brittle and is sold in sheets or small flakes. When wet, it can be used as a wrapper for other foods. Dry yuba is used in soup or fried in fat. Yuba is regarded as a very nutritious, high quality food but is produced in quite limited quantities. The proximate composition is given in Table 10.4; it is very similar to kori-tofu in analysis.

Kinako

The Japanese have a product which is similar to full-fat soy flour except that it is made from whole roasted soybean and thus contains the seed coat. It is called kinako (Watanabe 1969). To make kinako the clean whole seed is roasted for about 30 min, cooled, and ground to a powder. One of their favorite ways of using kinako is to spread it on rice or rice cakes. The rice cakes are made from rice which has been pounded. Kinako is an inexpensive source of good quality protein for supplementing Japanese foods.

FIG. 10.8. SHEETS OF YUBA, AN ORIENTAL FOOD HIGH IN PROTEIN AND OIL, CHINA, 1948

Soybean Sprouts

Bean sprouts, which have been used as food in the Orient since ancient times, are made either from soybeans or mung beans *(Phaseolus aureus)*. Most of the commercial sprouts produced in the United States are made from mung beans, perhaps because they make longer sprouts in relation to seed size. The soybean sprouts contain more protein and oil and less water than the mung beans and consequently are higher in food value. Commercial equipment for growing bean sprouts has been described by Chen and Chen (1956).

The equipment consists of an earthen jar about 3 ft high and 18 in. in diameter with several small holes in the bottom for draining the water; a cloth is placed over the bottom of the jar to prevent passage of the beans. Only selected sound whole beans having a high percentage of germination should be used. The beans are washed well, placed in a jar, covered to screen from light, and sprinkled with water three times a day. As discussed in Chap. 2, excessive moisture is unfavorable to rapid sprouting but it should be above 50%. The water should contain a small amount of calcium hypochlorite (1 tsp of calcium hypochlorite in 3 gal. of water) to discourage the growth of microorganisms. If the temperature is kept at near $30°$ C, the sprouts will grow to about 3 in. in 4 days (Drown 1943) when they will be ready for harvesting. For home use a flower pot may be used for sprouting. An alternate procedure is to sprout the beans between layers of wet cloth supported on a wire mesh in the absence of light. After the beans are harvested the sprouts may be considered as a fresh vegetable and can be used throughout the year, but are especially valuable in winter in areas where fresh vegetables are not available. The sprouts may be used uncooked in a salad, boiled in water with suitable seasoning, or they may be fried in fat and used alone or mixed with other vegetables. Fig. 10.9 shows the marketing of soybean sprouts in Canton, China in 1948.

Compositional Changes.—Block and Bolling (1944) found that for the first five days of germination or sprouting there was very little change in the amino acid composition of the system. However, in later experiments Dunn *et al.* (1948) observed a decrease of about 30% in total amino acids after germination and growth of 30 days.

In laboratory investigations, McKinney *et al.* (1958) determined the loss of dry matter during the soaking and germination, and changes in the oil, nitrogen, thiamine, and ascorbic acid content using the Hawkeye variety from the 1955 crop. In this experiment, 100 gm of beans were washed with a solution of chloride of lime (350 ppm Cl_2), soaked in 275 ml of tap water for 17 hr, and germinated in the dark between layers of cheese cloth at $24°-27°$ C.

The germinated beans were dried overnight at $70°$ C and then ground and dried to a constant weight at $100°$ C. The weight loss during the soaking period was 0.7%; and the sterile or spoiled beans, 1.4%. After 48 hr of germination the dry matter loss was 0.8%, and after 144 hr it was 2.6%. After 6 days of

FIG. 10.9. BASKET OF SOYBEAN SPROUTS AND IN-
FLATED   CURRENCY,   CANTON,   CHINA,
1948

germination the loss of total nitrogen was 2.6%, with a decrease in protein nitrogen of 15% for which there was a compensatory increase in nonprotein nitrogen. There was no significant loss in the petroleum ether extract (PEE) for the first 72 hr but after that period there was a rapid decrease in PEE which was partially replaced by free fatty acids.

The thiamine did not change in concentration during the first 4 days of germination which agrees with Sugimoto (1954) who reported no remarkable changes in thiamine for the first 13 days of germination. Ascorbic acid was absent in ungerminated beans, but appeared within the first 24 hr and reached a value of 290 mg per gm at the end of 3 days. Weakley and McKinney (1957) reported that the ascorbic acid occurs in the reduced form only. Thus, it appears that the vitamin content of soybeans is substantially increased by germination.

Lee *et al.* (1959) studied changes in the carbohydrate composition during germination and found a rapid disappearance of the galactose-containing oligo-saccharides (stachyose and raffinose) originally present in the normal seed, as well as a decrease in the sucrose. These reactions were reported to be in contrast to the production of glucose and fructose in the etiolated tissue.

Catsimpoolas *et al.* (1968A,B) using disc and immunoelectrophoresis techniques studied the protein subunits in dormant and germinating seed. They found that all of the protein components are subject to a degradation process but the major proteins (7S and 11S) disappeared at a slower rate than the smaller molecular weight proteins. They believe that the tertiary structure of each type of molecule acts as a limiting factor controlling the enzymatic degradation, since the availability of peptide bonds susceptible to cleavage will vary from one protein to another. Their data indicate that the 11S and 7S protein components were present in significant amounts after 16 days of germination.

BIBLIOGRAPHY

ADLER, L., and POMERANZ, Y. 1959. Use of lecithin in production of bread containing defatted soy flour as a protein supplement. J. Sci. Food Agr. *10*, 449–456.

ANDREGG, H. 1957. Process of making imitation nutmeats. U.S. Pat. 2,776,212. Jan. 1.

ANON. 1965. New foods from spun protein. Natl. Provisioner *153*, No. 15, 15–18.

ANON. 1967A. New route to texture, flavor, or color control of product uses "tailored proteins." Food Prod. Develop. *1*, No. 1, 35–36.

ANON. 1967B. Soybean protein products new entry in market. Food Technol. *21*, 176.

ANON. 1970A. FDA withdraws proposed imitation milk standards. Food Chem. News, June 8, 22–23.

ANON. 1970B. Improves properties of textured proteins. Food Eng. *42*, No. 7, 161.

ANON. 1970C. The industries: how they shape up. Snack Food *59*, No. 12, 35.

ANON. 1971. FDA studying repeal of filled milk, dry milk acts. Food Chem. News, Febr. 15, 17.

ANSON, M. L., and PADER, M. 1959. Method for preparing a meat-like product. U.S. Pat. 2,879,163. Mar. 24.

ANSON, M. L., and PADER, M. 1958. Protein food products and method of making same. U.S. Pat. 2,830,902. Apr. 15.

ARAI, S. *et al.* 1970. N-hexanal and some volatile alcohols; their distribution in raw soybean tissue and formation in crude soy protein concentrate by lipoxidase. Agr. Biol. Chem. (Tokyo) *34*, 1420–1423.

ARENSON, S. W. 1969. Imitation dairy products–their formulation, processing, quality control. Food Eng. *41*, No. 4, 76–79.

ATKINSON, W. T. 1969. Process for preparing a high protein snack. U.S. Pat. 3,480,442. Nov. 25.

ATKINSON, W. T. 1970. Meat-like protein food product. U.S. Pat. 3,488,770. Jan. 6.

BADENHOP, A. F., and HACKLER, L. R. 1970. Effects of soaking soybeans in sodium hydroxide solution as pretreatment for soy milk production. Cereal Sci. Today *15*, 84–88.

BADENHOP, A. E., and WILKENS, W. F. 1969. The formation of 1-Octen-3-ol in soybeans during soaking. J. Am. Oil Chemists' Soc. *46*, 179–182.

BECKEL, A. C., BELTER, P. A., and SMITH, A. K. 1949. A new soybean product–Gelsoy. Soybean Dig. *10*, No. 1, 17–18, 40.

BENNETT, O. L. 1948. Rapid method for determination of soya flour and cereal in sausages and similar meat products. J. Assoc. Offic. Agr. Chemists *31*, 513–517.

BERRY, R. E., BISSET, O. W., and LASTINGER, J. C. 1965. Method for evaluating foams from citrus concentrates. Food Technol. *19*, 1168–1171.

BLOCK, R. J., and BOLLING, D. 1944. Nutritional opportunities with amino acids. J. Am. Dietet. Assoc. *20*, 69–76.

BOYER, R. A. 1954. High protein food product and process for its preparation. U.S. Pat. 2,682,466. June 29.

BOYER, R. A. 1956. Method of manufacturing a high protein food product and the resulting product. U.S. Pat. 2,730,447. Jan. 10.

BOYER, R. A., ATKINSON, W. T., and ROBINETTE, C. E. 1945. Artificial fibers and manufacture thereof. U.S. Pat. 2,377,854. June 12.

BOYER, R. A., and SAEWERT, H. E. 1956. Method of preparing imitation meat products. U.S. Pat. 2,730,448. Jan. 10.
BROCK, F. H. 1951. Soy flours improve quality of many bakery products. Can. Baker. *54,* 36 (Oct.).
BOURNE, M. C. 1970. Recent advances in soybean milk processing technology. FAO/WHO /UNICEF Protein Advisory Group PAG Bull. *10.* 14–2.
BROWN, M. A. 1966. Soy flour in the modern U.K. bakery. Soybean Dig. *26,* 62–64.
BULLER, A. R., and KLIS, J. B. 1965. Spun soy protein foods get supermarket sales test. Food Process. *26,* No. 9, 115–117, 120.
BURNETT, R. S. 1951. Soybean food products. *In* Soybeans and Soybean Products, Vol. II, K. S. Markley (Editor). John Wiley & Sons, New York.
CATSIMPOOLAS, N., CAMPBELL, T. G., and MEYER, E. W. 1968A. Immunochemical study of changes in reserve proteins of germinating soybean seed. Plant Physiol. *43,* 799–805.
CATSIMPOOLAS, N., EKENSTAM, C., ROGERS, D. A., and MEYER, E. W. 1968B. Protein subunits in dormant and germinating soybean seed. Biochim. Biophys. Acta *168,* 122–131.
CHEN, P. S., and CHEN, H. D. 1956. Soybeans for Health, Longevity and Economy. The Chemical Elements, South Lancaster, Mass.
CIRCLE, S. J., and JOHNSON, D. W. 1958. Edible isolated soybean protein. *In* Processed Plant Protein Foodstuffs. A. M. Atschul (Editor). Academic Press, New York.
CIRCLE, S. J., MEYER, F. W., and WHITNEY, R. W. 1964. Rheology of soy protein dispersions. Effect of heat and other factors on gelation. Cereal Chem. *41,* 157–172.
COLEMAN, R. J., and CRESWICK, N. W. 1966. Method of dehydrating meat and product. U.S. Pat. 3,253,931. May 31.
CROSTON, C. B., EVANS, C. D., and SMITH, A. K. 1945. Zein fibers. Preparation by wet spinning. Ind. Eng. Chem. *37,* 1194–1198.
DEVOSS, L., BECKEL, A. C., and BELTER, P. A. 1950. Vegetable gel. U.S. Pat. 2,495,706. Jan. 31.
DISER, G. M. 1962. Soy flour and grits as protein supplements of cereal products. Proc. Conf. Soybean Products for Protein in Human Foods, USDA, Peoria, Ill. Sept. 13–15, 1961. USDA–ARS *52–72.*
DOWNEY, W. J., and EISERLE, R. J. 1970A. Substitutes for natural flavor. J. Agr. Food Chem. *18,* 983–987.
DOWNEY, W. J., and EISERLE, R. J. 1970B. Problems in the flavoring of fabricated foods. Food Technol. *24,* 1226–1229.
DROWN, M. J. 1943. Soybeans and soybean products as food. USDA Misc. Publ. *534.*
DUNN, M. S., CAMEIN, M. N., SHANKMAN, S., and BLOCK, H. 1948. Amino acids in lupine and soybean seeds and sprouts. Arch. Biochem. *18,* 195–200.
FEDERAL REGISTER. 1968. Imitation milks and creams. Vol. *33,* No. 98, 7456–7458.
FEDERAL REGISTER. 1969. Soda water, identity standard. Vol. *34,* No. 137, 12087.
FEDERAL REGISTER. 1970. Textured protein products. Vol. *35,* No. 236. 18530–18531.
FELDBERG, C. 1969. Extruded starch based snacks. Cereal Sci. Today. *14,* 211, 214.
FRANK, S. S., and CIRCLE, S. J. 1959. The use of isolated soybean protein for non-meat simulated sausage products, frankfurter and bologna types. Food Technol. *13,* 307–313.
FUJIMAKI, M., KATO, H., ARAI, S., and TAMAKI, E. 1968. Applying proteolytic enzymes on soybeans. I. Proteolytic enzyme treatment of soybean protein and its effect on flavor. Food Technol. *22,* 889–893.
FUJIMAKI, M. *et al.* 1969. Fundamental investigations of proteolytic enzyme application to soybean protein in relation to flavor. Dept. Agr. Chem., Univ. Tokyo. USDA Final Rept. *UR-A11-(40)-8* for period Oct. 1964–Mar. 1969.
FUJIMAKI, M., YAMASHITA, M., ARAI, S., and KATO, H. 1970. Plastein reaction. Its application to debittering of proteolyzates. Agr. Biol. Chem. (Tokyo) *34,* 483–484.
FUKUSHIMA, D., and VAN BUREN, J. P. 1970. Effect of physical and chemical processing factors on the redispersibility of dried soy milk proteins. Cereal Chem. *47,* 572–578. Mechanisms of protein insolubilization during drying of soy milk. Role of disulfide and hydrophobic bonds. *Ibid.* 687–696.

GENERAL FOODS. 1969. Whippable composition and manufacture thereof. Brit. Pat. 1,140,937. Jan. 22.

GIDDEY, C. 1960. Protein compositions and process of producing the same. U.S. Pat. 2,952,542. Sept. 13.

GIDDEY, C. 1965. Artificial edible structures from non-animal proteins. Cereal Sci. Today 10, 516–518, 514.

GLABE, E. F. et al. 1956. Uses of Gelsoy in prepared foods. Food Technol. 10. 51–56.

GUY, E. J., VETTEL, H. E., and PALLANSCH, M. J. 1969. Spray-dried cheese whey-soy flour mixtures. J. Dairy Sci. 52, 432–438.

HAAS, L. W. 1934A. Bleaching agent and process of preparing bleached bread dough. U.S. Pat. 1,957,334. May 1.

HAAS, L. W. 1934B. Method of bleaching flour. U.S. Pat. 1,957,337. May 1.

HAAS, L. W., and BOHN, R. M. 1934A. Bleaching agent and process of utilizing same for bleaching flour. U.S. Pat. 1,957,333. May 1.

HAAS, L. W., and BOHN, R. M. 1934B. Bleaching agent for flour and process of utilizing same in making bread. U.S. Pat. 1,957,336. May 1.

HACKLER, L. R. et al. 1965. Effect of heat treatment on nutritive value of soy milk protein fed to weanling rats. J. Food Sci. 30, 723–728.

HAN, T. B. 1958. Technology of Soybean Milk. A thesis. Lab. Technol. Agr. Univ. Netherlands.

HAND, D. B. et al. 1964. Pilot plant studies on soybean milk. Food Technol. 18, 139–142.

HANG, Y. D., and JACKSON, H. 1967A. Preparation of soybean cheese using lactic starter organism. I. General characteristics of the finished cheese. Food. Technol. 21, 1033–1034.

HANG, Y. D., and JACKSON, H. 1967B. Preparation of soybean cheese using lactic starter organism. II. Effects of addition of rennet extract and skim milk. Food Technol. 21, 1035–1038.

HARTMAN, W. E. 1966. Vegetarian protein foods. Food Technol. 21, No. 1, 39–40.

HAYWARD, J. H. 1941. Soybean flakes for brewing–a promising adjunct. Western Brewing World. 49, 7–8, 26.

HEINER, D. C., WILSON, J. P., and LAHEY, M. E. 1964. Sensitivity to cow's milk. Council Foods Nutr. J. Am. Med. Assoc. 189, 563–567.

HETRICK, J. H. 1969. Imitation dairy products, past, present, and future. J. Am. Oil Chemists' Soc. 46, 58A, 60A, 62A.

HONIG, D. H., RACKIS, J. J., and SESSA, D. J. 1971. Isolation of ethyl-D-galactopyranoside and pinitol from hexane-ethanol extracted soybean flakes. J. Agr. Food Chem. 19, 543–546.

HONIG, D. H., SESSA, D. J., HOFFMAN, R. L., and RACKIS, J. J. 1969. Lipids of defatted soybean flakes: extraction and characterization. Food Technol. 23, 95–100.

HOOVER, W. J., and TSEN, C. C. 1970. Add to high protein baked products. Southwestern Miller Oct. 13.

HUFFMAN, D. L., and POWELL, W. E. 1970. Fat content and soy level effect on tenderness of ground beef patties. Food Technol. 24, 1418–1419.

IRMITER, T. F. 1964. Foods from spun protein products. Nutr. Rev. 22, 33–35.

JENKINS, S. L. 1970. Expanded soybean products. U.S. Pat. 3,496,858. Febr. 24.

JOHNSON, D. W. 1970. Functional properties of oilseed proteins. J. Am. Oil Chemists' Soc. 47, 402–407.

KAMM, L. 1970. Immunochemical quantitation of soybean protein in raw and cooked meat products. J. Assoc. Offic. Anal. Chemists 53, 1248–1252.

KATZ, M. H. 1969. Edible dry mix for producing an aerated food product. U.S. Pat. 3,434,848. Mar. 25.

KELLEY, J. J., and PRESSEY, R. 1966. Studies with soybean protein and fiber formation. Cereal Chem. 43, 195–206.

KIRATSOUS, A. S. 1969. Meat analogs and flavor. Cereal Sci. Today 14, 147–149.

KNIGHTLY, W. H. 1969. The role of ingredients in the formation of coffee whiteners. Food Technol. 23, 171–173, 177, 180, 182.

KURAMOTO, S. WESTEEN, R. W., and KEEN, J. L. 1965. Process of preparing spun protein food products using sulfur dioxide. U.S. Pat. 3,177,079. Apr. 6.

KYD, G. H. 1963. Edible soy protein fibers promise new family of foods. Food Process. May, 123–126, 138.

LEARMONTH, E. M. 1952. The influence of soya flour on bread doughs. II. The inhibition of the proteolytic enzymes of malt and wheat flours. J. Sci. Food Agr. *3*, 54–59.

LEARMONTH, E. M. 1958. The influence of soya flour on bread doughs. III. The distribution of the papain-inhibiting factor in soya beans. J. Sci. Food Agr. *9*, 269–273.

LEARMONTH, E. M., and WOOD, J. C. 1960. Influence of soybean flour on bread doughs. IV. a-amylase of soybean. Cereal Chem. *37*, 158–169.

LEE, K. Y., LEE, C. Y., LEE, T. V., and KWON, T. W. 1959. Chemical changes during germination of soybean carbohydrate metabolites. National University, Seoul, Korea. Seoul Univ. J. *8*, 35. Chem. Abstr. *54*, 697d.

LEMANCIK, J. F., and ZIEMBA, J. V. 1962. Versatile soy flours. Food Eng. *34*, No. 7, 90–91.

LO, W. Y., STEINKRAUS, K. H., and HAND, D. B. 1968A. Concentration of soy milk. Food Technol. *22*, 1028–1030.

LO, W. Y., STEINKRAUS, K. H. and HAND, D. B. 1968B. Soaking soybeans before extraction as it affects chemical composition and yield of soy milk. Food Technol. *22*, 1188–1190.

LORANT, G. J. 1969. Frozen whipped topping composition. U.S. Pat. 3,431,117. Mar. 4.

LUKE, C. E. 1948. Manufacture of flaked cereal-soya product. U.S. Pat. 2,436,519. Febr. 24.

MCKINNEY, L. L., WEAKLEY, F. B., CAMPBELL, R. E., and COWAN, J. C. 1958. Changes in the composition of soybeans on sprouting. J. Am. Oil Chemists' Soc. *35*, 364–366.

MARTIN, R. E. 1969. Legal problems faced by soy proteins on state and national levels. Soybean Dig. *30*, No. 1, 19, 51. Nov.

MARTIN, R. E., and LECLAIR, O. V. 1967. For creating products–textured proteins. Food. Eng. *39*, No. 4, 66–69.

MATTICK, L. R., and HAND, D. B. 1969. Identification of a volatile compound in soybeans that contributes to the raw bean flavor. J. Agr. Food Chem. *17*, 15–17.

MELNYCHYN, P., and STAPLEY, R. B. 1969A. Acetylated soybean protein for coffee whiteners. S. African Pat. 68–07, 706. June. 27.

MELNYCHYN, P., and STAPLEY, R. B. 1969B. Plant protein purification for human consumption. Ger. Offen. 1,800,403. Nov. 27. (U.S. Pat. Appl. Oct. 2, 1967).

MILLER, H. W. 1962. Traditional method of processing and use of soy liquid and powdered milk. Proc. Conf. Soybean Products for Protein in Human Foods, Sept. 1961, Peoria, Ill. USDA–ARS, June, 149–156.

MIZRAHI, S., BERK, Z., and COGAN, U. 1967A. Isolated soybean protein as a banana spray-drying aid. Cereal Sci. Today, *12*, 322–325.

MIZRAHI, S., ZIMMERMANN, G., BERK, Z., and COGAN, U. 1967B. The use of isolated soybean protein in bread. Cereal Chem. *44*, 193–203.

MODLER, H. W., RIPPEN, A. L., and STINE, C. M. 1970. Physical and chemical stability of soybean oil-filled milk. J. Food Sci. *35*, 302–305.

MOSER, H. *et al.* 1967. Sensory evaluation of soy flour. Cereal Sci. Today, *12*, 296, 298, 299, 314.

MUSTAKAS, G. C. *et al.* 1969. Lipoxidase deactivation to improve stability, odor, and flavor of full fat soy flours. J. Am. Oil Chemists' Soc. *46*, 623–626.

MUSTAKAS, G. C. *et al.* 1970. Extruder processing to improve nutritional quality, flavor, and keeping quality of full-fat soy flour. Food Technol. *24*, 1290–1296.

NOYES, R. 1969. Protein Food Supplements. Noyes Development Corp. Park Ridge, N. J.

OBARA, T. 1968. Basic investigations on the development of foods from enzymatically treated protein concentrates. Dept. Agr., Tokyo Univ. Educ. Tokyo. USDA Public Law 480, Rep. Proj. *UR-A11-(40)-26.* Available from Natl. Agr. Library.

ODELL, A. D. 1967. Meat analogs from modified vegetable tissue. Proc. Int. Conf. Soybean Protein Foods. USDA–ARS *71–35*, 163–169.

OFELT, C. W., SMITH, A. K., EVANS, C. D., and MOSER, H. A. 1952. Soy flour bread wins its place. Food Eng. Dec. 145, 147–149.

OFELT, C. W., SMITH, A. K., and BELTER, P. A. 1953. Crumb softness from soy fractions. Food Technol. *7*, 432–434.

OFELT, C. W., SMITH, A. K., and DERGES, R. E. 1954A. Baking behavior and oxidation requirements of soy flour. I. Commercial full fat soy flours. Cereal Chem. *31*, 15–21.

OFELT, C. W., SMITH, A. K., and MILLS, J. M. 1954B. Baking behavior and oxidation requirements of soy flour. II. Commercial defatted soy flours. Cereal Chem. *31*, 23–21.

OFELT, C. W., SMITH, A. K., and MILLS, J. M. 1955. Effect of soy flour on amylograms. Cereal Chem. *32*, 48–52.

OFELT, C. W., and SMITH, A. K. 1955. Importance of oxidation on the use of soy flour with high extraction wheat flours. Trans. Am. Assoc. Cereal Chemists, *13*, 122–129.

PAULSEN, T. M. 1961. A study of macaroni products containing soy flour. Food Technol. *15*, 118–121.

PAULSEN, T. M. 1968. Process for treating soybean particulates. U.S. Pat. 3,361,574 (continuation in part, U.S. Pat. 3,100,709, Aug. 13 1963), also U.S. Pat. 3,361,575. Jan. 2.

PAULSEN, T. M., and HORAN, F. E. 1965. Functional characteristics of edible soy flour. Cereal Sci. Today *10*, 14–17.

PENTY, W. P. 1947. Making ready-to-east food. U.S. Pat. 2,421,216. Ready-to-eat food. U.S. Pat. 2,421,217. May 27.

POLLOCK, J. M., and GEDDES, W. F. 1960A. Soy flour as a white bread ingredient. I. Preparation of raw and heat treated soy flours and their effects on dough and bread. Cereal Chem. *37*, 19–29.

POLLOCK, J. M., and GEDDES, W. F. 1960B. Soy flour as a white bread ingredient. II. Fractionation of raw soy flour and the effect of the fractions on bread. Cereal Chem. *37*, 30–37.

POMERANZ, Y., and MILLER, G. D. 1960. Detection and estimation of soy flour in wheat flour. J. Assoc. Offic. Agr. Chemists, *43*, 442–444.

RACKIS, J. J., HONIG, D. H., SESSA, D. J., and STEGGERDA, F. R. 1970. Flavor and flatulence factors in soybean protein products. J. Agr. Food Chem. *18*, 977–982.

RAKOSKY, J. 1967. Soy proteins—their preparation and uses in comminuted meat products. *In* Meat Hygiene. USDA–CMS *8*, No. 6, 1–13.

RAKOSKY, J. 1969. Using soybeans for new improved candies. Mfg. Confectioner *49*, No. 2, 47–50.

RAKOSKY, J. 1970. Soy products for the meat industry. J. Agr. Food Chem. *18*, 1005–1009.

RALSTON PURINA. 1970. Soybean protein foodstuff. Japanese Pat. 16778/70. June 10.

ROCK, H., SIPOS, E. F., and MEYER, E. W. 1966. Soy protein—its role in processed meat production. Meat *32*, 52–56.

RUSOFF, I. I., GOODMAN, A. H., SOMMER, J., and CANTOR, S. M. 1961. Protein fortification of doughnuts. Publ. Central Lab. DCA Food Industries, New York.

SAIO, K., KAMIYA, M., and WATANBE, T. 1969. Food processing characteristics of soybean 11S and 7S proteins. Part I. Effect of different protein components among soybean varieties on formation of tofu-gel. Agr. Biol. Chem. (Tokyo) *33*, 1301–1308.

SANDERUDE, K. G. 1969. Continous cooking extrusion—benefits to the snack food industry. Cereal Sci. Today, *14*, 209–210, 214.

SCHAEDER, W. E., FAST, R. B., CRIMMINS, J. P., and DESROSIER, N. W. 1969. Evolving snack technology. Cereal Sci. Today, *14*, 203–204, 206.

SESSA, D. J., HONIG, D. H., and RACKIS, J. J. 1969. Lipid oxidation in full fat and defatted soybean flakes as related to soybean flavor. Cereal Chem. *46*, 675–686.

SMITH, A. K. 1958. Use of United States soybeans in Japan. USDA–ARS *71–12*.

SMITH, A. K. 1949. Oriental methods of using soybeans as food with special attention to fermented products. USDA. AIC-234.

SMITH, A. K., and BECKEL, A. C. 1946. Soybean or vegetable milk. A resume and bibliography. Chem. Eng. News, *24*, 54–56.

SMITH, A. K., NASH, A. M., and WILSON, L. 1961. Water absorption of soybeans. J. Am. Oil Chemists' Soc. *38*, 120–123.

SMITH, A. K., WATANABE, T., and NASH, A. M. 1960. Tofu from Japanese and United States soybeans. Food Technol. *14*, 332–336.

SUGIMOTO, K. 1954. Biochemical and nutritional studies of germinating soybeans. Osaka Med. School Bull. *1*, 1–16.

TAUBER, F. W., and LLOYD, J. H. 1947. Variation in composition of frankfurters with special reference to cooking changes. Food Res. *12*, 158–163.

THULIN, W. W., and KURAMOTO, S. 1967. Bontrae—A new meat-like ingredient for convenience foods. Food Technol. *21*, 168–171.

TURRO, E. J. 1970A. Baking formulation. U.S. Pat. 3,529,970. Sept. 22.

TURRO, E. J. 1970B. Protein base for bakery goods. U.S. Pat. 3,529,969. Sept. 22.

TURRO, E. J., and SIPOS, E. 1968. Effect of various soy protein products on bread characteristics. Bakers Dig. *42*, 44–50, 61.

TURRO, E. J., and SIPOS, E. 1970. Soy protein products in commercial cake formulations. Bakers Dig. *44*, 58–64.

USDA. 1962. Proc. Conf. Soybean Products for Protein in Human Foods. Sept. 13–15, 1961, USDA, Peoria, Ill. USDA–ARS.

USDA. 1963. Proc. Seed Protein Conf. January 21–23, 1963. USDA–ARS.

USDA. 1967. Proc. Intern Conf. Soybean Protein Foods, Oct. 17–19, 1966, Peoria, Ill. USDA–ARS *71-35.*

USDA. 1969A. Conf. on Protein-Rich Foods from Oilseeds, May 15–16, 1968, New Orleans. USDA–ARS *72-71.*

USDA. 1969B. Synthetics and substitutes for agricultural products. A compendium. USDA–ERS Misc. Publ. *1141.*

USDA. 1969C. The U.S. soybean market in the Republic of China, Taiwan. USDA–FAS Misc. Publ. *M-209.*

VAN BUREN, J. P. *et al.* 1964. Heat effects on soy milk. J. Agr. Food Chem. *12*, 524–528.

WAHL, A. S. 1944. Wahl Handbook of American Brewing Industry, Vol. II. Wahl Institute, Chicago.

WAHL, R., and WAHL, A. S. 1937. Wahl Handbook of American Brewing Industry, Vol. I. Wahl Institute, Chicago.

WATANABE, T. 1969. Industrial production of soybean foods in Japan. UN Ind. Develop. Organ. Meeting on Soya Bean Processing and Use, Peoria, Ill. Nov. 17–21. United Nations, N. Y.

WATTS, B. M. 1937. Whipping ability of soybean whey protein. Ind. Eng. Chem. *29*, 1009–1011 (1937). *Ibid. 31*, 1282–1283. (1939).

WEAKLEY, F. B., and MCKINNEY, L. L. 1957. Modified indophenol-xylene method for the determination of ascorbic acid in soybeans. J. Am. Oil Chemists' Soc. *34*, 281–284.

WILKENS, W. F., MATTICK, L. R., and HAND, D. B. 1967. Effect of processing methods on oxidative off-flavors of soybean milk. Food Technol. *21*, 1630–1633.

WOLF, W. J. 1970. Soybean proteins. Their functional, physical and chemical properties. J. Agr. Food Chem. *18*, 969–976.

WOOD, J. C. 1967. Soy flour in food products. Ingredient Survey. Food Manuf. *42*, 11, 12–15.

YAMASHITA, M. *et al.* 1970A. Enzymatic modification of proteins in foodstuffs. II. Nutritive properties of soy plastein and its bio-utility evaluation in rats. Agr. Biol. Chem. (Tokyo) *34*, 1333–1337.

YAMASHITA, M., ARAI, S., TSAI, S.-J., and FUJIMAKI, M. 1970B. Supplementing S-containing amino acids by plastein reaction. Agr. Biol. Chem. (Tokyo) *34*, 1593–1596.

YAMANAKA, Y., OKAMURA, O., and HASEGAWA, Y. 1970. Method of preparing a sour milk beverage. U.S. Pat. 3,535,117. Oct. 20.

ZICK, W. F. 1969. Lipid and protein derived flavors for snack food application. Cereal Sci. Today, *14*, 205–208.

ZIEMBA, J. V. 1966A. Create new foods with textured soy proteins. Food Eng. *38,* No. 4, 58–60.

ZIEMBA, J. V. 1966B. Let soy proteins work wonders for you. Food Eng. *38,* No. 5, 82–89.

ZIEMBA, J. V. 1969. Simulated meats–how they're made. Food Eng. *41,* No. 11, 72–75.

ZIEMBA, J. V. 1970. R&D's low-cost dairy-like cheese. Food Eng. *42,* No. 8, 48–49.

ZIEMBA, J. V. 1971. Showcase plant spins soy protein. Food Eng. *43,* No. 1, 66–69.

CHAPTER 11

C. W. Hesseltine
H. L. Wang

Fermented Soybean Food Products

INTRODUCTION

Soybeans have been an important source of protein, fat, and flavor for oriental people for thousands of years. While a large variety of foods were developed from soybeans, the four most important were miso, shoyu (soy sauce), tempeh, and tofu. While the fermented products, miso and shoyu, contribute amino acids to the diet, their contribution is more in flavor than in nutrition. Having a high content of protein and fat, tofu makes a substantial contribution to nutrition. Tempeh, which is used mostly in Indonesia, has a good flavor and is also rich in both protein and fat.

The development of fermented foods, which depends on the use of rather sophisticated microbiology, was a remarkable achievement in the early history of China. There is considerable evidence (Chap. 1) that the Buddhist priests, who taught the people to avoid eating meat, were largely responsible for the development of fermented foods in the Orient. The somewhat lesser known fermented products included in this chapter are natto, hamanatto, and sufu, as well as recent work on the development of an American cheeselike product. The ancient methods of making fermented foods are changing rapidly through the introduction of modern microbiology and technology. Today, Japan has become a leader in the field of industrial microbiology.

To make some of these products, fermentation is carried out directly on cooked soybeans with a selected organism under specific conditions. In making miso and shoyu, a koji is prepared in a preliminary fermentation that is used in a second stage to ferment a combination of cooked soybeans and cereal; the cereal is usually rice when making miso and wheat when making shoyu. Koji is used also in other fermentations such as distilled spirits (shocha), sake, wine, and other beverages where it serves in place of malt.

KOJI

The word "koji" (Tamiya 1958) is an abbreviation of kabi-tacki meaning "bloom of mold." The process, like much of Japanese culture, was introduced into Japan from China about 200 A.D. Koji is a source of enzymes for converting starch into fermentable sugars and proteins into peptides and amino acids. The koji molds are grown on rice to produce amylases, which convert rice starch to fermentable sugars, so that in the second fermentation stage sugar becomes available for the yeast growing in the moromi or mash.

The microorganisms used in koji are almost always fungi of the genus *Aspergillus*. When we examine the tane koji (seed or inoculum used to make koji), we may find a great deal of variation in the purity of the inoculum. Some starters contain a mixture of mold types, including mucoraceous molds, various aspergilli and penicillia, along with an assortment of bacteria and yeasts. Obviously, these starters were not prepared under controlled conditions and contain many "weed" microorganisms. Such preparations should be avoided because of the possibility of mycotoxins. On the other hand, in modern plants making koji starters, extreme care is taken to maintain vigorous tane koji that are pure cultures containing one or more purposely mixed strains whose spores have an extremely high and rapid rate of germination (98-99%). As far as we are aware, the molds that make koji belong to the *Aspergillus oryzae* group and include *A. oryzae* and *A. sojae*. Other fungi are employed to a certain extent for the conversion of starch to fermentable sugars in the amylo-koji process including strains of *Rhizopus* according to Inui *et al.* (1965). Some of the black aspergilli in the *A. niger* group may also be used.

Molds used for koji have been extensively investigated and are known to produce a great variety of enzymes including a-amylase, proteases (3 types are known, 1 is active at alkaline, 1 is active at acid and 1 is active at neutral pH's), nucleases, sulfatases, phosphatases, transglycosidases, peptidases, acylase, ribonucleo-depolymerases, mononucleotide phosphatase, adenyl-deaminases and purine nucleosidases.

The modern koji process begins with growing a selected *A. oryzae* strain on an agar slant in pure culture. The strain is selected for its special abilities by natural selection or by induced mutation and must have the ability to sporulate luxuriantly on rice. Selection is made to give a desirable koji for a particular fermentation. Spores from the mature slant culture are used to inoculate 1-1½ kg of moist, sterile, brown rice in a wooden tray, which is also sterile. To the rice is added 2% wood ash as a source of trace elements. Addition of ash of certain trees is an important step in the process because the RNA/DNA ratio is higher. A high RNA/DNA ratio gives spores with greater viability and vigor. After five days' incubation, spores from the first tray are used to inoculate all the sterile wooden trays of the tane production plant.

The entire production area is such that it may be sterilized, and workers entering areas devoted to spore production must go through a sterile entry room dressed entirely in sterile clothing to prevent contamination. In incubation rooms, both trays and incubator are sterilized with steam between each spore run. Spores are produced at a temperature of $30°$ C, dried at $50°$ C, and stored at $15°$ C for sale. Purity of the inoculation is regularly determined. Spores from one or more strains are blended together and are prepared for a particular fermentation. Thus, tane koji for the miso fermentation is composed of several strains mixed together in a definite proportion: 150 gm of inoculum is equal to 1×10^{12} viable spores.

If we plate out the tane koji used in red miso fermentation, we will encounter perhaps 3 distinct morphological strains: 1 tall and light sporing with a brownish tint; a second, which is rather short, heavily sporulating and yellow green; and an intermediate form sporulating vigorously and yellow green.

The use of tane koji to prepare specific types of koji needed in fermenting soybeans will be described under the appropriate fermentations in which the various substrates are rice or wheat as well as soybeans. Besides the type of substrate used, the appearance of the koji will be different. For example, in koji prepared from rice, the mold will have converted the rice into a solid cake bound together with the mold mycelium and will be harvested just before sporulation when the conidiophores of the mold are just forming. At this stage, the maximum amount of desirable enzymes will be formed, but the molds will not have produced any undesirable odors and flavors. On the other hand, koji for shoyu, composed of roasted cracked wheat and soybean flakes at the time of harvest, will be green as a result of the sporulation of the *Aspergillus* strains.

In the older methods of making koji, wooden trays were employed but today much of the koji is made by automatic equipment in rooms in which the substrate is loaded, inoculated, turned, and harvested mechanically; the conditions of moisture, temperature, and aeration all being carefully controlled with automatic equipment. Under these conditions practically no contamination occurs and the chances for any mycotoxins or harmful bacteria being present are excluded. Incidentally, all the certified strains of molds used in Japan have been tested for the absence of toxin.

MISO

Miso is a food prepared by the fermentation of soybeans and salt with or without a cereal. Innumerable variations are possible based on the ratios of substrate, salt, length of fermentation, and aging. This account will be general in nature and follows the procedure for making red miso. For a longer account of the process, see Shibasaki and Hesseltine (1962) and Watanabe (1969).

Miso is produced in a number of countries of the Orient. In China it is called chiang. Actually, the product is produced and consumed over the whole area of Japan, China, Indochina, Indonesia, and the East Indies.

In appearance, miso is a paste resembling peanut butter in consistency and smooth in texture. But in China, chiang may be unground so that individual particles of soybeans are present. Its color varies from a light, bright yellow to a very blackish brown. Generally speaking, the darker the color the stronger the flavor. The product is typically salty and has a distinctive pleasant aroma.

Usually, miso is not consumed by itself but dissolved in water as a base for various types of soups. Often the soup contains vegetables, algae, tofu, or fish. It can serve as a seasoning for cooked meat and vegetable dishes and may be mixed with fresh vegetables such as cucumbers. Sometimes it may be added to fish. In Western culture, its counterpart is catsup. Currently, miso is made in Hawaii and is available in the United States, being packaged like sausages in plastic tubing.

Watanabe (1969) summarized the statistics concerning traditional foods in Japan. We have abstracted some of his data on the fermented foods of Japan (Table 11.1). In Table 11.2, we have also abstracted Watanabe's data on the composition and cost of fermented foods. According to him, 180,000 metric tons of soybeans and defatted soybean meal are used annually in Japanese factories for miso manufacture and an estimated additional 60,000–70,000 metric tons of soybeans are turned into various types of miso in the home. Figures on annual production of miso are shown in Table 11.3. It must be remembered that miso contains water and cereals, as well as soybeans.

Miso may be classified in various ways. The substrates used with soybeans may be rice, barley, or only soybeans. About 80% is made from rice and soybeans and the remaining 20% is of the last two types. In 1965, the various ingredients to produce 492,650 metric tons of miso were, in metric tons: soybeans, 150,181; defatted soybeans, 12,504; rice, 77,250; barley, 14,413; salt, 65,393; and corn meal, 3,842. The last is a substitute for rice. As noted, miso may be classified on the basis of color and taste of the finished product which ranges from white and light yellow to red and from sweet to salty. Rice miso is the most popular and may be divided into five types. These are white miso, light-yellow sweet miso, light-yellow salty miso and yellow-red sweet miso and yellow-red salty miso. These types of rice miso are made by varying the ratio of ingredients and soaking and cooking conditions involving varying the time, temperature, and length of fermentation and ripening. Ebine (1967), for example, collected detailed information about three types of miso which is summarized in Table 11.4. Miso can also be classified on the basis of the area where it is produced; namely, Sendai, Shinshu, etc.

TABLE 11.1

DEMAND FOR WHOLE SOYBEANS IN JAPAN

Soybeans	1964	1966	1967
		(1000 Metric Tons)	
Whole			
Miso	145	158	169
Shoyu	16	15	15
Natto	30	38	47
Defatted			
Miso	15	10	8
Shoyu	165	162	154
Total used			
Miso	160	168	177
Shoyu	181	177	169
Natto	30	38	47

TABLE 11.2

CHEMICAL COMPOSITION OF SOYBEAN FOODS

Fermented Foods	Moisture (%)	Protein (%)	Fat (%)	Soluble Carbohydrate (%)	Fiber (%)	Ash (%)	Retail Cost in 1968 (¢/100 Gm)
Miso							
Salty light	50.0	12.6	3.4	19.4	1.8	12.8	3.9
Salty red	50.0	14.0	5.0	14.3	1.9	14.8	
Soybeans	47.5	16.8	6.9	13.6	2.2	13.0	
Natto	58.5	16.5	10.0	10.1	2.3	2.6	5.5
Soybeans	12.0	34.3	17.5	26.7	4.5	5.0	

TABLE 11.3

ANNUAL PRODUCTION OF MISO

Year	Tons	Year	Tons
1956	530,078	1962	453,955
1957	520,176	1963	476,533
1958	514,974	1964	473,846
1959	505,354	1965	492,650
1960	505,086	1966	510,304
1961	482,357	1967	520,510

TABLE 11.4

COMPOSITION OF MISO IN RELATION TO TIME OF
FERMENTATION AND RATIO OF SOYBEANS:RICE:SALT

Item	White Miso	Light-yellow Salty Miso	Yellow-red Salty Miso
Soybean:rice:salt	100:200:35	100:60:45	100:50:48
Duration and temperature of fermentation	2–4 days, 50° C	30 days, 30°–35° C	60 days, 30°–35° C
Color	Bright light yellow	Light yellow	Yellow red
Taste	Very sweet	Salty	Salty
NaCl (%)	5	12–13	12.5–13.5
Moisture (%)	43	48	50
Protein (%)	8	10	12
Sugar (%)	20	13	11
Shelf-life	Short	Fairly long	Long

In miso manufacture, methods differ from variety to variety, but basically the process is the same. Briefly, it involves the cleaning and cooking of soybeans; preparation of rice koji; mixing of soybeans, salt, koji, and inoculum; fermentation under anaerobic conditions; and blending and packaging of the product for market. The process is outlined on a laboratory scale in Figure 11.1

Miso manufacture is essentially two successive fermentations. The first involves the preparation of koji under aerobic conditions from strains of *A. oryzae* and *A. sojae.* The molded rice koji serves as source of enzymes and nutrients for the second fermentation. The second is an anaerobic fermentation involving yeasts and bacteria.

Preparation of Koji

Rice koji is prepared from polished rice since it is essential that mold mycelium quickly penetrate the rice kernels. Brown rice is therefore not suitable. Polished rice is soaked in cold water (about 15° C) overnight or until the moisture content is about 35%. Excess water is drained off and the swollen rice is cooked with steam at atmospheric pressure for 40–60 min, cooled to 35° C, and inoculated at the rate of 0.1% tane koji (*A. oryzae* inoculum). In modern plants the cooling and inoculation are carried out as a continuous process.

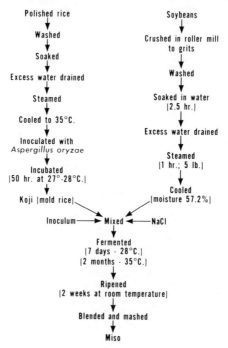

FIG. 11.1. PROCESS FOR MAKING RED MISO

The inoculated rice is then carried by closed conveyor tubes to the koji rooms which may be circular or rectangular. The rectangular rooms may be about 7 ft high and perhaps 15 ft wide and 30 ft or more in length. In the older process, the rice was placed in wooden trays. During development of the koji, temperature (35–40° C), moisture, and aeration are extremely important factors and must be rigidly controlled. Since fermentation is carried out in large rooms, contamination by bacteria is a problem. Bacterial growth is characterized by overheating, resulting in condensation of free water on the rice kernels which causes even more rapid growth of bacteria. Contamination is prevented by steaming the koji rooms, by air filtration, and by ultraviolet lamps to reduce weed microorganisms that might be present in the air introduced into koji rooms. An important factor, however, is to keep the rice kernels moist but not to permit free water on the surface of the kernels. Also during the preparation of koji, the rice is turned several times. During this stirring, the bottom rice must be brought to the top and the lumps broken apart to maintain uniform temperature, moisture, and aeration throughout the fermenting rice. In modern plants this turning of the rice is done with automatic mechanical devices. In the old process stirring was done by hand and this operation increased the amount of contamination.

To encourage mold growth, the humidity is maintained above 90%. In new plants, the air is washed and, hence, high in moisture. Ventilation must be adequate to supply sufficient oxygen and to remove the carbon dioxide.

In about 40–48 hr, the cooked rice is completely covered with white mycelium of the inoculated *A. oryzae* strains. The miso koji appears as a white felt of mycelia with the rice kernels bound together with mycelium. Harvesting is done while the koji is white and before any sporulation has occurred. At this time good koji has a pleasant smell, lacks any musty or moldy odors, and is quite sweet in taste. At this time the rice koji should be removed from the koji room, stirred, cooled, and then mixed with salt, cooked and cooled soybeans, and the yeast inoculum for the second fermentation.

Treatment of Soybeans

Simultaneous with koji production, soybeans are being prepared for fermentation. Although almost any variety may be used, manufacturers require that the soybeans should have a very light yellow, thin, and glossy seed coat; a light yellow or white hilum; a high protein content; absorb water uniformly and cook rapidly; and be large and even in size. The miso made with soybeans should be a product that is uniform in color and consistency.

Beans are cleaned, washed, and soaked in water overnight, absorbing about 1.2 times their weight of water. The beans are cooked under 10–15 lb of pressure until they are sufficiently soft to be mashed between the fingers for dark red miso but less cooking and pressure are needed for white or yellow miso. Cooking temperature and time influence color and flavor of the final product. Other factors that determine flavor and color depend on the proportion of soybeans to rice, the koji, amount of salt, and length of fermentation. Cooked soybeans are chopped or mashed. For white or light yellow miso, dehulled soybeans or soybean grits are required. A patent by Smith *et al.* (1961) covers the use of grits.

Mixing.—After cooling the beans are mixed with rice koji, salt, and water containing the inoculum. To obtain the proper moisture level, about 10% water is added with the inoculum. In the past, the inoculum was miso from a previous fermentation that contained the mixed inoculum of yeast and bacteria. Naturally, the inoculum from a previous fermentation is a selected flora of salt-tolerant yeast and bacteria capable of growing under anaerobic conditions. The dominant organisms are yeast *Saccharomyces rouxii* and certain lactic acid-producing bacteria. Actually, another yeast, *Torulopsis* sp., may also be present. In recent years, pure culture starters speed up the fermentation and reduce the influence of weed yeasts and bacteria.

The proportions of ingredients at the time of mixing, which is carried out in closed blending equipment, determine the type, color, flavor, odor, and appearance of the final product. Thus, white miso contains more rice than soybeans while darker types will contain 50-90% soybeans. Less expensive grades of miso are made from defatted soybean meal instead of whole soybeans. The true miso flavor is believed to come entirely from soybeans, whereas sweetness is derived from the sugars produced by enzymes from rice koji. White miso contains less salt (4-8%), which permits more rapid fermentation but gives the product a shorter shelf-life; red or brown miso contains 11-13% salt. The inoculum, on the other hand, is always the same regardless of the type of miso being made.

Miso contains 48-52% water. If it is less than 48%, the miso is too hard, and if more than 55%, the product becomes too soft. Naturally, moisture content greatly affects the velocity of the fermentation.

Fermentation.—After blending, the mixture is transferred to open tanks either made of wood or concrete. Weight is placed on top of the tanks to force liquid to the surface ensuring anaerobic conditions. Wooden tanks are used in the most modern plants because they can be picked up with fork lifts and moved into and out of incubator rooms. Also, they are easier to dump or unload at the time of harvest. These usually are quite large with a capacity of four tons or more. Wooden tanks last for years without damage due to moving and dumping. Since they can be moved, they may be washed easily between fermentations.

During fermentation the enzymes convert the rice into dextrin, maltose, and glucose, which serve as fermentable sugars for the yeasts and bacteria. The soybean protein is converted to peptides and amino acids. One of the chief amino acids produced is glutamic acid, which gives the delicious flavor to miso. Soybean oil is converted in part to fatty acids.

The fermentation temperature is carried out at $30°-38°$ C. The length of fermentation varies with the type of miso being made. For light yellow miso at least 1 month is required, whereas 3 months are necessary for red miso. The darker grades of miso, containing a high concentration of salt, may take from 3 to 12 months. These darker grades can be preserved for several months without refrigeration. The high salt concentration in these grades inhibits the growth and development of undesirable microorganisms including those which produce toxins. In some regions, miso is fermented at the temperature of locality and such miso is called "natural miso." This type requires at least one whole summer to complete the fermentation because little fermentation occurs during winter months.

Today, a revolution has occurred in the manufacture of miso, which has taken the form of automated equipment and continuous processing.

One of the big problems still remaining is the control of microorganisms because the fermentation is carried out under an open system. Thus, undesirable microorganisms can develop that can grow under anaerobic conditions and in a high salt concentration. This contamination is especially true when the fermentation and aging time is long.

According to Watanabe (1969), there is a trend towards miso being made in factories. There is a trend also towards larger factories; for instance, in 1959, there were 3000 factories while in 1965 there were 2400. The largest company has a capacity of 22,000 metric tons per year. The current price range of miso in Japan varies from 28¢ to $1.10 per kg, depending on the kind and grade of miso.

Ebine (1967) has outlined analytical methods used in the quality control during miso processing and those suitable in evaluation of the product.

SHOYU

Shoyu is the Japanese name for a dark-brown liquid, with a salty taste and sharp flavor, which is made by fermenting soybeans, wheat, and salt. It is an all-purpose seasoning agent used in preparation of foods as well as a table condiment in oriental and many other countries. The product is known as chiang-yu in China; tao-yu in Indonesia; and tayo in the Philippines. Shoyu is the only traditional oriental fermented product that has become well-known in the cookery of Western countries, and is often referred to as soy sauce. Among fermented products, chiang-yu is considered to be the most important one in China, whereas in Japan miso and shoyu are of nearly equal importance with shoyu production exceeding 1.2 million kiloliters (315 million gallons) a year.

The fermentation of shoyu is essentially a process of enzymatic hydrolysis of proteins, carbohydrates, and other constituents of soybeans and wheat to peptides, amino acids, sugars, alcohols, acids, and other low molecular compounds by the enzymes of mold, yeast, and bacteria. In addition to the fermentation technique, two other processes are followed. One is a chemical method in which acid hydrolyzes the proteins and the carbohydrates; and the other, a combination of the two. Traditionally, shoyu is made of whole soybeans; however, in recent years defatted soybean meals and flakes have taken its place. Table 11.1 indicates that 90% of the 170,000 metric tons of soybeans consumed in shoyu production in Japan per year are defatted soybean products.

Fermented shoyu.—The fermentation method involves two steps similar to that of miso fermentation except that more water is used in shoyu production and the filtered liquid, not the whole product, is consumed. The process flowsheet is shown in Fig. 11.2.

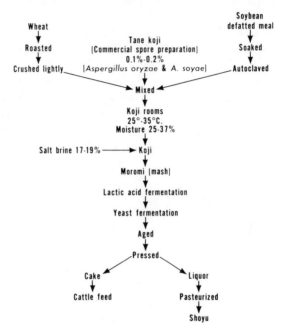

FIG. 11.2. FLOWSHEET   FOR   MANUFACTURE
OF SHOYU

The preparation of shoyu koji from defatted soybean meals or flakes was described by Umeda *et al.* (1969). Defatted soybean products are first moistened by spraying with water amounting to about 130% soybean weight, which are then steamed at 13 lb for 45 min. Whole soybeans are usually soaked overnight, drained, and then steamed at 10 lb for several hours. After steaming and cooling, the mass is mixed with wheat that has been roasted and cracked or very coarsely ground. The proportion of soybeans to wheat may vary from one manufacturer to another. In Japan, equal amounts of soybeans and wheat are strictly observed, whereas in China, soybeans and wheat are mixed in ratios from 4:1 to 1:1. The mixture having a high ratio of soybeans to wheat produces a product rich in nitrogen and taste; however, it is difficult to make koji of superior quality from such mixtures because koji made from protein-rich and moistened substrate is easily contaminated by undesirable bacilli. The mixture is next inoculated with tane koji (*A. oryzae* and *A. soyae*) at a rate of 0.1-0.2% and then incubated at 30° C. After 24 hr of incubation, the mixture is covered with a thin white growth of mold. As mold growth continues, the temperature of the mixture could rise above room temperature to 40° C or higher. The mixture, therefore, should be turned or stirred several times during koji making to maintain uniform temperature, moisture, and aeration. The mixture should also be free of lumps to minimize bacteria propagation. As the incubation time increases, the molds continue to grow and their growth turns yellow and dark green. The moisture of the mixture gradually decreases. After about 72 hr of incubation, the shoyu koji, which has a moisture content about 26%, is ready to be harvested. In recent years, automatic koji-making processes have been developed to replace the traditional way of making koji involving wooden trays and hand mixing. The new equipment includes a continuous cooker, automatic inoculator, automatic mixer, large perforated shallow vats in closed chambers equipped with forced air devices, temperature controls, and mechanical devices for turning the substrates during incubation.

Shoyu koji of superior quality has a dark green color, pleasant aroma, sweet but bitter, vigorous mold growth, high population of yeast, low bacteria count, and strong activities of proteases and amylases.

The second step in preparation of shoyu is brine fermentation. Shoyu koji is transferred to a deep vessel in which a salt solution of 22.6% is added to make a liquid mash of about 18% of salt. A typical example of raw materials used in shoyu fermentation reported by Umeda *et al.* (1969) is as follows: defatted soybean meal, 330 kg; wheat, 337.5 kg; water, 430 liters; salt solution of 22.6%, 1200 liters. *Lactobacillus delbrueckii* and strains of yeast *Hansenula* are added to the mash. The liquid mash, or moromi as it is called by Japanese, is stirred frequently to provide enough aeration for good growth of yeast, to prevent the growth of undesirable anaerobic microorganisms, to maintain uniform temperatures, and to facilitate the removal of carbon dioxide. In modern plants, this process is carried out in large concrete vats with aeration devices. The change of temperature is said to be important for normal progress of fermentation. Therefore, shoyu fermentation in Japan usually starts in April and it will take a year to complete. In general, low-temperature fermentation gives better results; because the rate of enzyme inactivation is slow, the enzymes remain active longer (Komatsu 1968). Watanabe (1969) indicated that good quality shoyu can be obtained by 6-month fermentation when the temperature of moromi is controlled as follows: starting at 15° C for 1 month, followed by 28° C for 4 months, and finishing the fermentation at 15° C for 1 month. Takeda and Nakayama (1968) found that the fermentation time can be reduced from 1 yr to about 2 months when koji is enriched with peptidase. Steamed, defatted soybean flakes are first hydrolyzed by commercial bacterial protease at 55° C for 20 hr, and then mixed with roasted wheat to make koji. The peptidase of koji so prepared is much higher than that of conventional koji.

A perfect fermented moromi should have a bright reddish brown color, pleasant aroma, and be salty but tasty. The matured moromi is pressed to remove the liquid. The liquid or raw shoyu is then pasteurized at $65°-80°$ C, filtered to remove precipitates, and bottled ready for market.

Umeda (1963) and Umeda *et al.* (1969) have reported analytical results of shoyu made from whole soybeans and defatted soybean meal by 11 factories in Japan. Their average results are given in Table 11.5.

TABLE 11.5

AVERAGE COMPOSITION OF SHOYU MADE FROM
WHOLE SOYBEANS AND DEFATTED
SOYBEAN MEAL

	Raw Material	
Conditions	Whole	Defatted Meal
Baumé°	22.7	23.4
NaCl (%)	18.5	18.0
Total nitrogen (%)	1.6	1.5
Amino nitrogen (%)	0.7	0.9
Reduced sugar (%)	1.9	4.4
Alcohol (%)	2.1	1.5
Acidity I (ml)	10.1	14.0
Acidity II (ml)	9.8	13.6
pH	4.8	4.6
Glutamic acid (%)	1.3	1.2
Nitrogen yield (%)	75.7	73.7

A good shoyu has a salt content of about 18%. Its pH is between 4.6 to 4.8; below that the product is considered too acid suggesting acid produced by undesirable bacteria. It is also generally recognized in Japan that the quality and price of shoyu are determined by nitrogen yield, total soluble nitrogen, and the ratio of amino nitrogen to total soluble nitrogen. The nitrogen yield is the percentage of nitrogen of raw material converted to soluble nitrogen showing the efficiency of enzymic conversion. The total soluble nitrogen is a measure of the concentration of nitrogenous material in the shoyu indicating a standard of quality. A ratio of greater than 50% of amino nitrogen to total soluble nitrogen is also evidence of quality. These results can be affected by many factors such as raw materials, steaming conditions, tane koji, and brine fermentation. Technology to improve these values is constantly being sought. Defatted soybean products have proved to be as good a raw material as whole soybeans for shoyu fermentation, whereas full-fat soybean meal is not a good substitute for whole soybeans. Not only does full-fat soybean meal give a low nitrogen yield, it also results in a product having too much acid taste. Alcohol-washed meal can produce a high quality shoyu as well as a high nitrogen yield; however, the cost of alcohol-washed meal makes it impractical as a raw material.

The high salt content of shoyu prevents the growth of most microorganisms; preservatives such as n-butyl-p-hydroxy-benzoate and sodium benzoate are also used in shoyu. n-butyl-p-hydroxy-benzoate is easily hydrolyzed by enzymes produced by Aspergilli, but Hanaoka (1962) reported that enzyme activity is greatly inhibited by high salt concentration, heat treatment, and a long period of fermentation.

The chemical changes in the production of shoyu and its flavor are complicated. Yokotsuka (1960) has written a complete review on these changes. He states that of the total nitrogen, about 40–50% are amino acids, 40–50% peptides and peptones, 10–15% ammonia, and less than 1% protein. Seventeen common amino acids are present, glutamic acid and its salts being the principal flavoring constituents. The organic bases, believed to be hydrolyzed products of nucleic acids, are adenine, hypoxanthine, xanthine, guanine, cytosine, and uracil. Sugars present are glucose, arabinose, xylose, maltose and galactose; also two sugar alcohols, glycerol and mannitol. Organic acids reported in shoyu are lactic, succinic, acetic and pyroglutamic.

According to Watanabe (1969) 1 metric ton of soybeans will produce about 5 kl of shoyu. Based on this estimate, the amount (170,000 metric tons) of soybeans going into shoyu production yearly in Japan is rather high (1.2 million kiloliters). The blend of chemical shoyu with fermented shoyu in the commercial products perhaps accounts for the difference.

Chemical Shoyu.—The chemical method of making shoyu is a process of acid hydrolysis. According to Watanabe (1969), defatted soybean products or other proteinous materials are hydrolyzed by heating with 18% hydrochloric acid for 8–10 hr. After hydrolysis, the hydrolysate is neutralized with sodium carbonate and filtered to remove the insoluble materials giving a clear dark-brown liquid, or chemical shoyu. Acid hydrolysis usually results in a more complete breakdown of substrates than enzymatic hydrolysis; however, acid hydrolysis cannot perform many of the other specific reactions or interreactions of hydrolyzed products as carried out by multiple enzymes produced by molds, yeasts, and bacteria. Chemical shoyu, therefore, is a solution of amino acids and salts which is tasty and, according to oriental taste, does not possess the flavor and odor of fermented shoyu. It is often blended with fermented shoyu before being sold.

Attempts have been made to combine the chemical method, which is convenient, with the fermentation process, which produces the more desirable flavor characteristics of shoyu. First, the substrates are subjected to partial acid hydrolysis. After neutralization, the partially hydrolyzed material is mixed with moromi and the fermentation is carried out for about a month. Tenbata and Morinaga (1968) report that the taste of chemical shoyu was improved by brewing acid hydrolysates of soybeans with 3–5 volumes of moromi for 10–30 days.

NATTO

In the natural fermentation of soybeans, molds usually dominate, but natto is one of the few products in which bacteria predominate during fermentation. *Bacillus natto,* identified as *Bacillus subtilis,* is claimed to be the organism responsible for natto fermentation. Consequently, natto possesses the characteristic odor and persistent musty flavor of this organism, and is also covered with viscous, sticky polymers that this organism produces. Because of its characteristic odor, flavor, and slimy appearance, natto, even though it is well known in Japan, is not so popular nor so widely consumed as miso.

In Japan, natto is eaten with a sauce or mustard and often used for breakfast and dinner along with rice. Making natto is a simple operation and can be easily done at home. Before the microorganism was isolated, natto was made by wrapping cooked soybeans in rice straw and setting in a warm place for 1-2 days. The quality of the product is then ascertained by the stickiness of the beans and their flavor. Rice straw was credited not only in supplying the fermenting organism, but also in providing the aroma of straw, which many consumers were fond of, and in absorbing the unpleasant odor of ammonia from natto.

Many papers have been published concerning the microorganisms in natto fermentation; however, it is now well established that bacilli are the most important ones. Based on Muramatsu's account (1912), Sawamura was first to give the name of *B. natto* to 1 of the 2 bacilli that he isolated from natto. He identified the other one as a variety of *B. mes. vulgatus.* He also believed that both bacilli were required to make good natto. *B. natto* produced natto with good taste and aroma and *B. mes. vulgatus* provided the needed stickiness. But Muto (also cited by Muramatsu 1912) found that only one bacterium, which belonged to the *B. subtilis* group, was necessary for the preparation of natto. Muramatsu (1912) also made a detailed study of the three bacilli he isolated from natto. He learned that the three bacilli were similar to those isolated by Sawamura, Muto, and others. He also supported Muto's account that only 1 bacillus was essential for natto fermentation, and either 1 of his 3 bacilli was suitable for making natto. He agreed with Sawamura that the organism similar to *B. natto* Sawamura did not yield enough viscosity, but Muramatsu discovered that whenever the fermentation was carried out at high temperature (45° C), the organism produced natto with high viscosity and good taste. All three bacilli produced trypsin-like proteolytic enzymes.

In 1960, Sakurai reconfirmed that *B. natto* is aerobic, Gram-positive rod, and classified as a related strain of *B. subtilis.* There are two types of *B. natto* in the laboratory of the Food Research Institute, Ministry of Agriculture and Forestry, Tokyo, Japan. One has optimum temperature from 30° to 45° C and the other, from 35° to 45° C. He recommended that the culture known as *B. natto* SB-3010 and having optimum temperature of 30°-45° C appeared to be the one most suitable for making natto.

Pure culture fermentation has been adopted for making natto ever since the isolation of *B. natto*. Soybeans are cleaned and soaked in water for 12–20 hr, depending on the temperature, until they approximately double in weight. The soaked beans are then cooked until tender, drained, cooled to 40° C, inoculated with a water suspension of *B. natto,* and packed in a wooden box or polyethylene bag; sometimes they are wrapped in a paper-thin sheet of pine wood. The packages, which contain about 1/3 lb of cooked beans, are then placed in an incubating room at 40°–43° C for 12–20 hr. The product has a short storage life: partly because it has a moisture content of more than 50%; partly because natto is usually prepared in small-scale plants of poor quality control. Hayashi (1959A) suggested that the addition of H_3PO_4 to soaking water (0.05–0.1%) seemed to increase the storage life of the product and yet did not affect natto quality.

To improve the keeping properties of natto and broaden its uses, dry powdered natto was developed (Sakurai and Nakano 1961). Fermentation time was reduced to 6–8 hr so that the product would be more suitable for general consumption as food. After fermentation, the beans are spread out on metal trays for drying at low temperatures, either in vacuum or aeration, until the moisture content is less than 5%; then the beans are milled. Arimoto (1961) reports that powdered natto can be added to biscuits, crackers, and soup. The addition of 15% powdered natto in biscuits, 20% in crackers, and 5% in curry soup was acceptable to school children.

There are many reports concerning the changes occurring in natto fermentation. Hayashi (1959A) made one of the most comprehensive studies on natto. His data indicated that there was no change in fat and fiber contents of soybeans during a 24-hr period of fermentation but that carbohydrate almost totally disappeared. A great increase in water-soluble and ammonia nitrogen was noted during fermentation as well as during storage. The amino acid composition remained the same. Boiling markedly decreased the thiamine level of soybeans; but fermentation by *B. natto* enhanced the thiamine content of natto approximately to the same level of soybeans. Riboflavin in natto greatly exceeded that in soybeans. Vitamin B_{12} in natto was found by Sano (1961) to be higher than in soybeans.

Conflicting results on the nutritive value of natto have been published by several investigators (Arimato 1961; Hayashi 1959B; Sano 1961). They disagreed on the nutritive value of natto protein as being superior to that of boiled soybeans. But they agreed that rats fed a diet containing natto and rice grew as well as rats receiving a complete laboratory diet; whereas, rats fed a diet of boiled soybeans and rice did not grow as well as the controls. However, Hayashi found that addition of thiamine to the diet of boiled soybeans and rice corrected the deficiency; the growth of rats receiving a thiamine-enriched diet of boiled soybeans and rice was comparable to that of rats fed natto and rice. Here is further evidence that the benefits of fermented foods are manifold.

HAMANATTO

Hamanatto is the Japanese name for a product made by fermenting whole soybeans with strains of *A. oryzae*. The fermented beans are made in the vicinity of Hamanatsu from which the name of the product was perhaps derived. However, similar products are widely produced and consumed in China, the Philippines, the East Indies, and probably in other countries of the Orient. The product is known as "toushih," (which means salted beans) by the Chinese; "tao-si" by the Filipinos; and "tao-tjo" by the East Indians. In the United States, such fermented beans are often referred to as black beans because of their color.

The methods of preparing soybeans for fermentation and the composition of the brine may vary from country to country, but the essential features are similar. Soybeans are soaked and steamed until soft, drained, cooled, mixed with parched wheat flour, and then inoculated with a strain of *A. oryzae*. After incubation, the beans are packed with the desired amount of salt, spices, wine, and water and aged for several weeks or months. The finished products are blackish. They have a salty taste and their flavor resembles that of shoyu. However, they may differ in salt and moisture contents. Hamanatto is rather soft, having a high moisture content. Toushih has a much lower moisture content than that of hamanatto and, therefore, is not so soft as hamanatto. Tao-tjo tends to have a sweet taste because sugar is often added to the brine.

A typical process for making hamanatto in Japan is outlined in Figure 11.3, based on information furnished by Dr. A. Kaneko of Nagoya University, Japan. According to Dr. Kaneko, the finished product has a salt content of 13% and a moisture content of 38%.

The fermented beans can be used as an appetizer to be consumed with bland foods, such as rice gruel, or they can be cooked with vegetables, meats, and seafoods as a flavoring agent.

TEMPEH

One of the most important fermented soybean foods, originating in Indonesia, is tempe kedelee, or tempeh. This is a cake-like product made by fermenting soybeans with *Rhizopus*. When fried in oil it has a pleasant flavor, aroma, and texture. Unlike most of the other fermented soybean foods, which are usually used as flavor agents or relishes, tempeh serves as a main dish in Indonesia. Because of its unusually good and mild flavor and because of its high protein content, tempeh has been suggested as a possible source of low-cost protein.

Making tempeh in Indonesia is a household art. The procedure may vary from one household to another, but the principal steps are as follows: Soybeans are soaked in tap water until the hulls can be easily removed by hand. The dehulled soybeans are boiled with excess water for 30 min, drained, and spread for surface drying. Small pieces of tempeh from a previous fermentation are mixed with the soybeans. The inoculated beans are wrapped with banana leaves and

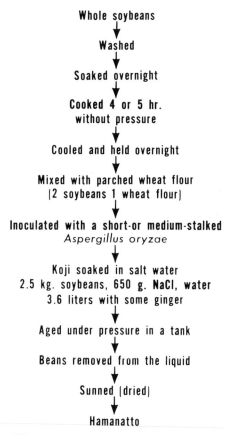

Whole soybeans
↓
Washed
↓
Soaked overnight
↓
Cooked 4 or 5 hr.
without pressure
↓
Cooled and held overnight
↓
Mixed with parched wheat flour
(2 soybeans 1 wheat flour)
↓
Inoculated with a short-or medium-stalked
Aspergillus oryzae
↓
Koji soaked in salt water
2.5 kg. soybeans, 650 g. NaCl, water
3.6 liters with some ginger
↓
Aged under pressure in a tank
↓
Beans removed from the liquid
↓
Sunned (dried)
↓
Hamanatto

FIG. 11.3 PROCESS FOR MAKING
HAMANATTO

allowed to ferment at room temperature for one day. By this time, the beans are covered with white mycelium and bound together by mycelium as a cake, which has a pleasant odor. Traditionally, the cake, which is consumed within a day, is cut into thin slices, dipped into a salt solution, and fried in coconut oil. Sliced tempeh can be baked or added to soup.

To understand the fermentation, it is necessary to produce tempeh under scientifically controlled conditions. It was not until the late 1950's that two groups of scientists in the United States began to study the tempeh fermentation: New York Agricultural Experiment Station, Geneva, N. Y., and the Northern Regional Research Laboratory, Peoria, Ill. As a result, a pure culture fermentation method on a laboratory scale was developed. Changes in soybeans during tempeh fermentation and nutritional value of tempeh were studied in detail. The physiology and biochemistry of the tempeh mold are also being investigated.

We have received cultures of *Rhizopus* and other fungi isolated from different lots of tempeh in Indonesia and found that only *Rhizopus* could make tempeh in pure culture fermentation. Of the 40 strains of *Rhizopus* received, 25 of them are *R. oligosporus;* others are *R. stolonifer* (Ehren) Vuill, *R. arrhizus* Fischer, *R. oryzae* Went and Geerligs, *R. formosaensis* Nakazawa, and *R. achlamydosporus* Takeda. Apparently, *R. oligosporus* is the principal species used in Indonesia for tempeh fermentation. The characteristics of this species have been described by Hesseltine (1965).

Hesseltine *et al.* (1963B) have described a petri dish procedure for making tempeh in the laboratory (Figure 11.4). The preparations of soybeans for fermentation are similar to the traditional manner. Mechanically dehulled soybean grits are also suitable to make good tempeh. Since soybean grits absorb water easily, the soaking time can be reduced to 30 min. The beans are inoculated with spores of *R. oligosporus,* which have been grown on potato-dextro-agar slants at 28° C for 5–7 days. The spore suspension is prepared by adding a few milliliters of sterilized distilled water to the slant. The inoculated beans are mixed, packed tightly into petri dishes, and placed in an incubator at 30°–31° C for about 20 hr. *R. oligosporus* does not require much aeration as do many other molds; as a matter of fact, too much aeration may cause spore formation. It is therefore important to pack the petri dishes tightly; even so, some sporulation may still occur at the edge of the dish, but it will not affect the product. This procedure can also be adopted for making tempeh either in shallow wooden or metal trays with perforated bottoms and covers or in perforated plastic bags and tubes (Martinelli and Hesseltine 1964).

Steinkraus and his co-workers (1960) suggested the use of 0.85% lactic acid as soaking water. The dehulled, soaked beans are also cooked in the acid solution. This treatment would bring the pH of the beans to a range of 4.0–5.0. At this pH range, the growth of contaminating bacteria will be inhibited, but not that of the tempeh mold. We have not, however, encountered bacteria growth in our process. Because *R. oligosporus* produces an antibacterial agent (Wang *et al.* 1969); and because this organism also has the unique characteristic of fast growing, probably there is little chance for bacteria to gain ground before the tempeh fermentation is complete. Djien (1970) has further investigated this matter. He purposely inoculated with different amounts of *Escherichia coli, B. mycoides, Pseudomonas pyocyanea, Proteus spec.* or *P. cocovenenaus* along with *R. oligosporus* in making tempeh as described by Hesseltine. His results indicated that the fermentation is not intervened by the presence of inoculated bacteria. Djien commented that prefermentation during soaking or addition of acid to the soaking water may not be very important in the process of tempeh fermentation.

To prevent the loss of water-soluble substances during preparation and cooking, soybeans were treated in a minimum amount of water, just enough to soak the beans thoroughly, or were sprayed with a certain volume of water before autoclaving. But Smith *et al.* (1964) found that when this procedure was

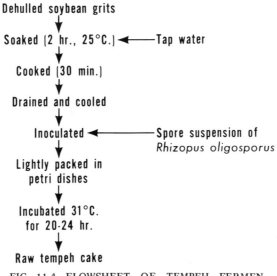

FIG. 11.4. FLOWSHEET OF TEMPEH FERMEN-
TATION ON A LABORATORY SCALE

followed, the tempeh showed less mold development and much sporulation. The product also had an unpleasant odor and poor flavor. The presence of a water-soluble and heat-stable mold inhibitor in soybeans was suggested by Hesseltine *et al.* (1963A). Later, Wang and Hesseltine (1965) found that the water-soluble and heat-stable fraction of soybeans also inhibited the formation of proteolytic enzymes by *R. oligosporus.*

When copra (pressed coconut cake) is substituted in the tempeh fermentation, it is then known as tempeh bongkrek. We have developed (Hesseltine *et al.* 1967) new tempeh-like products by fermenting cereal grains, such as wheat, oats, barley, or mixtures of cereals and soybeans with *Rhizopus.* Among the cultures known to make good soybean tempeh, *R. oligosporus* is the only species suitable to ferment cereal grains to make acceptable products. Other species either grow poorly on cereal grains or produce a product with unacceptable odor and taste.

As mentioned earlier, soybean tempeh is perishable and usually consumed the day it is made because the release of ammonia by enzymatic action causes the product to become obnoxious. Its shelf-life, however, can be prolonged by various methods. In Indonesia, they cut the tempeh into slices which are then dried under the sun. We found that the most satisfactory way to keep tempeh is first blanching to inactivate the mold and enzymes and then freezing. Steinkraus *et al.* (1961) developed a pilot-plant process to dehydrate tempeh by a hot air dryer at 93° C for 90–120 min. Recently, Iljas (1969) evaluated the acceptability and stability of tempeh preserved in a sealed can for ten weeks. There was no significant change in acceptability of the tempeh, when the can was sealed and immediately stored at $-29°$ C or when the can was filled with water, steam-vacuum sealed, heat-processed at $115°$ C for 20 min, and stored at room temperature. However, when tempeh was first air dried at $60°$ C for 10 hr and then sealed in a can which was stored at room temperature, acceptability of the tempeh tended to decrease as storage progressed.

The effects of *Rhizopus* on soybeans have been studied by several investigators. A number of interesting changes occur in soybeans during fermentation. Steinkraus *et al.* (1960) found that the temperature of fermenting beans rises to above that of the incubators as fermentation progresses, but that it falls as the growth of mold subsides. The pH increases steadily presumably because of protein breakdown. After 69 hr of incubation, soluble solids rise from 13 to 28%, soluble nitrogen also increases from 0.5 to 2.0%, whereas total nitrogen remains fairly constant, and reducing substances slightly decrease probably due to this utilization by the mold. Similar changes were observed when wheat is fermented by *R. oligosporus* (Wang and Hesseltine 1966). Wagenknecht *et al.* (1961) reported that 1/3 of the total ether-extractable soybean lipid is hydrolyzed by the mold after 69 hr of incubation, and among all the fatty acids, 40% of the linolenic acid is utilized by the mold. Niacin, riboflavin, pantothenic acid and vitamin B_6 contents of soybeans increase after fermentation, whereas thiamine slightly decreases (Roelofsen and Talens 1964). Wang and Hesseltine (1966) also noticed in fermenting wheat with *R. oligosporus* that the amount of niacin and riboflavin of the wheat tempeh greatly exceeds that of wheat alone, while thiamine appears to be less. Apparently, *R. oligosporus* has a great synthetic capacity for niacin, riboflavin, pantothenic acid, and vitamin B_6, but not for thiamine.

Although free amino acids in tempeh increase during fermentation, the amino acid composition of soybeans is not significantly changed by fermentation (Stillings and Hackler 1965). Perhaps the amount of mycelial protein present in tempeh is not high enough to alter greatly the amino acid composition of the soybeans, nor does the mold depend upon any specific amino acid for growth as suggested by Sorenson and Hesseltine (1966).

The nutritional value of fermented products has always been a controversial subject; it is no exception in regarding the nutritional value of tempeh. Indonesians consider tempeh to be a nourishing and easily digestible food. Van Veen and Schaefer (1950) observed beneficial effects of tempeh on patients with dysentery in the prison camps of World War II, and they suggested that tempeh was much easier to digest than soybeans. However, animal feeding experiments have not substantiated this conclusion (Hackler *et al.* 1964; Smith *et al.* 1964; Murata *et al.* 1967; Wang *et al.* 1968). Nevertheless, the superior nutritive value of tempeh over untreated soybeans has been noted by György (1961) on animals fed low protein diets. His results seem to resemble those obtained with animals fed antibiotics added to their protein source. Recently, we found that *R. oligosporus* indeed produces an antibacterial agent during tempeh fermentation as well as in submerged culture (Wang *et al.* 1969). The compound is especially active against some Gram-positive bacteria including both microaerophilic and anaerobic bacteria; e.g., *Streptococcus cremoris, B. subtilis, Staphylococcus aureus, Clostridium perfringens* and *C. sporogenes.* The compound contains polypeptides having high carbohydrate content. Its activity is not affected by pepsin or *R. oligosporus* proteases, and is slightly decreased by trypsin and peptidase. It is, however, rapidly inactivated by pronase.

It is well established that antibiotics, in addition to minimizing infections, elicit growth-stimulating effects in animals especially those whose diets are deficient in any one of several vitamins, or proteins, or some growth factors. Oriental people are constantly exposed to overwhelming sources of infection and their diets are frequently inadequate. The finding of antibacterial agents produced by *R. oligosporus,* therefore, offers a possible clearer understanding of the true value of tempeh in the diet of Indonesians, and perhaps of fermented foods in the diets of all Orientals. Furthermore, these antibacterial agents would minimize the bacterial contamination during tempeh fermentation.

Of the 40 strains of tempeh mold maintained at the Northern Regional Research Laboratory, *R. oligosporus* NRRL 2710 is a typical representative. Sorenson and Hesseltine (1966) studied carbon and nitrogen utilization by NRRL 2710. They found that such common sugars as glucose, fructose, galactose, and maltose provide excellent growth of the mold; whereas the soluble carbohydrates of soybeans i.e., stachyose, raffinose, and sucrose—are not utilized by the mold as a sole source of carbon. On the other hand, various vegetable oils can be substituted for sugars as sources of carbon. The strong lipase activity and utilization of linolenic acid by mold found in tempeh fermentation (Wagenknecht *et al.* 1961) would strongly suggest that lipid materials are the primary sources of energy. Sorenson and Hesseltine also disclosed that ammonium salts and some amino acids, such as proline, glycine, aspartic acid, and leucine, are excellent sources of nitrogen for *R. oligosporus,* but that the mold does not depend on any specific amino acid for growth.

Strains of tempeh molds produce various amounts of amylase, pectinase, lipase, and proteases. Although strains of *R. oryzae* produce high amounts of amylase, *R. oligosporus* forms little or no amylase. Since starch is seldom found in mature soybeans, it is not particularly important that this species produces amylase during tempeh fermentation. Certain species of *Rhizopus* are known to be active pectinase producers. Of strains suitable for tempeh fermentation, *R. arrhizus* NRRL 1526 appears to produce the highest amount of pectinase. All the strains of *R. oligosporus* have little or no pectinase activity (Hesseltine *et al.* 1963B). Lipase is also produced by molds in tempeh fermentation (Wang and Hesseltine 1966; Wagenknecht *et al.* 1961). Wagenknecht *et al.* (1961) found that fatty acids are liberated by hydrolysis of soybean lipids, but there is no further subsequent utilization of these fatty acids. They concluded that either the mold does not possess the enzyme systems to metabolize these fatty acids or these fatty acids are not permeable to the cytoplasmic membrane of *Rhizopus*. Proteases are, perhaps, much more important enzymes in tempeh fermentation. The ability of *Rhizopus* to produce proteolytic enzymes varied greatly between different strains of the same species as well as between species (Wang and Hesseltine 1965). The proteolytic enzyme systems have optimal pH at 3.0 and 5.5, with the pH 3.0 type predominating in submerged cultivation and pH 5.5 type predominating in tempeh fermentation. The enzymes are stable at pH 3–6 and have high milk-clotting activity.

SUFU

Sufu is a phonetic rendition of the Chinese words, putrid bean curd. It is also known as "fu-ju" or "tou-fu-ju" by many Chinese, "bean cake" by Chinese grocers of this country, and "Chinese cheese" by many scientists. Many more translated names can be found in English literature (Wang and Hesseltine 1970). Indeed, these various names give a good description of the product. Sufu, therefore, is a soft cheese-type product and made from cubes of soybean curd (tofu) by the action of microorganisms. Sufu has been widely consumed as a relish by all segments of the Chinese people and has been manufactured in China long before the Ching Dynasty. Chao of Vietnam, tahuri of the Philippines, and taokoan of the East Indies perhaps are similar to sufu. But no comparison will be made because of the sparse literature on similar products from other countries.

The process of making sufu was considered a natural phenomena. Not until 1929 was a microorganism believed to be responsible for sufu fermentation isolated and described by Wai (1929). He identified the microorganism to be an undescribed species of *Mucor* and proposed the name *Mucor sufu*. Wai also thought that this fungus inhabited originally on rice straw because rice straw was always used to cover the tofu cubes for fermentation in the traditional way. Almost 40 yr later, Wai (1968) reinvestigated the microorganism in sufu fermentation and as a result a pure culture fermentation for making sufu was developed. Wai and his co-workers collected molds from several plants in Taiwan and Hong Kong. They consistently obtained strains of *Actinomucor elegans*. However, strains of *M. hiemalis* and *M. silvaticus* were isolated from homemade sufu. We have received a number of cultures reported to have been isolated from sufu fermentation of various factories and found they all belong to the genus of *Mucor* or a related genus, *Actinomucor*.

In obtaining good quality fermented products, the fungus for sufu fermentation must have white or yellowish-white mycelium to warrant an attractive appearance. The texture of mycelial mat should also be dense and thick so that a firm film will be formed over the surface of the fermented tofu cubes to prevent any distortion in its shape. It is also important that the organism elaborate enzyme systems having high proteolytic activity because the mold grows on a protein-rich medium. Furthermore, the mold growth should develop neither a disagreeable odor nor an astringent taste. Wai and his co-workers confirmed that

A. elegans, M. hiemalis, M. silvaticus and *M. subtilissimus* possess all these characteristics and can produce sufu having good quality. But they indicated that *A. elegans* is the best 1 among the 4 molds for sufu fermentation and is the 1 adopted commercially.

Three steps are normally involved in making sufu: preparing tofu, molding, and brining (Fig. 11.5). To make tofu, soybeans are first washed, soaked overnight, and then ground with water. The finely ground mixture is strained through a coarse cloth to separate the soybean milk from the insoluble residue. After the soybean milk is heated to boiling, calcium or magnesium sulfate is added to coagulate the proteins. The coagulated milk is then transferred into a cloth-lined wooden box and pressed with weight on top to remove whey. A soft, but firm cake-like curd (tofu) forms. Tofu has a bland taste and a high content of water (about 90%). It can be consumed directly and is so eaten extensively throughout the Far East. But the water content of tofu for making sufu is lower than that of tofu consumed directly; otherwise, it is likely to be spoiled by bacterial growth.

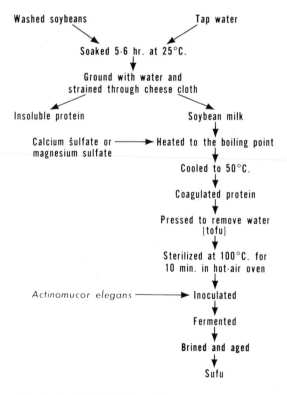

FIG. 11.5. FLOWSHEET FOR THE PREPARA-
TION OF SUFU

According to Wai, tofu for making sufu has a water content of 83%; insoluble protein, 9.1%; soluble protein, 0.4%; and lipid, 4.0%. To prepare for molding, tofu is cut into small cubes of about 2 × 2 × 4 cm. The cubes are immersed in an acid-saline solution of 6% sodium chloride plus 2.5% citric acid for 1 hr and then subjected to hot air sterilization at 100° C for 15 min. This treatment prevents the growth of contaminating bacteria but does not affect the growth of fungi needed in making sufu. Tofu cubes should be separated from one another in a tray with small openings in the bottom and top to facilitate the circulation of air, because mycelia are required to develop on all sides of the cubes. After cooling, the cubes are then inoculated over their surface with pure culture of an appropriate fungus grown on filter paper impregnated with a culture solution. The inoculated cubes are incubated at 20° C or lower for 3–7 days depending on the culture. The freshly molded cubes, known as pehtze, have a luxurious growth of white mycelium and no disagreeable odor. The pehtze has a water content of 74%; insoluble protein, 10.9%; soluble protein, 1.3%; and lipid, 4.3% as reported by Wai.

The last step in making sufu is brining and aging. The pehtze are placed in various types of brining solutions depending on the flavor desired. The most common brine would be one containing 12% sodium chloride and rice wine amounting to about 10% of ethyl alcohol. The immersed cubes are allowed to age for about 40–60 days. The product is then bottled with the brine, sterilized, and marketed as sufu.

Other additives, either to give color or flavor, are frequently incorporated into the brine. Red rice and soy mash are added to the brine to make a red product known as "red sufu." Fermented rice mash can be added to the brine so that the product has a more alcoholic fragrance; tsui-fang, which means drunk sufu, is so made. The addition of hot pepper to the brine would make hot sufu. Rose sufu is one aged in brine containing rose essence. Therefore, the taste and aroma of sufu, in addition to its own characteristics, can be easily enhanced or modified by changing the ingredients of the brining solution.

Changes occurring during the aging process were also studied by Wai and his co-workers. After 30 days of aging at room temperature, total soluble nitrogen increased from 1.00 to 2.74% and total insoluble nitrogen reduced from 7.89 to 6.05%, while total nitrogen changed slightly. The soluble nitrogenous compounds were reported to consist of soluble proteins, peptides, and amino acids, including aspartic acid, serine, alanine, leucine/isoleucine, and glutamic acid. Lipids in pehtze were also partially hydrolyzed through the aging period. Free fatty acids increased from 12.8 to 37.1% and total lipids remained unchanged.

Tofu has a high content of protein and lipids, 55 and 30%, respectively. Therefore, it is expected that the molds would produce high activities of proteolytic and lipolytic enzymes. The proteolytic enzymes elaborated by *M. hiemalis* 28 NRRL 3103 (Wang 1967) and *A. elegans* NRRL 3104 have been studied at the Northern Regional Research Laboratory. *M. hiemalis* grown in soybean medium produces protease having optimal pH 3.0-3.5, whereas *A. elegans* produces proteolytic enzymes having optimal pH at 3.0, 6.0, and 9.0. However, enzyme activity is very low in culture filtrates unless sodium chloride is added to the medium. Further work indicated that the enzyme is loosely bound to the mycelial surface, possibly by ionic linkage and can be easily eluted by sodium chloride or other ionizable salts. Apparently, sodium chloride in the brining solution not only retards mold growth and imparts a salty taste to the final product, but also releases the mycelium-bound proteases, which, in turn, penetrate the pehtze and act on the protein.

Lipolytic enzymes produced by sufu molds were studied by Wai (1968). He found that the enzymes can hydrolyze glycerol esters of fatty acids, as well as such phospholipids as lecithin, which is high in soybeans. A mixture of phospholipases, perhaps, is produced by the molds because choline and fatty acids, stearic acid, palmitic acid, and linoleic acid were formed in the reaction mixture when lecithin was the substrate.

In addition to its salty taste, sufu has its own characteristic flavor. Since the composition of substrate in sufu fermentation is rather a simple one, it is likely that the hydrolytic products of proteins and lipids provide the principle flavor constituents of sufu. The added wine, fermented rice mash, anise, or any synthetic flavor to the brining solution contributing esters, organic acids, sugars, and alcohols further enhance the flavor and aroma.

Traditionally, sufu is consumed directly as a relish or is cooked with vegetables or meats. Either way, sufu adds a zest to the bland taste of a rice-vegetable diet. Because sufu is a cream cheese-type product and has a mild flavor, it would be suitable to use in Western countries as a cracker spread or as an ingredient of dips and dressings.

NEW SOYBEAN PRODUCTS MADE BY FERMENTATION

Cheese-type Products

Soybean milk and soybean curd or tofu to the people of Asia have the same importance as cow's milk and cheese to the people of dairy countries. The relationship of soybean curd to soybean milk is like that of cottage cheese to cow's milk, except that the curdling of soybean milk is traditionally accomplished by the addition of calcium salts, or occasionally by acid. Asiatic people have preferred the method of salt precipitation for making curd. Not only does the salt precipitation yield a product having acceptable texture, but also the salt-precipitated curd serves as a good source of calcium. Unlike most types of cheese, soybean curd is usually consumed without the ripening process carried out by microorganisms. Sufu, as described, is the only traditional fermented soybean product made by a curdling process and a ripening process carried out with a mold.

In recent years, attempts have been made to develop a new cheese-type product from soybean milk through the use of cultures and technology employed in the making of cheese from cow's milk. Although these new products have not become available, these studies have resulted in several publications and patents. These studies show that in addition to the method of salt precipitation, the curdling of soybean milk can be brought about by the acid produced by lactic acid bacteria. The ripening process can be carried out by enzyme preparations, bacteria, or mold.

Hang and Jackson (1967) made a cheese from soybean milk using a cheese starter, *Streptococcus thermophilus.* Although the product had a clean and fresh flavor, the process did not change the substrate other than that ·the acid produced by the bacteria caused soybean milk to curdle. During the 63 days of ripening, the starter bacteria gradually disappeared; other lactobacilli of primary importance in cheesemaking were not found. The water-soluble nitrogen remained constant throughout the ripening process, a condition indicating the lack of proteolysis. However, they found that the addition of rennet extracts and skim milk to the soybean milk and starter stimulated proteolysis and yielded a product having a clean and mild flavor.

Kenkyusho (1965) of Japan used *S. faecalis* as a starter to produce a cheese-like product. Soybean milk was first supplemented with casein, glucose, butter fat, and vegetable oil. The mixture was emulsified, heated, and then treated with *S. faecalis,* rennet extract, and calcium chloride. The resulting curd was then treated according to the conventional method of cheese making.

Soybean or defatted soybean powder was also used to make a cheese-like product. The powder was mixed with water and then treated with takadiastase or other digestive enzymes (Nihon Koyu Kogyo Co. 1965) or inoculated with mold (Kikkoman Shoyu Co. 1965).

According to Obara of Tokyo University of Education, Japan, no acceptable product can be obtained from soybean milk by the conventional cheese-making process. Obara (1968) treated the curd obtained by salt precipitation with proteolytic enzymes. After aging, the finished product had a good flavor and texture. His procedure is outlined in Fig. 11.6. He found that calcium sulfate at the concentrations of 0.03–0.04 N was the best salt to curdle soybean milk. The starter culture was a mixture of *S. cremoris* and *S. lactis* (1:1). Among the proteolytic enzymes tested, papain at a concentration of 0.394% in respect to the protein content gave the best result. The product had an extremely smooth texture and desirable taste. Pronase and bioprase were also considered to be suitable. Trypsin and molsin, however, were unsatisfactory because of their poor digestion ability and the inferior taste of the product.

Fermented Soybean Milk

Soybean milk has a long history as a popular beverage in Asia, but fermented soybean milk has never been a traditional one. In recent years, attempts have

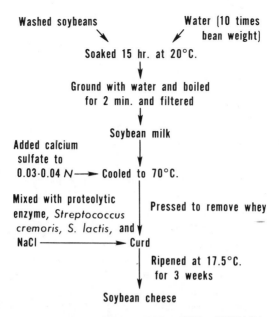

Washed soybeans   Water (10 times
                                    bean weight)

Soaked 15 hr. at 20°C.

Ground with water and boiled
for 2 min. and filtered

Soybean milk

Added calcium
sulfate to
0.03-0.04 N ⟶ Cooled to 70°C.

Mixed with proteolytic
enzyme, *Streptococcus*   Pressed to remove whey
cremoris, S. lactis, and
NaCl ⟶ Curd

Ripened at 17.5°C.
for 3 weeks

Soybean cheese

FIG. 11.6. FLOWSHEET FOR THE PREPARA-
TION OF SOYBEAN CHEESE

been undertaken to develop fermented soybean milk by the same cultures employed in making fermented cow's milk.

In Japan, when soybean milk was subjected to yogurt-type fermentation (Ariyama 1963), the resulting product had higher protein and mineral contents than that of yogurt made from cow's milk. To prepare soybean milk with a high-protein content, Ariyama suggested that the soybeans be first steamed, then soaked and ground with 0.1% NaOH solution, and filtered. According to Ariyama, the use of 0.1% NaOH to extract soybeans yielded a soybean milk having a higher protein content than that of soybean milk obtained by water extraction. Before fermentation, the milk was neutralized with hydrochloric or citric acid and supplemented with 15% cane sugar. The enriched soybean milk was then boiled, cooled, and inoculated with *Lactobacillus bulgaricus*. After incubating at 37–43° C for 4–6 hr, the soybean milk coagulated into a yogurt-like product. Ariyama's results indicate that soybean milk yogurt has a protein content of 9.8% as compared to 3.4% in yogurt made from cow's milk.

At the Northern Regional Research Laboratory, acidophilus-type soybean milk has been prepared. Freshly prepared soybean milk was sterilized at 120° C for 30 min, cooled, and inoculated with *L. acidophilus* NRRL B-629 and incubated at 37° C for 24 hr. The resulting fermented soybean milk tasted good after addition of sugar and vanilla flavoring.

An Ontjom-type Product

Ontjom is an Indonesian food made by fermenting peanut press cake with strains of mold belonging to *Neurospora*. A culture isolated from ontjom was identified as *Neurospora sitophila* and is now maintained in our USDA–ARS Culture Collection designated as NRRL 2884. In Indonesia, the ontjom fermentation is carried out much like tempeh and the product is also eaten much like tempeh. Coconut press cake and residue from soy milk product are occasionally used as substrates, but dehulled soybeans are not known to be used for ontjom fermentation in Indonesia.

Steinkraus *et al.* (1965) studied the ontjom mold to ferment soybeans. Their fermentation yielded a tempeh-like product having an acceptable flavor and texture. The dehulled beans were soaked for 17–18 hr at 25° C in water acidified with 85% lactic acid on a 1.5–2.0% vol/vol basis. The soaked beans were cooked for 90 min at 100° C, drained, cooled, and inoculated. They used a culture of *Neurospora* isolated from ontjom obtained from Indonesia. The inoculum was prepared by growing the mold on sterilized soybeans which were then freeze dried and ground. One gram of the inoculum was used to inoculate 1 kg of cooked beans. The inoculated beans were packed approximately 1 in. thick in a covered stainless steel pan (10 × 14 × 2-1/2 in.) and incubated at 30° C. After 35–40 hr incubation, the beans were bound together by the mold mycelia into a cake-like product similar to tempeh. When the bean cake was sliced and deep-fat fried, it had an almond-like characteristic flavor which was not quite like that of tempeh.

The studies of Steinkraus *et al.* also indicated that the ontjom mold, *Neurospora* sp., had a lower maximum growth temperature than does *R. oligosporus* and that the fermentation cannot be carried out at a temperature above 32° C. During fermentation, the pH gradually rose. The total solids and total nitrogen remained fairly constant, whereas soluble solids and soluble nitrogen progressively increased. Thus, the changes in soybeans during fermentation were similar to those in tempeh.

FUTURE OF FERMENTED SOYBEAN FOODS

If we concede that more and more people will have to use nonanimal protein, then what direction should the development of soybean foods and particularly soybean food fermentations take? The mistake is often made that what we in the West believe is acceptable food, should also be acceptable in the rest of the world. There is a better alternative to this approach. Each country or cultural group has developed a preference for certain food flavors, textures, mouth feel, and appearance. Why not recognize and accept these cultural preferences and develop foods on scientific grounds that do not alter the target people's preferences? Recently, we heard that bulgur being imported into the East Indies is not being used as it was intended but rather mixed with soybeans and then fermented into cakes with the tempeh mold. The product now is acceptable because it takes on a form familiar to the people. Native foods should be examined in detail and modern scientific methods used to develop a food technology on the product. Hence, acceptance would be assured by the native people toward which the market is aimed. If research were developed in this direction, problems of microbial spoilage, retention of vitamins, uniformity of product, and lowered cost could be solved. This concept applies as well to nonfermented foods. Therefore, the goal of fermented soybean food research for developing countries should be aimed at the preparation of a native food by modern technology to give a uniform product which is cheap, nutritious, free of dangerous microorganisms, and completely familiar to the native population.

BIBLIOGRAPHY

ARIMOTO, K. 1961. Nutritional research on fermented soybean products. *In* Meeting the Protein Needs of Infants and Preschool Children. Publ. *843*. Natl. Acad. Sci.–Natl. Res. Council, Washington, D. C.

ARIYAMA, H. 1963. Process for the manufacture of a synthetic yoghurt from soybean. U.S. Pat. 3,096,177. July 2.

DJIEN, K. S. 1970. Personal communication. Albardoweg 31, Wageningen, The Netherlands.

EBINE, H. 1967. Evaluation of dehulled soybean-grits from United States varieties for making miso. USDA Final Tech. Rept. Public Law 480. Project *UR-ALL-(40)-2*. Available at a cost from Natl. Agr. Library.

GYÖRGY, P. 1961. The nutritive value of tempeh. *In* Meeting the Protein Needs of Infants and Preschool Children. Publ. *843*. Natl. Acad. Sci.–Natl. Res. Council.

HACKLER, L. R., STEINKRAUS, K. H. VAN BUREN, J. P., and HAND, D. B. 1964. Studies on the utilization of tempeh protein by weanling rats. J. Nutr. *82*, 452–456.

HANAOKA, Y. 1962. Studies on preservation of soy sauce. I. On the esterase produced by *Aspergillus sojae*. J. Ferment. Technol. *40*, 610–614.

HANG, Y. D., and JACKSON, H. 1967. Preparation of soybean cheese using lactic starter organisms. Food Technol. *21*, 95–100.

HAYASHI, U. 1959A. Studies on natto. III. The quality of natto in its relation to the time of soaking soybeans in water and to the pH value of it. J. Ferment. Technol. *37*, 276–280.

HAYASHI, U. 1959B. Experimental studies of the nutritive value of "natto" (a fermented soybean). Japan J. Nation's Health *28*, 568–596.

HESSELTINE, C. W. 1965. A millennium of fungi, food and fermentation. Mycologia *57*, 149–197.

HESSELTINE, C. W., DECAMARGO, R., and RACKIS, J. J. 1963A. A mould inhibitor in soybeans. Nature *200*, 1226–1227.

HESSELTINE, C. W., SMITH, M., BRADLE, B., and DJIEN, K. S. 1963B. Investigations of tempeh, an Indonesian food. Develop. Ind. Microbiol. *4*, 275–287.

HESSELTINE, C. W., SMITH, M., and WANG, H. L. 1967. New fermented cereal products. Develop. Ind. Microbiol. *8*, 179–186.

ILJAS, N. 1969. Preservation and shelf life studies of tempeh. M.S. Thesis, Ohio State Univ.

INUI, T., TAKEDA, Y. and IIZUKA, H. 1965. Taxonomical studies on genus *Rhizopus*. J. Gen. Appl. Microbiol. *11*, 1–121.

KENKYUSHO, C. M. 1965. Preparation of soy cheese. Japanese Pat. 16,737. July 30.

KIKKOMAN SHOYU CO. 1965. Preparation of soybean-base mold cheese. Japanese Pat. 21,228. Sept. 20.

KOMATSU, Y. 1968. Changes of some enzyme activities in shoyu brewing. 1. Changes of the constituents and enzyme activities in shoyu fermentation after low-temperature mashing. Seasoning Sci. (in Japanese) *15*, No. 2, 10–20.

MARTINELLI, A. F., and HESSELTINE, C. W. 1964. Tempeh fermentation: package and tray fermentations. Food Technol. *18*, 167–171.

MURAMATSU, S. 1912. On the preparation of natto. J. Coll. Agr., Imp. Univ., Tokyo, *5*, 81–94.

MURATA, K., IKEHATA, H., and MIYAMOTO, T. 1967. Studies on the nutritional value of tempeh. J. Food Sci. *32*, 580–594.

NIHON KOYU KOGYO CO. 1965. Preparation of a soybean-base cheese-like protein food. Japanese Pat. 21,230. Sept. 20.

OBARA, T. 1968. Basic investigations on the development of foods from enzymatically treated soybean protein concentrates to increase the use of United States soybeans in Japan. USDA Final Tech. Rept. Public Law 480. Project *UR-A11-(40)-26*. Available at a cost from Natl. Agr. Library.

PEDERSON, C. S. 1971. Microbiology of Food Fermentations. Avi Publishing Co., Westport, Conn.

ROELOFSEN, P. A., and TALENS, A. 1964. Changes in some B-vitamins during molding of soybeans by *Rhizopus oryzae* in the production of tempeh kedelee. J. Food Sci. *29*, 224–226.

SAKURAI, Y. 1960. Report of the researches on the production of high-protein food from fermented soybean products. Food Res. Inst. Min. Agr. Forestry, Fukagawa, Tokyo, Japan.

SAKURAI, Y., and NAKANO, M. 1961. Production of high-protein food from fermented soybean products. *In* Meeting the Protein Needs of Infants and Preschool Children. Publ. *843*. Natl. Acad. Sci.–Natl. Res. Council.

SANO, T. 1961. Feeding studies with fermented soy products (natto and miso). *In* Meeting the Protein Needs of Infants and Preschool Children. Publ. *843*. Natl. Acad. Sci.–Natl. Res. Council, Washington, D. C.

SHIBASAKI, K., and HESSELTINE, C. W. 1962. Miso fermentation. Econ. Botany *16*, 1,80–195.

SMITH, A. K., HESSELTINE, C. W., and SHIBASAKI, K. 1961. Preparation of miso. U.S. Pat. 2,967,108. Jan. 3.

SMITH, A. K. *et al.* 1964. Tempeh: Nutritive value in relation to processing. Cereal Chem. *41*, 173–181.

SORENSON, W. G., and HESSELTINE, C. W. 1966. Carbon and nitrogen utilization by *Rhizopus oligosporus.* Mycologia *58*, 681–689.

STEINKRAUS, K. H., HAND, D. B., VAN BUREN, J. P., and HACKLER, L. R. 1961. Pilot plant studies on tempeh. Proc. Conf. Soybean Products for Proteins in Human Food, USDA, Peoria, Ill. USDA–ARS *71–22*.

STEINKRAUS, K. H. *et al.* 1960. Studies on tempeh–an Indonesian fermented soybean food. Food Res. *25*, 777–788.

STEINKRAUS, K. H., LEE, C. Y., and BUCK, P. A. 1965. Soybean fermentation by the ontjom mold *Neurospora.* Food Technol. *19*, 119–120.

STILLINGS, B. R., and HACKLER, L. R. 1965. Amino acid studies on the effect of fermentation time and heat processing of tempeh. J. Food Sci. *30*, 1043–1049.

TAKEDA, R., and NAKAYAMA, S. 1968. Peptidase of *Aspergilli.* III. Enrichment of koji with peptidase and a proposal of rapid fermentation of shoyu. Seasoning Sci. (in Japanese) *15*, No. 3, 19–25.

TAMIYA, H. 1958. The koji, an important source on enzymes in Japan. Proc. Intern. Symp. Enzyme Chemistry, Tokyo and Kyoto.

TENBATA, M., and MORINAGA, T. 1968. Fermenting ability and the refined degree of soy moromi by addition of chemical soy sauce. Hiroshima-ken Shokuhin Kogyo Shikensho Hokoku. *10*, 37–44.

UMEDA, I. 1963. Comparison of United States and Japanese soybeans for making shoyu. Final Tech. Rept., Public Law 480, Project *UR-A11-(40)-(C)*. Available at a cost from Natl. Agr. Library.

UMEDA, I., NAKAMURA, K., YAMATO, M., and NAKAMURA, Y. 1969. Investigations of comparative production of shoyu (soy-sauce) from defatted soybean meals obtained from United States and Japanese soybeans and processed by United States and Japanese methods. USDA Final Tech. Rept. Public Law 480. Project *UR-A11–(40)-21.* Available at a cost from Natl. Agr. Library.

VAN VEEN, A. G., and SCHAEFER, G. 1950. The influence of the tempeh fungus on the soybean. Doc. Neerl. Indonesian Morbis. Trop. *2,* 270–281.

WAGENKNECHT, A. C. *et al.* 1961. Changes in soybean lipids during tempeh fermentation. J. Food Sci. *26,* 373–376.

WAI, N. 1929. A new species of *Mono-Mucor, Mucor sufu,* on Chinese soybean cheese. Science *70,* 307–308.

WAI, N. 1968. Investigation of the various processes used in preparing Chinese cheese by the fermentation of soybean curd with mucor and other fungi. USDA Final Tech. Rept. Public Law 480. Project *UR-A6-(40)-1.* Available at a cost from Natl. Agr. Library.

WANG, H. L. 1967. Release of proteinase from mycelium of *Mucor hiemalis.* J. Bacteriol. *93,* 1794–1799.

WANG, H. L., and HESSELTINE, C. W. 1965. Studies on the extracellular proteolytic enzymes of *Rhizopus oligosporus.* Can. J. Microbiol. *11,* 727–732.

WANG, H. L., and HESSELTINE, C. W. 1966. Wheat tempeh. Cereal Chem. *43,* 563–570.

WANG, H. L., and HESSELTINE, C. W. 1970. Sufu and lao-chao. Agr. Food Chem. *18,* 572–575.

WANG, H. L., RUTTLE, D. I., and HESSELTINE, C. W. 1968. Protein quality of wheat and soybeans after *Rhizopus oligosporus* fermentation. J. Nutr. *96,* 109–114.

WANG, H. L., RUTTLE, D. I., and HESSELTINE, C. W. 1969. Antibacterial compound from a soybean product fermented by *Rhizopus oligosporus.* Proc. Soc. Exptl. Biol. Med. *131,* 579–583.

WATANABE, T. 1969. Industrial production of soybean foods in Japan. UNIDO Expert Group Meeting on Soybean Processing and Use, USDA, Peoria, Ill. Nov. 17–21. (Proc. in press.)

YOKOTSUKA, T. 1960. Aroma and flavor of Japanese soy sauce. Advan. Food Res. *10,* 75–134.

CHAPTER 12

D. E. Hooton
D. W. Johnson

Marketing of Soybeans and Their Protein Products

THE MARKETING FUNCTION

There is only one valid definition of business purpose: to create a customer (Drucker 1954).

Modern marketing has evolved from what was once a generalized concept to what now embodies a series of precisely defined activities. In spite of its evolutionary growth to its current state of refinement, the term "marketing" still has different meanings in the minds of the people who use the term in the course of their business. For over a decade, now, the "marketing view" has been widely publicized. It has even acquired a fancy name: The Total Marketing Approach.

Not everything that goes by that name deserves it. "Marketing" has become a fashionable term. But a gravedigger remains a gravedigger even when called a "mortician"—only the cost of the burial goes up (Drucker 1964). The process of bestowing marketing titles to men whose function is to sell will not bring about much of a change in what they do or how well they do it. But, it is almost certain that the budget will be expanded to fit the new sophistication.

This chapter makes no effort to survey or discuss the variety of marketing philosophies now in existence, for the list is long and the task, though possibly worthy, should be handled more appropriately as a separate subject and analysis. But whether marketing is viewed in the old sense of "pushing products" or in the new sense of "customer satisfaction engineering" it is almost always viewed and discussed as a business activity (Kotler and Levy 1969). For the purpose of guiding the reader, the authors will characterize, briefly, the marketing concept to which they subscribe.

Terms and Definitions

The heart of the marketing discipline lies in the various functions which operate within it. McGarry (1950), identified the general functions of marketing as contactual, merchandising, pricing, propaganda, physical distribution, and termination. However, the systematic identification of these factors was not the complete picture, and the need persisted to determine more explicitly what these functions are and to what they subdivide. The relevant question became: what are the unique and inherent purposes of these marketing functions?

Primarily, there are two purposes: to obtain demand and to service demand with the goal of acceptable levels of profit and investment return. All of the activities in which marketing people are engaged are organized means to achieve these primary ends. Recognizing the fact that the above-mentioned functions are inherent to marketing makes it possible to arrive at a delineation of what marketing does (its objective) and how it does it (its activities). The following diagram, (Fig. 12.1) adapted from Lewis and Erickson's (1969) model, incorporates the marketing functions into the major categories and shows their interrelationships.

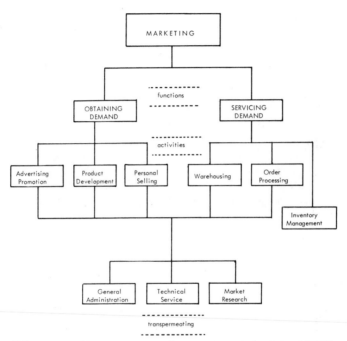

FIG. 12.1. SCOPE OF MARKETING FUNCTIONS AND ACTIVITIES

Marketing Philosophy

Setting down the marketing functions in the related order of Fig. 12.1 emphasizes their central objectives as being rooted to the basic purpose of the business: producing a customer for its goods and services. This was the great contribution of the new marketing concept promulgated in the 1950's, a concept that resulted from the analysis of marketing's place in business which occurred following World War II (Levitt 1960). This concept held that the objective of all business firms in an age of abundance and creativity is to develop customer loyalties and satisfactions. The key to achieving this objective is to focus on those things the customer needs and wants (Lewis and Erickson 1969).

A distinct clarity is evident in this kind of focusing, for it swings the company's planning away from product orientation, or, the assumption that the firm's manufacturing corps knows more about the kind of product a customer wants than does the customer. This latter view was often buttressed by the further assumption that the customer will always appreciate and pay for a well-made product (Oxenfelt 1966). The converse, which is just as extreme, is that once on the order books, the customer will accept a marginal version of the original model.

While strict product orientation with its obvious faults represents the nadir of modern marketing, it must be pointed out that its opposite, pure customer orientation, does not present the sure route to success that the field saleman would have his management believe. To produce economic results, there is more involved than just knowing the customer and giving him what he wants.

Today's operational and successful marketing programs go considerably beyond this to another central facet of the marketing task, as mentioned previously: that of profit orientation. In order to underwrite the risks and to insure that the business will have a future, the objective of earning a profit must be just as fervently espoused as knowing the customer's desires. From this position, what the customer wants matters only to the extent that the firm would profit by meeting and supplying those wants. To the extent that the business seeks to understand and accommodate its customers (including the cost of learning their preferences), it is in a position to pursue the course of profits with intelligence.

Conclusion

As we move through the discussion of the marketing of soybeans and their protein derivatives, this chapter will follow, at least topically, the marketing functions shown in the schematic (Fig. 12.1). As the case may demand, some functions will receive more attention and discussion than others. However, such stress does not imply an attempt to augment or diminish the importance of *all* marketing functions as essential parts of the whole subject.

A final objective of this chapter will be to bind together the information, ideas, opinions, and experiences constructed here with a central thread which will draw attention to the transformation of thought that has been taking place in a still heavily-structured commodity business which is developing into a complex and highly specialized enterprise. This kind of transformation is not, of course, unique to the soybean industry; but it is a fact of its history and its future, especially that which will be shaped by its marketing activity.

THE MARKETING OF BEANS

The Commodity Environment

The marketing of soybeans is much more complicated than for other agricultural products. This is due to the fact that not only do soybeans have a futures market but so do the two primary products resulting from the processing of beans: specifically, soybean oil and meal. While there does exist a general relationship between the prices for beans, meal, and oil, it is not a straight line relationship. Rather, it varies depending on the relative demand for oil and meal in relation to their counterpart commodities which compete for the same markets.

In the United States, soybeans, like cereal grains, are harvested over a relatively short period of time in the fall of the year. Therefore, the initial soybean marketing is a highly seasonal activity with the result that the beans are gathered and placed in storage for utilization over the entire year. Depending on circumstances, the grower may hold the beans in storage, either on the farm or in private storage space, or sell his beans for cash. The first commercial holder of beans (the storage elevator) and the second holder (the bean processor) are required to take in raw material which may have to be held in inventory for months.

Soybean Futures.—In order to alleviate the risk involved in such holdings, a futures market gradually developed for soybeans, meal and oil. In this way, if the futures price, at some point several weeks or months ahead, is more attractive to the farmer than the cash price at the time of harvest (plus storage and handling charges) a futures sale can be made. The same is true of the intermediate holder of beans and for the processor. Fortunately, there are a large number of companies or individuals who are willing to buy or sell soybeans, meal, and oil at any one time. However, if it were only the holder of beans who were the sellers and the processors and exporters who were the buyers, the marketing, at times, would be in an almost impossible situation through the resulting imbalance. In a free competitive market, there are always speculators who are willing to take the opposite side of a trading action at any one time. This has created a fair state of equilibrium, and has brought about a successful trading system both in the cash and futures market over the years. In view of the intricate nature of the futures market, a detailed analysis will not be attempted here. For further insight into its workings the reader is referred to the publications by Raclin (1963), Bartholomew (1968, 1969), Johnson (1969) and Merrill Lynch et al. (1969).

The Farmer as a Producer

The decision by the farmer to grow beans results from an overall appraisal of what dollar return he expects to net from beans per acre as opposed to estimates for other eligible crops. The decision considers crop restrictions imposed by the government and what options are open on unrestricted acreage.

The farmer generally is not in a position to be a direct factor in securing or creating demand for soybeans as a commodity. The total demand for beans is the result of a complex series of external events which are quite far removed from his control or influence. The point at which real demand begins is the level of the retail grocer where any number of consumer products, whose presence as a finished article is in some way related to soybean derivatives, have gained acceptance. The range is broad and extends from fresh meat and eggs to angel food cakes and casseroles. All of these represent a vast system of livestock feeding and food production with many intermediate points of refinement along the way.

As remote as the retail market is from the growers' sphere of influence, there is an area where they may play a role in creating demand for soybeans and their refined products. By taking an active part in agricultural trade associations such as The American Soybean Association and local and national agricultural groups, they have an opportunity to participate in the sponsoring of genetic research to develop new varieties with a view to improving crop yields, better ratios of protein to oil, a more desirable balance of essential amino acids, and more medically acceptable ratios of saturated to unsaturated fatty acids. Such innovations could bring about improved prices through an increased demand for the end products of oil and protein. It must be remembered that the U.S. farmer is relying on both the domestic and foreign market. In the latter, the various associations can cooperate with the federal government to stimulate the use of soybeans and soybean products abroad; but domestically, he must rely on the U.S. processors to generate the demand.

The Farmer as a Marketer

Once he has determined the number of acres he will devote to soybeans and has considered the risk due to weather and plant disease and has produced a crop, he must now choose how and when and where to market it. In a free competitive environment, this would be the economic line of transit for his particular set of circumstances and geographical location.

Options.—Participation in the loan program requires that the farmer either have storage space on the farm or secure space in an elevator. As a marketer and a businessman, the farmer considers a number of options that are open to him. He may sell his beans to the local elevator at a firm cash price even before the seed is in the ground. He may sell futures up to approximately 1 yr before the seed is planted if he thinks the board price is higher than the cash price will be at the time of harvest or sometime later. Another alternative would be to place the beans in storage and hold for a more favorable sale, or he may harvest and sell directly to the elevator at that time. If he feels the board price for beans will improve as time goes on, he may sell the cash beans at harvest and then buy futures. Finally, he may place the beans under a price support loan or purchase agreement for redemption when and if the cash price ever exceeds the loan value (Fiedler 1971).

If he elects the latter course, he must show evidence that he has prepaid storage. If he chooses to redeem the beans, he must repay the principal plus accumulated interest. If the beans are brought in for loan storage, there is an elevation charge which is returned to him by government only if he defaults. Elevation regulations, therefore, encourage default on off-farm stored beans.

Also, the premiums and discounts for moisture and foreign material in the loan program are presently more generous than those in normal commercial practice.

The Soybean Distributor

As with many other commodities, soybeans may be produced in areas where they are not totally consumed. Price differentials drive the flow of beans from surplus to deficit areas. Locational price factors are worked out by the bean merchandisers. When the price of beans at a consumption point is more than the full freight over the price at the country elevator point, the merchandisers buy at the latter and sell at the former. Beans first move by truck from the farm to the country elevator. This short truck haul is the most expensive move per mile the beans incur. From the elevator they then move by truck or rail to processing plants, subterminal elevators, or terminal elevators. If the latter, they may be shipped by rail or barge to an export elevator or to a processor.

In average years, the grower sells a relatively high proportion of the crop at harvest time to a nearby elevator. Generally, elevators store surplus beans from the time of harvest until demand arises in the areas they service. In deficit areas the storage period is quite short. The country elevator operator, as the second handler, has all of the options available to him which the farmer has with the exception of the support loan. The process by which soybeans are attracted into and out of storage is when the "basis" (the difference between the cash price and a futures price) is sufficiently weak to create a storage incentive. In this case soybeans are bought and futures sold. When the "basis" subsequently strengthens, so that storage is not attractive, cash beans are sold and futures are bought.

If prices are unsatisfactory, the grower usually does not sell his beans until harvest or later. If crop scares or other bullish developments force prices up, some growers sell a portion of their crop ahead. With the increasing tendency of U.S. and foreign processors to plan ahead, growing season purchases may become an even more important requirement for country and subterminal operators.

THE MARKETING OF MEAL AND OIL

The actions which characterize the buying and selling of meal are similar to those for beans. As a commodity, meal does not lend itself to either the kind or variety of marketing methods associated with the refined protein products. Meal is bought and sold on a cash basis or on some type of futures basis, e.g. nearby option. Its transactions are predicated on the demand which exists both in this country and abroad. It has its own system of economics, strategies, and risks which can be best understood by referring to the same authors previously suggested for soybean trading. The foregoing also applies to the commercial activities relating to unrefined soybean oil.

THE MARKETING OF REFINED PROTEINS

History

Flour and Grits.—Soy flour products have been produced in this country for almost 50 yr. A brief account of their development has been presented in Chap. 10. As reviewed in that chapter, flour products initially offered to the food industry were in rather crude form, and as such met with an indifferent and negative reception. The image projected by these products affected attitudes toward soy protein which persisted for many years, even when improved products later became available.

These early soy flours and grits epitomized bad marketing practices, and are often cited as how not to introduce new ingredient products. Generally, the failure of such products results from any one of several causes as reviewed here: (1) The fallacy of assuming that a need always constitutes a want, or, that a need always calls forth a market. (2) The failure to define precisely that segment of the total market where the products have the greatest value to users. (3) Underestimation of the kind and quantity of marketing effort required to generate demand. (4) Inadequate information about the buying process, particularly the pattern of influence and who makes the decision to buy. (5) The failure to anticipate the demands the new product makes on the customer's technical and operational skills. (6) The failure to accurately gauge the amount of new investment which will be required by the user and the extent to which their present technology is rendered obsolete. (7) Insufficient knowledge about the "result areas" of the user's business—or how his profits are generated.

Isolates.—Not all of the soy proteins had their inception characterized by poor planning. In documenting the commercial marketing of some of the other edible grades of protein, it will be appropriate to trace the origins of the industrial, inedible isolate which did so much to influence the pattern for the food grade product and its introduction to the trade some 20 yr later.

Although the technical development of soy protein isolates appeared in several early papers, even as far back as 1919, it did not make its appearance as a commercial product until 1937, when the first successful full-scale plant was put into operation by The Glidden Company. The product was geared to the industrial markets being served at that time by casein, which were primarily pigment binding in coatings for paper, folding carton- and box-board, and as a blanketing foam for fire fighting during World War II. In these applications, the protein functioned as an adhesive and as a protective colloid.

Strategies.—The marketing strategy for the industrial isolated protein was developed basically by Walter M. Bain, formerly a chemical engineer with Crown Zellerbach, who, in 1937, had been hired and appointed Sales Manager of Industrial Protein for The Glidden Company. Three important concepts came out of his efforts over a 28-yr period which solidly established the industrial grade of soy protein and gave shape to the marketing approach for the edible grade which entered the product stream in the mid-1950's. The first was to search out the changes in those markets where a particular protein could be modified or designed to perform in such a way that competitive products would yield below-standard results when substituted. The second was to recognize that with any product or ingredient the cost per unit of performance was the most realistic measure of to what degree a product would benefit a customer. And, third, the most effective way to establish this with the customer was through reformulation, actual plant trials, and continuing technical service.

An example of the first concept was the perception that the trend toward the development and conversion to trailing blade machines in the paper coating industry favored the properties of soy protein which did not "string out," leaving tracks in the coating which was characteristic of casein. An illustration of the second concept was the development of partially hydrolyzed soy protein for use on the trailing blade which made it possible to go to higher solids, lower moisture coatings. This resulted in faster machine speeds and a reduction in drier capacity and costs.

There are parallels to these examples in the field of edible isolates. The trend toward quick-service convenience foods requiring fabrication, dehydration, re-hydration brought the versatility of the textured soy proteins to the forefront in the mid and late 1960's. Likewise, the economic performance of soy protein in chopped emulsified meats, egg white replacement in cakes, and the substitution for nonfat dry milk in bread all point to a lower cost per unit of performance while, at the same time, maintaining the required level of consumer acceptance.

Major Use Classes of Proteins

The development of the soy flours, concentrates, isolates, and texturates that are here today was shaped largely by a creative and competititve U.S. food industry searching and pressing for better ways to structure and process foods and a desire to have an extensive array of basic ingredients with which to work. It is apparent that the flours, concentrates, and isolates are rarely, if ever, consumed in their original state, but as ingredients in some finished food product. The applied uses of these materials can be classified into three basic areas: (1) complete replacement for existing ingredients; (2) extenders for primary ingredients; and (3) a basic ingredient in a new food.

Substitutes.—The first category focuses on those uses of soy protein where a direct substitution is accomplished and value is advanced. The use of specialty soy flour in bread to replace nonfat dry milk is such an example. Another is the use of isolated soy protein in hypoallergenic infant milk formulas. In the first case, production economics is the motivator. In the second, customer well-being (the baby and its mother) is the objective and the expected sales prompt the manufacturer to make the replacement.

Extenders.—The next area, that of extenders, considers the advantages of enhanced performance at the same or lower cost—or—the same performance at a lower cost. Examples include such long-established applications as lecithinated soy flours to extend egg yolk in doughnuts, enzyme-modified soy proteins to partially replace egg white in angel food cake and in aerated candies, isolated soy proteins in finely chopped, cooked meats to extend the emulsifying and thermo-setting properties of salt-soluble muscle proteins.

Basic Ingredient.—The third category emphasizes the opportunities for proteins in tomorrow's food products, those now on the drawing board and those already hatched and in market test. Food products may be new in the sense that they offer a new taste, a new shape or consistency, a new motivation, or a new convenience but all within the dimension of a basically familiar food. Thus, the motivation to shed weight and slim down brought the measured-calories wafers and beverages containing soy protein isolates, concentrates, and flours into this market. The new convenience entered as dehydrated TV dinners and casseroles with isolates as a functional performer. The textured proteins bridge a multitude of use areas, offering both replacement and extender value to the processor, and present almost unlimited opportunities to put together familiar foods that provide improved consistency and greater convenience. Also in this third category are products that have had no previous history and have entered food lines as a uniquely fabricated entity. Here, one thinks of some of the "space" foods developed for the moon explorers. These might be in snack form or simply a new concoction designed to furnish a balanced meal that is a delightful eating experience but bears no resemblance to any of today's foods.

Nutrition as a Use Class.—Finally, the product planner considers another category, a separate but underlying thread which cross-stitches protein product development in current markets: nutrition. In some cases, it has been a required companion to physical function, as in infant milk and the controlled-calorie powdered drinks. In others, it has played a background role, present but not featured as with the flour-concentrate blends in baked goods.

We are now in the era where nutrition will play a more decisive part for three reasons: increasing awareness of protein nutrition on the part of the consumer who will consider it as an important factor to the improvement of his well-being; the positive action the government agencies have taken to set minimum standards for certain foods as well as what officially constitutes "lunch" and "dinner;" and the effect of progressive shortages of protein on a world scale.

Obtaining Demand

Referring again to Fig. 12.1, certain of the marketing activities are fairly specific for creating and maintaining business. Others identified here as transpermeating are a part of both, stimulating demand as well as servicing it. In discussing the role of these activities in terms of marketing soy protein products, attention will not be given to all of them. This is not to slight their importance but to better bring into view those which have undergone the most change and are exerting the greatest influence.

Product Planning.—What the producer or supplier thinks the most important feature of a product to be—what they mean when they speak of its "quality"—may be well be relatively unimportant to the customer. It is likely to be what is hard, difficult, and expensive to make. But the customer is not moved in the least by the manufacturer's troubles. His only question is and should be: What does this do for me (Drucker 1964)?

Value as Criterion.—The product planning activity addresses itself to determining "what products with what features for what uses by whom." To be effective, the planner must have a firm grasp of the market concept of value. Otherwise, the result of his effort will be a super product only he appreciates or a line of second-rate products which always lead to marginal business performance.

Value Defined.—Value, as understood by the planner, is based on the extent a product satisfies the user's sense of well-being as the user views it. Whereas all products possess a certain relation to well-being, there is a greater and a lesser degree in that relation. The lesser degree is present when a product is merely useful. But for the higher degree to exist it is necessary for the product not only to cause an enhancement of well-being, but also be an indispensable condition of it. Thus, all goods have usefulness; but not all goods have significant market value. In order for there to be significant value, utility must be paired with relative scarcity (Böhm-Bawerk 1960). For example, there is ample utility in soy flours in several use markets. But with the recent entry of at least two major firms into production of flour products, they are relatively plentiful, and the exchange value is pushed down. Similarly, there is a shortage of somewhere around 10 million tons of protein on a global basis, but proteins with utility in the use areas where a profitable exchange can be made are slow in development. Devotion to the "value" principle leads the product planner away from duplicates of what already exists and toward innovation in a specific area and forces him to concentrate his efforts.

Nutrition in Planning.—In combination, these factors exert a force which will bring more protein into the markets and go beyond quantity to insist that the protein (on either a weight or caloric basis) produces a physiological effect of a certain kind and magnitude. Just as protein nutrition has entered a more finely divided area of product proliferation in the mind of the marketer, so has the understanding of the effect that is expected from the use of soy proteins if they are to have a valid claim on the new markets and opportunities.

As new soy products have been and will be developed, their nutritional contribution must be tracked and measured. And, as soy proteins have been modified to achieve physical and organoleptic performance, these, too, must be evaluated should they find expanded uses in finished foods consumed for nutritional reasons. In the face of a recognized need, supplementation with essential amino acids is becoming an accepted and realistic approach to overcoming natural imbalances from product to product. Once the supplementation of soy protein with DL-methionine is officially established in the Federal Register by the US FDA, it will provide an impetus to new product development.

Summary.—The real task of the product planner in the protein field is not only to see the immediate desires of the market for particular protein concepts, e.g., high tensile strength in an edible protein fiber, or high nutritive value in a finished food system, but he also must perceive things that already have happened which constitute a major change and therefore irrevocably affect the future. Thus, trends are actually only variations in a pattern. Future opportunities that are of a real impact are those which result from a break in or departure from the pattern.

Personal Selling.—Today's protein salesman is a businessman who knows and understands his own firm's operation as well as the internal workings of other industrial businesses he calls on. His view, probably more so than anyone else in the company, is directed toward the action that lies outside the business, that area where the customer, the true profit center, exists. His job is to move the company's output, and in so doing, he assumes responsibility for one side of the revenue stream. Selling, as a part of marketing, is concerned with the direct attempts to persuade the customer to buy. A salesman, of course, is one who *sells.* As a salesman of nonstaple ingredients, his tactical plans take on the elements of an extended campaign, and involve a series of strategies designed to produce that ultimate buying decision which arises from a number of intermediate decisions.

Industrial Ingredients.—This chapter has been concerned largely with soy protein as an ingredient to the food industry, but at the same time recognizes that the advent of the textured proteins involves products which will enter the domain of consumer marketing. It is anticipated that these products will blossom steadily and bear mature fruit sometime in the mid-to-late 1970's. Also, proteins in the intermediate and rapidly-growing institutional and food service realm stand with a leg in both industrial and consumer marketing practices. These two areas are in their early stages of emergence, and the activity of selling to these markets will not be developed in detail here. Rather, this section will treat proteins mainly as industrial ingredients, and will give some attention to the relationship of selling to marketing. Going from the general to the specific, marketing is broader than selling and embraces all of the activities which are undertaken to win large, secure, profitable markets and which are intended to foster sales.

The Buying Decision.— From the marketing view, buying decisions for industrial ingredients as compared with retail products are more complex, involve more people, take longer, require the evaluation of more factors, and are less easily observable (Webster 1968). All of the elements which influence the decision to buy are widely varied and the sources of information involved exist in considerable variety and number. Basically, the flow of the buying decision process goes through the steps of awareness, interest, evaluation, trial, and adoption (Rogers 1962). The path of adoption of a soy protein requires the salesman to move in no predetermined or set pattern, but generally as follows: (1) Study of industrial prospect's business: how they produce their goods; identification of critical profit areas of their business; and determination of how this product can ultimately augment the prospect's economic situation. (2) Assess competitive status. (3) Make presentation to purchasing department. (4) In conjunction with technical service, make contact with and presentation to R & D, marketing, production, quality control, and other management. (5) Determine who make the decision to buy. (6) With technical service, demonstrate and establish through trial the value of the innovation. (7) Recontact purchasing to formalize and close the sale. (8) Continue contact and service.

The Time Lag.—At the point the salesman begins his study until the sale is made, a year or more may have transpired. During this time, he may have had personal contact with as many as 15–20 people who were directly or in some way involved in the adoption of the product. If the protein ingredient is part of a finished food product that is in some stage of its own development or market test, a 3-yr time elapse is considered to be a realistic end point.

Several factors are critical to the time element. One is the amount and the quality of information the salesman has available. Quality of information is a subjective concept, however, and relates to its ability to reduce uncertainty in the buyer's mind. The sum of buyer confidence comes from the degree of credibility which is a combination of the perceived expertise and trustworthiness of the company and the salesman. As more companies begin to use the protein innovation, "word gets around" and this, in conjunction with the prevailing advertising and promotion program, gives the prospect a chance to consider the product within his own industry, independently from the sales contact.

Another factor affecting time is the size of the customer. Larger firms are more likely to be able to tolerate risk, and can afford to assign people and other resources to evaluating the product. When these factors are dominant, the customer tends to reach a buying decision earlier. With smaller firms, their motivation to an early decision arises from the amount and relevance of the technical information plus the fact that their whole decision-making structure is less complex. In a small bakery or meat processing plant, one or two people may be all that are involved in the decision to buy.

The foregoing are general conclusions and concepts, and, as such, apply to whatever kind of selling force a firm elects to use, whether direct salesmen, agent-brokers, or jobber-distributors, and whether their end markets are situated in this country or overseas.

Generating and Servicing Demand

Market Research.—Considered in its broadest sense, market research is a knowledge task. In the scope of the marketing realm, it is described as a transpermeating type of activity, contributing to both the efforts of generating demand and servicing demand. It is servicing in the sense that it keeps on appraising the suitability of what is produced and sold in terms of what is wanted. On the other side, it is obtaining demand since its assumptions and conclusions direct both product planning and manufacturing to the opportunities that hold the most promise for economic results.

Market Reality.—It is well known that no single product or company is very important to the market. The market is a harsh employer who will dismiss even the most faithful servant without a penny of severance pay. The sudden disintegration of a big company would greatly upset employees, suppliers, banks, labor unions, and governments; but it would hardly cause a ripple in the market (Drucker 1964). Of all the facts he has to face, this one is the most difficult for even the seasoned market researcher. But once accepted as a reality, it starts his thinking from a new perspective and new insight into the customer's world and will enable him to develop the kind of "generating" and "servicing" information the business must have.

Its Role.—The striving for a reliable navigational fix on the company's position through the constant sifting and sorting of information characterizes much of how market research spend its time. Too often it becomes mired in the sheer weight of accumulated numbers and thereby loses its real value to the marketing function which is as a questioner rather than just a "questionnaire."

The Customer.—To make its contribution to the business significant, a marketing analysis has to be based on the assumption that the firm does not know who the customer really is but needs to find out. The view from outside has three dimensions rather than one. It asks not only "who buys?" but "where is it bought?" and "what is it being bought for?" Every business can be thus defined as serving either customers, or markets, or end uses (Drucker 1964). In the case of protein ingredients and materials, the markets represent the customer. But for the kind of analysis that will provide strategic information, it is necessary to superimpose the findings from a separate analysis of the end-uses and applications of these materials. Without a study in this kind of depth, the best that could be hoped for would be a vague notion that the "ABC" Company buys Soy Protein "X" and they are in dehydrated foods. So, dehydrated foods is outlined as a market opportunity. But until it is known why Protein "X" is used and not Protein "Z," which is also available, there is no way to get any insight as to the size of its future or if it even has a future.

The Questions.—In the external environment, the key questions are those not only concerned with who the customer is (that is, who makes the decision to buy and why he buys what he does), but also seeks answers to at least four other basic questions: (1) What could happen that could make our proteins no longer needed by the user? A new breed of cow that produces 25,000 lb of milk per year on a diet of waste paper and urea? (2) What could happen that would force the customer to drop it? A new food law or regulation? (3) Who also serves these markets but is not a competitor of ours? How long is this likely to remain at a status quo? (4) In these markets, whose noncompetitor are we? What did we perceive that others have missed or rejected? What have they "discovered" as an opportunity that we may have overlooked?

The Application.—Once the information is at hand concerning the hard numbers about actual and potential volume, growth, capacity, and expansion of suppliers, and where the opportunities lie, the second task of market research is to constantly challenge the validity of what it has so painstakingly documented, that is, the data and the assumptions made as they apply to the business. The point of emphasis is on the realization that what has been chronicled as a correct situation begins aging the minute it is put down on paper. Most likely it will change into a new situation. But how soon, and, is the change going to be favorable or detrimental to the future business?

That no business can afford to stay bound to the products and markets of "yesterday" seems too elementary to mention. Yet, it is a habit practiced by many who are fascinated and obsessed with trying to revive the past and get it breathing again. It is the duty of the market researcher to resist the temptation to nurse along with more figures what is actually not worth saving and to steer the company's planning toward tomorrow's prospects.

Technical Service.—In the marketing of soy proteins, technical service, by the nature of its involvement in customer-focused work, is the bridge which connects and brings together the two major marketing functions: obtaining demand and servicing demand. Although customarily thought of as largely service center-ed, the more effective groups are structured and staffed to generate sales as well as to service existing business.

Its Organization.—Functionally, technical service is a line performer, operating in an expanding sphere of selling actions. Organizationally, it appears as a staff, the wanted but somewhat awkward step-child of two departments, bound to one, working for both. Its tactical resource, the laboratory, is controlled by R&D. Its mission, product acceptance, is the province of marketing.

Figure 12.1 shows technical service as part of the marketing team, and identifies its transpermeating nature. Not all producers of soy protein follow this kind of organizational design. Some assign technical service to R&D. The choice may result from a number of causative factors within the business, including people resources and the objectives of the business. It is not unusual for a given company to go through cycles in which technical service is transferred back and forth between marketing and R&D at recurring intervals. More often this represents a search for effectiveness rather than an experiment in organizational structure. It may also reflect the fact that management has not yet discerned what technical service's proper role should be because it has not yet arrived at a firm understanding about the nature of its own business.

As mentioned earlier in this chapter, it is a point in the history of soy protein that technical service became the fulcrum on which leverage was applied to move industrial isolated soy protein from dead center to a profit earner. It is also the reason why isolated soy protein continues to be used in many paper-coating mills in preference to other adhesives even during periods of adverse pricing.

However, as Hegel observed: "We learn from history; we learn nothing from history." And, a number of years went by before the value of technical service to the inedible product was translated into the edible protein activities where similar results could be realized. The shift from strictly personal selling to that of selling plus technical service has established some edible protein companies more firmly in their markets, and has made it that much more difficult for the "outsiders" to move in on the basis of offering duplicate products.

Its Purpose.—From the standpoint of marketing's basic purpose, how a company describes what it produces in the form of a protein product is not the cardinal point on which its success and its future depend. Rather, it is what the customer has confirmed and believes he is buying that decides the matter. In keeping with this precept, the purpose of technical service is to concentrate its effort where the buying decision is made, to see the product as the customer does, and to communicate to its own company the customer's view.

This is accomplished in the two functional areas of marketing as follows:

Obtaining demand: (1) Demonstrates product performance to potential users. (2) Guides field salesmen to applications for other products which offer improved performance to the customer's business. (3) Constructs new product ideas through perceived needs in the end use markets which are or will become "wants." (4) Stimulates sales growth for the future through building confidence in his products and company.

Servicing demand: (1) "Troubleshoots" existing problems in customer plant. (2) Assists customer in maximizing formulations. (3) Transmits ideas to customer in other areas of their production which may or may not have a relationship to his own products. (4) Promotes feed-forward of knowledge and experience gained in the crucial use markets to people in the decision areas of his own company.

Its Personnel.—The existence of a technical service group in a company marks it as an innovator, a firm looking for leadership as measured by the size of its economic performance rather than the size of its sale volume or market share. The greater the market proliferation a company seeks, the larger the technical service staff required. to serve it. These are people with ability in at least the following five areas: (1) problem solving; (2) perceptiveness; (3) communication; (4) practical experience in the industries or markets to be covered; and (5) inventiveness. Essentially, they are knowledge workers.

The man with these qualifications is obviously prepared for work in important areas of his company's business. The assignment of skills of this order to technical service emphasizes again that this activity is a critical link to an important result area and follows the principle that first class or high caliber people must always be allocated to major opportunities where there is promise of maximum return per unit of effort. It is a staffing decision, and carried out as such will demonstrate whether the company marketing soy protein has a program for effectiveness or merely a statement of its good intentions. The placement of people with superior skills and high performance will be the primary factor in deciding how far world supplies of protein can be carried over the next 25 yr and beyond. It will be through their capabilities, largely, that the protein ingredients and materials developed by research will be transformed into finished food products, featuring what tomorrow's consumer will expect in the way of value.

Some Related Considerations

New Economics.—The fact that the soybean has progressed from a 33 million bushel crop in 1936 to over 1 billion bushels in 1970 can be accounted for in a number of ways. It is apparent, though, that this growth has not been due to the demand just for oil or just for protein, but to both. Even though soybean oil initially was the desired refined product and processing was designed to maximize its extraction, protein has also been an important economic factor in the crush and in stimulating expansion of the crop. As oil from other seeds entered the commodity stream and the demand for protein enlarged, processing values have shifted so that for the past few years meal has been in the dominant position, being around 1 1/2 times the value of the oil.

Research in seed genetics which would seek to increase protein content at the expense of the oil finds little support, if any, from the oil milling industry. Yet, a few visionaries have at least considered what would happen to their oil markets if soy protein should become the chief source of supply for the present world population. It would require the protein from about 7 billion bushels of soybeans to provide the FAO level of protein for 3.5 billion people. At present levels of oil content, a crop of that size would place approximately 84 billion pounds of oil in a market which presently consumes about 50 billion pounds of edible vegetable oil from all sources. Answers to the question "what do we do with it?" are not easily seen. While the example represents an extreme situation, it still points to the need to start thinking from a new premise. Old concepts about the crush and the markets may produce right answers, but they will be for the wrong questions.

A New Attitude.—The increase in emphasis on protein has put the soybean in an important, if not strategic, position. But it is a position which needs re-evaluation and a different attitude about what part it will have in the new markets. For at least the last 15 yr, it has served as the "most available" land crop protein to the food industry. This is in spite of the fact that some of its inherent flavor characteristics have limited its level of usage. Soy protein offers valuable functional properties. It is the most economical commercial source of vegetable protein per unit of nutrition, and it is firmly established in its agricultural economics as a crop.

In a sense, refined soy protein has taken on the characteristics of a "monopoly," at least as a vegetable protein ingredient source to the food processing markets. And, here is where the alarm should be rung: There is rarely one right way to develop a market; the monopolist, even when in an enlightened position, tends to be complacent about markets and customers that cannot turn to another supplier. In the absence of the challenge of competition, the alternative ways to develop and service a market are not likely to be thought of or vigorously explored. The sales of a product line do not begin to grow or even approach their potential as long as they are offered from only one source.

A New Result.—Whatever are the reasons or set of circumstances that carry a business to a position of leadership, it must recognize that others have access to the same markets and the same knowledge. Market dominance, often equated with monopoly and the leadership position, actually shares with them only one area of common ground: in a free market, all three represent a transitory status; all are vulnerable to a drift downward. This tendency can be reversed by the infusion of new energy and new ideas from within (internal analysis) or from the outside (competition).

The application of this observation to soy protein involves two aspects of its business future. The first has to do with the stimulatory effect of the introspective examination. Until another protein makes its way into the commercial supply lines, the motivation to innovate on the part of the soybean industry will have to come as a result of its own situation analysis. The second concerns the effect of a competitive entry. When proteins from other sources are introduced, as they are certain to be, for the first time these markets will begin to expand in response to the thrust of competition. The result will be more uses, more applications, more satisfactions sought, and a market which will have come more fully alive.

It will be out of this kind of environment with its inspirations and its frictions that the creative proteins will be engineered. If substantial quantities of soy protein are to find their way into the bread basket of the world, especially the empty areas, it will be out of the spawning grounds of this next decade and in the challenging presence of the new protein entries.

BIBLIOGRAPHY

BARTHOLOMEW, D. M. 1968. Profitable soybean hedging for farmers. J. Am. Oil Chemists Soc. *45*, 700A–704A.
BARTHOLOMEW, D. M. 1969. Hedging opportunities for soybean processors. J. Am. Oil Chemists Soc. *46*, 396A–398A.

BÖHM-BAWERK, E. VON. 1960. Value and Price. Libertarian Press, South Holland, Ill.

DRUCKER, P. F. 1954. The Practice of Management. Harper & Row, New York.

DRUCKER, P. F. 1964. Managing for Results. Harper & Row, New York.

FIEDLER, R. E. 1971. Economics of the soybean industry. J. Am. Oil Chemists' Soc. *48*, 43–48.

JOHNSON, I. B. 1969. Selling the crop—use of the futures market. Blue Book Issue Soybean Dig. *29*, No. 6, Mar. 32–39.

KOTLER, P., and LEVY, S. J. 1969. Broadening the concept of marketing. J. Marketing *33*, No. 1, 10–15.

LEWIS, R. J., and ERICKSON, L. G. 1969. Marketing functions and marketing systems: a synthesis. J. Marketing *33*, No. 3, 10–14.

LEVITT, T. 1960. Marketing myopia. Harvard Business Rev. *38*, 45–46.

MCGARRY, E. D. 1950. Some functions of marketing reconsidered. *In* Theory in Marketing, Cox and Anderson (Editors). Richard D. Irwin, Homewood, Ill.

MERRILL LYNCH *et al.* 1969. The Soybean. New York.

OXENFELT, A. R. 1966. Executive Action in Marketing. Wadsworth Publishing Co. Belmont, Calif.

RACLIN, R. L. 1963. Futures Market in Soybeans and Their Products. Address, Congress Intern. Assoc. Seed Crushers, Amsterdam, Neth.

ROGERS, E. M. 1962. Diffusion of Innovations. The Free Press of Glencoe, Glencoe, Ill.

WEBSTER, F. E., JR. 1968. On the application of communication theory to industrial markets. J. Marketing Res. *5*, 426–428.

APPENDIX

GLOSSARY OF SOYBEAN TERMS

TERMS USED IN CONJUNCTION WITH THE PROCESSING OF SOYBEANS AND
THE UTILIZATION OF SOY PRODUCTS

Soybean(s)   A legume, the botanical name of which is Glycine max (L.)
Merrill; a summer annual varying in height from less than a foot to more
than 6 feet and in habit of growth from stiffly erect to prostrate; the
cultivated plant may reach a height of 3 feet or more; the seeds (soybeans)
are borne in pods that grow in clusters of three to five with each pod usually
containing two or three or more seeds; the oil content of the soybean varies
from 13% to 26% (average 18% to 22%) and from 38% to 45% of protein
(on a moisture-free basis); grown for centuries in the Orient and first
introduced to the United States early in the 19th century; soybeans grow
best in areas having hot, damp summer weather but they can be grown
under a great variety of climatic conditions; the term "soybean" has been
accepted as standard in the United States in preference to "soya bean,"
"soy bean," "soyabean," or "soja bean"; it is incorrect to use the term
"beans" when referring to soybeans.

Soybean Processor   An individual, or a group of two or more individuals
working together as a company or firm, whose primary business is the
separation of the oil and meal in soybeans; the activities of a processor may
also include refining and/or distribution of the oil as well as distribution of
the soybean meal.

Soybean Processing   The procedures involved in the removing of oil from
soybeans; there are two types of processes presently used in the United
States for this purpose:

   (1) Solvent Extraction   The process whereby the oil is leached or
washed (extracted) from flaked soybeans by the use of hexane as the
solvent; the level of oil in the extracted flakes can be reduced to 1% or less
by this processing method; the products resulting from the use of this
process are designated in the trade as "solvent extracted," e.g., "solvent
extracted soybean meal," "dehulled solvent extracted soybean meal,"
"solvent extracted soy grits," "solvent extracted soy flour."

   (2) Mechanical Processing   A continuous pressing process, at elevated
temperatures, using expellers or screw presses which utilize a worm shaft
continuously rotating within a pressing cylinder or cage to express the oil
from soybeans after they have been ground and properly conditioned; the
oil content of the resulting press cake is reduced to from 4% to 6% by this
processing method; although technically incorrect, the products resulting
from this type of processing are often referred to as "expeller," e.g.,
"expeller soybean meal"; the term "Expeller" applies specifically to the
screw press manufactured by the V. D. Anderson Co. of Cleveland Ohio,
whereas the screw press manufactured by the French Oil Mill Machinery
Co. of Piqua, Ohio, is designated as a "mechanical screw press."

   (3) Pre-press Solvent Processing   A combination of mechanical press-

ing followed by solvent extraction, used in some mills equipped for other types of oilseeds in addition to soybeans.

A fourth type of processing, called Hydraulic Pressing, is not used at present in the United States to any appreciable extent but is used in some parts of the world for recovering the oil from soybeans and other oil-bearing materials; it is the oldest process known to the vegetable oil processing industry; it consists of an intermittent pressing operation carried out at elevated temperatures in a hydraulic press after the soybeans have been rolled into flakes and properly conditioned by heat treatment.

Soybean Oil   A mixture of triglycerides composed of unsaturated fatty acids (oleic, linoleic, linolenic) and saturated fatty acids, together with usually not more than 1.5% of the free fatty acids and from 1.8% to 3.2% of phospholipids, depending on the quality and kind of soybeans and the procedure used in the processing.

Crude Soybean Oil   Sometimes referred to as "crude raw soybean oil"; the unrefined oil produced by any one of the procedures described for the processing of soybeans; it is customary to filter the oil and/or allow it to settle after being processed from the soybeans as required by the standard trade specifications.

Edible Crude Soybean Oil   Soybean oil which shall be of any of the following designated types produced from domestic-grown, undamaged, mature yellow soybeans: (1) expeller pressed, (2) expeller pressed degummed, (3) hydraulic pressed, (4) hydraulic pressed degummed, (5) solvent extracted, (6) solvent extracted degummed, (7) mixtures of any of the above-described types; when the oil is produced by solvent extraction, the name of the solvent used in the process must be given.

Refined Soybean Oil   Oil produced from crude soybean oil which has been washed with water and centrifuged (degumming), treated with dilute alkali solution (refining), decolorized by treatment with absorbent clay materials (bleaching) and subjected to steam distillation at high temperatures under vacuum (deodorizing); depending upon the use for which they are intended, some refined soybean oils are not deodorized during the refining process.

Edible Refined Soybean Oil   Degummed soybean oil which has been subjected to special refining processes to adapt it specifically for use in food products; in addition to treatment with alkali, bleaching and deodorization, the oils may be also winterized; these oils are further classified as salad oils and cooking oils.

Hydrogenated Soybean Oil   During the process of hydrogenation, the soybean oil is exposed to hydrogen gas in the presence of heat and a catalyst (nickel) and the hydrogen combines with certain of the chemical components (unsaturated fatty acids) of the triglycerides with a resultant increase in the melting point of the oil; sometimes referred to as "hardening."

Degummed Soybean Oil   The product resulting from washing crude soybean oil with water and/or steam for a specified period of time and then centrifuging the oil-and-water mixture to remove the phosphatides; it shall not contain more than 0.03% of phosphorus.

Winterized Oil   Oil which has been treated in a manner to partially remove the saturated glycerides which have relatively high melting points and are soluble only to a limited extent in the unsaturated glycerides; the process consists of chilling the oil slowly and then maintaining it at 5° C. (41° F.) for a specified period of time and removing the crystallized glycerides from the liquid fraction of the oil by the use of filter presses; soybean oil does not require winterizing because it is made up primarily of unsaturated glycerides.

Technical Grade Refined Soybean Oil   Includes a wide variety of soybean oils that are specially refined and further processed to meet requirements and/or specifications for a definite industrial use.

Soybean Fatty Acids   The product obtained when glycerine is split off from the triglycerides in soybean oil by any method of hydrolysis; in the field of industrial usage, soybean fatty acids are usually further classified according to the treatment to which they are subjected subsequent to hydrolysis.

Soybean Soapstock   The byproduct that results from the alkali refining of soybean oil; averages about 6% of the volume of crude oil refined; usually referred to as "foots" since it accumulates at the bottom (foot) of the refining tank; contract grade should contain not less than 30% total fatty acid.

Acidulated Soybean Soapstock   The product that results from the complete acidulation of soybean soapstock; contract grade should contain not less than 85% total fatty acid.

Soybean Lecithin   The mixed phosphate product obtained from soybean oil by the degumming process; contains lecithin, cephalin and inositol phosphatides, together with glycerides of soybean oil and traces of tocopherols, glucosides and pigments; it shall be designated and sold according to conventional descriptive grades with respect to consistency and bleaching; the dehydrated emulsion of mixed phosphatides and soybean oil is further processed to produce the commercial grades which may be described as follows: plastic or firm consistency; soft consistency; fluid; unbleached; bleached; and double bleached; high-quality commercial lecithin contains 60% to 65% phosphatides and 35% to 40% soybean oil, depending on the consistency desired in the final product; two types of lecithin are generally recognized in the trade: crude soybean lecithin and refined lecithin.

Break Material   Flocculent material (precipitate) that appears in and can be separated from crude soybean oil that has been rapidly heated to temperatures between 250° C. and 300° C. (482° F. and 572° F.); found to be very high in ash content, rich in phosphorus, calcium and magnesium; presumed to be derived from phospholipids which are thermally decomposed at these high temperatures; sometimes referred to as "foots."

Sludge   The solid residue which settles to the bottom of the tank or car in which crude soybean oil is stored or shipped; cannot be pumped out or removed by rubber-faced scrapers from the surface to which it adheres; steam must be applied to facilitate its removal; sometimes referred to as "tank bottoms" or "foots."

Soybean Products

The following definitions have been adopted officially or tentatively by the American Feed Control Officials, Inc.:

Ground Soybeans   The product obtained by grinding whole soybeans without cooking or removing any of the oil; sometimes erroneously called "soybean meal." (Adopted 1933.)

Ground Soybean Hay   The ground soybean plant including the leaves and beans; it shall be reasonably free of other crop plants and weeds and shall contain not more than 33.0% crude fiber. (Adopted 1944, Amended 1964).

Soybean Hulls   The product consisting primarily of the outer covering of the soybean. (Adopted 1948).

Solvent Extracted Soybean Feed   The product remaining after the partial removal of protein and nitrogen-free extract from dehulled solvent extracted soybean flakes. (Adopted 1948, Amended 1960, 1964.)

Soybean Meal   The product obtained by grinding the cake, chips or flakes which remain after removal of most of the oil from soybeans by a mechanical or solvent extraction process; if solvent extracted, it shall be so designated; heat must be applied during the process; it shall contain not more than 7.0% crude fiber. (Proposed 1960, Adopted 1961, Amended 1964.)

Dehulled Solvent Extracted Soybean Meal   The product obtained by grinding the flakes remaining after removal of most of the oil from dehulled soybeans by a solvent process; heat must be applied during the process; it shall contain not more than 3.0% crude fiber. (Proposed 1960, Adopted 1961, Amended 1964.)

Soybean Mill Feed   The product composed of soybean hulls and the offal from the tail of the mill which results from the manufacture of soy grits or flour; it shall contain not less than 13.0% crude protein and not more than 32.0% crude fiber. (Proposed 1960, Adopted 1961, Amended 1964.)

Soybean Mill Run   The product composed of soybean hulls and such bean meats that adhere to the hulls which results from normal milling operations in the production of dehulled soybean meal; it shall contain not less than 11.0% crude protein and not more than 35.0% crude fiber. (Proposed 1960, Adopted 1961, Amended 1964).

Heat Processed Soybeans   The product resulting from heating whole soybeans without removing any of the component parts; it may be ground, pelleted, flaked or powdered; it shall be sold according to its protein content. (Proposed 1960, Adopted 1964.)

Nitrogen Free Extract (N.F.E.)   The figure obtained by subtracting from 100 percent the sum of the percentages of ash, moisture, protein, fat and fiber.

Standard Specifications

Have been adopted for certain soybean products as follows by the National Soybean Processors Association and made a part of their Trading Rules governing purchase and sale of soybean meal in the United States:

Soybean Chips, Soybean cake, 41% Protein Soybean Meal   Produced by cooking ground soybeans and reducing the oil content of the cooked product by pressure to 6% or less on a commercial basis; standard specifications are as follows:

Protein	Minimum 41.0%
Fat	Minimum 3.5%
Fiber (when loaded by seller)	Maximum 7.0%
N.F.E.	Minimum 27.0%
Moisture (when loaded by seller)	Maximum 12.0%

Soybean Flakes, 44% Protein Soybean Meal   Produced by conditioning ground soybeans and reducing the oil content of the conditioned product by the use of hexane or homologous hydrocarbon solvents to 1% or less on a commercial basis; standard specifications are as follows:

Protein	Minimum 44.0%
Fat	Minimum 0.5%
Fiber (when loaded by seller)	Maximum 7.0%

N.F.E.	Minimum 27.0%
Moisture (when loaded by seller)	Maximum 12.0%

Dehulled Soybean Flakes, 50% Protein Solvent Extracted Soybean Meal   Produced by cracking, heating, and flaking dehulled soybeans and reducing the oil content of the conditioned flakes by the use of hexane or homologous hydrocarbon solvents to 1% or less on a commercial basis; the extracted flakes are cooked and marketed as such or ground into meal; standard specifications are as follows:

Protein	Minimum 50.0%*
Fat	Minimum 0.5%
Fiber (when loaded by seller)	Maximum 3.0%
N.F.E.	Minimum 27.0%
Moisture (when loaded by seller)	Maximum 12.0%

*Revised to 49% in 1967 for trade purposes.

Soybean Proteins

The following terms and their definitions are quite widely used and accepted in the trade in conjunction with the production and utilization of soy flour and other soy products:

Soy Flour   The screened, graded product obtained after expelling or extracting most of the oil from selected, sound, clean, dehulled soybeans, except that full-fat soy flour is not subjected to expelling or extraction and contains all of the oil originally present in the soybeans; ground finely enough to pass through a 100-mesh or smaller screen; usually available in various grinds to meet specific requirements.

Soy Grits   The screened, graded product obtained after expelling or extracting most of the oil from selected, sound, clean, dehulled soybeans; usually ground to conform to the following range of granulations in terms of majority percent through respective U.S. standard screens:

Coarse	No. 10 to No. 20
Medium	No. 20 to No. 50
Fine	No. 50 to No. 80

Soybean Meal   The product resulting from grinding the material remaining after the removal of part of the oil by pressure or solvents from soybeans; its use in generally restricted to feeds for livestock and poultry as compared with soy flour and grits commonly used for human consumption.

Defatted Soy Flour   Flour produced by the nearly complete removal of the oil from soybeans by the use of hexane or other homologous hydrocarbon solvents; usually contains about 1% of fat.

Low-Fat Soy Flour   Flour produced either by partial removal of the oil from soybeans or by adding back soybean oil and/or lecithin to defatted soy flour to a specified level, usually in the range of 5% to 6%.

High-Fat Soy Flour   Flour produced by adding back soybean oil and/or lecithin to defatted soy flour to a specified level, usually in the range of 15%.

Full-Fat Soy Flour   Flour containing all of the oil originally present in the raw soybeans; usually contains 18% to 20% of fat.

Lecithinated Soy Flour   A type of low-fat or high-fat flour in which lecithin is added to defatted soy flour to a specified level, usually in a range up to 15%.

Protein   A naturally occurring combination of amino acids, containing the chemical elements carbon, hydrogen, oxygen, nitrogen and usually sulphur; one of the essential constituents of all living things and of the diet of the animal organism.

Isolated Protein   Protein which has been greatly concentrated from its native location, either by chemical or mechanical means; this is accomplished by removing a portion of the carbohydrate and mineral matter that is assoicated with the native protein; in most instances the proteinaceous material is then precipitated from solution, followed by subsequent separation from the liquid phase and drying.

Toasting   The term commonly but erroneously applied to the processes (wet) of cooking soybean meal, flour or grits by atmospheric or pressure methods for the purpose of increasing the protein efficiency of these soya products or improving their functional properties and/or improving physical texture.

Textured Protein Products (TPP)   Fabricated palatable food ingredients processed from edible protein sources, including among others soy grits, soy protein isolates and soy protein concentrates, with or without suitable optional ingredients added for nutritional or technological purposes. They are made up as fibers, shreds, chunks, bits, granules, slices or other forms. When prepared for consumption by hydration, cooking, retorting and other procedures, they retain their structural integrity and characteristic chewy texture.

Meat Analogs   Textured protein products fabricated to simulate muscle tissue foods from various animal sources in appearance, texture, flavor and behavior on cooking.

Definitions

The following tentative product definitions have been recommended by the National Soybean Processors Association:

Soy Grits and/or Soy Flour   Is the mechanically classified product obtained from high quality, sound, clean, dehulled soybeans by processing so as to yield products on a moisture-free basis in a protein range of 40%–60% (N × 6.25), fiber maximum of 3.5%, and a variable fat content.

Isolated Soy Protein   Is the major proteinaceous fraction of soybeans prepared from high quality, sound, clean, dehulled soybeans by removing a preponderance of the non-protein components and shall contain not less than 90% protein (N × 6.25) on a moisture-free basis.

Soy Protein Concentrate   Is the product prepared from high quality, sound, clean, dehulled soybeans by removing most of the oil and water soluble non-protein consitituents and shall contain not less than 70% protein (N × 6.25) on a moisture-free basis.

Vegetable Fats

Margarine   A fatty food product used as a substitute for butter and in similar applications; the composition is regulated by the U.S. Food and Drug Administration under a specific Standard of Identity, requiring a minimum fat content of 80%, blended with some form of milk (e.g. whole milk, skim milk or a mixture of dried nonfat milk solids and water), salt, vitamins A and D, and traces of other ingredients added for improvement of flavor and/or physical characteristics; either animal or vegetable fats may be used but refined edible vegetable oils, such as soybean, cottonseed, corn, peanut and safflower oils are most widely used; any or all of the fats used in the formulation may be hydrogenated to varying degrees or hydrogenated fats may be blended with unhydrogenated oils.

Vegetable Shortening   A semi-solid fatty food product made entirely from refined edible vegetable oils, such as soybean, cottonseed, corn, peanut and safflower oils, used in cooking, baking and frying; there are two general classes: (a) blended, classified on the basis of the raw materials used as ingredients, and (b) hydrogenated, which may be further identified on the basis of their intended use; hydrogenated shortenings are usually made from blends of two or more hydrogenated fats; the oils may be hydrogenated individually or in combination to the same or different degrees of saturation and then blended in the proportions necessary to give the desired characteristics in the finished products.

Oriental Foods

Over the centuries a great many foods based on the soybean have been developed and utilized by the Oriental peoples as a major part of their diet. Many of these foods have been adopted and are used to a considerable extent in other parts of the world. Some of the better-known of these soy-based foods are defined and/or described as follows:

Soy Sauce (Shoyu)   A hydrolysate of soybean protein mixed with wheat flour, resulting from the action of molds, yeasts and bacteria as prepared by the Oriental method; the fermentation or enzymatic action is permitted to progress for up to 1 1/2 years, at which time the extract is heated and processed to produce the liquid for edible uses; may also be prepared by chemical (acid) hydrolysis of soybean protein; used as a seasoning in the preparation of foods and as a table condiment.

Soy Milk   When prepared by the Oriental technique, it is a suspension of ground soybeans in water in that soaked soybeans are finely ground and mixed with water and the resultant mass poured into cheesecloth through which the liquid phase passes and is recovered as the milk; in the western world and other areas where it is becoming more widely used as food for infants that are suffering from malnutrition and for individuals afflicted with certain allergies, diabetes and other diseases associated with diet, soy milk is made as a dispersion of soy flour or soy protein isolate in water with added vitamins, minerals, carbohydrates and, in many instances, a flavoring compound.

Miso   Made from soaked, steam-heated soybeans which are inoculated with cultures of microorganisms grown on rice or barley and then allowed to ferment; one method of fermentation is called the "natural brewing process" in which the soybeans are allowed to ferment for approximately 9 months; the other method is known as "the quick brewing process" and the material is produced in a short time by reducing the length of time for processing (heating) the soybeans and the fermentation; a sweet type or a salty type may be produced; used as a soup which is mixed with tofu or vegetables or seaweed, for preserving fish, meat and vegetables and for various kinds of Japanese-style cooking with vegetables, meat or fish.

Tofu   A cheese-like product made from soy milk; the protein is precipitated by the use of calcium sulfate or a comparable coagulating agent and then placed into molding boxes or forms and allowed to cool; when the curd has properly cooled, it is cut into appropriate sized portions for cooking; may be eaten seasoned with shoyu, put into miso soup or cooked with miso, vegetables, fish and meat; also used in western-style cooking as a component of omelets, croquettes, soups, stews or in the fried form; a method for making soft tofu has been recently introduced, a process for which American-grown soybeans are especially well suited.

Dried Tofu   Tofu frozen for several weeks and dried; serves to preserve the product for extended periods of storage; soaked in hot water for about 5 minutes and then used in the same manner as ordinary tofu.

Aburaage   Properly drained tofu that is fried lightly in soybean oil or other edible oils; consumed chiefly in the cooked state with proper seasoning.

Kinako   Ground toasted soybeans; used for making Japanese-style cakes.

Namaage   A kind of aburaage, produced by frying thick slices of tofu in soybean oil or other edible oils; eaten in the cooked form seasoned with shoyu or ginger.

Ganmodoki   Drained tofu is crushed and mixed with minced vegetables, the resultant mixture is molded into a round shape and fried; popularly used for home cooking and is eaten in the cooked state with appropriate seasoning.

Tempeh   A soy food product developed in Indonesia in which soybeans are soaked overnight and then cooked for a short time; the cooked soybeans are inoculated with the fungus Rhizopus oryzae and allowed to stand for 18 to 48 hours to permit optimum growth of the mycelium of the organism; the product is roasted, cooked in soup or fried in oil; may also be sliced and dried.

Natto   A fermented soy product produced in Japan in essentially the same manner as tempeh, except that a different organism is used to bring about the desired effect on the soy protein.

Yuba   The name given to the protein film that forms on the surface of soy milk when it is heated nearly to the boiling point; the film is removed and dried, requiring soaking in water before it is used as food; eaten in the cooked form with appropriate seasoning.

Moyashi (Soybean Sprouts)   Produced by germinating soybeans in the dark for 10 days or more until the sprouts reach about 1.5 centimeters (about 0.6 inch) in length; eaten in salads and in cooked form.

Vanaspati   An edible, semi-solid fat used as a food in India, Pakistan and other areas of western Asia; generally manufactured from edible refined vegetable oils (soybean, peanut, cottonseed and sesame oils); the refined oil is processed into a semi-solid fat by partial hydrogenation in combination with careful filtration, additional refining and deodorization, refined, deodorized sesame oil and vitamin A are blended into the hydrogenated oil so as to conform to the relevant government regulations; adequate amounts of vitamin D are also added; the hot, liquid fat is poured into tin containers which are then sealed and subjected to a controlled cooling process to insure the development of uniform crystallization of the fat and the characteristic granular texture in the finished product.

Ghee   A semi-fluid clarified form of butter fat, commonly referred to as "butter oil," made in India and neighboring countries; usually made from buffalo or cows' milk but the milk from sheep and goats is used in some areas; the fat is separated from the milk by heating and the subsequent formation of a curd; the curd is churned or centrifuged to recover the butter which is then heated to remove water to give the desired physical characteristics and improve the keeping qualities.

Source: Anon. (1966).

OFFICIAL STANDARDS OF THE UNITED STATES FOR SOYBEANS

Revised Effective
  Sept. 1, 1955

Sec.
26.601 Terms defined.
26.602 Principles governing application of standards.
26.603 Grades, grade requirements, and grade designations.

AUTHORITY: § § 26.601 to 26.603 issued under sec. 8, 39 Stat. 485; 7 U. S. C 84. Interpret or apply sec. 2, 39 Stat. 482, as amended; 7 U. S. C. 74.

§ 26.601 **Terms defined.** For the purposes of the official Grain Standards of the United States for soybeans:

(a) **Soybeans.** Soybeans shall be any grain which consists of 50% or more of whole or broken soybeans which will not pass readily through an 8/64 sieve and not more than 10% of other grains for which standards have been established under the United States Grain Standards Act.

(b) **Classes.** Soybeans shall be divided into the following five classes: Yellow soybeans, green soybeans, brown soybeans, black soybeans, and mixed soybeans.

(c) **Yellow soybeans.** Yellow soybeans shall be any soybeans which have yellow or green seedcoats, and which in cross section are yellow or have a yellow tinge, and may include not more than 10% of soybeans of other classes.

(d) **Green soybeans.** Green soybeans shall be any soybeans which have green seedcoats, and which in cross section are green, and may include not more than 10% of soybeans of other classes.

(e) **Brown soybeans.** Brown soybeans shall be any soybeans with brown seedcoats, and may include not more than 10% of soybeans of other classes.

(f) **Black soybeans.** Black soybeans shall be any soybeans with black seedcoats, and may include not more than 10% of soybeans of other classes.

(g) **Mixed soybeans.** Mixed soybeans shall be any mixture of soybeans which does not meet the requirements of the classes yellow soybeans, green soybeans, brown soybeans, or black soybeans. Bicolored soybeans shall be classified as mixed soybeans.

(h) **Grades.** Grades shall be the numerical grades, sample grade, and special grades provided for in § 26.603.

(i) **Bicolored soybeans.** Bicolored soybeans shall be soybeans with seedcoats of two colors, one of which is black or brown.

(j) **Splits.** Splits shall be pieces of soybeans that are not damaged.

(k) **Damaged kernels.** Damaged kernels shall be soybeans and pieces of soybeans which are heat-damaged, sprouted, frosted, badly ground-damaged, badly weather-damaged, moldy, diseased, or otherwise materially damaged.

(l) **Heat-damaged kernels.** Heat-damaged kernels shall be soybeans and pieces of soybeans which are materially discolored and damaged by heat.

(m) **Foreign material.** Foreign material shall be all matter, including soybeans and pieces of soybeans, which will pass readily through an 8/64 sieve and all matter other than soybeans remaining on such sieve after sieving.

(n) **Stones.** Stones shall be concreted earthy or mineral matter and other substances of similar hardness that do not disintegrate readily in water.

(o) **8/64 sieve.** An 8/64 sieve shall be a metal sieve 0.032 inch thick perforated with round holes 0.125 (8/64) inch in diameter with approximately 4,736 perforations per square foot.

§ 26.602 **Principles governing application of standards.** The following principles shall apply in the determination of the classes and grades of soybeans:

Maximum limits of

Grade	Minimum test weight per bushel Pounds	Moisture %	Splits %	Damaged kernels		Foreign material %	Brown, black, and/or bicolored soybeans in yellow or green soybeans %
				Total %	Heat damaged %		
1	56	13.0	10	2.0	0.2	1.0	1.0
2	54	14.0	20	3.0	0.5	2.0	2.0
3†	52	16.0	30	5.0	1.0	3.0	5.0
4‡	49	18.0	40	8.0	3.0	5.0	10.0

Sample grade: Sample grade shall be soybeans which do not meet the requirements for any of the grades from No. 1 to No. 4, inclusive; or which are musty, sour, or heating; or which have any commercially objectionable foreign odor; or which contain stones; or which are otherwise of distinctly low quality. † Soybeans which are purple mottled or stained shall be graded not higher than No. 3 ‡ Soybeans which are materially weathered shall be graded not higher than No. 4.

(a) **Basis of determination.** Each determination of class, splits, damaged kernels, and heat-damaged kernels, and of black, brown, and/or bicolored soybeans in yellow or green soybeans, shall be upon the basis of the grain when free from foreign material. All other determinations shall be upon the basis of the grain as a whole.

(b) **Percentages.** All percentages shall be upon the basis of weight. The percentage of splits shall be expressed in terms of whole percents. All other percentages shall be expressed in terms of whole and tenths percents.

(c) **Moisture.** Moisture shall be ascertained by the air-oven method prescribed by the United States Department of Agriculture, as described in Service and Regulatory Announcement No. 147, issued by the Agricultural Marketing Service, or ascertained by any method which gives equivalent results.

(d) **Test weight per bushel.** Test weight per bushel shall be the weight per Winchester bushel as determined by the method prescribed by the United States Department of Agriculture, as described in Circular No. 921 issued June 1953, or as determined by any method which gives equivalent results.

§ 26.603 **Grades, grade requirements, and grade designations.** The following grades, grade requirements, and grade designations are applicable under these standards:

(a) **Grades and grade requirements for soybeans** (see also paragraph (c) of this section).

(b) **Grade designation.** The grade designation for soybeans shall include in the order named the number of the grade or the words "Sample grade," as the case may be; the name of the class; and the name of each applicable special grade. In the case of mixed soybeans the grade designation shall also include, following the name of the class, the approximate percentages of yellow, green, brown, black, and bicolored soybeans in the mixture.

(c) **Special grades, special grade requirements and special grade designation for soybeans—**

(1) **Garlicky soybeans—(i) Requirements.** Garlicky soybeans shall be soybeans which contain 5 or more garlic bulblets in 1,000 grams.

(ii) **Grade designation.** Garlicky soybeans shall be graded and designated according to the grade requirements of the standards applicable to such soybeans if they were not garlicky and there shall be added to and made a part of the grade designation the word "garlicky."

(2) **Weevily soybeans—(i) Requirements.** Weevily soybeans shall be soybeans which are infested with live weevils or other live insects injurious to stored grain.

(ii) **Grade designation.** Weevily soybeans shall be graded and designated according to the grade requirements of the standards applicable to such soybeans if they were not weevily, and there shall be added to and made a part of the grade designation the word "weevily."

Interpretations (Added)

§ 26.901 **Interpretation with respect to the term "distinctly low quality."**

The term "distinctly low quality", when used in the official grain standards of the United States, shall be construed to include grain which contains more than two crotalaria seeds (Crotalaria spp.) in 1,000 grams of grain.

§ 26.902 **Interpretation with respect to the term "purple mottled or stained."**

The term "purple mottled or stained" when used in the official grain standards of the United States for soybeans (see § 26.603(a)) shall be construed to include soybeans which are discolored by the growth of a fungus; or by dirt; or by a dirtlike substance including nontoxic inoculants; or by other nontoxic substances.

§ 26.903 **Interpretation with respect to the term "bicolored soybeans."**

The term "bicolored soybeans," when used in the Official Grain Standards of the United States for Soybeans (see § 26.601(i)), shall be construed to include any soybeans with seedcoats of two colors, one of which is black or brown, when the black and/or brown color covers 50 percent or more of the seedcoat. The hilum of a soybean shall not be considered a part of the seedcoat.

Source: Anon. (1971).

SOY FLOUR STANDARDS

	Full-fat soy flour	Low-fat soy flour	Defatted soy flour
Protein (N X 6.25)*	40.0% Min.	45.0% Min.	50.0% Min.
Fat (Ether Extract)*	18.0% Min.	4.5% Min. 9.0% Max.	2.0% Max.
Fibre*	3.0% Max.	3.3% Max.	3.5% Max.
Moisture	8.0% Max.	8.0% Max.	8.0% Max.
Ash*	5.5% Max.	6.5% Max.	6.5% Max.

*Moisture free basis.
Screen—97% through No. 100 U. S. Standard Screen for each of the above.
Approved and adopted by the Soya Food Research Council.

Definition

The products shall be processed from high quality, sound, clean, dehulled yellow soybeans as defined in the United States Grain Standards Act. The soybeans shall be subjected to a thorough initial cleaning operation, that shall substantially remove all foreign material. Disagreeable flavors and odors shall be removed by subjecting the soy material to adequate processing.

Soy flour shall be prepared and packaged under modern sanitary conditions. The soy flour shall be free from burnt, musty, rancid or other undesirable flavors or odors; free from burnt, scorched, or grayish color; and free from insects, insect webbing, dirt or other extraneous matter.

Source: Anon. (1971).

ANALYTICAL DATA RANGE OF COMMERCIAL SOY PROTEIN FOOD INGREDIENTS

Soy Protein Isolates

	(%)
Protein (N X 6.25)	86 to 94
Moisture	4 to 7
Fat	0.1 to 0.3
Ash	3 to 7
Crude fiber	0.1 to 0.3
Color	Cream

Soy Protein Concentrates

	(%)
Protein (N X 6.25)	65 to 72
Moisture	3 to 7
Fat	2
Ash	4 to 7
Crude fiber	2.7 to 5
Color	Cream

Extruded Textured Protein Products (No Additives)

	(%)
Protein (N X 6.25)	48 to 53
Moisture	5 to 9
Fat	1 to 1.3
Ash	5.5 to 7
Crude fiber	2.7 to 3
Color	Light to dark tan

Spun Meat Analogs (Moist Form)

	(%)
Protein (N X 6.25)	16 to 22
Moisture	55 to 61
Fat	2 to 18
Ash	2 to 4

Enzyme-modified Soy Proteins

	(%)
Protein (N X 6.25)	60 to 65
Moisture	4 to 5
Ash	12 to 15
Color	Light tan

SOME U.S. COMPANIES MARKETING SOY PROTEIN FOOD INGREDIENTS

Archer Daniels Midland Co., Specialty Div., Decatur, Ill. 62525: extruded textured protein products; soy flours and grits.

Borden, Inc., Industrial Food Products Div., Elgin, Ill. 60122: enzyme-modified soy proteins.

Cargill, Inc., Soy Special Products Div., Minneapolis, 55402: Extruded textured protein products; soy flours and grits.

Carnation Co., Industrial Products Div., Van Nuys, Calif. 91412: soy protein isolates.

Central Soya, Chemurgy Div., Chicago, 60639: soy protein isolates; soy protein concentrates; textured protein products; soy flour and grits.

Crest Products, Inc., Park Ridge, Ill. 60068: specialty soy protein products.

Dairyland Products, Inc., Savage, Minn. 55378: processed soybean "nuts."

Deltown Chemurgic Corp., Yonkers, N.Y. 10701: dairy whey-soy blends.

Erie Casein Co., Erie, Ill. 61250: dairy whey-soy blends.

Far-Mar-Co., Inc., Hutchinson, Kans. 67501: extruded textured protein products.

General Mills, Inc., Food Service and Protein Products Div., Minneapolis, 55440: spun textured protein products; meat analogs.

Griffith Laboratories, Chicago, 60609: soy protein concentrates; textured protein products.

B. Heller & Co., Chicago, 60653: soy protein concentrates.

Land-O-Lakes, Inc., Minneapolis, 55413: dairy whey-soy blends.

Loma Linda Foods, Mt. Vernon, Ohio 43050: soy flours; soy milks; soy cheeses; canned soybeans; meat analogs.

Miles Laboratories, Marschall Div., Elkhart, Ind. 46514; Worthington Foods Div., Worthington, Ohio 45085: spun textured protein products; meat analogs; soy milks.

Ralston Purina Co., Food Protein Dept., St. Louis, Mo. 63199: spun and extruded textured protein products; soy protein isolates.

A. E. Staley Mfg. Co., Decatur, Ill. 62525: extruded textured protein products; soy flours and grits. Gunther Products Div., Galesburg, Ill. 61404: enzyme-modified soy proteins; soy protein isolates; soy protein concentrates.

Swift Chemical Co., Div. of Swift & Co., Vegetable Protein Products Dept., Oak Brook, Ill. 60521: extruded textured protein products; soy protein concentrates; soy flours and grits; dairy whey-soy blends.

H. B. Taylor Co., Div. of National Can Corp., Chicago, Ill. 60632: textured protein products.

Source: Anon. (1971).

A.O.C.S. Method Ba 11–65
Revised 1969

NITROGEN SOLUBILITY INDEX (NSI)

Definition: This method determines the dispersible nitrogen in soybean products under the conditions of the test. In contrast to the alternate fast stir method for Protein Dispersibility Index (PDI), No. Ba 10–65, the slower stirring technique used in this method will give generally lower results than those obtained by the fast stir method.

Scope: Applicable to ground soybeans, whole or ground full-fat or extracted flakes, full-fat and defatted soy flours and grits, and soybean meal.

A. Apparatus:
1. 400 ml. beaker.
2. 200 ml. graduated cylinder.
3. Glass stirring rod.
4. 250 ml. volumetric flask.
5. Stirring apparatus*, adustable up to 120 rpm, and having a segment of a circle, the extreme diameter of the paddle being 50 mm. (this type of paddle is commercially available; e.g. E. H. Sargent Cat. No. S-76670).
6. Balance (0.01 g. accuracy).
7. Water bath to maintain a temperature of 30° C.
8. Wash bottle.
9. Centrifuge with 50 ml. round bottom centrifuge tubes.
10. Standard Kjeldahl equipment. See A.O.C.S. Method Ac 4–41.
11. Small glass funnels.
12. 100 ml. beakers.
13. 25 ml. pipet.
   *Most of the commercial electrical stirring devices can be adjusted to 120 rpm with a variable Powerstat. Cone drive-type stirrers are particularly adapted for slow speed stirring.

B. Reagents:
1. Distilled Water.
2. Standard Reagents as used for protein determination in A.O.C.S. Method Ac 4–41.
3. Silicone antifoam (Dow Corning AF, Hodag F-2, FD-82, or equivalent).
4. Glass Wool (Fiberglas–Corning No. 800.)

C. Preparation of Sample:
1. If the sample is not in form of a fine flour, grind it so that at least 95% of the sample will go through a 100-mesh screen. Use an impact-type laboratory mill such as a MIKRO-PULVERIZER or MIKRO-SAMPLMILL (Pulverizing Machinery Division, 26 Chatham Road, Summit, N.J.) equipped with a No. 3428 Mikro-Screen (tweed type). To obtain sufficient impact, rotor speed should be at least 10,000 rpm (never exceed 16,000 rpm).
2. In case of a full-fat sample, use a screen (such as No. 027RD26G or equivalent) that will grind the material to a fine flour so that at least 80% of the material will go through an 80-mesh screen and 90% through a 60-mesh screen.

3. In order to avoid heat denaturation during grinding, mix an equal amount of dry ice with the sample prior to grinding. At least 25 g. of sample should be used for the mill. (This grinding step is optional; see E., Calculations).

D. Procedure:
1. Weigh 5 g. of the sample into a 400 ml. beaker. Measure 200 ml. of distilled water at 30° C. Add a small portion of the water at a time and disperse it thoroughly with a stirring rod. Stir in the remainder of the water, using the last of it to wash off the rod.
2. Stir the mixture at 120 rpm with a mechanical stirrer for 120 minutes at 30° C. with the beaker immersed in a water bath. Transfer the mixture to a 250 ml. volumetric flask, by carefully washing out the contents of the beaker into the flask. Add 1 or 2 drops of antifoam, dilute to mark with distilled water, and mix the contents of the flask thoroughly.
3. Allow to stand for a few minutes and decant off about 40 ml. into a 50 ml. centrifuge tube. Centrifuge 10 minutes at 1500 rpm and decant supernatant through a funnel containing a plug of glass wool (being careful not to transfer any of the centrifuged solids to the filter). Collect the clear filtrate in a 100 ml. beaker.
4. Pipet out 25 ml. of the clear liquid into a Kjeldahl flask and then proceed according to standard practice for protein determination (A.O.C.S. Method Ac 4-41).
5. Determine total nitrogen on the sample using A.O.C.S. Method Ac 4-41.

E. Calculation:

$$\% \text{ Water Soluble Nitrogen} = \frac{(B-S) \times N \times .014 \times 100}{\text{Wt. of sample}}$$

Where B = ml. of alkali back titration of blank
  S = ml. of alkali back titration of sample
  N = Normality of alkali
Nitrogen Solubility Index (NSI) =

$$\frac{\% \text{ Water soluble nitrogen} \times 100}{\% \text{ Total nitrogen}}$$

F. Precision:

Two single determinations performed in one laboratory should not differ by more than 3.735.

Agreement between laboratories: Two single determinations performed in different laboratories should not differ by more than 7.224.

NOTE: Nitrogen Solubility Index (NSI) may be reported on an "as is" basis or on a "ground sample" basis, and this should be so specified: "Nitrogen Solubility Index (NSI as is)" or "Nitrogen Solubility Index (NSI ground sample)"

Source: Anon. (1970).

A.O.C.S. Method Ba 10–65
Revised 1969

PROTEIN DISPERSIBILITY INDEX (PDI)

Definition: This method determines the dispersible protein in soybean products under the conditions of the test. In contrast to the alternate slow stir method for Nitrogen Solubility Index (NSI), No. Ba 11-65, the faster stirring technique used in this method will give generally higher results than those obtained by the slow stir method.

Scope: Applicable to ground soybeans, whole or ground full-fat or extracted flakes, full-fat and defatted soy flours and grits, and soybean meal.

A. Apparatus:
1. Hamilton Beach Drinkmaster No. 30. Modified to accommodate Waring Blendor blade and cup.
2. Blade Assembly. Cenco-Pinto blades. Central Scientific Company, No. 17251-L55. Use two blades, one horizontal, and one with tips pointing down with the cutting edge in the direction of rotation.
3. Waring Blendor Cup. 1 qt. capacity with bottom sealed with No. 3 stopper.
4. Glassware. 300 ml. volumetric flask, 15 ml. pipet, 600 ml. beaker.
5. Centrifuge. International Type SB size 1, 2,700 rpm, with 50 ml. tubes or any equivalent, capable of delivering 1,400 relative centrifugal force at the tip.
6. Balance. 0.1 g. accuracy, important.
7. Timer. Interval, alarm.
8. Variable Transformer.
9. Standard Kjeldahl Equipment. See A.O.C.S. Method Ac 4-41.
10. Tachometer. Range to 10,000 rpm.
11. Voltmeter. (Use optional).

B. Reagents:
1. Distilled Water. Neutral.
2. Standard Reagents as used for protein determination, A.O.C.S. Method Ac 4-41.

C. Preparation of Sample:
1. No preparation necessary; use sample as received.

D. Standardization of Blendor:
1. Measure 300 ml. of distilled water into the Blendor cup, and place in position on the mixer.
2. Remove chrome cap, which covers the top of the drive shaft. Using the proper tip, place tachometer in position on the rotating shaft.
3. With the switch in high position, gradually increase the transformer setting until the shaft shows 8,500 rpm on the tachometer. Note voltmeter reading and transformer setting, and use for blending of sample.
4. Standardization of the Blendor should be done before each series of tests to eliminate errors on account of fluctuation in line voltage.

E. Procedure:
1. Weigh 20 ± 0.1 g. of soy product.
2. Fill a 300 ml. volumetric flask with distilled water at $25° C \pm 1° C$. Pour about 50 ml. of the water into the Blendor cup. (Water-dispersible protein is related to temperature so the Blendor cup should be at room temperature). Transfer the weighed sample quantitatively to the Blendor cup. Stir with a spatula to form a paste. Add remainder of water in increments, with stirring, to form a smooth slurry. Use last of water to rinse spatula and Blendor cup walls. Place cup in position for blending.

3. Turn Blendor on with switch in high position, and gradually adjust the variable transformer to the point indicated by the water standard at 8,500 rpm. Blend at this speed for 10 min.

4. Remove the Blendor cup, and pour the slurry into a 600 ml. beaker. After the slurry has separated, decant or pipet a portion into a 50 ml. centrifuge tube, and centrifuge 10 min. at 2,700 rpm.

5. Pipet 15 ml. of supernatant liquid into a Kjeldahl flask, and determine protein by using A.O.C.S. Method Ac 4-41 (15 ml.=1.0 g. sample).

F. Calculation of Results

$$\% \text{ Water Dispersible Protein} = \frac{(B-S) \times N \times 0.014 \times 100 \times 6.25}{\text{Wt. of sample}}$$

Where B = ml. of alkali back titration of blank.

S = ml. of alkali back titration of sample.

N = Normality of alkali.

Protein Dispersibility Index (PDI) =

$$\frac{\% \text{ Water dispersible protein} \times 100}{\% \text{ Total protein}}$$

G. Precision:

Two single determinations performed in one laboratory should not differ by more than 4.375.

Agreement between laboratories: Two single determinations performed in different laboratories should not differ by more than 9.664.

Source: Anon. (1970).

A.O.C.S. Tentative Method Ba 9-58

UREASE ACTIVITY

Definition: This method determines the activity of the residual urease in the soybean products under the conditions of the test.

Scope: Applicable to soybean meals, soy flour and to soybean mill feeds except when urea has been added.

A. Apparatus:

1. Water bath capable of being maintained at a temperature of $30° \pm .5°$ C.

2. pH meter equipped with glass and calomel electrodes and with provision for testing 5 ml. of solutions. It should be a precision instrument with a temperature compensator having a sensitivity of ± 0.02 pH units or better. Follow manufacturer's instructions for operation of the instrument and determination of pH. Calibrate the meter with standard buffers with values at or near the range at which measurements are to be made.

3. Test tubes, 20 mm. $\times$ 150 mm., fitted with rubber stoppers.

B. Solutions:

1. Phosphate buffer solution, 0.05M. Dissolve 3.403 g. of monobasic potassium phosphate (KH_2PO_4, AR grade) in approximately 100 ml. of freshly distilled water. Dissolve 4.355 g. of dibasic potassium phosphate (K_2HPO_4, AR grade) in approximately 100 ml. of water. Combine the two solutions and make to 1000 ml. If reagents are pure, pH should be at 7.0.

If it is not, adjust to 7.0 with a solution of a strong acid or base before using. The useful life of the buffer solution, prepared as described, is less than 90 days.

2. Buffered urea solution. Dissolve 15 g. urea (AR grade) in 500 ml. of the phosphate buffer solution. Add 5 ml. of toluene to serve as a preservative and to prevent mold formation. Adjust the pH of the urea solution to 7.0 as in B,1.

C. Preparation of Sample:

1. Grind the sample as fine as possible without raising the temperature and mix. At least 60% of the sample should pass a No. 40 U.S. Standard sieve. Soy flour requires no grinding but make certain it is well mixed.

D. Procedure:

1. Weigh 0.200 g (± 0.001 g.) of sample into a test tube and add 10 ml. of the buffered urea solution. Stopper, mix and place in water bath at 30° C. Do not invert the tube during the process of mixing.

2. Prepare a blank by weighing 0.200 g (± 0.001 G.) sample into a test tube and to this add 10 ml of the phosphate buffer solution. Stopper, mix and place in water bath at 30° C. Allow a time interval of 5 minutes between the preparation of the test and the blank portions. Agitate the contents of each tube at 5 minute intervals.

3. Remove the test and blank portions from the water bath after 30 minutes. Transfer the supernatant liquids to a 5.0 ml. beaker, maintaining the 5-minute interval between the test and blank. Determine the pH of the supernatant liquids at exactly 5 minutes after removal from the bath. (See Note 1.)

E. Calculations:

1. The difference between the pH of the test and the pH of the blank is an index of urease activity.

F. Note:

1. Care must be exercised to prevent contamination of all glassware or electrodes. Should the pH instrument fail to deliver a prompt and stable reading, investigate. Frequently, the flow of electrolyte through the porous fibers in the calomel electrode may be retarded by a coating of the soluble fraction from soybean.

2. This method is a modification of the procedure of Caskey, C. D., and Knapp, F. C., Ind. Eng. Chem., Anal. Ed. *16*, 640 (1944).

Source: Anon. (1970).

WATER ABSORPTION OF SOY FLOUR

A. Apparatus:

1. Centrifuge, International No. 1 with 8 inch head, or equivalent.
2. Centrifuge tube, 50 ml.
3. Burette, 50 ml.

B. Procedure:

1. Weigh 5 g of sample into a 50 ml centrifuge tube and add 40 ml of distilled water from a burette.

2. Stir with a stirring rod until the mixture is homogeneous (usually ca 1 minute).
3. Centrifuge for 5 minutes at 2000 r.p.m.
4. Decant the clean liquor back into the burette which must contain distilled water to at least the lowest graduation point. Determine the volume of decanted liquor.

C. **Calculation:**

Ml of water absorbed = 40 − ml decanted liquor.

$$\% \text{ water absorption} = \frac{\text{ml water absorbed} \times 100}{\text{Weight of sample}}$$

Source: Anon. (1946).

BIBLIOGRAPHY

ANON. 1946. Handbook of analytical methods for soybeans and soybean products. Natl. Soybean Processors Assoc.

ANON. 1966. Blue Book Issue. Soybean Dig. *26,* No. 6, 18–21.

ANON. 1970. Official and Tentative Methods of the American Oil Chemists' Society, 3rd Edition. American Oil Chemists' Society, Chicago.

ANON. 1971. Blue Book Issue. Soybean Dig. *31.* No. 6, 52–53, 110–116.

Index